THE NATURE AND PROPERTIES OF SOILS

THE NATURE AND
PROPERTIES OF
SOILS *8th Edition*

NYLE C. BRADY

Professor of Soil Science
New York State College of Agriculture and Life Science
Cornell University
and Director, International Rice Research Institute,
Philippines

MACMILLAN PUBLISHING CO., INC.
New York
COLLIER MACMILLAN PUBLISHERS
London

MACMILLAN PUBLISHING CO., INC.
866 Third Avenue, New York, New York 10022

COLLIER-MACMILLAN CANADA, LTD.

Library of Congress Cataloging in Publication Data

Brady, Nyle C
  The nature and properties of soils.

  Sixth-7th ed. by H. O. Buckman and N. C. Brady.
    1. Soil science.   I.   Buckman, Harry Oliver,
  (date)   The nature and properties of soils.
  II.   Title.
  S591.B79   1974   631.4   73–1046
  ISBN 0–02–313350–3

Printing: 9 10 11 12   Year: 9 0 1 2 3 4

PREFACE

CHANGE is the name of the game in modern soil science. Not only have new discoveries dictated subject matter changes but society has forced changes by demanding solutions to pressing human problems. In addition, teaching methods have become more informal. Students are increasingly involved in independent study, each student being given an opportunity to move at his or her pace. Simultaneously, the science background of college students has broadened, which makes it easier for them to comprehend the constitution of soils and the biological, chemical, and physical processes occurring in soils. The changes made in this eighth edition reflect both increased knowledge of soils and the use of this knowledge to solve emerging societal problems.

Growing knowledge of and concern for environmental pollution have dictated major changes from previous editions. Soil is being used more and more as a recipient of organic and inorganic wastes from both the farm and the city. These wastes include pesticides, animal manures, industrial chemicals, domestic sewage, and nuclear contaminants. Their effects on soil processes and properties and on plants grown in these soils are a public as well as a scientific concern.

This eighth edition responds to the concern for the soil as a recipient of wastes in two ways. First, a new chapter on soil pollutants has been added. This chapter provides a general discussion of the pollutants, their reactions in soils, and their effects on soil organisms including plants. Second, the chapter on animal manures has been rewritten to reflect changes in the concepts of animal wastes in modern agriculture, and other chapters, such as those concerned with the loss of nutrients from soils, have been revised to take into account aspects concerned with environmental quality.

Another marked change over previous editions relates to the treatment of soil water. In recent years, scientists have developed a unified energy concept of water as it enters and moves in soils, is absorbed and translocated through plants, and is finally evaporated from the leaves into the atmosphere. This concept of a soil–plant–atmosphere continuum (SPAC) is the basis for major changes in the three chapters on soil moisture. The energy relations of water as it moves through the SPAC have been presented in terms easily understood by students having their initial exposure to soils.

The introductory chapter has been expanded to include a glimpse of the historical development of soil science. Similarly, some historical perspective of the evolution of current concepts of soil classification is included in Chapter 12. The comprehensive classification system of the U.S. Department of Agriculture is emphasized, only general reference being made to other systems.

Minor revisions have been made in all chapters of the book. The treatment of soil organisms and organic matter has been brought up to date. Chapter 13 has been revised to incorporate the new comprehensive classification system as it relates to organic soils.

The readability of the text has been improved by reducing the number of footnotes and by placing references at the ends of chapters rather than at the bottoms of pages and in figures and tables. Changes were made to reduce the formality of expression throughout the text, and many new figures and tables have been added.

The author is indebted to the many soil scientists who made suggestions for improving the text; to his wife, Martha, for typing, proofreading, and checking references; to Frances Geherin of his office for editing the manuscript; and to Patricia Oplinger, Roberta Reniff, Grace Saatman, and Mildred Townsley for typing it.

N. C. B.

CONTENTS

1. The Soil in Perspective   1

AII

   1:1   What Is Soil?   2
   1:2   Evaluation of Modern Concepts of Soil   2
   1:3   The Approach—Edaphological Versus Pedological   6
   1:4   A Field View of Soil   7
   1:5   The Soil Profile   9
   1:6   Subsoil and Surface Soil   10
   1:7   Mineral (Inorganic) and Organic Soils   11
   1:8   General Definition of Mineral Soils   12
   1:9   Four Major Components of Soils   12
   1:10 Mineral (Inorganic) Constituents in Soils   13
   1:11 Soil Organic Matter   14
   1:12 Soil Water—A Dynamic Solution   15
   1:13 Soil Air—Also a Changeable Constituent   16
   1:14 The Soil—A Tremendous Biological Laboratory   17
   1:15 Clay and Humus—The Seat of Soil Activity   18

2. Supply and Availability of Plant Nutrients in
   Mineral Soils   19

   2:1   Factors Controlling the Growth of Higher Plants   19
   2:2   The Essential Elements   20
   2:3   Essential Elements from Air and Water   20
   2:4   Essential Elements from the Soil   21
   2:5   Macronutrient Contents of Mineral Soils   23
   2:6   Forms of Macronutrients in Soils   26
   2:7   Transfer of Plant Nutrients to Available Forms   29
   2:8   Soil Solution   32
   2:9   Nutritional Importance of Soil pH   35
   2:10 Forms of Elements Used by Plants   35
   2:11 Soil and Plant Interrelations   37
   2:12 Soil Fertility Inferences   38

3. Some Important Physical Properties of Mineral
AII Soils   40

   3:1   Classification of Soil Particles and Mechanical Analysis 40
   3:2   Physical Nature of the Soil Separates   42
   3:3   Mineralogical and Chemical Compositions of the Soil
        Separates   44
   3:4   Soil Textural Classes   45

3:5 Determination of Soil Class 48
3:6 Particle Density of Mineral Soils 49
3:7 Bulk Density of Mineral Soils 50
3:8 Pore Space of Mineral Soils 53
3:9 Structure of Mineral Soils 55
3:10 Aggregation and Its Promotion in Arable Soils 58
3:11 Structural Management of Soils 62
3:12 Soil Consistence 64
3:13 Tilth and Tillage 66

4. Soil Colloids: Their Nature and Practical Significance 71

4:1 General Constitution of Silicate Clays 71
4:2 Adsorbed Cations 75
4:3 Fundamentals of Silicate Clay Structure 76
4:4 Mineralogical Organization of Silicate Clays 78
4:5 Source of the Negative Charge on Silicate Clays 83
4:6 Chemical Composition of Silicate Clays 86
4:7 Genesis of Silicate Clays 88
4:8 Mineral Colloids Other Than Silicates 91
4:9 Geographic Distribution of Clays 92
4:10 Organic Soil Colloids—Humus 94
4:11 Colloids—Acid Salts 96
4:12 Cation Exchange 97
4:13 Cation Exchange Capacity 99
4:14 Cation Exchange Capacity of Whole Soils 103
4:15 Percentage Base Saturation of Soils 103
4:16 Cation Exchange and the Availability of Nutrients 105
4:17 Other Properties of Colloids—Plasticity, Cohesion, Swelling, Shrinkage, Dispersion, and Flocculation 106
4:18 Conclusion 109

5. Organisms of the Soils 111

5:1 Organisms in Action 111
5:2 Organism Numbers, Biomass, and Metabolic Activity 114
5:3 Earthworms 116
5:4 Soil Microanimals 118
5:5 Roots of Higher Plants 121
5:6 Soil Algae 123
5:7 Soil Fungi 123
5:8 Soil Actinomycetes 126
5:9 Soil Bacteria 128
5:10 Conditions Affecting the Growth of Soil Bacteria 129

5:11 Injurious Effects of Soil Organisms on Higher Plants   130
5:12 Competition Among Soil Microorganisms   132
5:13 Effects of Agricultural Practice on Soil Organisms   133
5:14 Activities of Soil Organisms Beneficial to Higher Plants   135

6. Organic Matter of Mineral Soils   137
6:1   Sources of Soil Organic Matter   137
6:2   Composition of Plant Residues   137
6:3   Decomposition of Organic Compounds   139
6:4   Energy of Soil Organic Matter   142
6:5   Simple Decomposition Products   143
6:6   The Carbon Cycle   143
6:7   Simple Products Carrying Nitrogen   145
6:8   Simple Products Carrying Sulfur   146
6:9   Mineralization of Organic Phosphorus   146
6:10 Humus—Genesis and Definition   146
6:11 Humus—Nature and Characteristics   148
6:12 Direct Influence of Organic Compounds on Higher
      Plants   150
6:13 Influence of Soil Organic Matter on Soil Properties   150
6:14 Carbon–Nitrogen Ratio   151
6:15 Significance of the Carbon–Nitrogen Ratio   151
6:16 Amount of Organic Matter and Nitrogen in Soils   154
6:17 Factors Affecting Soil Organic Matter and Nitrogen   156
6:18 Regulation of Soil Organic Matter   161

7. Soil Water: Characteristics and Behavior   164
7:1   Structure and Related Properties of Water   164
7:2   Soil Water Energy Concepts   167
7:3   Soil Moisture Content Versus Suction   173
7:4   Measuring Soil Moisture   175
7:5   Capillary Fundamentals as They Relate to Soil Water   178
7:6   Types of Soil Water Movement   181
7:7   Saturated Flow Through Soils   182
7:8   Unsaturated Flow in Soils   184
7:9   Water Movements in Stratified Soils   185
7:10 Water Vapor Movement   186
7:11 Retention of Soil Moisture in the Field   189
7:12 Conventional Soil Moisture Classification Schemes   192
7:13 Factors Affecting Amount and Use of Available Soil
      Moisture   195
7:14 How Plants Are Supplied with Water—Capillarity and
      Root Extension   197
7:15 Conclusion   199

8. Vapor Losses of Soil Moisture and Their Regulation   200

8:1   Interception of Rain Water by Plants   200
8:2   The Soil–Water–Plant Continuum   202
8:3   Evapo-transpiration   204
8:4   Magnitude of Evaporation Losses   207
8:5   Efficiency of Water Use   208
8:6   Evaporation Control: Mulches and Cultivation   210
8:7   Vaporization Control in Humid Regions   215
8:8   Vaporization Control in Semiarid and Subhumid Regions   215
8:9   Evaporation Control of Irrigated Lands   216

9. Liquid Losses of Soil Water and Their Control   220

9:1   Percolation and Leaching—Methods of Study   220
9:2   Percolation Losses of Water   221
9:3   Leaching Losses of Nutrients   223
9:4   Land Drainage   227
9:5   Open Ditch Drainage   228
9:6   Tile Drains   229
9:7   Benefits of Land Drainage   232
9:8   Runoff and Soil Erosion   234
9:9   Accelerated Erosion—Mechanics   236
9:10   Accelerated Erosion—Causes and Rate Factors   237
9:11   Types of Water Erosion   240
9:12   Sheet and Rill Erosion—Losses Under Regular Cropping   240
9:13   Sheet and Rill Erosion—Methods of Control   243
9:14   Gully Erosion and Its Control   245
9:15   Wind Erosion—Its Importance and Control   245
9:16   Conservation Treatment Needs in the United States   251
9:17   Summary of Soil Moisture Regulation   251

10. Soil Air and Soil Temperature   253

10:1   Soil Aeration Defined   253
10:2   Soil Aeration Problems in the Field   254
10:3   Composition of Soil Air   256
10:4   Factors Affecting the Composition of Soil Air   258
10:5   Effects of Soil Aeration on Biological Activities   260
10:6   Other Effects of Soil Aeration   264
10:7   Aeration in Relation to Soil and Crop Management   265
10:8   Soil Temperature   266
10:9   Absorption and Loss of Solar Energy   268

10:10   Specific Heat of Soils   270
10:11   Heat of Vaporization   271
10:12   Movement of Heat in Soils   272
10:13   Soil Temperature Data   272
10:14   Soil Temperature Control   274

11.   Origin, Nature, and Classification of Parent
      Materials   277

11:1   Classification and Properties of Rocks   277
11:2   Weathering—A General Case   279
11:3   Mechanical Forces of Weathering   281
11:4   Chemical Processes of Weathering   282
11:5   Factors Affecting Weathering of Minerals   285
11:6   Weathering in Action—Genesis of Parent Materials   287
11:7   Geological Classification of Parent Materials   287
11:8   Residual Parent Material   288
11:9   Colluvial Debris   290
11:10   Alluvial Stream Deposits   290
11:11   Marine Sediments   292
11:12   The Pleistocene Ice Age   293
11:13   Glacial Till and Associated Deposits   295
11:14   Glacial Outwash and Lacustrine Sediments   298
11:15   Glacial-Aeolian Deposits   299
11:16   Agricultural Significance of Glaciation   300

12.   Soil Formation, Classification, and Survey   303

12:1   Factors Influencing Soil Formation   303
12:2   Weathering and Soil Profile Development   310
12:3   The Soil Profile   312
12:4   Concept of Individual Soils   315
12:5   Soil Classification in the United States   317
12:6   Soil Classification—New Comprehensive System   318
12:7   Soil Orders   322
12:8   Soil Suborders, Great Groups, and Subgroups   333
12:9   Soil Families, Series, Phases, Associations, and Catenas   337
12:10   Soil Classification—1949 System   340
12:11   Soil Survey and Its Utilization   343
12:12   Land Capability Classification   347

13.   Organic Soils (Histosols): Their Nature, Properties,
      and Utilization   353

13:1   Genesis of Organic Deposits   353
13:2   Area and Distribution of Peat Accumulations   355

13:3   Peat Parent Materials   356
13:4   Uses of Peat   358
13:5   Classification of Organic Soils   359
13:6   Physical Characteristics of Field Peat Soils   361
13:7   Colloidal Nature of Organic Soils   362
13:8   Chemical Composition of Organic Soils   364
13:9   Bog Lime—Its Importance   367
13:10   Factors That Determine the Value of Peat and Muck
        Soils   367
13:11   Preparation of Peat for Cropping   368
13:12   Management of Peat Soils   368
13:13   Organic Versus Mineral Soils   370

14.   Soil Reaction: Acidity and Alkalinity   372

14:1   Source of Hydrogen Ions   372
14:2   Colloidal Control of Soil Reaction   378
14:3   Major Changes in Soil pH   379
14:4   Minor Fluctuations in Soil pH   381
14:5   Hydrogen Ion Heterogeneity of the Soil Solution   381
14:6   Active Versus Exchange Acidity   382
14:7   Buffering of Soils   384
14:8   Buffer Capacity of Soils and Related Phases   385
14:9   Importance of Buffering   386
14:10   Soil-Reaction Correlations   387
14:11   Relation of Higher Plants to Soil Reaction   391
14:12   Determination of Soil pH   393
14:13   Soil Acidity Problems   394
14:14   Methods of Intensifying Soil Acidity   395
14:15   Reaction of Soils of Arid Regions   396
14:16   Reaction of Saline and Sodic Soils   396
14:17   Growth of Plants on Halomorphic Soils   399
14:18   Tolerance of Higher Plants to Halomorphic Soils   399
14:19   Management of Saline and Sodic Soils   400
14:20   Conclusion   402

15.   Lime and Its Soil–Plant Relationships   404

15:1   Liming Materials   404
15:2   Chemical Guarantee of Liming Materials   406
15:3   Fineness Guarantee of Limestone   409
15:4   Changes of Lime Added to the Soil   410
15:5   Loss of Lime from Arable Soils   412
15:6   Effects of Lime on the Soil   413
15:7   Crop Response to Liming   414

15:8   Overliming   415
15:9   Shall Lime Be Applied?   415
15:10   Form of Lime to Apply   416
15:11   Amounts of Lime to Apply   417
15:12   Methods of Applying Lime   420
15:13   Lime and Soil Fertility Management   421

16.   Nitrogen and Sulfur Economy of Soils   422

16:1   Influence of Nitrogen on Plant Development   422
16:2   Forms of Soil Nitrogen   423
16:3   The Nitrogen Cycle   423
16:4   Ammonia Fixation   427
16:5   Nitrification   428
16:6   Soil Conditions Affecting Nitrification   429
16:7   Fate of Nitrate Nitrogen   431
16:8   Gaseous Losses of Soil Nitrogen   431
16:9   Fixation of Atmospheric Nitrogen by Legume Bacteria   434
16:10   Amount of Nitrogen Fixed by Legume Bacteria   436
16:11   Fate of Nitrogen Fixed by Legume Bacteria   437
16:12   Do Legumes Always Increase Soil Nitrogen?   437
16:13   Fixation by Organisms in Symbiosis with Nonlegumes   438
16:14   Nonsymbiotic Fixation of Atmospheric Nitrogen   439
16:15   Amount of Nitrogen Fixed by Nonsymbiosis   440
16:16   Addition of Nitrogen to Soil in Precipitation   441
16:17   Reactions of Nitrogen Fertilizers   441
16:18   Practical Management of Soil Nitrogen   443
16:19   Importance of Sulfur   444
16:20   Natural Sources of Sulfur   446
16:21   The Sulfur Cycle   449
16:22   Behavior of Sulfur Compounds in Soils   449
16:23   Sulfur and Soil Fertility Maintenance   453

**17.   Supply and Availability of Phosphorus and
        Potassium   456**

17:1   Importance of Phosphorus   456
17:2   Influence of Phosphorus on Plants   457
17:3   The Phosphorus Problem   457
17:4   Phosphorus Compounds in Soils   458
17:5   Factors That Control the Availability of Inorganic Soil
        Phosphorus   460
17:6   pH and Phosphate Ions   460
17:7   Inorganic Phosphorus Availability in Acid Soils   462

17:8   Inorganic Phosphorus Availability at High pH Values   466
17:9   pH for Maximum Inorganic Phosphorus Availability   466
17:10   Availability and Surface Area of Phosphates   467
17:11   Phosphorus-Fixing Power of Soils   468
17:12   Influence of Soil Organisms and Organic Matter on the
         Availability of Inorganic Phosphorus   469
17:13   Availability of Organic Phosphorus   469
17:14   Practical Control of Phosphorus Availability   470
17:15   Potassium—The Third "Fertilizer" Element   472
17:16   Effects of Potassium on Plant Growth   472
17:17   The Potassium Problem   473
17:18   Forms and Availability of Potassium in Soils   475
17:19   Factors Affecting Potassium Fixation in Soils   479
17:20   Practical Implications in Respect to Potassium   480

18.   Micronutrient Elements   484

18:1   Deficiency Versus Toxicity   484
18:2   Role of the Micronutrients   486
18:3   Source of Micronutrients   487
18:4   General Conditions Conducive to Micronutrient
         Deficiency   489
18:5   Factors Influencing the Availability of the
         Micronutrient Cations   490
18:6   Chelates   493
18:7   Factors Influencing the Availability of the
         Micronutrient Anions   496
18:8   Need for Nutrient Balance   498
18:9   Soil Management and Micronutrient Needs   499

19.   Fertilizers and Fertilizer Management   503

19:1   The Fertilizer Elements   503
19:2   Three Groups of Fertilizer Materials   503
19:3   Nitrogen Carriers—Two Groups   504
19:4   Inorganic Nitrogen Carriers   504
19:5   Phosphatic Fertilizer Materials   510
19:6   Fertilizer Materials Carrying Potassium   513
19:7   Sulfur in Fertilizers   515
19:8   Micronutrients   515
19:9   Mixed Fertilizers   517
19:10   Effect of Mixed Fertilizers on Soil pH   520
19:11   The Fertilizer Guarantee   520
19:12   Fertilizer Inspection and Control   521
19:13   Fertilizer Economy   522

19:14   Movement of Fertilizer Salts in the Soil   524
19:15   Methods of Applying Solid Fertilizers   525
19:16   Application of Liquid Fertilizers   526
19:17   Factors Influencing the Kind and Amount of Fertilizers
        to Apply   527
19:18   Kind of Crop to Be Fertilized   528
19:19   Chemical Condition of the Soil—Total Versus Partial
        Analyses   529
19:20   Tests for Available Soil Nutrients—Quick Tests   530
19:21   Broader Aspects of Fertilizer Practice   531

20.   Animal Manures and Green Manures   534

20:1   Quantity of Manure Produced   534
20:2   Chemical Composition   538
20:3   Storage, Treatment, and Management of Animal
        Manures   541
20:4   Utilization of Animal Manures   544
20:5   Long-Term Effects of Manures   546
20:6   Green Manures—Defined   546
20:7   Benefits of Green Manures   546
20:8   Plants Suitable as Green Manures   548
20:9   Problems with Green Manures   549
20:10   Practical Utilization of Green Manures   550

21.   Soils and Chemical Pollution   551

21:1   Chemical Pesticides—Background   551
21:2   Kinds of Pesticides   553
21:3   Behavior of Pesticides in Soils   554
21:4   Effects of Pesticides on Soil Organisms   559
21:5   Contamination with Toxic Inorganic Compounds   560
21:6   Behavior of Inorganic Contaminants in Soils   563
21:7   Prevention and Elimination of Inorganic Chemical
        Contamination   567
21:8   Organic Wastes   568
21:9   Use of Organic Wastes for Crop Production   570
21:10   Soils as Organic Waste Disposal Sites   572
21:11   Soil Salinity   573
21:12   Radionuclides in Soils   575
21:13   Three Conclusions   576

22.   Soils and the World's Food Supply   578

22:1   Expansion of World Population   578
22:2   Factors Influencing World Food Supplies   579

22:3   The World's Land Resources   580
22:4   Potential of Broad Soil Groups   582
22:5   Problems and Opportunities in the Tropics   585
22:6   Requisites for the Future   590

Glossary of Soil Science Terms   593

Index   623

Chapter 1

THE SOIL IN PERSPECTIVE

MAN is dependent on soils—and to a certain extent good soils are dependent upon man and the use he makes of them. Soils are the natural bodies on which plants grow. Man enjoys and uses these plants because of their beauty and because of their ability to supply fiber and food for himself and his animals. Man's standard of living is often determined by the quality of soils and the kinds and quality of plants and animals grown on them.

But soils have more meaning for man than as a habitat for growing crops. They underlay the foundations of houses and factories and determine whether these foundations are adequate. They are used as beds for roads and highways and definitely influence the length of life of these structures. In rural areas soils are often used to absorb domestic wastes through septic sewage systems. They are being used more and more as recipients of other wastes from municipal, industrial, and animal sources. The deposition of undesirable silt in municipal reservoirs makes the protection of soils in upstream watersheds as important to the city dweller as to his counterpart on the farm or in the forest. Obviously, soils and their management are of broad societal concern.

Great civilizations have almost invariably had good soils as one of their chief natural resources. The ancient dynasties of the Nile were made possible by the food-producing capacity of the fertile soils of the valley and its associated irrigation systems. Likewise, the valley soils of the Tigris and Euphrates rivers in Mesopotamia and of the Indus, Yangtse, and Hwang-Ho rivers in India and China were habitats for flourishing civilizations. Subject to frequent replenishing of their fertility by natural flooding, these soils provided continued abundant food supplies. They made possible stable and organized communities and even cities, in contrast to the nomadic, shifting societies associated with upland soils and with their concomitant animal grazing. It was not until man discovered the value of manures and crop residues that he was able to make extensive use of upland soils for sustained crop culture.

Soil destruction or mismanagement was associated with the downfall of some of the same civilizations which good soils had helped to build. The cutting of timber in the watersheds of these rivers encouraged erosion and topsoil loss. In the Euphrates and Tigris valleys, the elaborate irrigation and drainage systems were not maintained. This resulted in the accumulation of harmful salts, and the once productive soils became barren and useless. The proud cities which had occupied selected sites in the valleys disintegrated and the people migrated elsewhere.

History provides lessons which modern man has not always heeded. The wasteful use of soil resources in the United States during the white man's first century of intensive agricultural production provides such an example. Even today there are many who do not fully recognize the long-term significance of soils. This may be due in part to widespread ignorance as to what soils are, what they have meant to past generations, and what they mean to us and future generations.

1:1. WHAT IS SOIL?

CONCEPTS OF SOIL. Part of the lack of concern for soils may be attributed to different concepts and viewpoints concerning this important product of nature. For example, to a mining engineer the soil is the debris covering the rocks or minerals which he must quarry. It is a nuisance and must be removed. To the highway engineer the soil may be the material on which a roadbed is to be placed. If its properties are unsuitable, he must remove it and replace it with rock and gravel.

The average homeowner also has a concept of soil. It is good if the ground is mellow or loamy. The opposite viewpoint is associated with "hard clay" which resists being spaded up into a good seedbed for a flower garden. The homemaker can differentiate among variations in the soil, especially those relating to its stickiness or tendency to cling to shoe soles and eventually to carpets.

The farmer, along with the homeowner, looks upon the soil as a habitat for plants. He makes a living from the soil and is thereby forced to pay more attention to its characteristics. To him the soil is more than useful—it is indispensable.

A prime requisite for learning more about the soil is to have a common concept of what it is. This concept must encompass the viewpoints of the engineer, the homeowner, and the farmer. In developing this concept, brief consideration will be given to the practical and scientific discoveries of the past.

1:2. EVALUATION OF MODERN CONCEPTS OF SOIL

There are two basic sources of our current knowledge of soils. First, there is the practical knowledge gained by farmers through centuries of trial and error. This was the only information available before the advent of modern science, which now provides a second source of facts about soils and their management.

EXPERIENCE OF THE CULTIVATOR. The earliest recorded history contains evidence that, through trial and error, man learned to distinguish differences in soils. He also learned the value of treating soils with plant and animal wastes. More than 42 centuries ago the Chinese used a schematic soil map as a basis for taxation. Homer, in his Odyssey, said to have been written about 1000 B.C., makes reference to the use of manure on the land. Biblical references are made to the "dung hill" and to the beneficial practice of "dunging" around plants. Greek and Roman writers describe a reasonably elaborate system of farming which involved leguminous plants and the use of ashes and sulfur as soil amendments. These evidences suggest that by the time of the early Roman civilization many of the practical principles governing modern agriculture, including soil management, had been discovered and put to use by farmers and animal husbandmen.

Further development and application of these principles were held in abeyance following the barbaric invasions of Rome. Even so, Roman agriculture was the foundation for most European agriculture during the feudal dark ages. The cultural practices were passed from generation to generation even though the farmers were ignorant as to why the practices were necessary. When in the seventeenth and eighteenth centuries intellectual activity was again revived, the stage was set for the application of science to the improvement of agricultural systems including those involving soils.

EARLY SCIENTIFIC INVESTIGATIONS AND SOIL PRODUCTIVITY. From the seventeenth century until the middle of the twentieth century the primary occupation of soil scientists has been to increase the production of crop plants. Jan Baptista van Helmont, a Flemish chemist, in his famous five-year willow tree experiment concluded that 164 pounds of dry matter came primarily from the water supplied since the soil lost no weight while producing the tree. This concept was altered by John Woodward, an English researcher, who later found muddy water to produce more plant growth than rain water or river water, leading to the conclusion that the fine earth was the "principle" of growth. Others concluded that the principle was humus taken in by the plants from soil. Still others assumed that the principle was somehow passed on from dead plants or animals to the new plant. Jethro Tull early in the eighteenth century demonstrated the benefits of cultivation, thinking erroneously that stirring the soil would make it easier for plants to absorb small quantities of fine earth.

It remained for the French agriculturalist J. B. Boussingault, through a series of field experiments starting in 1834, to provide evidence that the air and rain were the primary sources of carbon, hydrogen, and oxygen in plant tissues. But his investigations were largely disregarded until 1840 when the eminent German chemist Julius von Liebig reported findings that crop yields were directly related to the content of "minerals" in the manures applied to the soil. Liebig's reputation as an outstanding physical chemist was instrumental in convincing the scientific community that the old theories were wrong. He gave birth to the concept that mineral elements in the soil and in added manures and fertilizers are essential for plant growth.

Liebig's work revolutionized agricultural theory and opened the way for numerous other investigations. Those of J. B. Lawes and J. H. Gilbert at the Rothamsted Experiment Station in England are most noteworthy since they put Liebig's theories to test in the field. While they confirmed much of his findings, they identified two errors. Liebig had theorized that nitrogen came primarily from the atmosphere rather than the soil. He further assumed that salts could be fused prior to their being added as fertilizers; apparently he gave no thought to what this drastic action would do to the solubility and "availability" to plants of the nutrient elements. The Rothamsted field research proved that nitrogen applications to the soil markedly benefited plant growth. Further they showed that mineral elements must be in an "available" form for optimum uptake by plants. Their investigations led to the development of acid-treated phosphate rock, or "superphosphate," which is still the mainstay of commercial fertilizer sources of phosphorus.

While the work of Gilbert, Lawes, and others pinpointed weaknesses in components of his theory, Liebig's primary concept is still considered basically sound. For example, he stated what has since been called the Law of the Minimum: "by the deficiency or absence of one necessary constituent, all the others being present, the soil is rendered barren for all those crops to the life of which that one constituent is indispensable." The significance of this finding will be more apparent later as soil fertility and plant nutrition interactions are considered.

After Liebig, unraveling the complexities of nitrogen transformations in soil awaited the emergence of soil bacteriology. J. T. Way discovered in 1856 that nitrates are formed in soils from ammonia-containing fertilizers. Twenty years later R. W. Warington demonstrated that this process was biological, and in 1890 S. Winogradski isolated the two groups of bacteria responsible for the transformation of ammonia to nitrate. Coupled with the discovery in the 1880s that nitrogen-assimilating bacteria grow in nodules of the roots of legumes, these findings provided background information for sound soil and crop management practices.

The European investigations on soil fertilizers were found to be quite applicable to the United States as they were tested in the late nineteenth century. Upon being tilled, the soils along the eastern and southern seaboards were easily depleted of nutrients by the percolation of rainfall and by crop removal. Edmund Ruffin, a Virginia farmer, grasped the concepts of nutrient depletion and in his writings was especially critical of those who did not properly care for and replenish their soil. Unfortunately, the abundance of open land to the west encouraged the abandonment of the "worn-out" soils in the east rather than the adoption of more realistic management systems.

The establishment of the U. S. Department of Agriculture in 1862 and the State Agricultural Experiment Stations in 1886 gave a decided boost to both field and laboratory investigations on soils. Numerous field trials were initiated to test the applicability of findings of the European investigators. Unfortunately, in most of the trials soil was not considered to be a dynamic medium for crop growth. Instead the soil was considered merely as a "storage bin" in keeping with Liebig's concepts. Exceptions existed, such as the work of men like F. H. King of Wisconsin, who studied the movement and storage of water in soils in relation to root penetration and crop growth. Also C. G. Hopkins of Illinois developed effective soil-management systems based on limestone, rock phosphate, and legumes. Milton Whitney of the U. S. Department of Agriculture urged greater consideration of properties of soils in the field and initiated the first national soil survey system.

FIELD SOIL INVESTIGATIONS. Liebig's concepts thoroughly dominated the thinking of soil scientists in the late nineteenth and early twentieth centuries. Furthermore, except for the field testing of crop response to fertilizer, much of the research was done in the laboratory and greenhouse. Inadequate attention was given to the variable characteristics of the soils as found in the field. Nor was much significance given to the differences in soils as related to the climate in which they were found. Soils were considered as geological residues on the one hand and as reservoirs of nutrients for plant growth on the other.

As early as 1860, E. W. Hilgard, then in Mississippi, published his findings, which called attention to the relationships among climate, vegetation, rock materials, and the kinds of soils which develop. He conceived soils not merely as media for plant growth but as dynamic entities subject to study and classification in their field setting. Unfortunately, Hilgard was ahead of his time. It was necessary for many of his concepts to be rediscovered before they were accepted.

Parallel to Hilgard's investigations were those made by the brilliant team of soil scientists in Russia led by V. V. Dokuchaev. These scientists found unique horizontal layerings in soils—layerings associated with the climate, vegetation, and underlying soil material. The same sequence of layering was found in widely separated geographical areas provided the climate and vegetation were similar among the areas. The concept of soils as natural bodies was well developed in the Russian studies, as were concepts of soil classification based on field soil characteristics.

The Russian studies were underway as early as 1870. Unfortunately, language barriers prevented effective communication of the Russian concepts to scientists from Western Europe, Asia, and the Americas until 1914, when they were published in German by K. D. Glinka, one of the team of Russian soil scientists. These concepts were quickly grasped by C. F.

Marbut of the U. S. Department of Agriculture, who had been placed in charge of the U. S. National Soil Survey by Dr. Whitney. Marbut and his associates developed a nationwide soil classification system based to a great extent on the Russian concepts. Consideration of soils as natural bodies has led to further modifications in soil classification systems which will receive attention in later chapters.

Even a brief background of the history of the development of current concepts of soils would be incomplete without reference to the work of many scientists who have contributed to knowledge of soil characteristics and soil processes. The concept of soil acidity, its causes and remedies, has resulted from such studies, as has a knowledge of how nutrient elements are held in and released from soils. Characterization and identification of clays of different types in soils and an understanding of the interaction of soil and water are further examples. Although much is yet to be learned about soil characteristics and processes, much knowledge has been gained in the past century.

1:3. The Approach—Edaphological Versus Pedological

The previous section suggests that two basic concepts of soil have evolved through two centuries of scientific study. The first considers soil as a natural entity, a biochemically weathered and synthesized product of nature. The second conceives of the soil as a natural habitat for plants and justifies soil studies primarily on that basis. These conceptions illustrate the two approaches that can be used in studying soils—that of the *pedologist* and that of the *edaphologist*.

Certain aspects, such as the origin of the soil, its classification, and its description, are involved in *pedology* (from the Greek word *pedon*, which means soil or earth). Pedology considers the soil as a natural body and places minor emphasis on its immediate practical utilization. The pedologist studies, examines, and classifies soils as they occur in their natural environment. His findings may be as useful to highway and construction engineers as to the farmer.

Edaphology (from the Greek word *edaphos*, which also means soil or ground) is the study of the soil from the standpoint of higher plants. It considers the various properties of soils as they relate to plant production. The edaphologist is practical, having the production of food and fiber as an ultimate goal. At the same time, he must be a scientist to determine reasons for variation in the productivity of soils and to find means of conserving and improving this productivity.

In this textbook the dominant viewpoint will be that of the edaphologist. Pedology will be used, however, to the extent that it gives a general understanding of soils as they occur in nature and are classified. Furthermore, since studies of the basic physical, chemical, and biological characteristics of soils contribute equally to the edaphology and pedology, it is not possible to clearly separate these approaches. This is illustrated in the following section, which deals with soils as they are found in the field.

1:4. A FIELD VIEW OF SOIL

Someone has said that the soil is to the earth as the peel is to the orange. This analogy is acceptable but should be modified to stress the great variability in soil from site to site on the earth's surface. Even a casual examination of road cuts from one geographic area to another suggests differences in soil depth, color, and mineral makeup. The trained eye of the soil scientist, however, can identify common properties of soils from areas as distant as Hawaii, India, and the United States. The common properties will receive immediate attention, leaving the variations for succeeding chapters.

SOIL VERSUS REGOLITH. Views such as that shown in Fig. 1:1, in which unconsolidated materials are found on underlying rocks, are quite familiar. Above bedrock some unconsolidated debris is present almost universally. This material, known as *regolith*, may be negligibly shallow or hundreds of

FIGURE 1:1. Relative positions of the regolith, its soil, and the underlying country rock. Sometimes the regolith is so thin that it has been changed entirely to soil, which, in such a case, rests directly on bedrock. [*Photo courtesy Tennessee Valley Authority.*]

feet in thickness. It may be material which has weathered from the underlying rock or it may have been transported by the action of wind, water, or ice, and deposited upon the bedrock. As might be expected, the regolith tends to vary in composition from place to place.

An examination of the upper 3 to 6 feet of the regolith shows that it differs from the material below. Being nearer to the atmosphere, this upper zone has been subject to the weathering actions of wind, water, and heat. Furthermore, it is the zone in which most of the plant roots are found. Plant residues deposited originally on the surface have become incorporated by earthworms into the surface layers and have been disintegrated and partially decomposed by microorganisms.

The presence of some undecomposed organic matter and the weathering of minerals in the soil have resulted in the formation of characteristic horizontal layering. This upper and biochemically weathered portion of the regolith is the soil and is distinguished from the material below[1] by (a) a relatively high organic matter content, (b) an abundance of the roots of higher plants and of soil organisms, (c) more intense weathering, and (d) the presence of characteristic horizontal layers. The characteristics which differentiate the upper few feet of the regolith from the underlying material have led scientists to the concept of the soil as a *natural body* or a collective group of natural bodies. The soil is considered as a three-dimensional product of nature which has resulted from both destructive and synthetic forces. Weathering and microbial decay of organic residues are examples of destructive processes, whereas the formation of new minerals, such as certain clays, and the development of characteristic layer patterns are synthetic in nature. These forces have given rise to a distinctive entity in nature called the soil, which, in turn, is comprised of a large number of individual soils.

THE SOIL VERSUS A SOIL. Characteristics of the soil vary widely from place to place. For example, on steep slopes the soil is generally not deep and productive as when it is formed on gentle slopes. Where it has developed from sandstone it is more sandy and less fertile than where it has formed from rocks such as shale. Its properties are quite different when it has developed under tropical climates as compared to temperate or arctic conditions.

Scientists have recognized these soil variations from place to place and have set up classification systems in which the soil is considered as composed of a large number of individual soils, each having its distinguishing characteristics. Therefore, *a* soil, as distinguished from *the* soil, is merely a well-defined subdivision having recognized limits in its characteristics and properties. Thus, a Cecil clay loam, a Marshall silt loam, and a Norfolk sand are

---

[1] In cases where the regolith was relatively uniform originally, the material below the soil is considered to have a composition similar to the parent material from which the soil formed.

examples of *specific soils* which collectively make up the overall *soil* covering the world's land areas. The term "soil" is a collective term for all soils just as "vegetation" is used to designate plants.

1:5. THE SOIL PROFILE

Examination of a vertical section of a soil in the field shows the presence of more or less distinct horizontal layers (see Fig. 1:2). Such a section is called a *profile* and the individual layers are regarded as *horizons*. These horizons above the parent material are collectively referred to as the *solum*, from the Latin legal term meaning soil, land, or parcel of land. Every well-developed, undisturbed soil has its own distinctive profile characteristics which are utilized in soil classification and survey and are of great practical importance. In judging a soil its whole profile must be considered.

SOIL HORIZONS. The upper layers or horizons of a soil profile generally contain considerable amounts of organic matter and are usually darkened

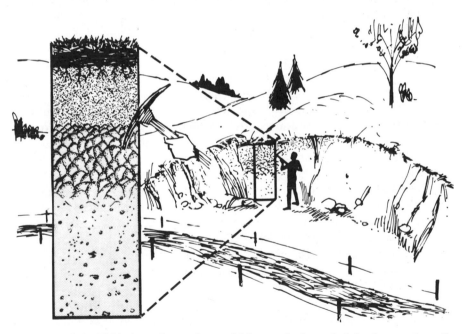

FIGURE 1:2. Field view of a road cut which reveals the underlying layers of a soil. The closeup emphasizes soil layering and the distinctive character of the *soil profile*. The surface layer is darker in color because of its higher organic matter content. One of the subsurface horizons (point of pick) is characterized by a distinctive structure. The existence of layers such as those shown is used to help differentiate one soil from another.

appreciably because of such an accumulation. Layers thus characterized are referred to as the major zone of organic-matter accumulation. When a soil is plowed and cultivated, these layers are included in the familiar surface soil, which is sometimes referred to as the *furrow slice* because it is the portion of the soil turned or "sliced" by the plow.

The underlying subsoil contains comparatively less organic matter than the upper layers. The various subsoil layers, especially in mature, humid-region soils, present two very general belts: (a) an upper zone of transition, and (b) a lower zone of accumulation of compounds, such as iron and aluminum oxides, clays, gypsum, and calcium carbonate.

The solum thus described extends a moderate depth below the surface. A depth of 3 or 4 feet is representative for temperate region soils. Here, the noticeably modified lower subsoil gradually merges with the less weathered portion of the regolith whose upper portion is geologically on the point of becoming a part of the lower subsoil and hence of the solum.

The various layers comprising a soil profile are not always distinct and well defined. The transition from one to the other is often so gradual that the establishment of boundaries is rather difficult. Nevertheless, for any particular soil the various horizons are characteristic and greatly influence the growth of higher plants.

1:6. SUBSOIL AND SURFACE SOIL

The productivity of a soil is determined in no small degree by the nature of its subsoil. This is of practical significance since the subsoil normally is subject to little field alteration except by drainage. Even when roots do not penetrate deeply into the subsoil, the permeability and chemical nature of the subsoil influence the surface soil in its role as a medium for plant growth (see Fig. 1:3).

The situation in respect to the surface soil is somewhat different. In the first place, it is the major zone of root development, it carries much of the nutrients available to plants, and it supplies a large share of the water used by crops. Second, as the layer which is plowed and cultivated, it is subject to manipulation and management. By proper cultivation and the incorporation of organic residues, its physical condition may be modified. It can be treated with chemical fertilizers and limestone and it can be drained. In short, its fertility and to a lesser degree its productivity[2] may be raised, lowered, or satisfactorily stabilized at levels consistent with economic crop production.

---

[2] The term "fertility" refers to the inherent capacity of a soil to supply nutrients to plants in adequate amounts and in suitable proportions. Productivity is related to the ability of a soil to yield crops. "Productivity" is the broader term since fertility is only one of a number of factors that determine the magnitude of crop yields.

FIGURE 1:3. Root system of an 8-year-old juniper tree. The roots extend more than 16 feet into the soil. Moisture and nutrients are likely absorbed to great depths in this soil. [*Photo courtesy U. S. Soil Conservation Service.*]

This explains why much of the soil investigation and research has been expended upon the surface layer. Plowing, cultivation, liming, and fertilization are essentially furrow-slice considerations. The term "soil" in practice usually denotes the surface layer, the "topsoil," or, in practical terms, the furrow slice.

1:7. MINERAL (INORGANIC) AND ORGANIC SOILS

The profile generalizations just described relate to soils which are predominantly *mineral* or *inorganic* in composition. Even in their surface layer organic matter contents of mineral soils are comparatively low, generally ranging from 1 to 10 percent. In contrast, soils in swamps, bogs, and marshes commonly contain 80–95 percent organic matter. These *organic* or *muck* soils when drained and cleared are most productive, especially for high-value crops such as fresh market vegetables. Organic deposits may also be excavated, bagged, and sold as organic supplements for home gardens and potted plants. The economic significance of organic soils is considerable in localized regions.

Since they occupy such a high proportion of the total land area, mineral soils are of greater importance than organic soils. Consequently, mineral soils deservedly receive the major attention in this text, the origin, character, and agricultural use of organic soils being considered as a unit in Chapter 13. Until then the discussion will be concerned almost exclusively with mineral soils.

1:8. GENERAL DEFINITION OF MINERAL SOILS

Mineral soils have already been denoted as the "upper and biologically weathered portion of the regolith." When expanded by profile and horizon study, this statement presents a pedological concept of soil origin and characterization. If, however, the production of higher plants is to be a part of the picture, a broader and more inclusive statement must be offered. In the light of the ideas respecting the general function of the soil which have been presented, *soil* may be defined as "a collection of natural bodies which has been synthesized in profile form from a variable mixture of broken and weathered minerals and decaying organic matter, which covers the earth in a thin layer, and which supplies, when containing the proper amounts of air and water, mechanical support and sustenance for plants."

1:9. FOUR MAJOR COMPONENTS OF SOILS

The definition just cited leads logically to the question of soil composition. Mineral soils consist of four major components: mineral materials, organic matter, water, and air. These exist mostly in a fine state of subdivision and are so intimately mixed that satisfactory separation is rather difficult.

VOLUME COMPOSITION OF MINERAL SOILS. Figure 1:4 shows the approximate volume composition of a representative silt loam surface soil in optimum condition for plant growth. Note that it contains about 50 percent pore space (air and water). The solid space is made up by volume of about 45 percent mineral matter and 5 percent organic matter. At optimum moisture for plant growth, the 50 percent of pore space possessed by this representative soil is divided roughly in half—25 percent water space and 25 percent air. The proportion of air and water is subject to great fluctuations under natural conditions, depending on the weather and other factors (see Fig. 1:4).

In presenting such an arbitrary volume representation of a surface mineral soil, it must be emphasized that the four major components of the normal soil exist mainly in an intimately mixed condition. This encourages both simple and complex reactions within and between the groups and permits an ideal environment for the growth of plants.

The volume composition of subsoils is somewhat different from that just described. Compared to topsoils they are lower in organic matter content, are

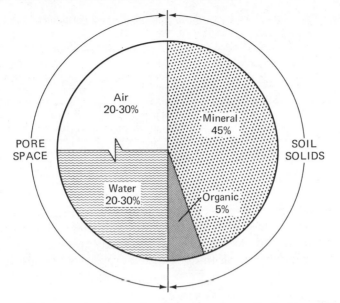

FIGURE 1:4. Volume composition of a silt loam surface soil when in good condition for plant growth. The air and water in a soil are extremely variable, and their proportion determines in large degree its suitability for plant growth.

somewhat more compact, and contain a higher percentage of small pores. This means they have higher percentage of minerals and water and a considerably lower content of organic matter and air.

1:10. MINERAL (INORGANIC) CONSTITUENTS IN SOILS

A casual examination of a sample of soil illustrates that the inorganic portion is variable in size and composition. It is normally composed of small rock fragments and minerals[3] of various kinds. The rock fragments are remnants of massive rocks from which the regolith and in turn the soil have been formed by weathering. They are usually quite coarse (see Table 1:1). The minerals, on the other hand, are extremely variable in size. Some are as large as the smaller rock fragments; others, such as colloidal clay particles, are so small that they cannot be seen without the aid of an electron microscope.

Quartz and some other "primary minerals" have persisted with little change in composition from the original country rock. Other minerals such as the silicate clays and iron oxides have been formed by the weathering of less resistant minerals as the regolith developed and soil formation progressed.

[3] The word "mineral" is used in this book in two ways: first, as a general term to describe soils dominated by inorganic constituents, and second, as a more specific term to describe distinct minerals found in nature, such as quartz and feldspars. A more detailed discussion of the common soil-forming minerals and the rocks in which they are found is given in Chapter 11.

TABLE 1:1. *Four Major Size Classes of Inorganic Particles and Their General Properties*

Size Fraction	Common Name	Means of Observation	Dominant Composition
Very coarse	Stone, gravel	Naked eye	Rock fragments
Coarse	Sands	Naked eye	Primary minerals
Fine	Silt	Microscope	Primary and secondary minerals
Very fine	Clay	Electron microscope	Mostly secondary minerals

These minerals are called secondary minerals. In general, the primary minerals dominate the coarser fractions of soil, whereas secondary minerals are most prominent in the fine materials, especially in clays. Clearly, mineral particle size will have much to do with the properties of soils in the field.

1:11. SOIL ORGANIC MATTER

Soil organic matter represents an accumulation of partially decayed and partially synthesized plant and animal residues. Such material is continually being broken down as a result of the work of soil microorganisms. Consequently, it is a rather transitory soil constituent and must be renewed constantly by the addition of plant residues.

The organic matter content of a soil is small—only about 3 to 5 percent by weight in a representative mineral topsoil. Its influence on soil properties and consequently on plant growth, however, is far greater than the low percentage would indicate. Organic matter functions as a "granulator" of the mineral particles, being largely responsible for the loose, easily managed condition of productive soils. Also, it is a major soil source of two important mineral elements, phosphorus and sulfur, and essentially the sole source of nitrogen. Through its effect on the physical condition of soils, organic matter also increases the amounts of water a soil can hold and the proportion of this water available for plant growth (see Fig. 1:5). Finally, organic matter is the main source of energy for soil microorganisms. Without it, biochemical activity would come practically to a standstill.

Soil organic matter consists of two general groups: (a) original tissue and its partially decomposed equivalents, and (b) the humus. The original tissue includes the undecomposed roots and the tops of higher plants. These materials are subject to vigorous attack by soil organisms, both plant and animal, which use them as sources of energy and tissue-building material.

The gelatinous, more resistant products of this decomposition, both those synthesized by the microorganisms and those modified from the original plant tissue, are collectively known as *humus*. This material, usually black

FIGURE 1:5. Soils high in organic matter are darker in color and have higher water-holding capacities than do soils low in organic matter. The same amount of water was applied to each container (*bottom photo*). Because of the higher water-holding capacity of the soil on the left, depth of water penetration was less.

or brown in color, is colloidal in nature. Its capacity to hold water and nutrient ions greatly exceeds that of clay, its inorganic counterpart. Small amounts of humus thus augment remarkably the soil's capacity to promote plant production.

1:12. SOIL WATER—A DYNAMIC SOLUTION

Two major concepts concerning soil water emphasize the significance of this major component of the soil in relation to plant growth: (a) water is held within the soil pores with varying degrees of tenacity depending on the amount of water present; and (b) together with its dissolved salts, soil water makes up the *soil solution*, which is so important as a medium for supplying nutrients to growing plants.

The tenacity with which water is held by soil solids determines to a marked degree the movement of water in soils and its use by plants. For example, when the moisture content of a soil is optimum for plant growth (see Fig. 1:4), plants can readily absorb the soil water, much of which is present in pores of intermediate size. As some of the moisture is removed by the growing plants, that which remains is present in only the tiny pores and as thin films around the soil particles. The attraction of the soil solids for this water is great and they can compete successfully with higher plants for it. Consequently, not all the water soils can hold is available to plants. Much of it remains in the soil after plants have used some water and have wilted or died as a consequence of water shortage.

The soil solution contains small but significant quantities of dissolved salts, many of which are essential for plant growth. There is an exchange of nutrients between the soil solids and the soil solution and then between the soil solution and plants. These exchanges are influenced to a degree by the concentration of salts in the solution, which in turn is determined by the total salts in the soil and by the content of soil water. Such is the dynamic nature of this solute-bearing water and its importance to plant life.

1:13. SOIL AIR—ALSO A CHANGEABLE CONSTITUENT

Soil air differs from the atmosphere in several respects. First, the soil air is not continuous but is located in the maze of soil pores separated by soil solids. This fact accounts for its variation in composition from place to place in the soil. In local pockets, reactions involving the gases can greatly modify the composition of the soil air. Second, soil air generally has a higher moisture content than the atmosphere; the relative humidity of soil approaches 100 percent when the soil moisture is optimum. Third, the content of carbon dioxide is usually much higher and that of oxygen lower than that found in the atmosphere. Carbon dioxide is often several hundred times more concentrated than the 0.03 percent commonly found in the atmosphere. Oxygen decreases accordingly, and in extreme cases may be no more than 10 to 12 percent as compared to about 20 percent for normal atmosphere.

The content and composition of soil air is determined to a large degree by the soil–water relationships. A mixture of gases, the air simply moves into those soil pores not occupied by water. Following a rain, large pores are the first vacated by the soil water, followed by medium-sized pores as water is removed by evaporation and plant utilization. Thus the soil air ordinarily occupies the large pores and, as the soil dries out, those intermediate in size.

This explains the tendency for soils with a high proportion of tiny pores to be poorly aerated. In such soils, water dominates and the soil-air content and composition is unsatisfactory for best plant growth. The dynamic nature of soil air is apparent. The tendency for rapid changes in air content and composition has marked effects not only on the growth of economic plants but also upon the soil organisms, plant and animal, which occupy the soil.

1:14. THE SOIL—A TREMENDOUS BIOLOGICAL LABORATORY

The representative mineral soil harbors a varied population of living organisms which comprise the life of the soil. Representatives of both animals and plants are abundant in soils. The whole range in size from the larger rodents, worms, and insects to the tiniest bacteria commonly occurs in normal soils. Moreover, most organisms vary so much both in numbers and in amounts as to make precise statements impossible. For example, the number of bacteria alone in 1 gram of soil may range from 100,000 to several billion, depending on conditions.

The total weight of living matter, including plant roots, in an acre–furrow slice of a representative mineral soil may vary from 5,000 pounds to 10,000 or 20,000 pounds (see Fig. 1:6). In any case, the quantity of living organic matter is sufficient to influence profoundly the physical and chemical trend

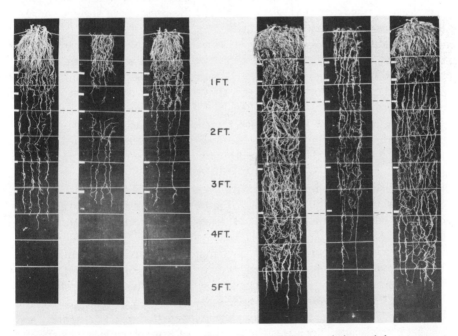

FIGURE 1:6. Plant roots tell us something about soil characteristics and the treatment the soil has received. The crop was grown on an Illinois (Cisne) soil which received no fertilizers or crop residues (*left*) and which received both fertilizers and crop residues (*right*). [*Photo courtesy J. B. Fehrenbacher, University of Illinois.*]

of soil changes. Virtually all natural soil reactions are directly or indirectly biochemical in nature.

Activities of soil organisms vary from the largely physical disintegration of plant residues by insects and earthworms to the eventual complete decomposition of these residues by smaller organisms such as bacteria, fungi, and actinomycetes. Accompanying these decaying processes is the release of several nutrient elements, including nitrogen, phosphorus, and sulfur from organic combination. By contrast, conditions in nature are such that organisms need these elements for their growth and a reversal occurs; that is, the elements are converted again into organic combinations not available to higher plants.

Humus synthesis, exclusively a biochemical phenomenon, results from soil-organism activity. This material is certainly one of the most useful products of microbial action.

1:15. CLAY AND HUMUS—THE SEAT OF SOIL ACTIVITY

The dynamic nature of the finer portions of the soil—clay and humus—has been indicated. Both these constituents exist in the "colloidal state," wherein the individual particles are characterized by extremely small size, large surface area per unit weight, and the presence of surface charges to which ions and water are attracted.

The chemical and physical properties of soils are controlled largely by clay and humus. They are centers of activity around which chemical reactions and nutrient exchanges occur. Furthermore, by attracting ions to their surfaces, they temporarily protect essential nutrients from leaching and then release them slowly for plant use. Because of their surface charges, they also act as "contact bridges" between larger particles, thus helping to maintain the stable granular structure which is so desirable in a porous, easily worked soil.

On a weight basis, the humus colloids have greater nutrient- and water-holding capacities than does clay. However, clay is generally present in larger amounts, and its total contribution to the chemical and physical properties will usually equal that of humus. The best agricultural soils contain a balance of the properties of these two important soil constituents.

COLLOIDO-BIOLOGICAL CONCEPT. Two major concepts must be emphasized before moving into a more detailed study of soil components. First, most of the chemical activity in soils is associated with a relatively small proportion of the total soil components—the colloidally active clay and humus. Second, there is a vigorous soil organism population associated with humus and plant residues which largely controls the turnover of organic materials, including humus, and regulates the supply of several nutrient elements. These two ideas lead logically to the fact that a study of soils from the standpoint of plants can best be approached from a colloido-biological viewpoint.

Chapter 2

SUPPLY AND AVAILABILITY OF PLANT NUTRIENTS IN MINERAL SOILS

THE research of nineteenth century scientists reviewed in Chapter 1 clearly identified certain elements as essential for the growth of plants. The capacity of soils to supply these essential elements is a fundamental edaphological problem. To the extent that they cannot be supplied by soils, these elements must be added in fertilizers, manures, and crop residues. In this chapter the essential elements as they exist and function in soils will be discussed. Means of supplementing the soil's capacity to supply nutrients will be considered in Chapters 14 through 20.

2:1. FACTORS CONTROLLING THE GROWTH OF HIGHER PLANTS

The essential elements are only one of the environmental factors influencing the growth of plants. In addition to the absence of disease and freedom from insect pests, six such external factors are generally recognized: (a) light, (b) mechanical support, (c) heat, (d) air, (e) water, and (f) nutrients. The soil is an agent in supplying, either wholly or in part, all of these external factors with the exception of light.

It is well to remember that plant growth is dependent upon a favorable combination of these factors and that any one of them, if out of balance with the others, can reduce or even entirely prevent the growth of plants. Furthermore, the factor which is *least* optimum will determine the level of crop production (see Fig. 2:1). This principle, sometimes called the *principle of limiting factors,* may be stated as follows: *the level of crop production can be no greater than that allowed by the most limiting of the essential plant growth factors.*

This concept is a most important one and is applicable with nutrient elements. We must be concerned not only with the supply of a given element but also with the relationship of this supply to all other factors which may affect plant growth.

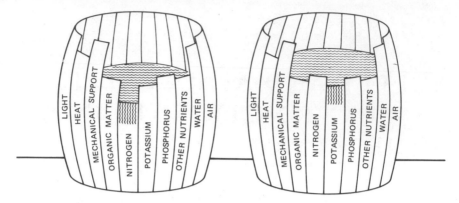

FIGURE 2:1. An illustration of the principle of limiting factors. The level of water in the barrels above represents the level of crop production. (*Left*) Nitrogen is represented as being the factor that is most limiting. Even though the other elements are present in more adequate amounts, crop production can be no higher than that allowed by the nitrogen. When nitrogen is added (*right*) the level of crop production is raised until it is controlled by the next most limiting factor, in this case, potassium.

2:2. THE ESSENTIAL ELEMENTS

Starting with the investigations of Liebig and other nineteenth century scientists discussed in Chapter 1, research has shown that certain elements are necessary for the normal growth of plants. These *essential elements* must be present in forms usable by plants and in concentrations optimum for plant growth. In addition, there must be a proper balance among the concentrations of the various soluble nutrients in the soil. Too much calcium, for example, may interfere with phosphorus and boron nutrition or may encourage chlorosis because of a reduction in the availability of the soil iron, zinc, or manganese.

Seventeen elements have been demonstrated to be essential for plant growth. They are listed in Table 2:1 and are classified according to sources— whether primarily from air and water or from soil solids—and usage— whether used by plants in relatively large or small amounts. It is not uncommon for a crop such as corn to remove 150 pounds of nitrogen per acre. The amount of boron carried away by the same crop might be as little as 10 or 15 grams.

2:3. ESSENTIAL ELEMENTS FROM AIR AND WATER

Higher plants obtain most of their carbon and oxygen directly from the air by photosynthesis, the process by which sugar is synthesized in plant leaves from water and carbon dioxide. The hydrogen is derived, either directly or indirectly, from the water of the soil. All of the other essential elements

TABLE 2:1. *Essential Nutrient Elements and Their Sources*[a]

Essential Elements Used in Relatively Large Amounts		Essential Elements Used in Relatively Small Amounts	
Mostly from Air and Water	From Soil Solids	From Soil Solids	
Carbon	Nitrogen	Iron	Copper
Hydrogen	Phosphorus	Manganese	Zinc
Oxygen	Potassium	Boron	Chlorine
	Calcium	Molybdenum	Cobalt
	Magnesium		
	Sulfur		

[a] Other minor elements, such as sodium, fluorine, iodine, silicon, strontium, and barium, do not seem to be universally essential, as are the seventeen listed here, although the soluble compounds of some may increase crop growth.

except certain supplies of the nitrogen acquired from the soil air indirectly by legumes are obtained from the soil solids.

It must not be inferred from this that the bulk of the plant tissue is synthesized from the soil nutrients. Quite the reverse is true. Ordinarily from 94 to 99.5 percent of fresh plant tissue is made up of carbon, hydrogen, and oxygen; only from 0.5 to perhaps 5 or 6 percent is from soil constituents. In spite of this, it is the nutrient elements obtained from the soil that usually limit crop development. Except in cases of drought, cold weather, poor drainage, or disease, plant growth in the field is not seriously retarded by a lack of carbon, hydrogen, and oxygen. Hence, there is justification for the nutrient emphasis that is placed on the soil and the fourteen elements that it supplies.

2:4. ESSENTIAL ELEMENTS FROM THE SOIL

MACRONUTRIENTS. Of the fourteen essential elements obtained from the soil by plants, six are used in relatively large quantities and are thus referred to as *macronutrients*. They are *nitrogen, phosphorus, potassium, calcium, magnesium,* and *sulfur*. Plant growth may be retarded because these elements are actually lacking in the soil, because they become available too slowly, or because they are not adequately balanced by other nutrients. Sometimes all three limitations are operative.

Nitrogen, phosphorus, and potassium are commonly supplied to the soil as farm manure and as commercial fertilizers. They are called *primary* or *fertilizer* elements. In the same way, calcium, magnesium and sulfur are referred to as *secondary* elements. Calcium and magnesium are added to acid soils in limestone and are called *lime* elements. Sulfur, other than that present in rain water, usually goes into the soil as an ingredient of such fertilizers as farm manure, superphosphate, and sulfate of ammonia, or is applied alone as flowers of sulfur.

MICRONUTRIENTS. The other nutrient elements (*iron, manganese, copper, zinc, boron, molybdenum, chlorine,* and *cobalt*) are used by higher plants in very small amounts, thereby justifying the name *micronutrients* or *trace elements.* Such a designation does not mean that they are less essential than the macronutrients but merely that they are needed in small quantities (see Fig. 2:2).

FIGURE 2:2. Some essential elements are required in very small quantities. Only 5 ounces of molybdenum *per acre–furrow slice* nearly doubled the yield of the alfalfa in this experiment. [*Photo courtesy W. K. Kennedy, Cornell University.*]

Except for iron and in some cases manganese, trace elements are found sparingly in most soils (see Table 2:2), and their availability to plants is often very low. Consequently, even though their removal by plants is small, the cumulative effects of crop production over a period of years may rapidly reduce the limited quantities of these elements originally present in soils. The three general soil situations where micronutrients are most apt to be a problem are (a) sandy soils, (b) organic soils, and (c) very alkaline soils. This is due to the relatively small quantities of micronutrients in sands and organic soils and to the low availability of most of these elements under very alkaline conditions.

TABLE 2:2. *Range in Micronutrient Content Commonly Found in Soils and a Suggested Analysis of a Representative Surface Soil*

Nutrient	Normal Range		Suggested Analysis of a Representative Surface Soil (ppm)
	Percent	ppm [a]	
Iron	0.500 –5.000	5,000–50,000	25,000
Manganese	0.020 –1.000	200–10,000	2,500
Zinc	0.001 –0.025	10– 250	100
Boron	0.0005 –0.015	5– 150	50
Copper	0.0005 –0.015	5– 150	50
Chlorine	0.001 –0.1	10– 1,000	50
Cobalt	0.0001 –0.005	1– 50	15
Molybdenum	0.00002–0.0005	0.2– 5	2

[a] ppm = parts per million. These estimates are based on published data from a number of sources, especially Mitchell (4).

The deficiencies of micronutrients discovered in many of our soils in recent years have highlighted the practical significance of these elements. Because macronutrient fertility problems are more widespread, however, they will be given our first attention. Micronutrients will be considered in more detail in Chapter 18.

FOUR NUTRIENT QUESTIONS. To arrive at a logical conclusion as to why nutrient deficiencies often occur in soils, four phases must be examined. They are (a) the macronutrient contents of mineral soils, (b) their forms of combination, (c) the processes by which these elements become available to plants, and (d) the soil solution and its pH. These phases will be considered in order.

2:5. MACRONUTRIENT CONTENTS OF MINERAL SOILS

The chemical composition for representative surface soils of humid temperate and arid temperate regions is given in Table 2:3. It should be noted that such figures do not fit any particular soil but present a very rough average of the data available for top soils of these two regions. These data suggest that soils of arid regions are in general higher in all of the important constituents except organic matter and nitrogen. An exception even to this is found in the black earth soils (Mollisols) of subhumid regions, which sometimes range as high as 16 percent of organic matter and 0.70 to 0.80 percent of nitrogen.

ORGANIC MATTER, NITROGEN, AND PHOSPHORUS. The percentage of organic matter exceeds that of any other constituent listed in Table 2:3. Yet its

TABLE 2:3. *Total Amounts of Organic Matter and Primary Nutrients Present in Temperate Region Mineral Surface Soils* [a]

		Representative Analyses			
		Humid Region Soil		Arid Region Soil	
Constituents	Ranges That Ordinarily May Be Expected (%)	Percent	lb/acre– furrow slice	Percent	lb/acre– furrow slice [b]
Organic matter	0.40–10.00	4.00	80,000	3.25	65,000
Nitrogen (N)	0.02– 0.50	0.15	3,000	0.12	2,400
Phosphorus (P)	0.01– 0.20	0.04	800	0.07	1,400
Potassium (K)	0.17– 3.30	1.70	34,000	2.00	40,000
Calcium (Ca)	0.07– 3.60	0.40	8,000	1.00	20,000
Magnesium (Mg)	0.12– 1.50	0.30	6,000	0.60	12,000
Sulfur (S)	0.01– 0.20	0.04	800	0.08	1,600

[a] As a supplement to the generalized figures of Table 2:3 the analyses of eight representative United States surface soils are presented as published by Marbut (3):

Constituents	Norfolk Fine Sand, Florida (%)	Sassa-fras Sandy Loam, Virginia (%)	Ontario Loam, New York (%)	Loam from Ely, Nevada (%)	Hagers-town Silt Loam, Tennes-see (%)	Cascade Silt Loam, Oregon (%)	Marshall Silt Loam, Iowa (%)	Summit Clay from Kansas (%)
SiO_2	91.49	85.96	76.54	61.69	73.11	70.40	72.63	71.60
TiO_2	0.50	0.59	0.64	0.47	1.05	1.08	0.63	0.81
Fe_2O_3	1.75	1.74	3.43	3.87	6.12	3.90	3.14	3.56
Al_2O_3	4.51	6.26	9.38	13.77	8.30	13.14	12.03	11.45
MnO	0.007	0.04	0.08	0.12	0.44	0.07	0.10	0.06
CaO	0.01	0.40	0.80	5.48	0.37	1.78	0.79	0.97
MgO	0.02	0.36	0.75	2.60	0.45	0.97	0.82	0.86
K_2O	0.16	1.54	1.95	2.90	0.91	2.11	2.23	2.42
Na_2O	Trace	0.58	1.04	1.47	0.20	1.98	1.36	1.04
P_2O_5	0.05	0.02	0.10	0.18	0.16	0.16	0.12	0.09
SO_3	0.05	0.07	0.08	0.12	0.07	0.21	0.12	0.11
Nitrogen	0.02	0.02	0.16	0.10	0.27	0.08	0.17	0.09

[b] The furrow slice of a representative mineral soil is considered to contain approximately 2 million pounds of dry earth to the acre.

amount in most surface soils usually is critical. It is of prime importance in keeping the soil loose and open and is an essential source of several nutrient elements. The addition and subsequent decay of organic matter in the soil is thus highly significant both physically and chemically.

Nitrogen and phosphorus are almost always present in comparatively small amounts in mineral soils. Moreover, a large proportion of these elements at any one time is held in combinations unavailable to plants. For example, even the more simple compounds of phosphorus are relatively insoluble in most soils. As a result, this element is doubly critical—low total amounts and very low availability to plants.

POTASSIUM, CALCIUM, AND MAGNESIUM. The total quantity of potassium, in marked contrast to phosphorus, is usually plentiful except in sandy soils. The main problem is one of availability. Calcium shows great variation but it is generally present in lesser amounts than is potassium. When it is lacking, soils tend to be acid. Calcium-containing limestones are generally added to correct this condition.

Magnesium, besides its importance as a nutrient, functions in the soil much as does calcium. Its deficiency in some soils has long been suspected. Until recently, however, it has not been considered especially critical because it is carried by most limestones, sometimes in large amounts. Where liming[1] is practiced, the lack of magnesium often is automatically rectified. In spite of this, magnesium deficiency is a major problem in many areas in the eastern United States.

SULFUR. Although it is usually no more plentiful than phosphorus, sulfur is more readily available. This is because its simple inorganic compounds are not rendered insoluble by reacting with certain other soil constituents as is the case with phosphorus. As already suggested, the addition of sulfur in farm manure, rain water, and fertilizers tends in an automatic way to relieve a possible deficiency in humid temperate regions. In certain areas of the west and south, however, specific additions of sulfur-containing compounds are required.

CRITICAL CONSTITUENTS. The above discussion seems to indicate that three constituents are likely to be critical in almost all mineral soils. Two—*organic matter* and *nitrogen*—merit particular attention because of the small amounts originally present and because of their ready loss through oxidation, leaching, or crop removal. The third, *phosphorus*, faces a double handicap as already explained—an exceptionally small amount present and a low availability to higher plants.

Under humid conditions *calcium* by all means must be included in the above list because it is sure to be much depleted by leaching. Consequently, it is needed not only as a nutrient but also as a means of controlling soil acidity. In arid regions, where the leaching of calcium usually is negligible, this nutrient is likely to be present in abundance, especially in the subsoil.

---

[1] Liming refers to the application to agricultural soils of basic calcium- and magnesium-containing materials with the objective of reducing soil acidity. Ground limestones are most commonly used, although burned lime (CaO) or slaked lime [$Ca(OH)_2$] are sometimes added. Collectively, these materials are referred to as *lime* and the practice of adding them as *liming* (see Chapter 15).

It is not to be inferred from the preceding generalizations, however, that *potassium, magnesium,* and *sulfur* may not be lacking in certain soils or that the problem of their supply may not at times be critical. The ever-increasing use of potash fertilizers, the demand for magnesium-containing dolomitic limestone, and the emphasis placed on sulfur additions are evidence of this.

2:6. FORMS OF MACRONUTRIENTS IN SOILS

The nutrient elements generally exist in two conditions: (a) complex and rather insoluble compounds, and (b) simple, more soluble forms readily available to higher plants. The latter are termed to be present in the *soil solution.* As a result of the chemical and biochemical processes at work, the general trend of the elements in the soil is from the complex to the simpler forms (see Table 2:4). The reverse process—that of synthesis and increased complexity—does occur, however. The building of proteins from simple nitrogen salts and the reversion of soluble phosphates to complex and insoluble compounds are examples.

SIMPLE AVAILABLE FORMS. The simpler and more soluble constituents of soils, especially those of humid regions, are subject to loss by leaching or are used by microorganisms and higher plants. Consequently, the greater proportion of the macronutrients exists in the soil in complex conditions. From thence they gradually become available through various processes of simplification. As a result, the productive capacity of a soil depends not so much upon the total amounts of the various nutrients present as upon the ease with which transfer is made to simple and available forms. Such a situation indicates why a *total* chemical analysis is likely to be of uncertain value in deciding the fertilizer needs of a soil. It is rather discouraging that the total amount of a nutrient may be determined with great accuracy, whereas its *availability* may be known in only a general way.

ORGANIC COMBINATIONS. The decomposition of soil organic matter allows nutrients held in this complex form to be released and simple compounds appear which are available to higher plants. Practically all of the nitrogen and much of the sulfur and phosphorus are held in organic combinations (see Table 2:4). Since phosphorus in complex mineral forms usually is very slowly available, there is some advantage of the organic association. Even so, organic phosphorus does not become available as easily and quickly as do organic sulfur and nitrogen.

INORGANIC COMBINATIONS. Most of the potassium, calcium, and magnesium exists in the soil in strictly inorganic forms. There is quite a difference, however, in the degree of availability of the three elements. For example, a much larger amount of calcium is held in an easily replaceable or *exchangeable*

TABLE 2:4. *Forms in Which the Macronutrients Occur in Mineral Soils*

Group 1 More Complex and Less Active Forms	Group 2 Some of the Simpler and More Available Forms and Their Ionic Equivalents	
Nitrogen		
Organic combinations: proteins, amino acids, and similar forms; colloidal and subject to decomposition	Ammonium salts Nitrite salts Nitrate salts	NH_4^+ NO_2^- NO_3^-
Phosphorus		
Apatite, an original source; secondary Ca, Fe, Al phosphates Organic; phytin, nucleic acid, and other combinations	Phosphate of Ca, K, Mg, etc. Soluble organic forms	$\left\{\begin{array}{l} HPO_4^{--} \\ H_2PO_4^- \end{array}\right.$
Potassium		
Original minerals such as feldspars and mica Complex secondary aluminum silicates such as clays, especially illite	Potassium ions adsorbed by colloidal complex Potassium salts such as sulfates, carbonates, etc.	$\left.\begin{array}{l} \\ \\ \\ \end{array}\right\} K^+$
Calcium		
Minerals such as feldspars, hornblende, calcite, and dolomite	Calcium ions adsorbed by colloidal complex A variety of simple calcium salts	$\left.\begin{array}{l} \\ \\ \\ \end{array}\right\} Ca^{++}$
Magnesium		
Minerals such as mica, hornblende, dolomite, and serpentine Secondary aluminum silicates such as clays, especially montmorillonite, chlorite, and vermiculite	Magnesium ions adsorbed by colloidal complex Numerous simple salts of magnesium	$\left.\begin{array}{l} \\ \\ \\ \end{array}\right\} Mg^{++}$
Sulfur		
Mineral combinations such as pyrite and gypsum Organic forms; colloidal and subject to decomposition	Various sulfites and sulfates of Ca, K, Mg, etc.	SO_3^{--} SO_4^{--}

condition by the colloidal fractions of the soil (see Fig. 2:3). In this form it is quite readily available to plants. The quantity of this element in an exchangeable or available form far exceeds that of any other macronutrient in the soil. This is reflected in humid regions in the significant losses of calcium in drainage water and the consequent need for its replacement by liming (application of limestone).

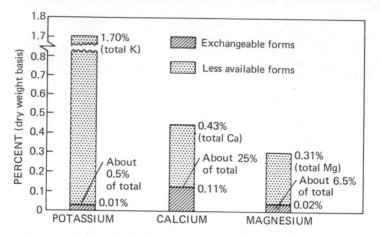

FIGURE 2:3. Percentages of total exchangeable (replaceable) K, Ca, and Mg present in a representative humid temperate regional mineral soil. Note that the amount of exchangeable Ca is approximately 10 times greater than the exchangeable K and 5 times that of the exchangeable Mg. About 25 percent of the total Ca is represented as exchangeable; the corresponding figures for K and Mg are 0.5 and 6.5 percent, respectively.

The situation in respect to potassium and to a lesser extent for magnesium is quite different. A very high percentage of the total quantities of these elements is held in the less available forms (see Fig. 2:3). The exchangeable or readily available potassium usually makes up less than 1 percent of the total quantity of this element in the soil. Although the relative content of exchangeable magnesium is usually somewhat higher, it is still lower than that of calcium.

Although sulfur is held in both mineral and organic forms (see Table 2:4), the latter combinations tend to predominate in humid region surface soils. The availability of sulfur depends on the rate of organic decomposition. Its transfer from organic forms to inorganic salts which can be used by plants seldom gives trouble. In soils of arid regions, considerable sulfur may occur in the sulfate form as well as in organic combination. Gypsum is a common carrier of inorganic sulfur in such soils. The presence of calcium sulfate in the lower horizons of the Mollisols of the Dakotas is an example (Fig. 12:7).

The quantity of simple inorganic salts such as potassium chloride, sodium chloride, and sodium sulfate in humid region soils is generally quite small. As one moves into more arid climates these salts are present in a somewhat higher concentration, especially in the lower horizons. Where drainage and leaching are restricted and rainfall is low, salts may be found in the plow layer or even at the soil surface in quantities sufficiently large to hinder or even prevent crop production.

2:7. TRANSFER OF PLANT NUTRIENTS TO AVAILABLE FORMS

NITROGEN. Since most of the soil nitrogen is found in organic matter, the latter must decompose if the nitrogen is to appear in simple forms. This decomposition is a very complex, biochemical process carried out by soil microorganisms. The nitrogen finally emerges as an ammonium compound which, if conditions are favorable, is oxidized to the nitrite and then to the nitrate form. The two oxidation changes are spoken of as *nitrification* and are brought about by two special purpose bacterial groups. Because most of the nitrogen utilized by higher plants is absorbed in the ammonium and nitrate forms, the importance of these processes is obvious. The transformations may be outlined simply as follows:

Organic nitrogen $\rightarrow$ Ammonium salts $\rightarrow$ Nitrate salts $\rightarrow$ Nitrate salts
Proteins, amino acids, etc. NH_4^+ NO_2^- NO_3^-

Mineralization   Nitrification

These microbiological transformations are influenced profoundly by soil conditions. When the soil is cold, waterlogged, or excessively acid, these changes do not progress rapidly. The nitrifying organisms are especially sensitive to these conditions.

When organic matter containing a large amount of carbon compared to nitrogen is added to a soil, the above processes may be reversed temporarily. The soil microorganisms, having large amounts of energy-producing materials at their disposal, multiply rapidly and use the nitrogen themselves, thus interfering with its simplification and appearance as ammonium and nitrates. In such cases, the soil organisms are competing directly with the higher plants. Thus, the simplification of the nitrogen is not always easy, rapid, or in proportion to the amounts present. This must be taken into consideration in the practical management of any soil.

PHOSPHORUS. When soil phosphorus is held in organic combination, decay will encourage its simplification, as was the case with nitrogen. The mineral phosphorus, however, presents a much more difficult problem. The various native soil phosphates are present in small quantities and are usually rather insoluble in water. Even when plant rootlets, aided by carbon dioxide and other root exudates, are in intimate contact with the mineral phosphates, the rate of solution is slow. A simple example illustrative of the solvent influence of carbon dioxide and water is as follows, where tricalcium phosphate represents the various insoluble soil phosphates:

$$Ca_3(PO_4)_2 + 4H_2O + 4CO_2 \rightarrow Ca(H_2PO_4)_2 + 2Ca(HCO_3)_2$$

Insoluble phosphate   Water-soluble phosphate   Soluble calcium bicarbonate

By this means growing plants encourage an availability of phosphorus that might otherwise be almost negligible. Thus, a soil may supply a crop with appreciable quantities of phosphorus and yet the soil solution and drainage water may contain very small amounts of this element (see Fig. 2:4).

FIGURE 2:4. Photomicrograph of dicalcium phosphate dihydrate crystals bonded to a root hair. This illustrates the intimate relationship between plant roots and some chemical fertilizers. [*Photo courtesy J. R. Lehr, Tennessee Valley Authority.*]

Simplification of phosphorus, like that of nitrogen, may be reversed. Microorganisms readily appropriate simple and soluble phosphorus compounds and build them up into complex organic forms. Soluble fertilizer compounds such as $Ca(H_2PO_4)_2$ and $NH_4H_2PO_4$ may also be changed to insoluble calcium phosphates or to the equally complex and insoluble iron and aluminum combinations. Such forms liberate their phosphorus very reluctantly.

POTASSIUM AND CALCIUM. Some of the calcium and most of the potassium occur as components of complex soil minerals. These forms slowly succumb through the years to the solvent action of water charged with carbonic and other acids. The ease with which the essential elements are rendered soluble depends upon the complexity of the soil minerals and on the intensity of weathering. The general reaction may be illustrated as follows, assuming the potash feldspar microcline to represent the complex minerals and carbonic acid, the soil acids:

$$2KAlSi_3O_8 + H_2CO_3 + H_2O \rightarrow H_4Al_2Si_2O_9 + K_2CO_3 + 4SiO_2$$

Microcline   Hydrated   Soluble
feldspar   silicate   carbonate

The potassium released through such a reaction may be taken up by plants, lost in drainage, or held by the negatively charged soil colloids. The latter form of nutrient combination will now receive our attention.

A small proportion of the potassium and much of the calcium present in soils is held on the surfaces of the colloids as *adsorbed*[2] cations. These cations are easily released to the soil solution by exchanging with other positively charged ions, giving rise to the term *exchangeable* cations. The general reaction may be shown as follows, assuming that hydrogen from an acid such as carbonic acid replaces calcium from the soil colloids:

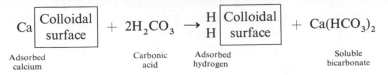

Adsorbed   Carbonic   Adsorbed   Soluble
calcium   acid   hydrogen   bicarbonate

A great deal of the calcium becomes mobile by this type of reaction. The replacement, called *cation exchange*, takes place with surprising ease and rapidity and is one of the most important types of reactions occurring in soils. Because of this situation, soils which contain considerably more total potassium than total calcium (see Fig. 2:3) release the latter element to the soil solution and to leaching much more lavishly. This has a direct bearing on calcium nutrition, soil acidity, liming, and other practical considerations.

The colloidal complex of the soil carries other cations besides calcium and hydrogen in a replaceable condition. Aluminum, magnesium, potassium, ammonium, sodium, and other ions are present in varying amounts.

MAGNESIUM. Magnesium is held in the soil in much the same conditions as is calcium and is released by the exchange reaction cited above. Some magnesium also comes directly from the soil minerals by weathering as does potassium. Hence, the illustrations cited for the transfer of potassium and calcium will serve for magnesium also. It is only necessary to visualize the release of soluble magnesium by the breakdown of minerals in one case and by ionic exchange in the other.

   [2] *Adsorption* refers to the adhesion of substances to the surfaces of solids. In soils it has to do with the attraction of ions and of water molecules to colloidal particles. The ions are not too tightly held, being replaceable by or exchangeable with ions of a like charge. *Absorption*, in contrast, refers to surface penetration such as takes place when nutrients and water enter plant roots. Thus, calcium ions are absorbed as they are taken in by plant roots but are adsorbed by soil colloids.

Like nitrogen and phosphorus, the release of potassium, calcium, and magnesium to the soil solution may easily be reversed. Thus, when soluble compounds of potassium, calcium, or magnesium are added to a soil, the colloidal matter adsorbs large quantities of the metallic ions. Potassium may be even more firmly fixed as a molecular part of the mineral colloids.

SULFUR. Transformations of sulfur, as of nitrogen, are largely biological. They go on readily in most soils and while subject to marked retardation at times, such influences are apparently not as serious as in the case of nitrogen. The transformations may be indicated as follows:

$$\text{Organic sulfur} \rightarrow \text{Decay products} \rightarrow \text{Sulfites} \rightarrow \text{Sulfates}$$

| Proteins and other organic combinations | Of which H_2S and other sulfides are simple examples | SO_3^{--} | SO_4^{--} |

$$\underbrace{\qquad\qquad}_{\text{Mineralization}} \qquad \underbrace{\qquad\qquad\qquad}_{\text{Sulfur oxidation}}$$

Sulfur oxidation, like nitrification, is brought about largely by specific types of bacteria. The sulfate compounds that result (see Table 2:4) are the source of most of the sulfur acquired by higher plants.

Soil organisms, especially bacteria and fungi, utilize sulfur as well as the metal cations already mentioned. The synthetic activities of these microbes are excellent examples of a practical reversal of the simplifying processes. While such transpositions temporarily compete with higher plants, they tend to conserve nutrients by reducing the loss of valuable constituents in drainage water.

2:8. SOIL SOLUTION

The soil solution is simply soil water in which are dissolved the ionic forms of plant nutrients discussed in the previous section. Because of its segregation in the large and small pore spaces of the soil, the soil solution is not always continuous nor does it move freely in the soil (see Section 1:12). Also, the soil solution is exceedingly changeable, varying as to the gross amount present as well as to the amount and proportion of its soluble constituents.

CONCENTRATION OF SOIL SOLUTION. As the soil moisture content is reduced by evaporation, the concentration of soluble salts in the soil solution rises. Moisture fluctuation in a humid region mineral soil is of sufficient range to permit a variation in the concentration of the soil solution from a few parts per million to 30,000 parts per million. Under ordinary conditions the acre–furrow slice of an arable humid region mineral soil contains from 500 to 1,000 pounds of soluble salts.

In arid and semiarid regions, the soil solution is usually more concentrated than where the rainfall is heavier. Under conditions of low rainfall and restricted drainage, salt concentrations are so high as to interfere at times with the growth of plants. The presence of even 0.5 percent total soluble salts is considered serious. This would mean about 10,000 pounds to an acre–furrow slice.

REACTION OF THE SOIL SOLUTION. Another important property of the soil solution is its reaction—that is, whether it is *acid, neutral,* or *alkaline.* Some soil solutions possess a preponderance of hydrogen over hydroxyl ions and therefore are acid. Some show the reverse and are alkaline, while others, which have an equal concentration of hydrogen and hydroxyl ions, are neutral (see Fig. 2:5).

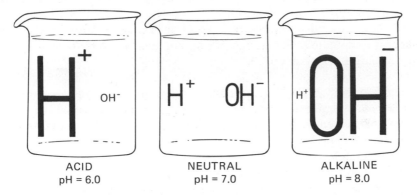

| ACID | NEUTRAL | ALKALINE |
| pH = 6.0 | pH = 7.0 | pH = 8.0 |

FIGURE 2:5. Diagrammatic representation of neutrality, acidity, and alkalinity. At neutrality (*center*) the H ions and OH ions of a solution are balanced, their respective numbers being the same (pH 7). (*Left*) At pH 6, the H ions are dominant, being 10 times greater, while the OH ions have decreased proportionately, being only $\frac{1}{10}$ as numerous. The solution, therefore, is acid, there being 100 times more H ions than OH ions present. (*Right*) At pH 8, the exact reverse is true, the OH ions being 100 times more numerous than the H ions. Hence, the solution is alkaline. This mutually inverse relationship must always be kept in mind in using pH data.

The exact relationship in any particular case is ordinarily evaluated in terms of hydrogen ion concentration, which is usually expressed in terms of pH. The pH value of a solution is the logarithm of the reciprocal of the hydrogen ion concentration. It may be stated conveniently as follows:

$$pH = \log \frac{1}{[H^+]}$$

Also, the concentration of the hydrogen ion (and consequently the pH) is related mathematically to the concentration of the hydroxyl ion. In any solution in which water is the solvent, the product of the concentration of these two ions is approximately 10^{-14} at 25°C. Thus,

$$[H^+] \times [OH^-] = 10^{-14}$$

If, for example, pH = 6, the hydrogen ion concentration is 10^{-6} equivalent per liter and the hydroxyl ion concentration is

$$\frac{10^{-14}}{10^{-6}} = 10^{-8} \text{ equivalent per liter}$$

It is not to be inferred that the distribution of the hydrogen and hydroxyl ions in the soil solution is homogeneous. Hydrogen ions usually are adsorbed by soils to a much greater degree than are the hydroxyl ions. Consequently, the hydrogen ions are seen (a) as being especially concentrated at and near the colloidal interfaces, and (b) as becoming less numerous as the outer portions of the water films are approached (see Fig. 14:4). Since hydroxyl ions vary in number inversely with hydrogen ions, this causes a higher pH in the outer moisture zones. Such a situation has many nutritional consequences in respect to both microorganisms and higher plants and will receive more attention later (see pp. 390–91).

RANGES IN PH. In Fig. 2:6 are shown the ranges of soil pH encountered as well as the relationship between pH values and terms commonly used to describe soil reaction. For mineral soils the extreme range in pH extends from near 3.5 to perhaps 10 or above. It is to be noted in this connection that certain peat soils may show a pH of less than 3. At the other extreme are alkali soils, some of which may reach a pH near 11.

The common ranges in pH shown by soils of humid regions and arid regions are sharply in contrast with the extreme spread just noted. That

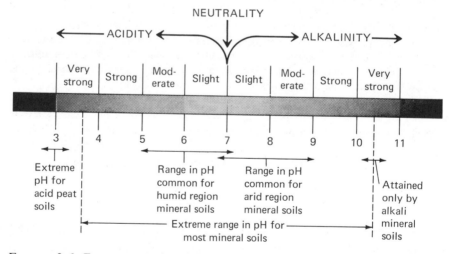

FIGURE 2:6. Extreme range in pH for most mineral soils and the ranges commonly found in humid region and arid region soils. The maximum alkalinity for alkali soils is also indicated, as well as the minimum pH likely to be encountered in very acid peat soils.

for soils of humid regions extends roughly from somewhat below 5 to above 7. Also, the latter figure overlaps the range common to soils of arid regions, whose usual pH spread is from a little below 7 to approximately 9.

2:9. NUTRITIONAL IMPORTANCE OF SOIL pH

The soil pH may influence nutrient absorption and plant growth in two ways: (a) through the *direct* effect of the hydrogen ion; or (b) *indirectly*, through its influence on nutrient availability and the presence of toxic ions. In most soils the latter effect is of great significance. Although at extreme pH values the direct toxic effect of the hydrogen ion can be demonstrated, most plants are able to tolerate a wide range in the concentration of this ion so long as a proper balance of the other elements is maintained. Unfortunately, the availability of several of the essential nutrients is drastically affected by soil pH, as is the solubility of certain elements that are toxic to plant growth. Examples of these effects will be cited.

Several essential elements tend to become less available as the pH is raised from 5.0 to 7.5 or 8.0. Iron, manganese, and zinc are good examples. Molybdenum availability, on the other hand, is affected in the opposite way, being higher at the higher pH levels. Phosphorus is never readily soluble in the soil, but it seems to be held with least tenacity in a pH range centering around 6.5 (see p. 462 and Fig. 17:3). Here, most plants seem to be able to extract it from the soil with least difficulty.

At pH values below about 5.0, aluminum, iron, and manganese are often soluble in sufficient quantities to be toxic to the growth of some plants. At very high pH values, the bicarbonate ion is sometimes present in sufficient quantities to interfere with the normal uptake of other ions and thus is detrimental to optimum growth. These few examples of the indirect effects of soil pH show why much importance must be placed on this characteristic in the diagnosis of fertility problems.

2:10. FORMS OF ELEMENTS USED BY PLANTS

There are two general sources of readily available nutrients in the soil. These are (a) nutrients adsorbed on the colloids, and (b) salts in the soil solution. In both cases the essential elements are present as ions, such as K^+, Ca^{++}, Cl^-, SO_4^{--}. The positively charged ions (cations), such as K^+, are mostly adsorbed by colloids, whereas the negatively charged ions (anions) and a small fraction of the cations are found in the soil solution.[3] The more

---

[3] Some soil solids have positive charges to which negatively charged ions (anions) are attracted. Such solids adsorb phosphates, sulfates, and nitrates. The exchange of negatively charged ions with the soil solution will be presented later (see p. 92).

important ions present in the soil solution or on the soil colloids may be tabulated as follows:

Nitrogen	NH_4^+,	NO_2^-, NO_3^-	Calcium	Ca^{++}
Phosphorus	HPO_4^{--},	$H_2PO_4^-$	Magnesium	Mg^{++}
Potassium	K^+		Sulfur	SO_3^{--}, SO_4^{--}
Iron	Fe^{++},	Fe^{3+}	Zinc	Zn^{++}
Molybdenum	MoO_4^{--}		Boron	BO_3^{3-}
Manganese	Mn^{++},	Mn^{4+}	Chlorine	Cl^-
Copper	Cu^+,	Cu^{++}	Water	H^+, OH^-

NITROGEN, PHOSPHORUS, AND SULFUR. Most of the nitrogen absorbed by plants is in either the ammoniacal or the nitrate form, depending on the conditions of the soil, the kind of plant, and its stage of growth. In general, the presence of both ions seems most favorable. The nitrite ion is generally present only in small quantities because it is so readily oxidized to the nitrate form. This is fortunate since any appreciable concentration of nitrite nitrogen is likely to be toxic to plants.

The particular phosphate ion presented to higher plants is determined to a considerable extent by the pH of the soil. When the pH is distinctly alkaline, the HPO_4^{--} ion is the form in which soluble phosphorus occurs. As the pH is lowered and the soil becomes slightly to moderately acid, both HPO_4^{--} and $H_2PO_4^-$ ions prevail; at high acidities the phosphorus is present largely as $H_2PO_4^-$ ion. Both these forms are believed to be absorbed by higher plants. Soluble organic forms of phosphorus cannot be used to any extent directly by higher plants but must undergo mineralization and appear in the mineral forms before appreciable utilization takes place.

The uptake of sulfur by higher plants apparently is largely as the SO_4^{--} ion. This is the final product of oxidation, and if the sulfur organisms are vigorous, few SO_3^{--} ions can accumulate. The situation in some respects resembles that already cited regarding nitrite oxidation.

OTHER ELEMENTS. Little need be said at this point regarding potassium, calcium, magnesium, zinc, boron, and chlorine since they occur in the soil solution in only one ionic form. But iron, manganese, and copper are in a slightly different category. The oxidation–reduction condition of the soil is a factor here. If the soil is well aerated, the ion of higher valence in each case tends to predominate. But if drainage is poor, reduction may occur and the lower-valent forms will be present. Thus, aeration is of considerable nutritive importance.

The carbon dioxide of the atmosphere in its photosynthetic role is the direct source of most of the carbon acquired by higher plants but not necessarily of all. In the soil solution carbonate and bicarbonate ions occur, and there is some evidence that these may be adsorbed by higher plants although the bulk of carbon present in plant tissue comes from the atmosphere through photosynthesis.

2:11. SOIL AND PLANT INTERRELATIONS

The uptake of nutrients by plants is determined not only by the "availability" of soil-held nutrients but by the supply of these nutrients to the plant root surfaces and by the nutrient absorption rates at these surfaces (1). Nutrients are supplied to the root surfaces in three ways. First, the roots and root hairs penetrate the soil and by so doing come in direct contact with the soil colloids. In other words, the roots move to the colloids and to the nutrients they hold. This is termed *root interception*. Second, some nutrients move to the roots along with water which the plants absorb for normal growth. Such movement is called *mass flow*. Third, as nutrients are absorbed by roots, a concentration gradient is set up between the zone immediately surrounding the root and the soil zones farther away. In response to this gradient, *diffusion* of ions toward the root surfaces takes place. For cations such as K^+ and Ca^{++} diffusion is by far the most important means of supplying nutrients to plant roots. Diffusion is also important for anions such as NO_3^-, although mass flow can also be quite significant for these negatively charged ions.

NUTRIENT ABSORPTION. Nutrient solubility and availability are not strictly soil phenomena with the plant simply absorbing in a passive way that which is presented to it. Nutrient solubility is markedly affected by root exudates and by microbial activity in the vicinity of the roots (the rhizosphere). Furthermore, once nutrients are solubilized their entrance into root cells is determined to a large extent by reactions associated with the plant. Aerobic respiration of root cells supplies energy for nutrient absorption, and reactions within the root cell membrane determine the rate at which any given element can be absorbed.

Ions move across a root cell membrane with the help of "nutrient carriers" (Fig. 2:7). The carrier for a given element (C) is "energized" by the process of respiration and in this state selectively binds ions from the soil solution. The carrier, coupled with the ion, moves across the cell membrane and releases the ion into the interior of the cell. This process makes possible the movement of ions from a dilute soil solution into a more concentrated solution in the cell. Furthermore, because a carrier is specific for one ion or a group of ions, it permits the ions of one element to be preferentially absorbed over others present in the soil solution.

Nutrient uptake by plants requires intimate root–soil association. It is accentuated by (a) root exudates, and (b) microbial activity in the immediate neighborhood of the absorbing surfaces. Plant roots give off large amounts of carbon dioxide and other acid-forming substances. These speed up interchange to a remarkable degree. Also, organic excretions from plant roots provide food and energy for microorganisms. The concentration of microbial activity within the plant-root zone (rhizosphere) is ample proof of this. Such

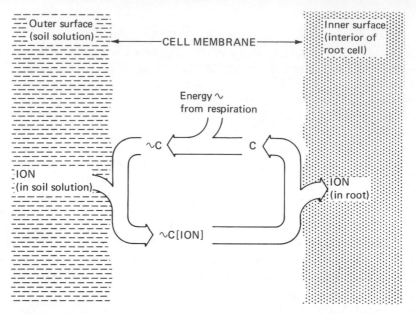

FIGURE 2:7. An illustration of how nutrient elements are thought to be transported from the soil solution into the root cell. An organic carrier, C, is specific for a given ion or group of ions. It is energized within the cell membrane through the process of respiration. In this energized state, $\sim$C, it binds ions of a specific element in the soil solution, $\sim$C[ion], and transports them across the membrane. At the inner surface of the membrane the carrier releases the ion into the interior of the root cell. The carrier may then be re-energized and the process is repeated. [*Modified from Hanson* (2).]

biochemical phenomena increase greatly the rate and ease of transfer of nutrients from soil to plant.

2:12. SOIL FERTILITY INFERENCES

Certain practical conclusions are inescapable in respect to the plant–nutrient relationships just presented. An adequate supply of each nutrient must be maintained in the soil. In addition, provision must be made for a rate of availability suitable to normal crop growth. This involves a complex transfer to the soil solution and to the plant, which participates in ways other than those of mere absorption.

Moreover, an adequate nutrient proportion is requisite, the total concentration of available nutrients being vital. Such a balance tends to ensure the desirable physiological conditions necessary for successful plant production. The pH of the soil solution, since it influences profoundly many of the important soil and plant processes, plays a critical role in such an adjustment. Soil management, to be successful, must encompass all of these phases.

REFERENCES

(1) Barber, S. A., "Mechanism of Potassium Absorption by Plants," in Kilmer, V. J., Younts, S. E., and Brady, N. C., Eds., *The Role of Potassium in Agriculture* (Madison, Wis.: American Society of Agronomy, Soil Science Society of America, and Crop Science Society of America, 1968).

(2) Hanson, J. B., "Roots—Selectors of Plant Nutrients," *Plant Food Review,* p. 8, Spring 1967.

(3) Marbut, C. F., "Soils of the United States," *Atlas of American Agriculture*, Part 3 (Washington, D.C.: U. S. Department of Agriculture, 1935).

(4) Mitchell, R. L., "Trace Elements," in Bear, F. E., Ed., *Chemistry of the Soil* (New York: Reinhold, 1955).

Chapter 3

SOME IMPORTANT PHYSICAL PROPERTIES OF MINERAL SOILS

PHYSICALLY, a mineral soil is a porous mixture of inorganic particles, decaying organic matter, and air and water. The larger mineral fragments usually are embedded in and coated over with colloidal and other fine materials. In some cases, the larger mineral particles predominate and a gravelly or sandy soil results. In other cases, the mineral colloids are more prevalent, which gives the soil clayey characteristics. All gradations between these extremes are found in nature. Organic matter acts as a binding agent to encourage the individual particles to cluster into clumps or aggregates.

Two very important physical properties of soils will be considered in this chapter: _soil texture_ and _soil structure_. Soil texture is concerned with the size of mineral particles. Specifically it refers to the relative proportion of particles of various sizes in a given soil. No less important is soil structure, which is the arrangement of soil particles into groups or aggregates. Together, these properties help determine not only the nutrient-supplying ability of soil solids but also the supply of water and air so important to plant life.

3:1. CLASSIFICATION OF SOIL PARTICLES AND MECHANICAL ANALYSIS

The size of particles in mineral soil is not subject to ready change. Thus, a sandy soil remains sandy and a clay soil remains a clay. For this reason the proportion of various size groups in a given soil (the texture) assumes added significance. It cannot be altered and thus is considered a basic property of a soil.

To study successfully the mineral particles of a soil, scientists usually separate them into convenient groups according to size. The various groups are spoken of as _separates_. The analytical procedure by which the particles are separated is called a _mechanical analysis_. It is a determination of the particle size distribution.

As might be expected, a number of different classifications have been devised. The size ranges for four of these systems are shown in Fig. 3:1. The

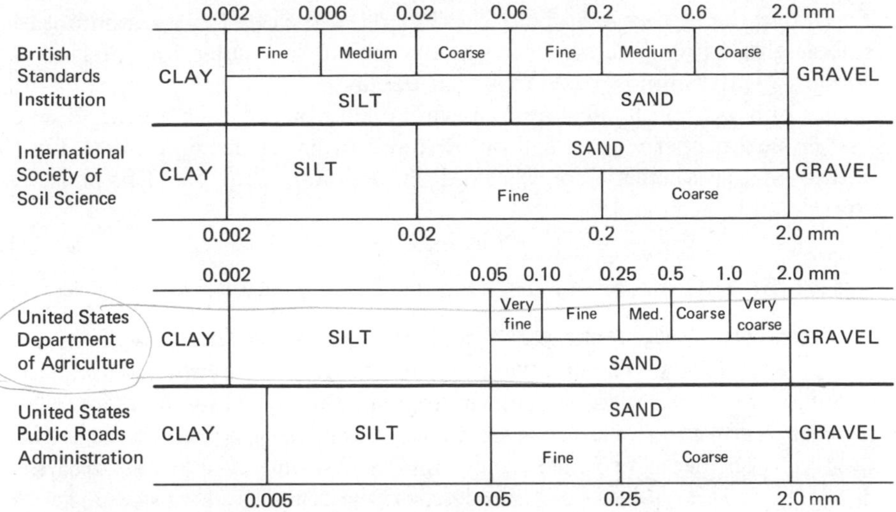

	0.002	0.006	0.02	0.06	0.2	0.6	2.0 mm
British Standards Institution	CLAY	Fine	Medium	Coarse	Fine	Medium	Coarse	GRAVEL
		SILT			SAND			

International Society of Soil Science	CLAY	SILT		SAND			GRAVEL
				Fine		Coarse	

| 0.002 | 0.02 | 0.2 | 2.0 mm |

	0.002	0.05	0.10	0.25	0.5	1.0	2.0 mm
United States Department of Agriculture	CLAY	SILT	Very fine	Fine	Med.	Coarse	Very coarse	GRAVEL
			SAND				

United States Public Roads Administration	CLAY	SILT	SAND			GRAVEL
			Fine		Coarse	

| 0.005 | 0.05 | 0.25 | 2.0 mm |

FIGURE 3:1. Classification of soil particles according to size by four systems. The U.S. Department of Agriculture system is used in this text. (Particle diameter in logarithmic scale.)

classification established by the U. S. Department of Agriculture will be used in this text.

MECHANICAL ANALYSES. To make a mechanical analysis, a sample of soil is broken up and the very fine sand and larger fractions are separated into the arbitrary groups by sieving. The silt and clay percentages are then determined by methods which depend upon the rate of settling of these two separates from suspension. The principle involved in the method is simple. When soil particles are suspended in water they tend to sink, and rapidity of settling is roughly proportional to their size. The suspension of a sample of soil is therefore the first step; the second step is that of settling and the withdrawal by some means of successive grades; and the third step is the determination of the percentage of each group of particles based on the original sample.

Although stone and gravel figure in the practical examination and evaluation of a field soil, they do not enter into the analysis of the fine earth. Their amounts are usually rated separately. The organic matter, ordinarily comparatively small in quantity, is normally removed by oxidation before the mechanical separation.

Sand, when dominant, yields a coarse-textured soil which has properties known to everyone as *sandy*. Such soils are sometimes referred to as *light* since they are easily tilled and cultivated. On the other hand, a fine-textured soil is made up largely of silt and clay, and its plasticity and stickiness indicate that it is likely to be difficult to till or cultivate and is therefore termed *heavy*.

The use of the terms "light" and "heavy" refer to ease of tillage and not to soil weight. As we shall see later, the dry weight of a cubic foot of sand is actually greater in most cases than that of clay.

Not only is a mechanical analysis valuable in picturing in a general way the physical properties of a soil but it is also of use in deciding the textural name—that is, whether a soil is a sand, sandy loam, loam, etc. This phase is considered in Section 3:5.

3:2. Physical Nature of the Soil Separates

COARSE SEPARATES. Stone, gravel, and sand, because of their sizes, function as separate particles. The first two range in size from 2 mm upward and may be more or less rounded, irregularly angular, or even flat. The distinction between gravel and stone is now technically based on size. Gravel, chert, and slate are considered to range from 2 mm to 3 inches along their greatest diameter. Stone, cobbles, and boulders, on the other hand, exceed 3 inches in respect to their greatest dimension.

Sand grains may be rounded or quite irregular depending on the amount of abrasion that they have received (see Fig. 3:2). When not coated with

FIGURE 3:2. Sand grains from soil. Note that the particles are irregular in size and shape. Quartz usually predominates, but other minerals may occur. Silt particles have about the same shape and composition, differing only in size.

clay and silt, such particles are not sticky even when wet. They do not possess the capacity to be molded (plasticity) as does clay. Their water-holding capacity is low, and because of the large size of the spaces between the separate particles, the passage of percolating water is rapid. Hence, they facilitate drainage and encourage good air movement. Soils dominated by sand or gravel, therefore, possess good drainage and aeration, and are usually in a loose, open condition.

CLAY AND SILT. Surface area is the characteristic most affected by the small size and fine subdivision of silt and especially clay. A grain of fine colloidal clay has about 10,000 times as much surface area as the same weight of medium-sized sand. The specific surface (area per unit weight) of colloidal clay ranges from about 10 to 1,000 square meters per gram. The same figures for the smallest silt particles and for fine sand are 1 and 0.1 square meter per gram. Since the adsorption of water, nutrients, gas, and the attraction of particles for each other are all surface phenomena, the significance of the very high specific surface for clay is obvious. This relationship is shown graphically in Fig. 3:3.

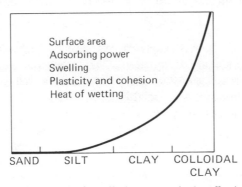

FIGURE 3:3. The finer the texture of a soil, the greater is the effective surface exposed by its particles. Note that adsorption, swelling, and the other physical properties cited follow the same general trend and that their intensities go up rapidly as the colloidal size is approached.

Clay particles commonly are micalike in shape and highly plastic when moist. When clay is wetted it tends to be sticky and is easily molded. On drying it absorbs considerable heat energy, and, on wetting again, evolves this same amount of heat. This is called the *heat of wetting*. This, too, is related to particle size (see Fig. 3:3).

In contrast with the platelike clay, silt particles are irregularly fragmental, diverse in shape, and seldom smooth or flat (see Fig. 3:2). They really are microsand particles, quartz being the dominant mineral. The silt separate, because it has an adhering film of clay, possesses some plasticity, cohesion (stickiness), and adsorption, but to a much lesser degree than the clay separate itself. The influence of silt is such as to make it a rather unsatisfactory soil constituent physically unless supplemented by adequate amounts of sand, clay, and organic matter.

The presence of silt and especially clay in a soil imparts to it a *fine texture* and slow water and air movement. Such a soil is highly plastic, becoming sticky when too wet, and hard and cloddy when dry unless properly handled.

The expansion and contraction on wetting and drying usually are great and the water-holding capacity of clayey and silty soils generally is high. As already mentioned, such soils are spoken of as *heavy* because of their difficult working qualities, markedly in contrast with *light*, easily tilled sandy and gravelly surface soils.

3:3. MINERALOGICAL AND CHEMICAL COMPOSITIONS OF THE SOIL SEPARATES

Although at this point our interest in soil particles is largely a physical one, a glance at their mineralogical makeup and chemical composition may be in order.

MINERALOGICAL CHARACTERISTICS. As already suggested, the coarsest sand particles often are fragments of rocks as well as minerals. Quartz commonly dominates the finer grades of sand as well as the silt separate (see Fig. 3:4). In addition, variable quantities of other primary minerals usually occur, such as the various feldspars and micas. Gibbsite, hematite, and limonite minerals also are found, usually as coatings on the sand grains. Hematite and limonite, because of their iron content, impart various shades of red and yellow if present in sufficient quantities. The soils of the southeastern part of the United States and well-oxidized tropical earths are good examples.

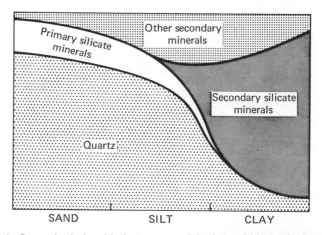

FIGURE 3:4. General relationship between particle size and kinds of minerals present. Quartz dominates the sand and coarse silt fractions. Primary silicates such as the feldspars, hornblende, and micas are present in the sands but tend to disappear as one moves to the silt fraction. Secondary silicates dominate the fine colloidal clay. Other secondary minerals, such as the oxides of iron and aluminium, are prominent in the fine silt and coarse clay fractions.

Some clay particles, especially those in the coarser clay fractions, are composed of minerals such as quartz and the hydrous oxides of iron and aluminum. Another is the complex aluminosilicates. Three main mineral types—*kaolinite, illite,* and *montmorillonite*—are at present recognized, although others are known to occur in significant quantities. These groups vary markedly in plasticity, cohesion, and adsorption, kaolinite being lowest in each case and montmorillonite highest. It is therefore important to know which clay type dominates or codominates any particular soil. These clays will be considered in greater detail in Chapter 4.

CHEMICAL MAKEUP. Since sand and silt are dominantly quartz (SiO_2), these two fractions are generally quite inactive chemically. Even the primary minerals which may contain nutrient elements in their chemical makeup are generally so insoluble as to make their nutrient-supplying ability essentially nil. An exception to this general rule is the silt fraction of certain potassium-bearing minerals such as the micas which have been known to release this element at a sufficiently rapid rate to partially supply plant requirements.

Chemically, kaolinite and the other members of that particular group are aluminum silicates. The same is true for montmorillonite and other clays of the same crystal pattern, but they carry in addition sodium, iron, or magnesium, as the case may be. Illite, often referred to as hydrous mica, is a potassium aluminum silicate. Its high potash content gives it a special nutrient significance. Obviously, the word "clay" is a term covering substances differing widely in their mineralogical and chemical compositions.

In well-weathered soils, especially those in the hot, humid tropics, oxides of iron and aluminum are prominent if not dominant, even in the clay-size fraction. Thus, climate can have a profound effect on the chemical and mineralogical composition of soil separates.

Since soil separates which range from very coarse sand to ultrafine clay differ so markedly in crystal form and chemical composition, it is not surprising that they also show a contrast in respect to mineral nutrients. Logically we would expect the sands, being mostly quartz, to be lowest and the clay separate to be highest. This inference is substantiated by the data in Table 3:1. The general relationships shown by these data hold true for most soils, although some exceptions may occur.

3:4. SOIL TEXTURAL CLASSES

As soils are composed of particles varying greatly in size and shape, specific terms are needed to convey some idea of their textural makeup and to give some indication of their physical properties. For this, *soil textural class* names are used, such as sand, sandy loam, and silt loam. These names originated through years of soil study and classification and gradually have become more or less standardized. Three broad and fundamental groups of

TABLE 3:1. *Phosphorus, Potassium, and Calcium Contents of Separates from Various United States Surface Soils* [a]

	Soils Developed from Indicated Materials				
Separate	Crystalline Residual	Limestone Residual	Coastal Plain	Glacial and Loessial	Arid
	Percent P				
Sand	0.03	0.12	0.03	0.07	0.08
Silt	0.10	0.10	0.10	0.10	0.10
Clay	0.31	0.16	0.34	0.38	0.20
	Percent K				
Sand	1.33	1.21	0.31	1.43	2.53
Silt	2.0	1.52	1.10	2.00	3.44
Clay	2.37	2.17	1.34	2.55	4.20
	Percent Ca				
Sand	0.36	8.75	0.05	0.91	2.92
Silt	0.59	7.83	0.14	0.93	6.58
Clay	0.67	7.08	0.39	1.92	5.73

[a] From Failyer et al. (3). Data in respect to the distribution of other elements through the various separates as reported by Joffe and Kunin (4) are as follows:

Chemical Composition of Montalto Silt Loam Topsoil

Separate	SiO_2 (%)	Fe_2O_3 (%)	Al_2O_3 (%)	TiO_2 (%)	CaO (%)	MgO (%)
Sand	86.3	5.19	6.77	1.05	0.37	1.02
Coarse silt	81.3	3.11	7.21	1.05	0.41	0.82
Fine silt	64.0	9.42	12.00	1.05	0.32	2.22
Coarser clay	45.1	13.50	21.10	0.96	0.38	2.09
Finer clay	30.2	17.10	22.80	0.88	0.08	1.77

soil textural classes are recognized: *sands, loams,* and *clays.* On the basis of these, additional class names have been devised.

SANDS. The *sand* group includes all soils of which the sand separates make up 70 percent or more of the material by weight. The properties of such soils are therefore characteristically sandy in contrast with the stickier nature of the heavier groups of soil. Two specific classes are recognized—*sand* and *loamy sand.*

CLAYS. A soil to be designated a clay must carry at least 35 percent of the clay separate and in most cases not less than 40 percent. In such soils the characteristics of the clay separate are distinctly dominant and the class name is *sandy clay, silty clay,* or, the most common of all, simply *clay.* It is well to note that sandy clays may contain more sand than clay. Likewise, the silt content of silty clays usually exceeds that of the clay fraction itself.

LOAMS. The *loam* group, which contains many subdivisions, is more difficult to explain. An ideal loam may be defined as a mixture of sand, silt, and clay particles which exhibits light and heavy properties in about equal proportions. Roughly it is a half-and-half mixture on the basis of properties.

Most soils of agricultural importance are some type of loam. They may possess the ideal makeup described above and be classed simply as *loam*. In most cases, however, the quantities of sand, silt, or clay present require a modified textural class name. Thus, a loam in which sand is dominant is classified as a *sandy loam* of some kind; in the same way there may occur *silt loams, silty clay loams,* and *clay loams.*

VARIATIONS IN THE FIELD. It can be seen readily that the textural class names already established—*sand, loamy sand, sandy loam, loam, silt loam, silty clay loam, clay loam, sandy clay, silty clay,* and *clay*—form a more or less graduated sequence from soils that are coarse in texture and easy to handle to the clays that are very fine and difficult to manage (see Table 3:2). It is also obvious that these textural class names are a reflection not only of particle-size distribution but also of tillage characteristics and other physical properties. For some soils, qualifying factors such as stone, gravel, and the various grades of sand become part of the textural class name. Even silt and clay become qualifying terms in practice.

TABLE 3:2. *General Terms Used to Describe Soil Texture*
in Relation to the Basic Soil Textural Class Names

U. S. Department of Agriculture Classification System

Common Names	Texture	Basic Soil Textural Class Names
Sandy soils	Coarse	Sandy Loamy sands
Loamy soils	Moderately coarse	Sandy loam Fine sandy loam
	Medium	Very fine sandy loam Loam Silt loam Silt
	Moderately fine	Clay loam Sandy clay loam Silty clay loam
Clayey soils	Fine	Sandy clay Silty clay Clay

3:5. DETERMINATION OF SOIL CLASS

"FEEL" METHOD. The common field method of determining the class name of a soil is by its *feel*. Probably as much can be judged about the texture and hence the class name of a soil merely by rubbing it between the thumb and fingers as by any other superficial means. Usually it is helpful to wet the sample in order to estimate plasticity more accurately. The way a wet soil "slicks out" or develops a continuous ribbon when pressed between the thumb and fingers gives a good idea of the amount of clay present. The sand particles are gritty, the silt has a floury or talcum-powder feel when dry and is only moderately plastic and sticky when wet. Persistent cloddiness generally is imparted by silt and clay.

The method as outlined is used in field operations such as soil survey and land classification. Accuracy in such a determination is of great practical value and depends largely on experience. Facility in class determination is one of the first things a field man should develop.

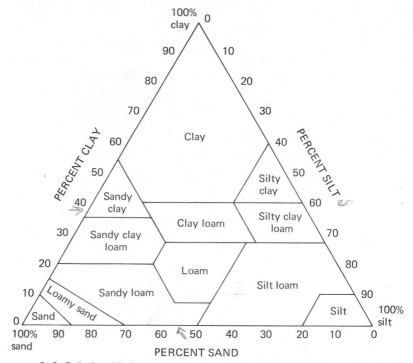

FIGURE 3:5. Relationship between the class name of a soil and its particle size distribution. In using the diagram the points corresponding to the percentages of silt and clay present in the soil under consideration are located on the silt and clay lines, respectively. Lines are then projected inward, parallel in the first case to the clay side of the triangle and in the second case parallel to the sand side. The name of the compartment in which the two lines intersect is the class name of the soil in question.

LABORATORY METHOD. A more accurate and fundamental method has been devised by the U. S. Department of Agriculture for the naming of soils based on a mechanical analysis. The relationship between such analyses and class names is shown diagrammatically in Fig. 3:5. The diagram re-emphasizes that a soil is a mixture of different sizes of particles. It illustrates how mechanical analyses of field soils can be used to check on the accuracy of the soil surveyor's class designations as determined by feel. A working knowledge of this method of naming soils is essential. The legend of Fig. 3:5 explains the use of this soil texture triangle.

The curves in Fig. 3:6 illustrate the particle size distribution in soils representative of three textural classes. Note the gradual change in percentage composition in relation to particle size. This figure emphasizes that there is no sharp line of demarcation in the distribution of sand, silt, and clay fractions which also suggests a gradual change of properties with change in particle size.

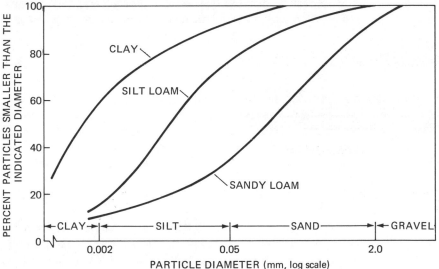

FIGURE 3:6. Particle size distribution in three soils varying widely in their textures. Note that there is a gradual transition in the particle size distribution in each of these soils.

3:6. PARTICLE DENSITY OF MINERAL SOILS

One means of expressing soil weight is in terms of the density of the solid particles making up the soil. It is usually defined as the mass (or weight) of a unit volume of soil solids and is called the *particle density*. In the metric system, particle density is usually expressed in terms of grams per cubic centimeter. Thus, if 1 cubic centimeter of soil solids weighs 2.6 grams, the particle density is 2.6 grams per cubic centimeter.

Although considerable range may be observed in the density of the individual soil minerals, the figures for most mineral soils usually vary between the narrow limits of 2.60 and 2.75. This occurs because quartz, feldspar, and the colloidal silicates with densities within this range usually make up the major portion of mineral soils. When unusual amounts of heavy minerals such as magnetite, garnet, epidote, zircon, tourmaline, and hornblende are present, the particle density may exceed 2.75. It should be emphasized that the *fineness* of the particles of a given mineral and the arrangement of the soil solids have nothing to do with the particle density.

Organic matter weighs much less than an equal volume of mineral solids having a particle density of 1.2 to 1.5. Consequently, the amount of this constituent in a soil markedly affects the particle density. This accounts for the fact that surface soils usually possess lower particle densities than do subsoils. Some mineral top soils high in organic matter may drop as low as 2.4 or even below in particle density. Nevertheless, for general calculations the average arable surface soil may be considered to have a particle density of about 2.65.

3:7. Bulk Density of Mineral Soils

BULK DENSITY. This is a second and different method of expressing soil weight. In this case, the *total soil space* (space occupied by solids and pore spaces combined) is considered. Bulk density is defined as the mass (weight) of a unit volume of dry soil. This volume includes both solids and pores. The comparative calculations of bulk density and particle density are shown in Fig. 3:7. A careful study of this figure should make clear the distinction between these two methods of expressing soil weight.

FACTORS AFFECTING BULK DENSITY. Bulk density is a weight measurement by which the entire soil volume is taken into consideration. Unlike particle density, which is concerned with the solid particles only, bulk density is determined by the quantity of pore spaces as well as soil solids. Thus, soils that are loose and porous will have low weights per unit volume (bulk densities) and those that are more compact will have high values. Since the particles of sandy soils generally lie in close contact, such soils have high bulk densities. The low organic matter content of sandy soils further encourages this. On the other hand, the particles of the finer-textured surface soils, such as silt loams, clay loams, and clays, ordinarily do not rest so close together. This occurs because these surface soils are comparatively well granulated, a condition encouraged by their relatively high content of organic matter. Granulation encourages a fluffy, porous condition which results in low bulk-density values. Consequently, the bulk density of a well-granulated

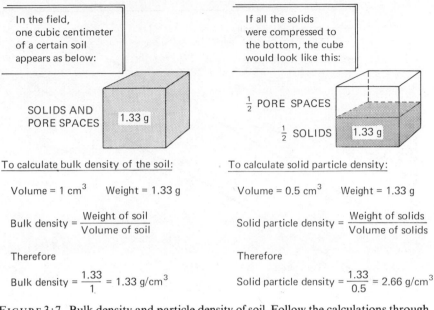

To calculate bulk density of the soil:

Volume = 1 cm^3 Weight = 1.33 g

$$\text{Bulk density} = \frac{\text{Weight of soil}}{\text{Volume of soil}}$$

Therefore

$$\text{Bulk density} = \frac{1.33}{1} = 1.33 \text{ g/cm}^3$$

To calculate solid particle density:

Volume = 0.5 cm^3 Weight = 1.33 g

$$\text{Solid particle density} = \frac{\text{Weight of solids}}{\text{Volume of solids}}$$

Therefore

$$\text{Solid particle density} = \frac{1.33}{0.5} = 2.66 \text{ g/cm}^3$$

FIGURE 3:7. Bulk density and particle density of soil. Follow the calculations through carefully and the terminology should be clear. In this particular case the bulk density is one half that of the particle density and the percentage pore space is 50.

silt loam surface soil is sure to be lower than that of a representative sandy loam.

The bulk densities of clay, clay loam, and silt loam surface soils normally may range from 1.00 to as high as 1.60 grams per cubic centimeter, depending on their condition. A variation from 1.20 to 1.80 may be found in sands and sandy loams. Very compact subsoils, regardless of texture, may have bulk densities as high as 2.0 grams per cubic centimeter or even greater. The relationship among texture, compactness, and bulk density is illustrated in Fig. 3:8.

Even in soils of the same surface texture, great differences in bulk density are to be expected when similar horizon levels are compared. This is clearly shown by data on Wisconsin silt and clay loams in Table 3:3. Moreover, there is a distinct tendency for the bulk density to rise with profile depth. This apparently results from a lower content of organic matter, less aggregation and root penetration, and a compaction caused by the weight of the overlying layers.

The system of crop and soil management employed on a given soil is likely to influence its bulk density, especially of the surface layers. The addition of farm manure in large amounts tends to lower the weight figure of surface soils, as does a bluegrass sod. Intensive cultivation operates in the opposite direction.

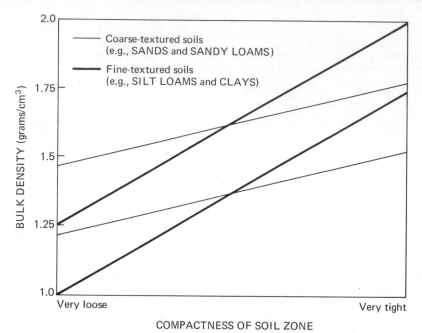

FIGURE 3:8. Generalized relationship between compactness and the range of bulk densities common in sandy soils and in those of finer texture. Sandy soils generally are less variable in their degree of compactness than are the finer-textured soils. For all soils, the surface layers are more likely to be medium to loose in compactness than are the subsoils.

TABLE 3:3. *Bulk Density Data for Certain Wisconsin Profiles* [a]

Horizon	Marathon Silt Loam	Miami Silt Loam	Spencer Silt Loam	Superior Clay Loam
Plow layer	1.34	1.28	1.38	1.46
Upper subsoil	1.49	1.41	1.55	—
Lower subsoil	1.59	1.43	1.66	1.66
Parent material	1.72	1.49	1.63	—

[a] From Nelson and Muckenhirn (7).

Data presented in Table 3:4 show this relationship very well. These data are from long-time experiments in four different states, the soils having been under cultivation for from 40 to 150 years. Cropping increased the bulk density of the topsoils in all cases.

TABLE 3:4. *Bulk Density and Pore Space of Certain Cultivated Topsoils and of Nearby Uncropped Areas* [a]

		Bulk Density		Percent Pore Space	
Soil Type	Years Cropped	Cropped Soil	Uncropped Soil	Cropped Soil	Uncropped Soil
Hagerstown loam (Pa.)	58	1.25	1.07	50.0	57.2
Marshall silt loam (Iowa)	50+	1.13	0.93	56.2	62.7
Nappanee silt loam (Ohio)	40	1.31	1.05	50.5	60.3
Av. 19 Georgia soils	45–150	1.45	1.14	45.1	57.1

[a] From Lyon et al. (6).

OTHER WEIGHT FIGURES. When the bulk density of a soil is known in terms of grams per cubic centimeter, its approximate dry weight in pounds per cubic foot may be found by multiplying by 62.42, the standard weight of a cubic foot of water. Clayey and silty surface soils may vary from 65 to 100 pounds to the cubic foot; sands and sandy loams may show a variation of 75 to 110 pounds. The greater the organic content, the less is this weight. Very compact subsoils, regardless of texture, may weigh as much as 125 pounds per cubic foot. The figures quoted are on a dry weight basis.

The actual weight of a soil also may be expressed in terms of an acre-foot, referring to a volume of soil 1 acre in extent and 1 foot deep. The weight of an acre-foot of surface mineral soil may range from 3 to 4 million pounds of dry substance. The figure most commonly used, however, is 2 million or sometimes 2.5 million pounds as the weight of average surface soil to a depth of 6 to 7 inches. This is considered an *acre–furrow slice*. Comparable weight figures expressed in metric units are 2.2 to 2.8 million kilograms per hectare.

3:8. PORE SPACE OF MINERAL SOILS

The pore space of a soil is that portion occupied by air and water. The amount of this pore space is determined largely by the arrangement of the solid particles. If they lie close together as in sands or compact subsoils, the total porosity is low. If they are arranged in porous aggregates, as is often the case in medium-textured soils high in organic matter, the pore space per unit volume will be high.

The validity of the above generalizations may readily be substantiated by the use of a very simple formula involving particle density and bulk density figures. The derivation of the formula used to calculate the percentage of total pore space in soil follows:

$$\% \text{ solid space} = \frac{\text{bulk density}}{\text{particle density}} \times 100$$

Since,

$$\% \text{ pore space} + \% \text{ solid space} = 100$$

and

$$\% \text{ pore space} = 100 - \% \text{ solid space}$$

then

$$\% \text{ pore space} = 100 - \frac{\text{bulk density}}{\text{particle density}} \times 100$$

Using this formula, a sandy soil having a bulk density of 1.50 and a particle density of 2.65 will be found to have 43.4 percent pore space. A silt loam in which the corresponding values are 1.30 and 2.65 possesses 50.9 percent air and water space. The latter value is close to the pore capacity of a normally granulated silt loam or clay loam surface soil.

FACTORS INFLUENCING TOTAL PORE SPACE. Considerable difference in the total pore space of various soils occurs depending upon conditions. Sandy surface soils show a range of from 35 to 50 percent, whereas medium- to fine-textured soils vary from 40 to 60 percent or even more in cases of high organic matter and marked granulation. Pore space also varies with depth; some compact subsoils drop as low as 25 to 30 percent. This accounts in part for the inadequate aeration of such horizons.

The handling of a soil exerts a decided influence upon pore space of the furrow slice. For instance, the continuous bluegrass sod of the Hagerstown loam of Pennsylvania cited in Table 3:4 had a total porosity of 57.2 percent, whereas the comparable rotation plot showed only 50 percent. Additional data presented in this table from three other states show that cropping tends to lower the total pore space below that of the virgin or uncropped soils. This reduction usually is associated with a decrease in organic matter content and a consequent lowering of granulation. Pore space in the subsoil has been found to decrease with cropping, although to a lesser degree.

SIZE OF PORES. Two types of individual pore space in general occur in soils—*macro* and *micro*. Although there is no sharp line of demarcation, the macropores characteristically allow the ready movement of air and percolating water. In contrast, in the micropores air movement is greatly impeded and water movement is restricted primarily to slow capillary movement. Thus, in a sandy soil, in spite of the low total porosity, the movement of air and water is surprisingly rapid because of the dominance of the macrospaces.

Fine-textured soils allow relatively slow gas and water movement despite the unusually large amount of total pore space. Here the dominating micropores often maintain themselves full of water. Aeration, especially in the subsoil, frequently is inadequate for satisfactory root development and desirable microbial activity. Therefore, the size of the individual pore spaces rather than their combined volume is the important consideration. The loosening and granulating of fine-textured soils promotes aeration not so much by increasing the total pore space as by raising the proportion of the macrospaces.

It has already been indicated (p. 12 and Fig. 1:4) that in a well-granulated silt loam surface soil at optimum moisture for plant growth, the total pore space will be near 50 percent and is likely to be shared equally by air and water. Soil aeration under such a condition is satisfactory, especially if a similar ratio of air to water extends well into the subsoil.

CROPPING AND SIZE OF PORES. Continuous cropping, particularly of soils originally very high in organic matter, often results in a reduction of large or macropore spaces. Data from a fine-textured soil in Texas presented in Table 3:5 show this effect very strikingly.

The amount of macropore space so necessary for ready air movement was reduced about one half by cropping the soil. This severe reduction in pore size also extended into the 6- to 12-inch layer. In fact, samples taken as deep as 42 inches showed the same trend.

A further examination of Table 3:5 shows that cultivation and cropping have appreciably reduced the total pore space. This was accompanied by a proportional rise in the micropore space. The decrease in pore size was associated with a corresponding decrease in organic matter content.

TABLE 3:5. *Effect of Continuous Cropping for at Least 40–50 Years on Total Pore Space and Macro- and Micropore Spaces in a Houston Black Clay from Texas* [a]

Sampling Depth (in)	Soil Treatment	Organic Matter (%)	Pore Space		
			Total (%)	Macro (%)	Micro (%)
0–6	Virgin	5.6	58.3	32.7	25.6
	Cultivated	2.9	50.2	16.0	34.2
6–12	Virgin	4.2	56.1	27.0	29.1
	Cultivated	2.8	50.7	14.7	36.0

[a] From Laws and Evans (5).

3:9. STRUCTURE OF MINERAL SOILS

Although *texture* undoubtedly is of great importance in determining certain characteristics of a soil, it is evident that the grouping or arrangement of particles must exert considerable influence also. The term *structure* is used to refer to such groupings. Structure is strictly a field term descriptive of the gross, over-all aggregation or arrangement of the primary soil separates.

A profile may be dominated by a single structural pattern. More often, a number of types of aggregation are encountered as progress is made from horizon to horizon. It is at once apparent that soil conditions and characteristics such as water movement, heat transfer, aeration, bulk density, and porosity will be much influenced by structure. In fact, the important physical changes imposed by the farmer in plowing, cultivating, draining, liming, and manuring his land are structural rather than textural.

TYPES OF SOIL STRUCTURE. Four primary types of soil structure are recognized: *platy, prismlike, blocklike,* and *spheroidal.* The last three have two subtypes each. A brief description of each of these structural types with schematic drawings will be found in Fig. 3:9. A more detailed description of each follows:

1. Platelike—*platy*—In this structural type the aggregates or groups are arranged in relatively thin horizontal plates, leaflets, or lenses. Platy structure is most noticeable in the surface layers of virgin soils but may characterize the subsoil horizons as well.

   Although most structural features are usually a product of soil-forming forces, the *platy* type is often inherited from the parent materials, especially those laid down by water or ice.

2. Prismlike—(*columnar* and *prismatic* subtypes)—These subtypes are characterized by vertically oriented aggregates or pillars which vary

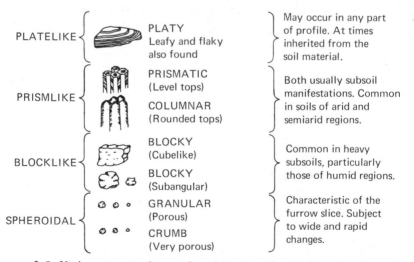

FIGURE 3:9. Various structural types found in mineral soils. Their location in the profile is suggested. In arable topsoils, a stable granular structure is prized.

in length with different soils and may reach a diameter of 6 or more inches. They commonly occur in subsoil horizons in arid and semiarid regions and when well developed are a very striking feature of the profile.

When the tops are rounded, the term *columnar* is used. This may occur when the profile is changing and certain horizons are degrading. When the tops of the prisms are still plane, level, and clean cut, the structural pattern is designated *prismatic*. Both the prismatic and columnar types of aggregation are divided into classes depending on the size or horizontal diameter of the prisms.

3. Blocklike—(*blocky* and *subangular* blocky subtypes)—In this case the original aggregates have been reduced to blocks, irregularly six-faced, and with their three dimensions more or less equal. In size these fragments range from a fraction of an inch to 3 or 4 inches in thickness. In general, the design is so individualistic that identification is easy.

   When the edges of the cubes are sharp and the rectangular faces distinct, the type is designated *blocky*. When subrounding has occurred, the aggregates are referred to as *subangular blocky*. These types usually are confined to the subsoil and their stage of development and other characteristics have much to do with soil drainage, aeration, and root penetration.

4. Spheroidal—(*granular* and *crumb subtypes*)—All rounded aggregates may be placed in this category, although the term more properly refers to those not over $\frac{1}{2}$ inch in diameter. These rounded complexes usually lie loosely and are readily shaken apart. When wetted, the intervening spaces generally are not closed so readily by swelling as may be the case with a blocky structural condition. Ordinarily the aggregates are called granules and the pattern *granular*. However, when the granules are especially porous, the term *crumb* is applied.

   Granular and crumb structures are characteristic of many surface soils, especially those high in organic matter. They are the only types of aggregation that are commonly influenced by practical methods of soil management.

As already emphasized, two or more of the structural conditions listed usually occur in the same soil solum. In humid temperate regions, a granular aggregation in the surface horizon with a blocky, subangular blocky or platy type of some kind in the subsoil is usual, although granular subhorizons are not uncommon. In soils of regions of lower rainfall the blocky type in the subsoil may be replaced by a columnar or prismatic arrangement.

GENESIS OF SOIL STRUCTURE. The mechanics of structure formation are exceedingly complicated and rather obscure. The nature and origin of the parent material are important factors as are the physical and biochemical processes of soil formation, particularly those resulting in the synthesis of clay and humus. Climate is also a prime consideration. Soluble salts probably play an important role, particularly in the soils of arid regions. Nor should the downward migration of clay, iron oxides, and lime be overlooked. Undoubtedly, the accumulation of organic matter and its type of decay play a major role, especially in the development of the granular structure so common in the surface soils of grasslands. The preservation and encouragement of this particular structural type are among the most important soil problems of cultivated lands.

3:10. AGGREGATION AND ITS PROMOTION IN ARABLE SOILS

In a practical sense we are concerned with two sets of factors in dealing with soil aggregation: (a) those responsible for aggregate formation, and (b) those which give the aggregates stability once they are formed. Since both sets of factors are operating simultaneously, it is sometimes difficult to separate relative effects on stable granule development in soils.

GENESIS OF GRANULES. Although there is some uncertainty about the exact mechanism by which granules form, several specific factors are known to influence their genesis. These include (a) wetting and drying, (b) freezing and thawing, (c) the physical activity of roots and soil animals, (d) the influence of decaying organic matter and of the slimes from the micro-organisms and other forms of life, (e) the modifying effects of adsorbed cations, and (f) soil tillage.

Any action that will develop lines of weakness, shift the particles back and forth, and force contacts that otherwise might not occur, encourages aggregation. Alternate wetting and drying, freezing and thawing, the physical effects of root extension, and the mixing action of soil organisms and of tillage implements encourage such contacts and, in turn, aggregate formation. The benefits of fall plowing on certain types of soil and the slaking of clods under the influence of a gentle rain have long been known and utilized in seedbed preparation. And the granulating influences of earthworms and other soil organisms should not be overlooked.

INFLUENCE OF ORGANIC MATTER. The major agency in the encouragement of granular-type aggregates in surface soil horizons is organic matter (see Fig. 3:10), which not only binds but also lightens and expands, making possible the porosity so characteristic of individual soil aggregates. Plant roots promote this granulation by their decay in the soil and by the disruptive action of their roots as they move through the soil. The electrochemical properties of the humus and clay are probably effective in the organization and the later stabilization of the aggregates. Moreover, slime and other

FIGURE 3:10. Puddled soil (*left*) and well-granulated soil (*right*). Plant roots and especially humus play the major role in soil granulation. For that reason a sod tends to encourage development of a granular structure in the surface horizon of cultivated land. [*Photo courtesy of U.S. Soil Conservation Service.*]

viscous microbial products probably encourage crumb development and exert a stabilizing influence. Granulation thus assumes a highly biological aspect.

Organic matter is of much importance in modifying the effects of clay. An actual chemical union may take place between the decaying organic matter and the silicate molecules. Moreover, the high adsorptive capacity of humus for water intensifies the disruptive effects of temperature changes and moisture fluctuations. The granulation of a clay soil apparently cannot be promoted adequately without the presence of a certain amount of humus. Therefore, the maintenance of organic matter, including synthesized products, is of great practical concern not only chemically and biologically but also physically (see Fig. 3:10).

EFFECT OF ADSORBED CATIONS. Aggregate formation is definitely influenced by the nature of the cations adsorbed by soil colloids (see p. 75). For instance, when sodium is a prominent adsorbed ion, the particles are dispersed and a very undesirable soil structure results. By contrast, the adsorption of calcium may encourage granulation by a phenomenon called *flocculation*. This occurs when the colloidal matter is brought together in floccules and encourages a type of structure which is quite desirable. When such means are present and active, flocculation assumes considerable practicable significance. Flocculation in itself, however, is not granulation because it usually does not provide for the *stabilization* of the aggregates.

While many surface soils highly charged with calcium are well granulated, it must not be inferred that this effect is due entirely to the direct influence of the adsorbed calcium. Exchangeable cations merely modify the influence of the other factors, especially the overall effects of decaying organic matter. The addition of limestone is effective as a granulating agent largely through its influence on biotic forces.

INFLUENCE OF TILLAGE. Tillage has both favorable and unfavorable effects on granulation. The short-time effect is generally favorable because the implements break up clods, incorporate the organic matter into the soil, and make a more favorable seedbed. Some tillage is thus considered necessary in normal soil management.

Over longer periods, tillage operations have detrimental effects on surface soil granules. In the first place, by mixing and stirring the soil, tillage generally hastens the oxidation of organic matter from soils. Second, tillage operations, especially those involving heavy equipment, tend to break down the stable soil aggregates. Compaction occurs from repeatedly running over fields with heavy farm equipment. An indication of the effect of such traffic upon bulk density is given in Fig. 3:11. These data explain the increased interest in techniques of drastically reducing tillage operations where possible.

AGGREGATE STABILITY. The stability of aggregates is of great practical importance. Some granules readily succumb to the beating of rain and the rough and tumble of plowing and fitting of the land. Others resist disintegration, thus making the maintenance of a suitable soil structure comparatively easy (see Fig. 3:12).

Apparently there are three major factors influencing aggregate stability. *First* is the temporary mechanical binding action of microorganisms, the fungi with their mycelia being especially effective. These effects are pronounced when fresh organic matter is added to soils and are at a maximum a few weeks or months after this application. *Second* is the cementing action of the intermediate products of microbial synthesis and decay, such as microbially produced gums and certain polysaccharides. These compounds are sometimes called "pre-humus constituents" and are clearly effective as aggregate stabilizers for at least several months. *Third* is the cementing action of the more resistant stable humus components, which provide most of the

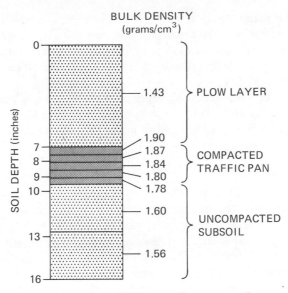

FIGURE 3:11. Tillage and heavy equipment traffic compacted a zone just below the plow layer of this Norfolk profile. Cotton roots would not penetrate layers with bulk densities of more than 1.8. [*From Camp and Lund* (1).]

long-term aggregate stability. These compounds are broken down only very slowly by microorganisms.

It should be emphasized that aggregate stability is not entirely an organic phenomenon. Particularly in the second and third phases identified above, there is interaction between organic and inorganic components. Polyvalent

FIGURE 3:12. The aggregates of soils high in organic matter are much more stable than are those low in this constituent. The low organic matter soil aggregates fall apart when they are wetted; those high in organic matter maintain their stability.

inorganic cations (for example, Ca^{++}, Mg^{++}, Fe^{++}, and Al^{3+}) are thought to act as bridges between the organic matter and soil clays, encouraging the development of clay–organic matter complexes. Generally speaking, soils in which kaolinite and the hydrous oxide clays are prominent have aggregates with the highest stability.

As a general rule, the larger the aggregates present in any particular soil, the lower is their stability. This is why it is difficult to build up soil aggregation beyond a certain size of granule or crumb in cultivated land.

3:11. STRUCTURAL MANAGEMENT OF SOILS

COARSE-TEXTURED SOILS. Looseness, good aeration and drainage, and easy tillage are characteristics of sandy soils. On the other hand, such soils are commonly too loose and open and lack the capacity to adsorb and hold sufficient moisture and nutrients. They are likely to be droughty and lacking in fertility. They need granulation. There is only one practical method of improving the structure of such a soil—the addition of organic matter. Organic material will not only act as a binding agent for the particles but will also increase the water-holding capacity. The addition of farm manures and the growth of sod crops are practices usually followed to improve the structural condition of sandy soils.

FINE-TEXTURED SOILS. The structural management of a clay soil is not such a simple problem as that of a sandy one. In clays and similar soils of temperate regions the potential plasticity and cohesion are always high because of the presence of large amounts of colloidal clay. When such a soil is tilled when wet, its pore space becomes much reduced, it becomes practically impervious to air and water, and it is said to be *puddled*. When a soil in this condition dries, it usually becomes hard and dense. The tillage of clay soils must be carefully timed. If plowed too wet, the structural aggregates are broken down and an unfavorable structure results. On the other hand, if plowed too dry, great clods are turned up which are difficult to work into a good seedbed. In sandy soils and the hydrous-oxide clays of the tropics such difficulties usually are at a minimum.

The granulation of fine-textured soils should be encouraged by the incorporation of organic matter. In this respect sod crops should be utilized fully and the crop rotation planned to attain their maximum benefits. The data in Table 3:6 from an experiment in Iowa show the degranulating influence commonly attributed to corn. The effect is less rapid, of course, when the crop is grown in a suitable rotation. The aggregating tendency of sod, whether it is a meadow mixture or a bluegrass sward, is likewise obvious. The data also suggest the degrees of soil granulation normally expected and disclose the rapidity with which aggregation may decline.

TABLE 3:6. *Percentage Water-Stable Aggregation of a Marshall Silt Loam near Clarinda, Iowa, under Different Cropping Systems* [a]

Crop	Percent of the Water-Stable Aggregates Which Were:	
	Large (1 mm and above)	Small (less than 1 mm)
Corn continuously	8.8	91.2
Corn in rotation	23.3	76.7
Meadow in rotation	42.2	57.8
Bluegrass continuously	57.0	43.0

[a] From Wilson et al. (9).

RICE SOILS. The detrimental effects of puddling the soil on most upland crops are not evident where lowland rice is grown. In fact the opposite is true—that is, puddling the soil is generally beneficial to the production of rice. In preparation for the planting of rice, the soil is flooded with water, either by irrigation or heavy rains, and is then puddled by intensive tillage, which essentially destroys the structural aggregates. Rice seeds are then sown or seedlings transplanted into the freshly prepared mud. Such soil management helps control weeds and also reduces the rate of water movement down through the soil (see Fig. 3:13). This is important since it is the common practice to maintain standing water in the rice through much of the growing season. By reducing water percolation, puddled soil markedly decreases the amount of water needed to produce a rice crop.

Unique characteristics of the rice plant account for its positive response to a type of soil management that destroys aggregate stability. Rice survives flooded conditions because oxygen moves downward inside the stem of the plant to supply the roots. This characteristic permits rice to compete well with all but a few aquatic weeds and grasses. The advantage possessed by the rice plant is its response, not to puddling per se, but to a flooded soil condition. Rice can be grown successfully on unpuddled but flooded soil. The unpuddled soil maintains at least some structural aggregation. After water is withdrawn and the rice harvested, the soil condition is then quite favorable for the growth of a crop such as corn or an edible legume that may follow rice in a cropping sequence.

The response of rice to soil structural management is important in two ways. It calls attention to a soil management system commonly followed in the tropics and subtropics where the majority of the world's population lives. Also, it illustrates the interaction between the plants and soil in determining the appropriate type of soil structure required.

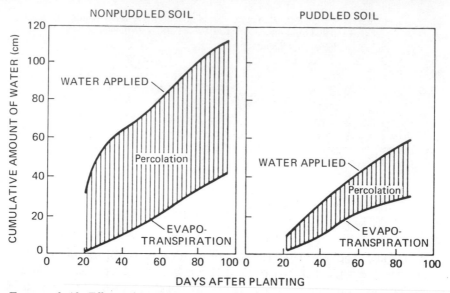

FIGURE 3:13. Effects of puddling a rice soil on the loss of water by percolation and evapo-transpiration. Note that in this case puddling greatly reduced percolation. [*From De Datta and Kerim* (2).]

3:12. SOIL CONSISTENCE

Soil *consistence* is a term used to describe the physical condition of a soil at various moisture contents, as evidenced by the behavior of that soil toward mechanical stresses or manipulations. This stress is commonly evidenced by feeling the soil, manipulating it by hand, or by pulling a tillage instrument through it. Soil consistence is considered a combination of soil properties dependent upon the forces of attraction between soil particles as influenced by soil moisture. The consistence of soils is generally described at three soil moisture levels: wet, moist, and dry. Also cementation of some soil horizons appears to be independent of soil moisture level.

WET SOILS. For wet soils, consistence is described in terms of *stickiness* and *plasticity*. In turn, the degree of stickiness is identified by the following terms: *nonsticky, slightly sticky, sticky,* and *very sticky.*

Plasticity is the capacity of soil to be molded, that is, to change shape in response to stress and to keep that shape when the stress is removed. It is evidenced when a thin rod is formed by rolling soil between the thumb and forefinger. Terms used to describe the degree of plasticity are *nonplastic, slightly plastic, plastic,* and *very plastic.*

MOIST SOILS. The consistence of moist soils is most important since it best describes the condition of soils when they are tilled in the field. In practice it is a general measure of the resistance of the soil to crushing between the thumb and forefinger. Consistence of a moist soil is described in the following terms, going from the material with least coherence to that which adheres so strongly as to resist crushing between the thumb and forefinger:

Loose: noncoherent
Very friable: coherent but very easily crushed
Friable: easily crushed
Firm: crushed under moderate pressure
Very firm: crushes only under strong pressure
Extremely firm: resists crushing between thumb and forefinger

Since the consistence of moist soils is quite dependent on the soil moisture level, the accuracy of field measurement of this soil characteristic is most dependent on the estimate of the soil moisture level. Therefore, coarse sands would be expected to have a loose consistence. Well-granulated loams and silt loams would be very friable, friable, or perhaps firm. Clays, silty clays, and silty clay loams are more likely to be firm or very firm, especially if they are low in organic matter. Such generalizations must be used with caution, however, since soil consistence is influenced by factors such as type of clay and the kind and amount of humus present.

DRY SOILS. When dry, soils resist crushing or other manipulation. The degree of this resistance is related to the attraction of the particles for each other and is expressed in such terms as rigidity and brittleness. In describing the consistence of dry soils, the following terminology is used starting with a condition of little interparticle attraction and moving to a state of high cohesive forces:

Loose: noncoherent
Soft: breaks under slight pressure between thumb and forefinger to a powdery mass
Slightly hard: breaks under moderate pressure
Hard: breaks with difficulty under pressure
Very hard: very resistant to pressure; cannot be broken between thumb and forefinger
Extremely hard: extreme resistance to pressure; cannot be broken in the hand

CEMENTATION. Some soil horizons exhibit cementation quite independent of soil moisture level. The cementing agents are compounds such as oxides of iron and aluminum, calcium carbonate, and silica. The consistence of the horizons is expressed in terms of the degree of cementation as follows:

Weakly cemented: cemented units can be broken in the hand

Strongly cemented: units cannot be broken in the hand but can be broken easily with a hammer

Indurated: units breakable only with sharp blows of a hammer

FIELD EXAMPLE. One explanation of how soil consistence varies with moisture content is to use the above terms in describing the variation of this soil property in the field. Fine-textured clay soon after a rain is quite high in soil moisture and is *sticky* to feel. It obviously is too wet to work and if manipulated will tend to puddle or run together. Even when this soil is allowed to dry somewhat it may still be slightly sticky and will be *plastic* in nature. It can be molded into various forms by applying pressure. Although it can be plowed at this moisture content, the furrow slice thus turned will likely form clods when the soil is allowed to dry. The consistence of these clods would then be considered as *hard* or *very hard*.

At a moisture content slightly below that required for plastic consistence, a soil is in optimum condition for working. If it has a high organic matter content, it probably has the appropriate properties to be termed *friable*. The exact moisture range over which this condition occurs will vary for different soils. In general, this range is much wider for medium-textured soils such as loams and some silt loams than for finer textured clays.

All soils should not be expected to behave like the clay soil just described. Sandy soils, for example, do not become plastic or sticky when wet, or hard when dry. They have a tendency to stay quite *loose* throughout their normal field moisture range. Loams and silt loams will be intermediate in their behavior between the clays and sands.

Consistence is important in determining the practical utilization of soils. The terms used to describe this soil property are meaningful as applied to soil tillage, compaction by farm machinery, and so on. These subjects are covered in the next section.

3:13. TILTH AND TILLAGE

Although frequent mention has been made of plowing and cultivation in relation to soil structure, attention must be given to seedbed preparation and maintenance of its granulation throughout the season. A convenient term—tilth—will greatly facilitate such a discussion.

TILTH DEFINED. Simply defined, tilth refers to "the physical condition of the soil in its relation to plant growth" and hence must take cognizance of all soil physical conditions that influence crop development.

Tilth depends not only on granulation and its stability but also on such factors as moisture content, degree of aeration, rate of water infiltration, drainage, and capillary-water capacity. As might be expected, tilth often changes rapidly and markedly. For instance, the working properties of fine-textured soils may be altered abruptly by a slight change in moisture.

One of the objectives of plowing and cultivation is the encouragement and maintenance of good tilth. Unfortunately, when improperly administered these operations may seriously impair tilth directly or set the stage for later deterioration.

TILTH AND PLOWING. The mold-board plow is a tillage implement designed to accentuate the granulation of the soil by its lifting, twisting, and shearing actions. At the same time, it turns under any organic residues that may be on or in the surface. It is a powerful tool for *good* when properly used—for *ill* if applied too frequently or under unsatisfactory conditions. Undoubtedly, it is the best implement for the preparation of sod land for cultivated crops. It cuts through the matted roots, turns the sod into the furrow, and exposes the shattered subsurface layers for further preparation.

On fine-textured soils, the plow is indispensable if maximum crop yields are to be attained. On sandy soils, however, the advantage of the mold-board plow is not so great because disking or some other type of cultivation may be as good or even better, especially if a stubble mulch is desired, as is often the case in areas of low rainfall.

Undoubtedly, many farmers plow too much and at times when soil aggregation is seriously impaired (see Fig. 3:14). This is likely to occur when clayey soils are plowed too wet or too dry. Yet in spite of its misuse and the criticism leveled at it, the plow continues to be an important factor in the structural management of land and the maintenance of crop yields. This is shown by the data of Table 3:7. Decisions as to when and how to plow in this age of heavy, mechanized farm implements are more important than ever before.

TABLE 3:7. *Influence of Various Methods of Seedbed Preparation upon Soil Porosity in Ohio* [a]

Seedbed Preparation	After Seven Years	
	Total Porosity (%)	Air Space Porosity (%)
Standard (plow, harrow)	56.3	25.9
Sod plow	56.0	24.9
Rotary tillage	56.9	24.2
Subsurface tillage	54.8	19.2
Surface tillage only	49.1	14.2
Standard plus straw mulch	55.8	26.9

[a] From Page et al. (8).

FIGURE 3:14. Cotton root development under two soil conditions. (*Left*) A "plow pan" has developed in this soil, and the cotton roots have not penetrated it. (*Right*) Another soil without "plow pan" with normal root penetration to a depth of 26 inches. [*Photos courtesy U.S. Soil Conservation Service.*]

TILTH AND CULTIVATION. Tillage of the soil after it is plowed is often more likely to impair tilth than is plowing. Such cultivation is employed frequently, is performed by many types of implements, and drastically influences the upper furrow slice, whose structure is so susceptible to destruction. Hence, in the preparation of a seedbed, only the minimum of cultivation should be applied after plowing, leaving the soil with a granular structure suitable for seeding yet coarse enough at the surface to resist erosion and the puddling effects of beating rains (see Fig. 3:15).

Cultivation during the growing season serves mainly to break up crusts induced by dashing rains, to ensure adequate aeration, and to kill weeds. When herbicides are used, the control of weeds by cultivation ceases to be so important. With intertilled crops, cultivation should be kept at a minimum and ordinarily should not extend much beyond midseason. This is because serious root pruning (see Fig. 8:9) may occur as well as a loss of water by evaporation from the exposed surfaces without a compensating improvement in tilth. Directly and indirectly, cultivation may impair soil aggregation. The decline of tilth should not be accelerated by too much cultivation.

FIGURE 3:15. A practice which is receiving some acceptance is that of mulch tillage. The furrow is opened with a disc lister or middle buster and the crop is planted in the furrow with no other seedbed preparation. Corn and soybeans are being grown using this technique on sandy and sandyloam soils. [*Photo courtesy U. S. Soil Conservation Service.*]

REFERENCES

(1) Camp, C. R., and Lund, J. F., "Effects of Soil Compaction on Cotton Roots", *Crops and Soils,* **17**:13–14, Nov. 1964.

(2) De Datta, S. K., and Kerim, A. A. A., "Water and Nitrogen Economy in Rainfed Rice on Puddled and Nonpuddled Soils," *Soil. Sci. Soc. Amer. Proc.* (in press).

(3) Failyer, G. H., et al., *The Mineral Composition of Soil Particles,* Bull. 54 (Washington, D.C.: U. S. Department of Agriculture, 1908).

(4) Joffe, J. S., and Kunin, R., "Mechanical Separates and Their Fraction in the Soil Profile: I. Variability in Chemical Composition and Its Pedogenic and Agropedogenic Implications," *Soil. Sci. Soc. Amer. Proc.,* **7**:187–93, 1942.

(5) Laws, W. D., and Evans, D. D., "The Effects of Long-Time Cultivation on Some Physical and Chemical Properties of Two Rendzina Soils," *Soil. Sci. Soc. Amer. Proc.,* **14**:15–19, 1949.

(6) Lyon, T. L., Buckman, H. O., and Brady, N. C., *The Nature and Properties of Soils* (New York: Macmillan, Inc., 1952), p. 60.

(7) Nelson, L. B., and Muckenhirn, R. J., "Field Percolation Rates of Four Wisconsin Soils Having Different Drainage Characteristics," *Jour. Amer. Soc. Agron.,* **33**:1028–36, 1941.

(8) Page, J. B., Willard, C. J., and McCuen, G. W., "Progress Report on Tillage Methods in Preparing Land for Corn," *Soil. Sci. Soc. Amer. Proc.,* **11**:77–80, 1945.

(9) Wilson, H. A., Gish, R., and Browning, G. M., "Cropping Systems and Season as Factors Affecting Aggregate Stability," *Soil. Sci. Soc. Amer. Proc.,* **12**:36–43, 1947.

Chapter 4

SOIL COLLOIDS: THEIR NATURE AND PRACTICAL SIGNIFICANCE

THE colloidal state refers to a two-phase system in which one material (or materials) in a very finely divided state is dispersed through a second. Good examples of the colloidal state are milk and cheese, clouds and fog, starch, gelatin, rubber, blood, proteins, plant and animal cells, and, of course, soil. Obviously, the colloidal state in nature is the rule rather than the exception. The upper limit in size of the mineral colloidal particles is less than 0.001 mm, or one micron (μ), values as low as 0.5 or even 0.2 microns being commonly accepted. Because the maximum size limit of the *clay* fraction of a soil is considered to be 0.002 mm or 2 microns, not all the clay is strictly colloidal.

It has already been emphasized that the most active portions of the soil are those in the colloidal state and that the two distinct types of colloidal matter, inorganic and organic, exist in intimate intermixture. The inorganic is present almost exclusively as clay minerals of various kinds; the organic is represented by humus. Attention will be focused initially on the inorganic fraction, leaving that of organic origin for later consideration (see pp. 94 and 153).

In a broad way, two groups of clays are recognized—the *silicate clays* so characteristic of temperate regions, and the *iron* and *aluminum hydrous oxide clays* found in the tropics and semitropics. The silicates will be discussed first because they are dominant in the most developed agricultural regions of the world.

4:1. GENERAL CONSTITUTION OF SILICATE CLAYS

SHAPE. Early students of colloidal clays visualized the individual particles as more or less spherical. It is now definitely established, however, that the particles are laminated, that is, made up of layers of plates or flakes (see Fig. 4:1). Their individual sizes and shapes depend upon their mineralogical organizations and the conditions under which they have developed. Some of these particles are micalike and definitely hexagonal; others are irregularly

FIGURE 4:1. Crystals of four silicate clay minerals found in soils. (*Above*) Kaolinite from Illinois magnified about 1,600 times (note hexagonal crystal upper right). (*Below*) Dickite from Kansas magnified about 9,000 times. (*Opposite above*) Illite from Wisconsin magnified about 15,000 times. (*Opposite below*) Montmorillonite from Wyoming magnified about 18,000 times. [*Scanning electron micrographs courtesy Dr. Bruce F. Bohor, Illinois State Geological Survey.*]

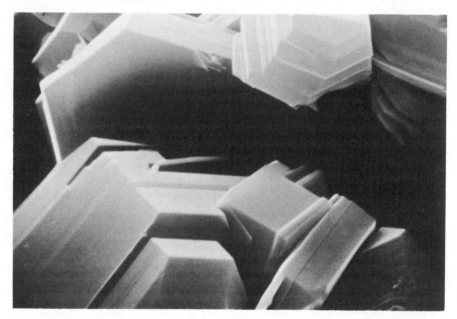

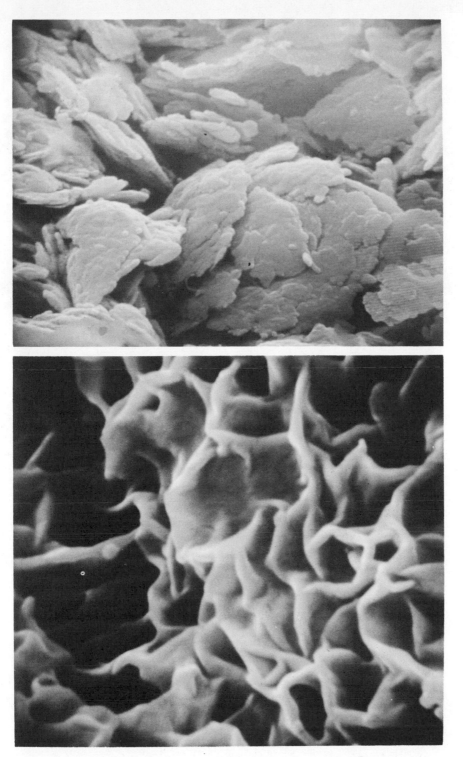

plate or flakelike; still others seem to be lath-shaped blades or even rods. The edges of some particles seem to be clean cut while the appearance of others is indistinctly frayed or fluffy. In all cases, the horizontal extension of the individual particles greatly exceeds their vertical dimension.

SURFACE AREA. All clay particles, merely because of their fineness of division, must expose a large amount of *external* surface. In some clays there are extensive *internal* surfaces as well. This internal interface occurs between the platelike crystal units that make up each particle (see Fig. 4:2). Thus, the

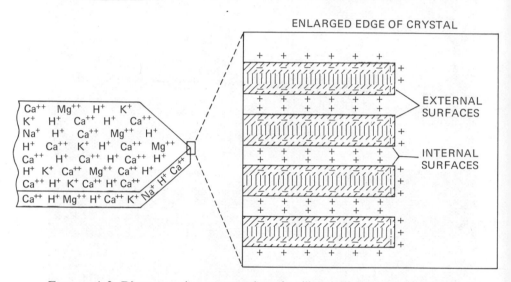

FIGURE 4:2. Diagrammatic representation of a silicate clay crystal (micelle) with its sheetlike structure, its innumerable negative charges, and its swarm of adsorbed cations. An enlarged schematic view of the edge of the crystal illustrates the negatively charged internal surface of this particular particle, to which cations and water are attracted. Note that each crystal unit has definite mineralogical structure.

tremendous surface area that characterizes clay is accounted for not only by fineness of division but also by the platelike structure of the fine particles. As a conservative estimate, it is suggested that the active interface due to the clay fraction of an acre–furrow slice of a representative silt or clay loam soil probably exceeds the land area of Illinois or Florida at least 40 or 50 times. The external surface area of 1 gram of colloidal clay is at least 1,000 times that of 1 gram of coarse sand.

ELECTRONEGATIVE CHARGE AND ADSORBED CATIONS. The minute silicate-clay colloid particles, referred to as *micelles* (microcells), ordinarily carry *negative* charges. Consequently, thousands of positively charged ions or *cations* are attracted to each colloidal crystal. This gives rise to what is known as an ionic double layer (see Fig. 4:2). The colloidal particle constitutes the *inner* ionic layer, being essentially a huge *anion*, the surfaces of which are highly negative in charge. The *outer* ionic layer is made up of a swarm of rather loosely held cations which surround and in some cases penetrate the particle. Thus, a clay particle is accompanied by an enormous number of adsorbed cations.

Associated with the layer of cations that throng the adsorptive surfaces of clay particles is a large number of water molecules. Part of these water molecules are carried by the adsorbed cations mentioned above since most of them are definitely hydrated. In addition, some silicate clays hold numerous water molecules packed between the plates that make up the clay micelle. These various types of water *in toto* are referred to when the hydration of clays is under consideration.

4:2. ADSORBED[1] CATIONS

Although all cations may be adsorbed by clay micelles, certain ones are especially prominent under natural conditions. For humid region colloids, these in the order of their numbers are H^+, Al^{3+}, and Ca^{++} first, Mg^{++} second, and K^+ and Na^+ third (see Table 4:1). For well-drained, arid and semiarid region soils, the order of the exchangeable ions is usually Ca^{++} and Mg^{++} first, Na^+ and K^+ next, and H^+ last. The humid region clays are considered to have a calcium–hydrogen complex; those in the drier regions are dominated by calcium and magnesium.

TABLE 4:1. *Relative Proportion of Adsorbed Metallic Cations Present in Certain Surface Soils of the United States* [a]

The percentage figures in each case are based on the sum of the metallic cations taken as 100. Note the geographic distribution of the soil samples.

Soil	Ca	Mg	K	Na
Penn loam (N.J.)	60.8	15.8	19.0	4.4
Mardin silt loam (N.Y.)	90.7	5.0	3.1	1.2
Webster series soil (Iowa)	76.8	20.4	1.2	1.6
Sweeney clay loam (Calif.)	76.1	21.3	1.3	1.3
Red River Valley soil (Minn.)	73.9	21.5	4.2	0.4
Keith silt loam (Nebr.)	77.1	13.3	7.1	2.5
Holdrege silt loam (Nebr.)	66.5	20.9	11.1	1.5

[a] Data compiled from various sources by Lyon et al. (4).

When the drainage of an arid region soil is impeded and alkaline salts accumulate, adsorbed sodium ions are likely to become prominent and may

[1] For a definition of adsorption see the Glossary.

equal or even exceed those of the adsorbed calcium. Here, then, would be a sodium or a sodium–calcium complex. By the same rule, in humid regions the displacement of the metallic cations by aluminum and hydrogen ions gives an aluminum–hydrogen clay. Since the cation, or cations, preponderant in a colloidal system has much to do with its physical and chemical properties as well as its relationship to plants, this phase of the subject is of much practical importance.

4:3. FUNDAMENTALS OF SILICATE CLAY STRUCTURE

For many years clays were thought to be amorphous. However, the use of X-rays, electron microscopy, and other techniques has shown that the silicate clay particles, in spite of their smallness, are definitely crystalline. The enlarged schematic view of the edge of a clay mineral crystal presented in Fig. 4:2 illustrates that individual layers or crystal units within a crystal have definite structures. The mineralogical organization of these crystal units varies from one type of clay to another and markedly affects the properties of the mineral. For this reason, some attention will be given to the fundamentals of silicate clay structure before consideration of specific silicate clay minerals.

SILICA TETRAHEDRAL AND ALUMINA OCTAHEDRAL LAYERS. Most silicate clays are *aluminosilicates*—that is, there are both aluminum and silicon components of the clay structure. These two basic molecular components are shown schematically in Fig. 4:3. One silicon atom surrounded by four oxygen atoms makes up the *silica tetrahedron*, so called because of its four-sided configuration. The *aluminum octrahedron* is an eight-sided building block consisting of a core aluminum atom surrounded by six hydroxyls or oxygens. An interlocking plane of a series of silica tetrahedra tied together by shared oxygen atoms gives a sheetlike tetrahedral layer. Similarly, large numbers of alumina octahedra, bound to each other by shared oxygen atoms in an octahedral layer, are arranged in a plane. These two basic layers, in different stacking arrangements and combinations, provide the fundamental structural units of silicate clays. The layers are bound to each other within the clay crystals by shared oxygen atoms.

ISOMORPHOUS SUBSTITUTION. The silicon in the tetrahedral layer and the aluminum in the octahedral layer are subject to replacement or substitution by other ions of comparable size. The atomic radii of a number of ions common in clays are listed in Table 4:2 to illustrate this point. Note that aluminum is only slightly larger than silicon. Consequently, aluminum can fit into the center of the tetrahedron in the place of the silicon atom and does so in some clays. As some silicates form, part of the silicon atoms in the layer are replaced by aluminum by a process called *isomorphous substitution*.

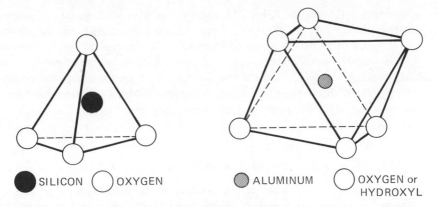

SILICON ○ OXYGEN   ● ALUMINUM ○ OXYGEN or
                                        HYDROXYL

FIGURE 4:3. Diagrammatic sketch of the two basic molecular components of silicate clays. (*Left*) A single *silica tetrahedron*, a four-sided molecular building block with a silicon atom surrounded by four oxygen atoms. When several silica tetrahedra are associated in the same plane, a silica *sheet* is formed. (*Right*) A single *alumina octahedron* showing one aluminum atom surrounded by six hydroxyls or oxygens. An alumina sheet is composed of a large number of these eight-sided molecular units tied together through shared oxygen atoms. (For ease of visualization, the oxygen atoms are shown as being about the same size as the silicon and aluminum. Actually the oxygens are much larger in radius, as shown in Table 4:2.)

As we shall see later (pp. 84–85), this substitution of a three-valent ion (Al^{3+}) for one with four valences (Si^{4+}) is responsible for the negative charge in an otherwise neutral silicate layer. The extent of such substitution helps determine the net negative charge. The importance of this will become obvious as we consider the properties of specific minerals.

TABLE 4:2. *Ionic Radii of Elements Common in Silicate Clays and an Indication of Which Are Found in the Tetrahedral and Octahedral Layers*

Note that aluminum can fit in either layer.

Ion	Radius (Å)[a]	Found in
Si^{4+}	0.41 ⎫	
Al^{3+}	0.50 ⎬	Silica tetrahedra
Fe^{3+}	0.64	
Mg^{++}	0.65	Alumina octahedra
Zn^{++}	0.70	
Fe^{++}	0.75 ⎫	
Ca^{++}	0.94 ⎬	Exchange sites
Na^+	0.98	
K^+	1.33 ⎭	
O^{--}	1.45	Both layers

[a] An Ångström unit (Å) is 10^{-8} centimeter.

Further reference to Table 4:2 shows that ions such as iron, zinc, and magnesium are not too greatly different in size from aluminum. As a result, these ions can fit in the place of aluminum as the central ion in the units making up the octahedral layer. The isomorphous substitution of a two-valent ion such as Mg^{++} for the three-valent Al^{3+} leaves unsatisfied negative charges from the oxygen atoms in the layer. As we shall see, this type of substitution helps account for the overall negative charge associated with several silicate clays.

4:4. MINERALOGICAL ORGANIZATION OF SILICATE CLAYS

On the basis of the number and arrangement of tetrahedral (silica) and octahedral (alumina) layers contained in the crystal units, silicate clays may be classified into four different groups: (a) 1:1 (silica:alumina)-type minerals; (b) 2:1-type minerals, which expand between crystal units; (c) 2:1-type nonexpanding minerals; and (d) 2:2-type minerals. Each of these is discussed briefly using the main member of a given group for illustrative purposes.

1:1-TYPE MINERALS. The crystal units of the 1:1-type minerals are made up of one silica (tetrahedral) layer alternating with one alumina (octahedral) layer—hence, the terminology 1:1-type crystal lattice (see Fig. 4:4). In soils, *kaolinite* is the most prominent member of this group which includes *halloysite, anauxite,* and *dickite.*

The two sheets of each crystal unit of kaolinite are held together by oxygen atoms which are mutually shared by the silicon and aluminum atoms in their respective sheets. These units, in turn, are held together rather rigidly by hydrogen bonding (Fig. 4:4). Consequently, the lattice is *fixed* and no expansion ordinarily occurs between units when the clay is wetted. Cations and water do not enter between the structural units of the micelle. The effective surface of kaolinite is thus restricted to its outer faces. Also, there is little isomorphic substitution in this mineral. Along with the relatively low surface area of kaolinite, this accounts for its low capacity to adsorb cations.

Kaolinite crystals usually are hexagonal with clean-cut edges (see Fig. 4:1). In comparison with montmorillonite particles, they are large in size, ranging from 0.10 to 5 microns across, with the majority falling within 0.2 to 2 microns. Because of the tightness with which their structural units are held together, kaolinite particles are not readily broken down into extremely thin sheets.

In contrast with the other silicate groups, the plasticity (capacity of being molded), cohesion, shrinkage, and swelling properties of kaolinite are very low. Its restricted surface and limited adsorptive capacity for cations and water molecules suggest that kaolinite does not exhibit colloidal properties of a high order of intensity.

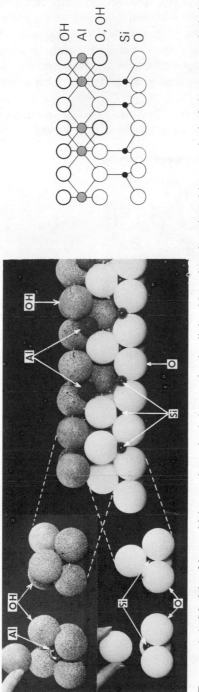

FIGURE 4:4. Models of ions which constitute the 1:1-type clay kaolinite. Note that the mineral is comprised of alternate octahedral (alumina) and tetrahedral (silica) sheets; thus, the designation "1:1." Aluminum ions surrounded by six hydroxyls make up the octahedral sheet (*upper left*). Smaller silicon ions associated with four oxygen ions constitute the tetrahedral sheet. These are coupled together (*center*) to give layers with hydroxyls on one surface and oxygens on the other. A schematic drawing of the ions (*right*) shows an end view of the crystal unit or layer.

2:1-TYPE EXPANDING MINERALS. The crystal units of these minerals are characterized by an alumina (octahedral) sheet sandwiched between two silica (tetrahedral) layers. Two general groups have this basic lattice structure, the *montmorillonite* group and the *vermiculite* group. In addition to montmorillonite, the former group includes *beidellite, nontronite*, and *saponite*.

The mineral montmorillonite is the prominent member of its group in soils. The flakelike crystals of this mineral (see Fig. 4:1) are composed of 2:1-type crystal units as shown in Fig. 4:5. In turn, these crystal units are loosely held together by very weak oxygen to oxygen linkages. Water molecules (as well as

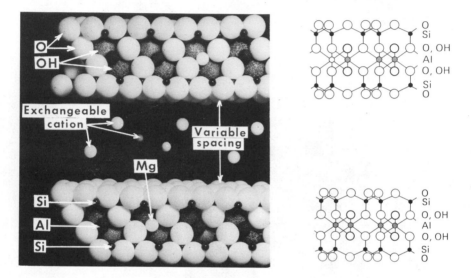

FIGURE 4:5. Model of two crystal units of an expanding lattice 2:1-type clay mineral, montmorillonite. Each layer is made up of an octahedral (alumina) sheet sandwiched in between two tetrahedral (silica) sheets. There is little attraction between oxygen ions in the bottom sheet of one unit and those in the top sheet of another. This permits a ready and variable expansion between layers. Water and exchangeable cations occupy space between the layers. This internal surface far exceeds the surface around the outside of the crystal. Note that magnesium has replaced aluminum in some sites of the octahedral sheet. This gives rise to a negative charge, which accounts for the high cation exchange capacity of this clay mineral.

cations) are attracted between crystal units, causing *expansion* of the crystal lattice. Consequently, montmorillonite crystals may be easily fractured artificially to give particles that approach the size of single crystal units. Commonly, however, montmorillonite crystals range in width from 0.01 to 1 microns. They are thus much smaller than the average kaolinite micelle.

The movement of water and cations in between montmorillonite crystal units exposes a very large *internal surface*, which greatly exceeds the external surface area of this mineral. The combined internal and external surfaces of montmorillonite (specific surface) in turn greatly exceeds the total surface area of kaolinite (see Table 4:3).

TABLE 4:3. *Comparative Properties of Three Major Types of Silicate Clay*

Property	Type of Clay		
	Montmorillonite	Illite	Kaolinite
Size (μm)	0.01–1.0	0.1–2.0	0.1–5.0
Shape	Irregular flakes	Irregular flakes	Hexagonal crystals
Specific surface (m^2/g)	700–800	100–120	5–20
External surface	High	Medium	Low
Internal surface	Very high	Medium	None
Cohesion, plasticity	High	Medium	Low
Swelling capacity	High	Medium	Low
Cation exchange capacity (meq/100 g)	80–100	15–40	3–15

Isomorphous substitution of magnesium for some of the aluminum in the octahedral layer and to a lesser extent of aluminum for silicon in the tetrahedral sheet leaves montmorillonite crystals with a high net negative charge. This charge is satisfied by a swarm of cations (H^+, Al^{3+}, Ca^{++}, K^+, etc.) which are attracted to both the internal and external surfaces (Fig. 4:2). Montmorillonite commonly shows a high cation adsorption capacity—perhaps 10–15 times that of kaolinite.

Montmorillonite also is noted for its high plasticity and cohesion and its marked shrinkage on drying. Wide cracks commonly form as soils dominated by montmorillonite are dried out (see Fig. 4:13). The dry aggregates or clods are very hard, making such soils difficult to till.

Vermiculites have structural characteristics similar to those of montmorillonite in that an octahedral layer is found between two tetrahedral sheets. In some vermiculites, however, the octahedral layer is dominated by magnesium rather than aluminum, three magnesium ions being in the place of two aluminums. In others, aluminum dominates the octahedral with only minor magnesium substitution. In the tetrahedral layer of vermiculites there has been considerable substitution of aluminum for silicon. This accounts for most of the very high net negative charge associated with this mineral.

Water molecules along with magnesium ions are strongly adsorbed between crystal units. However, they act more as bridges holding the units together than as wedges driving the units apart. The degree of swelling is, therefore, considerably less for vermiculite than for montmorillonite. For this reason vermiculite is considered a *limited-expansion* clay mineral having some internal surface but much less than is characteristic of montmorillonite.

The cation adsorption capacity of vermiculite commonly exceeds that of all other silicate clays, including montmorillonite. This is due to the high negative charge in the tetrahedral layer. Vermiculite crystals are larger than those of montmorillonite but are much smaller than those of kaolinite.

2:1-TYPE NONEXPANDING MINERALS. Often associated with montmorillonite clays is a rather indefinite group called the *hydrous micas*, of which *illite* is the most important in soils. Like montmorillonite, illite has a 2:1-type lattice. However, its particles are much larger than those of montmorillonite, and its source of charge is in the tetrahedral rather than the octahedral layer. About 15 percent of the tetrahedral silicon sites are occupied by aluminum atoms. This results in a high net negative charge in the tetrahedral layer. To satisfy this charge, potassium ions are strongly attracted between the crystal units. These K ions are just the right size to fit snugly into certain spaces in the adjoining tetrahedral layers (see Fig. 4:6). The potassium thereby acts as a binding agent, preventing much of the expansion of the crystal. Hence, illite is relatively nonexpansive.

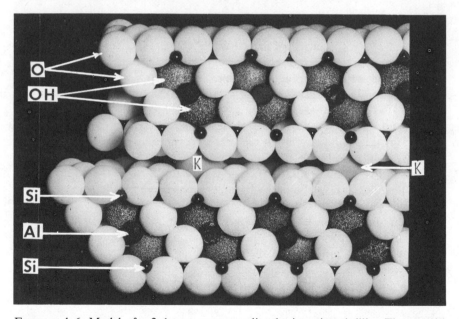

FIGURE 4:6. Model of a 2:1-type nonexpanding lattice mineral, illite. The general constitution of the layers is similar to that for montmorillonite—one octahedral (alumina) sheet between two tetrahedral (silica) sheets. However, potassium ions are tightly held between sheets giving the mineral a more or less rigid type of structure. This prevents the movement of water and cations into the space between layers. The internal surface and exchange capacity of illite are thus far below those of montmorillonite.

Such properties as hydration, cation adsorption, swelling, shrinkage, and plasticity are less intensively developed in illite than in montmorillonite. Nevertheless, illite exceeds kaolinite in respect to these characteristics. As to size, illite crystals are intermediate in size between those of montmorillonite and kaolinite (see Table 4:3).

2:2-TYPE MINERALS. This silicate group is represented by *chlorites*, which are common in some soils. Chlorites are basically silicates of magnesium with some iron and aluminum present. A typical chlorite crystal unit is composed of alternate talc (similar to a montmorillonite crystal unit) and brucite [$Mg(OH)_2$] layers. Magnesium dominates the octahedral position in the talc layer. Thus, the crystal unit contains two silica tetrahedral sheets and two magnesium octahedral sheets, giving rise to the term "2:2" structure sometimes used to describe this mineral.

The cation exchange capacity of chlorites is about the same as that of illite and considerably less than that of montmorillonite or vermiculite. Particle size and surface area for chlorite are also about the same as for illite. There is relatively little water adsorption between the chlorite crystal units, which accounts for the relatively nonexpansive nature of this mineral.

MIXTURES OF SILICATE CLAYS. In obtaining a general concept of clay minerals one should recognize that the specific groups do not occur independent of the other. In a given soil, one is likely to find several clay minerals in an intimate mixture. Furthermore, minerals having properties and composition intermediate between those of two well-defined minerals just described will be found. Such minerals are termed *mixed layer* or *interstratified* because within a given crystal the individual crystal units may well be of more than one type. Terms such as "chlorite–illite" and "illite–montmorillonite" are used to describe mixed-layer minerals. In some soils they are more common than the single-structured minerals such as montmorillonite.

4:5. SOURCE OF THE NEGATIVE CHARGE ON SILICATE CLAYS

EXPOSED CRYSTAL EDGES. There are at least two sources of the negative charges associated with silicate clay particles. The first involves unsatisfied valences at the broken edges of the silica and alumina sheets. Also, the flat external surfaces of minerals such as kaolinite have some exposed oxygen and hydroxyl groups which act as negatively charged sites. These groups are attached to silicon and aluminum atoms within their respective sheets. Especially at high pH, the hydrogen of these hydroxyls dissociates slightly and the colloidal surface is left with a negative charge carried by the oxygen. The loosely held hydrogen is readily exchangeable. The situation is illustrated in the diagram in Fig. 4:7.

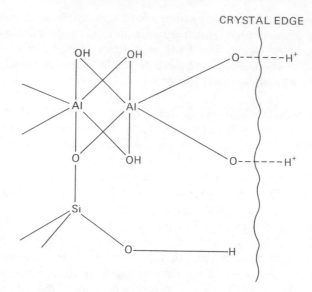

FIGURE 4:7. Diagram of a broken edge of a kaolinite crystal, showing oxygens as the source of negative charge. At high pH values the hydrogen ions tend to be held loosely and can be exchanged for other cations.

The presence of thousands of such groups gives the kaolinite clay particles a definite electronegativity. Consequently, they are surrounded by hydrogen ions and other cations that may have replaced such hydrogen. This phenomenon apparently accounts for most of the adsorbing capacity of 1:1-type colloidal clays. It also is of some significance with the 2:1 types, especially at broken crystal edges.

The charge sites at crystal edges are thought to hold hydrogen by covalent bonding. They are at least in part responsible for what has been termed the *pH-dependent* charge of inorganic colloids (see pp. 99–100 and Fig. 4:11). In moderately to strongly acid soils the hydrogen is apparently tightly held and not subject to ready replacement by other cations. Thus, the surface charge is not apparent. At pH values of 6 and above, hydrogen can be replaced by other cations such as calcium which, along with magnesium, tends to dominate these exchange sites in neutral and alkaline soils.

The magnitude of this pH-dependent charge varies with the type of colloid. It accounts for most of the charge of the 1:1-type minerals and up to one fourth of that of some 2:1 types. As will be seen later, it is the dominant type of charge for organic colloids.

IONIC SUBSTITUTION. The phenomenon of the isomorphic substitution of one ion for another in the crystal lattice was mentioned earlier as one of the fundamentals of silicate structure (see p. 76). We are now ready to see in a quantitative way the mechanism by which this substitution results in a second source of net negative charge in the clay crystal.

In Fig. 4:8 the structural arrangement in a segment of the octahedral layer is shown with and without the substitution of a magnesium ion for an aluminum ion. Without substitution, the positive and negative charges are in balance. The three positive charges of aluminum are fully satisfied by an equivalent of three negative charges from the surrounding oxygens or

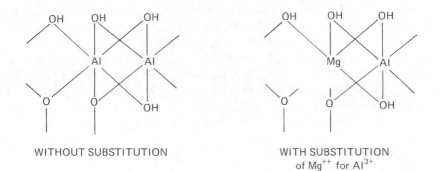

WITHOUT SUBSTITUTION   WITH SUBSTITUTION
                                          of Mg^{++} for Al^{3+}

FIGURE 4:8. Atomic configuration in the octahedral sheet of silicate clays without substitution (*left*) and with a magnesium ion substituted for one aluminum (*right*). Note that where no substitution has occurred, the three positive valences of aluminum are satisfied by an equivalent of half-valences from each of six oxygens or hydroxyls. With magnesium in the place of one aluminum only four of the half-valences are satisfied, leaving half-valences on each of two oxygen atoms unsatisfied. These two half-valences provide a net of one negative charge that must be satisfied by an adsorbed cation.

hydroxyls. There is no net negative or positive charge. However, when a magnesium ion having about the same diameter as an aluminum ion replaces one of the aluminums by isomorphous substitution, an imbalance occurs. The magnesium ion with only two positive charges cannot satisfy the three negative charges associated with the surrounding oxygens and hydroxyls. Consequently, the octahedral layer assumes one negative charge for each magnesium-for-aluminum substitution. This negative charge must be balanced by a positively charged cation such as Na^+ or K^+ which is adsorbed by the clay surface.

The situation just described exists in montmorillonite and related minerals and in some vermiculites. The resultant negative charge is far in excess of that resulting from broken crystal edges of these minerals.

Similar reasoning accounts for the net negative charge in the tetrahedral layer of silicate clay when aluminum, a three-valent ion, has become iso-morphically substituted for silicon, which has four positive charges. This can be shown simply as follows if we recall that each silicon is associated with an equivalent of two oxygens:

Tetrahedral layer (No substitution)	Tetrahedral layer (Al substituted for Si)
$O^{--}Si^{4+}O^{--}$	$O^{--}Al^{3+}O^{--}$
No charge	One excess negative charge

The 2:1-type silicate clays exhibit the substitution phenomenon just explained to a marked degree. Substitution may have occurred in either or both the octahedral and tetrahedral layers. In montmorillonite, most of the substitution has occurred in the octahedral layer, but some may also have taken place in the tetrahedral layer as well. Similarly, substitution in both layers has occurred in illite and vermiculite although the substitution and the source of charge lie largely in the tetrahedral layer. In addition to the magnesium and aluminum ions, iron, manganese, and other ions may substitute for aluminum or silicon in the lattice of certain minerals. The ionic diameter largely governs the substitutions which occur (see Table 4:2).

Unlike the charge associated with the exposed crystal edges, those resulting from ionic substitution are not dependent on pH. With the exception of certain complex aluminum ions, the cations attracted to these charges are subject to replacement at all common pH levels. Therefore, these charges are commonly referred to as *permanent charges* (see Fig. 4:11). This will be dealt with in greater detail as we consider soil acidity (see p. 374).

ANION EXCHANGE. It should be noted that some clay minerals exhibit positive as well as negative charges. This makes possible *anion exchange* between surface hydroxyl units and anions such as phosphate, sulfate, chloride, and nitrate. Since this property is associated most notably with the hydrous oxides of iron and aluminum, its mechanism will be considered later (see p. 92).

4:6. CHEMICAL COMPOSITION OF SILICATE CLAYS

Because of the extensive lattice modifications and ionic substitutions common in clays, chemical formulas cannot be used to identify specifically the clays in a given soil. However, "type" formulas can be used to illustrate differences in composition. Examples of these formulas, which are sometimes referred to as *unit layer formulas,* are shown in Table 4:4.

It is interesting to compare the unit layer formulas of clay minerals with those of the mineral pyrophyllite, the crystals of which have no net positive or negative charge. Referring to Table 4:4, compare the pyrophyllite formula, $Al_4Si_8O_{20}(OH)_4$, with that of montmorillonite in which $\frac{1}{2}$ mole of aluminum has been replaced with magnesium: thus, $(Al_{3.5}Mg_{0.5}) Si_8O_{20}(OH)_4$. The unit layer negative charge of montmorillonite, all of which emanates from the

TABLE 4:4. *Unit Layer Formulas of Important Clay and Other Minerals Showing the Most Prominent Substitution in the Al and Si Sheets as Well as the Molecules Between Crystal Units* [a]

Readily exchangeable ions shown in brackets.

Clay Mineral	Unit Layer Formula				Unit Layer Charge
	Octahedral (Al Sheet)	Tetrahedral (Si Sheet)	Numbers of Oxygen and Hydroxyl	Between Crystal Units	
Kaolinite	Al_4	Si_4	$O_{10}(OH)_8$		0
Pyrophyllite	Al_4	Si_8	$O_{20}(OH)_4$		0
Montmorillonite	$Al_{3.5}Mg_{0.5}$ $[Na_{0.5}]$	Si_8	$O_{20}(OH)_4$		0.5
Vermiculite	Mg_6	Si_7Al $[Mg_{0.5}]$	$O_{20}(OH)_4$	xH_2O, Mg^{++}	1.0
Chlorite	Mg_6	Si_6Al_2	$O_{20}(OH)_4$	$Mg_6(OH)_{12}$	2.0
Illite	Al_4	Si_7Al $[K_{0.2}]$	$O_{20}(OH)_4$	$K_{0.8}$	1.0
Muscovite	Al_4	Si_6Al_2	$O_{20}(OH)_4$	K_2	2.0

[a] Note that the substitution of Mg for Al or Al for Si is compensated for by either exchangeable or intercrystal unit ions (e.g., Na). (In some vermiculites and chlorites the octahedral layer is filled with four aluminum atoms rather than six magnesium atoms as shown.)

octahedral (alumina) layer, is said to be 0.5 since 0.5 equivalent of some cation is required to neutralize it. The negative charge is balanced by a chemically equivalent quantity of cations shown as $(Na_{0.5})$ in Table 4:4.

The unit layer charge for other silicate clay minerals is shown in Table 4:4. Note that kaolinite and pyrophyllite have no such charge since no isomorphous substitution has occurred in either mineral. The isomorphous substitution of 1 mole of aluminum for one of the 8 silicon atoms in each of vermiculite and illite formulas results in a unit layer charge of 1 for these minerals. This high negative charge accounts for the tenacity with which potassium is held between illite crystal units, and magnesium ions and water between crystal units of vermiculite. The high unit layer negative charge of muscovite (2) is satisfied by 2 moles of potassium bound tightly between crystal units. Chlorite's high charge is satisfied partly by positive charges resulting from the substitution of Al^{3+} for Mg^{++} in the brucite $(Mg_6(OH)_{12})$ intercrystal layers and partly by Mg^{++} ions held between crystal units.

The sodium, potassium, and magnesium ions (Table 4:4) are considered as mostly exchangeable; the others are part of the lattice structure. Only the major substitutions have been shown. It will be noted that the differences among the formulas for montmorillonite, illite, chlorite, and vermiculite largely reflect differences in the ions dominant in the tetrahedral and octahedral layers.

4:7. GENESIS OF SILICATE CLAYS

The silicate clays are developed most abundantly from such minerals as the feldspars, micas, amphiboles, and pyroxenes. Apparently, the transformation of these minerals into silicate clays has taken place in soils and elsewhere by at least two distinct processes: (a) a comparatively slight physical and chemical *alteration* of the primary minerals, and (b) a *decomposition* of the original minerals with the subsequent *recrystallization* of certain of their decomposition products into the silicate clays. These processes will each be given brief consideration.

ALTERATION. Alteration of the minerals may be encouraged by chemical action involving the removal of certain soluble constituents and the substitution of others within the crystal lattice. The changes which occur as muscovite is altered to hydrous micas may be used as an example. Muscovite is a 2:1-type primary mineral with a rigid-lattice structure. As the weathering process begins, some potassium is lost from the crystal structure and water molecules enter into the lattice to give a more loose and less rigid crystal. Also, there is a relative increase in the silica content as compared to aluminum in the silica sheet. Some of these changes, perhaps oversimplified, can be shown as follows:

$$K_2Al_4(Al_2Si_6)O_{20}(OH)_4 + Si^{4+} \xrightarrow{H_2O}$$

$$\{K_{0.2}\}(K_{0.8})Al_4(AlSi_7)O_{20}(OH)_4 + K^+ + Al^{3+}$$

Muscovite (Rigid lattice)   Illite (Semirigid lattice)

$\{K_{0.2}\}$ represents exchangeable potassium, and $(K_{0.8})$ represents potassium held semirigidly between crystal units.

There has been a release of potassium and aluminum, a minor change in the chemical makeup, a loosening of crystal lattice, and an initiation of exchangeable properties with little basic change in the crystal structure of the original mineral. It is still a 2:1 type, only having been *altered* in the process of weathering. Continued removal of potassium and substitution of magnesium for some of the aluminum in the alumina layer would result in the formation of montmorillonite (see Fig. 4:9).

These examples illustrate the structural similarity among some of the silicate clay minerals. They also illustrate the earlier reference to the gradual transition from one mineral to another and to the *intermediate* minerals with properties and characteristics in between those of the distinct groups (see

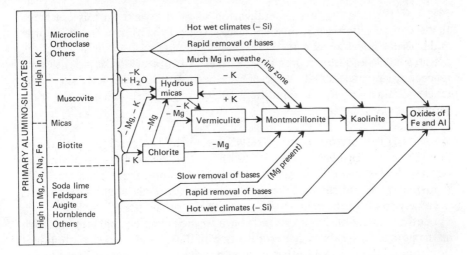

FIGURE 4:9. General conditions for the formation of the various silicate clays and oxides of iron and aluminum. Hydrous micas and chlorite are formed through rather mild weathering of primary alumino-silicate minerals, whereas kaolinite and oxides of iron and aluminum are products of much more intense weathering. Conditions of intermediate weathering intensity encourage the formation of vermiculite and montmorillonite. In each case, silicate clay genesis is accompanied by the removal of soluble elements such as K, Na, Ca, and Mg.

p. 83). The presence of "mixed-layer" minerals with such names as "illite–montmorillonite," "chlorite–illite," and "illite–vermiculite" suggests that a given colloidal crystal may contain crystal units of one mineral in between crystal units of another. Certainly they emphasize the complexity of clay mineralogy and of the soils of which these minerals are a part.

RECRYSTALLIZATION. The crystallization of silicate clays from soluble weathering products of other minerals is perhaps even more important in clay genesis than is alteration. A good example is the formation of kaolinite from solutions containing soluble aluminum and silicon. This process of recrystallization involves a complete change from the structural makeup of the original minerals and is usually the result of much more intense weathering than that required by the alteration process described above.

Moreover, such crystallization makes possible the formation of more than one kind of clay from a given original mineral. The exact silicate colloid which forms apparently depends upon the condition of weathering and the ions present in the weathering solution as crystallization occurs.

RELATIVE STAGES OF WEATHERING. The more specific conditions resulting in the formation of one or more of the important types of clay are shown in

Fig. 4:9. Perhaps the first generalization to be drawn from this outline is that there is a difference in the weathering stage of the minerals. The chlorite and hydrous micas apparently represent the younger weathering stages of the silicates, while kaolinite represents the oldest. Montmorillonite is considered to occupy an intermediate stage of weathering. With Fig. 4:9 as a guide, let us consider briefly the conditions which might yield each of the three groups of clays.

GENESIS OF INDIVIDUAL CLAYS. Hydrous micas represented by illite are thought to be formed by the alteration of the micas. This is postulated because illite is so similar to muscovite in makeup and general characteristics. Apparently, as indicated earlier, only a comparatively slight alteration is necessary to effect the changes from one to the other.

In other cases, illite has apparently been formed from orginal minerals such as the potash feldspars by recrystallization under conditions of an abundant potassium supply. In still other instances, illite may be formed from mont-morillonite if the latter is in contact with abundant potassium. More common, however, is the reverse reaction by which illite weathers to mont-morillonite by the loss of much of its potassium.

Chlorite is formed by the alteration of the magnesium- and iron-rich mica biotite. This change is accompanied by a loss of some magnesium, potassium, and iron. Further alteration and weathering may yield illite (hydrous micas) or vermiculite, either of which can be altered to form montmorillonite.

Montmorillonite may be formed by recrystallization from a variety of minerals provided conditions are appropriate. Apparently, mild-weathering conditions (usually slightly acid to alkaline), a relative abundance of magnesium, and an absence of excess leaching are conducive to the formation of this mineral. Alteration of other silicate clays, such as chlorite, illite, and vermiculite, may also yield montmorillonite.

As already stated, kaolinite represents a more advanced stage of weathering than does any of the other major types of silicate clays. It is formed from the decomposition of silicates under conditions of moderately to strongly acid weathering which results in the removal of the alkali and alkaline earth metals. The soluble aluminum and silicon products that are released may recrystallize under proper conditions to form kaolinite. This mineral, in turn, is subject to decomposition, especially in the tropics, with the formation of iron and aluminum oxides and soluble silica.

As weathering of primary and secondary minerals occurs, ions of several elements are released. The more soluble ions, such as sodium and potassium, are usually removed in leaching waters. Others, such as aluminum, iron, and silicon, either may recrystallize into new silicate-clay minerals or, more commonly, may form insoluble minerals, such as the hydrous oxides of iron and aluminum. As shown in Fig. 4:9, these compounds represent the most advanced stages of weathering. They usually dominate soils only under tropical or semitropical conditions where intense weathering has removed most of the silica (see p. 331).

4:8. MINERAL COLLOIDS OTHER THAN SILICATES

HYDROUS OXIDE CLAYS OF IRON AND ALUMINUM. The discussion so far has dealt only with the silicate clays. Hydrous oxide clays also deserve attention for a least two reasons: (a) they occur in temperature regions intermixed with silicate clays; and (b) this type of colloidal matter is commonly dominant in the soils of the tropics and semitropics. The red and yellow soils of these regions are controlled in large degree by iron and aluminum hydrous oxides of various types.

As their name suggests, *hydrous oxides* are oxides containing associated water molecules. The exact mechanism by which these water molecules are held is somewhat uncertain. For simplicity, the hydrous oxides are often shown as actual iron and aluminum hydroxides, $Al(OH)_3$ and $Fe(OH)_3$. Probably more correct general formulas are as follows: $Fe_2O_3 \cdot xH_2O$ and $Al_2O_3 \cdot xH_2O$. The x indicates that the quantity of associated water of hydration is different for different minerals. In soils, gibbsite ($Al_2O_3 \cdot 3H_2O$) is probably the dominant aluminum oxide; goethite ($Fe_2O_3 \cdot H_2O$) and limonite ($Fe_2O_3 \cdot xH_2O$) are the most prominent iron hydrous oxides.

Although relatively less is known about the hydrous oxide clays, they seem to have some properties in common with the silicates. For example, at least some of them are thought to have definite crystalline structure. The small particles may carry negative charges and thus serve as a central micelle around which a swarm of cations are attracted. The same general constitution described for the silicates may be visualized. Because of the much smaller number of negative charges per micelle, however, cation adsorption is even lower than for kaolinite. Also, most hydrous oxides are not as sticky, plastic, and cohesive as are the silicates. This accounts for the much better physical condition of soils dominated by hydrous oxides.

ALLOPHANE AND OTHER AMORPHOUS MINERALS. In many soils significant quantities of noncrystalline colloidal matter are found. For example, part of the iron and aluminum hydrous oxides in some soils are amorphous. The same is true of part of the silica, especially in soils formed from volcanic ash. In most cases, the properties of the amorphous mineral do not differ greatly from those of the crystalline materials.

Perhaps the most significant amorphous mineral matter in soils is *allophane*, the somewhat poorly defined combinations of silica and aluminum sesquioxide. Having a composition approximating $Al_2O_3 \cdot 2SiO_2 \cdot H_2O$, this material is found as a constituent in many soils. It is most prevalent in soils developed from volcanic ash.

The presence of allophane in a soil cannot be ignored because it has a high cation exchange capacity. This capacity is apparently pH dependent. Allophane also has a considerable anion exchange capacity, a property which will now be considered.

ANION EXCHANGE. Under certain conditions hydrous oxides of iron and aluminum, allophane and even kaolinite show evidence of having positive charges on their crystal surfaces. These charges are thought to have two sources: (a) the protonation or adding of hydrogen ions to hydroxyl groups on the edge of these minerals; and (b) the exchange of the hydroxyl groups for other anions, such as phosphate. These mechanisms can be illustrated by the appropriate equations. First we show the protonation reaction:

$$>Al-O\cdots H \ + \ H^+ \ \rightleftharpoons \ >Al-OH_2^+$$

Surface of mineral at high pH   Surface of mineral at low pH

At high pH values (left) the hydrogen ion tends to dissociate from the oxygen, leaving a negative charge on the surface. As the pH is lowered, an additional hydrogen ion associates with the hydroxyl, leaving a net positive charge. This charge will attract anions such as $H_2PO_4^-$, SO_4^{--}, NO_3^-, and Cl^-. These ions can exchange with each other, giving rise to the term *anion exchange*.

The second mechanism that shows a positive charge on certain colloids is that in which phosphates and similar ions exchange for hydroxyl:

$$>Al-OH \ + \ H_2PO_4^- \ \rightleftharpoons \ >Al-H_2PO_4 \ + \ OH^-$$

Surface of mineral   Surface of mineral

This reaction also occurs primarily at low pH values, the negative hydroxyl ion being replaced from positively charged aluminum ions in the crystal.

By one of the two of these mechanisms of anion exchange, SO_4^{--}, Cl^-, and NO_3^- ions as well as phosphates become adsorbed by allophane, the hydrous oxides, and to a degree by the kaolinite group. The reactions are most significant for phosphates since these ions tend to be quite tightly adsorbed. These reactions emphasize once again the importance of the type of clay in determining the basic reactions in soils.

4:9. GEOGRAPHIC DISTRIBUTION OF CLAYS

The clay of any particular soil is generally made up of a mixture of different colloidal minerals. In a given soil the mixture may vary from horizon to horizon. This occurs because the kind of clay that develops depends not only upon climatic influences and profile conditions but also upon the nature of the parent material. The situation may be further complicated by the presence in the parent material itself of clays that were formed under a preceding and perhaps an entirely different type of climatic regime. Nevertheless, some very general deductions seem possible, taking advantage of the relationships shown in Fig. 4:9.

REGIONAL DIFFERENCES. The well-drained soils of humid and subhumid tropics tend to be dominated by the hydrous oxides of iron and aluminum. These clays are also prominent in the warmer humid regions of the temperate zone such as are found in the southeastern part of the United States. Kaolinite is commonly the dominant silicate mineral in these soils (see Table 4:5) and is also found along with the hydrous oxide clays in more tropical areas.

TABLE 4:5. *Dominant Clay Minerals Found in Different Soil Orders of the Comprehensive Classification System* [a]

Soil Order	General Weathering Intensity	Typical Location in U. S.	Dominant Clay Minerals						
			Hydrous Oxides	Kaolinite	Montmorillonite	Illite	Vermiculite	Chlorite	Intergrades
Entisol	Low	Variable			×	×			
Inceptisol		Variable				×			×
Aridisols		Desert				×	×	×	×
Vertisols		Ala., Tex.			×				
Mollisols		Central U. S.			×	×	×	×	
Alfisols		Ohio, Pa., N.Y.		×	×	×		×	×
Spodosols		New England	×			×			×
Ultisols		Southeast U. S.	×	×			×		×
Oxisols	High	Tropical zones	×	×					×

[a] Adapted from Jackson (3).

As one might expect, montmorillonite, vermiculite, and illite are more prominent in cooler than in warm climates. Intensity of weathering is less there than in tropical and subtropical regions. In the northern part of the United States, in Canada, and in similar temperate regions throughout the world these clay minerals are common. The particular minerals which form depend largely on the parent materials and on soil–water regime. Where either the parent materials or the soil solution surrounding the weathering minerals is high in potassium, illite and related minerals are apt to be formed. Parent materials high in bases, particularly magnesium, or a soil drainage situation which discourages the leaching of these bases, encourages montmorillonite formation. For this reason, illite and montmorillonite are more likely to dominate soils in the semiarid and arid parts of the United States than in the humid East or South.

The strong influence of parent material on the geographic distribution of clays can be seen in the "black-belt" Vertisols of Alabama, Mississippi, and Texas. These soils have developed from base-rich parent materials and are dominated by montmorillonitic clays. They are surrounded by soils high in kaolinite and hydrous oxides which are more representative of this warm, humid region. Similar situations exist in central India and the Sudan.

Data in Table 4:5 show the dominant clay minerals in different soil orders, the descriptions of which are given in Chapter 12. These data tend to substantiate the generalization just discussed. For example, Oxisols and Ultisols are characteristic of areas of intense weathering, while Aridisols are found in desert areas. The dominant clay minerals for these areas are as expected on the basis of Fig. 4:9.

While a few broad generalizations relating to the geographic distribution of clays are possible, these examples suggest that local parent materials and weathering conditions tend to dictate the kinds of clay minerals found in soils.

4:10. ORGANIC SOIL COLLOIDS—HUMUS

Because the clays of surface soils usually carry an appreciable admixture of humus, a brief word about organic colloids is necessary at this point. Otherwise, the soil significance of the colloidal state of matter cannot be fully visualized.

COLLOIDAL ORGANIZATION. Humus may be considered to have a colloidal organization similar to that of clay shown in Fig. 4:10. A highly charged

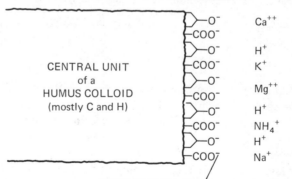

FIGURE 4:10. Adsorption of cations by humus colloids. The phenolic hydroxyl groups (⟩—O⁻) are attached to aromatic rings; the carboxyl groups (—COO⁻) are bonded to other carbon atoms in the central unit. Note the general similarity to the adsorption situation in silicate clays. In this case only surface adsorption is shown, but adsorption occurs within the micelle as well.

anion (micelle) is surrounded by a swarm of adsorbed cations. As later sections will show, the reactions of these cations are essentially the same whether they are adsorbed by clay or by humus.

Some important differences should be noted, however, between humus and the inorganic micelles. First, the complex humus micelle is composed basically of carbon, hydrogen, and oxygen rather than of aluminum, silicon, and oxygen as are the silicate clays. Also, the cation exchange capacity of humus per unit weight far exceeds even that of montmorillonite. The humus micelle is not considered crystalline, and the size of the individual particles, although extremely variable, may be at least as small as montmorillonite. Last, humus is not as stable as clay and is thus somewhat more dynamic—being formed and destroyed much more rapidly than clay.

Because of their complexity, relatively less is known about the specific structure of humus colloids than of the silicate clays. However, it is known that humus is not a specific compound nor does it have a single structural makeup. The major sources of negative charge are thought to be partially

neutralized carboxylic (—COOH) and phenolic (⟨ ⟩—OH) groups

associated with central units of varying size and complexity. This relationship is illustrated in a general way in Fig. 4:10.

The charge on humus colloids is pH dependent, as is the case for part of the silicate clays (see p. 84). Under strongly acid conditions hydrogen is tightly bound and not easily replaceable by other cations. The colloid therefore exhibits a low negative charge. With the addition of bases and the consequent rise in alkalinity, first the hydrogen from the carboxyl groups and then the hydrogen from the phenolic groups ionizes and is replaced by calcium, magnesium and other cations.

On the basis of solubility in acids and alkalis, humus is thought to be made up of three classes: (a) *fulvic acid*, lowest in molecular weight and lightest in color, soluble in both acid and alkali; (b) *humic acid*, medium in molecular weight and color, soluble in alkali but insoluble in acid; and (c) *humin*, highest in molecular weight, darkest in color, and insoluble in both acid and alkali. In spite of differences in chemical and physical makeup, the three classes tend to be very similar with regard to such properties as cation adsorption and nutrient release. Consequently, they will all be considered under the general term "humus."

The information on colloids presented thus far has emphasized the complexity of the compounds and crystals of which the individual micelles are made. In spite of this complexity, however, the general organization is about the same for each of the colloidal groups. That is, a central negatively charged micelle is surrounded by a swarm of cations. Consideration will now be given in somewhat more detail to this organization and the reactions modifying it.

4:11. COLLOIDS—ACID SALTS

Colloidal particles, regardless of their composition, are made up of a complex negative radical, the micelle, and a miscellany of adsorbed cations. In a humid region, those cations of calcium, aluminum, and hydrogen are by far the most numerous. Consequently, a colloidal complex may be represented in the following simple and convenient way:

The significance of the Ca, Al, and H of the formula is obvious because in a humid region these cations jointly dominate the complex. The M stands for the small amounts of metallic and other "base-forming" cations—Mg, K, Na, etc.— that are usually present in the ionic outer layer. The a, b, c, and d indicate that the numbers of cations are variable.

The above illustration emphasizes that clays and their associated exchangeable ions can be considered in perhaps an oversimplified way as *complex acid salts*. This can be verified by comparing the above formula with that of $NaHSO_4$, a well-known acid salt. In both cases negatively charged radicals (anions) are associated with metallic cations and hydrogen. The only difference is one of size and charge, the clay anion (micelle) being much larger than the sulfate and, of course, having many more negative charges per anion.

When by the proper laboratory manipulations the metallic cations are replaced entirely by hydrogen ions, a clay "acid" or H $\boxed{\text{Micelle}}$ results. In a like manner calcium ions may be given dominance and a calcium "salt" Ca $\boxed{\text{Micelle}}$ comes into being. It should be noted that several different acids and calcium salts are possible depending upon the nature of the micelle— that is, whether we are dealing with kaolinite, montmorillonite, humus, or some other colloid. Of course, in nature, soil colloids are very seldom wholly acids or wholly salts. As illustrated by the general formula at the beginning of this section, they are actually mixtures of *complex acid salts*.

WHY ADSORBED CALCIUM, ALUMINUM, AND HYDROGEN IONS ARE SO PROMINENT. In the early stages of clay formation the solution surrounding the decomposing silicates contains calcium, magnesium, potassium, sodium, iron, and aluminum ions, which have been liberated by weathering. These ions are not all held with equal tightness by the soil colloids. The order of strength of adsorption when they are present in equivalent quantities is Al > Ca > Mg > K > Na. Consequently, one would expect the quantity of these ions in the exchangeable form to be in the same order, with aluminum and calcium being the most dominant cation and sodium the least dominant. This is generally the case in most well-drained moderately acid humid-region soils.

As organic matter gradually accumulates and decomposes, organic and other acids are generated. These supply hydrogen, which influences the cation adsorption in two ways. First, it helps solubilize or keep in solution aluminum ions, which are quickly adsorbed by the colloids. Second, the hydrogen ions are rather tightly adsorbed themselves by both organic and inorganic micelles. The aluminum and hydrogen ions are very important because between them they characterize soil acidity and determine the amount of lime needed for optimum plant growth. Their presence is favored by high rainfall levels, which tend to leach out the other ions, most of which are held with relatively less tenacity.

In arid and semiarid regions, the calcium and other bases do not leach from the soil. These metallic cations therefore tend to dominate the adsorptive sites and pH values of 7.0 and above result. Under these conditions the aluminum ions form insoluble compounds and the adsorbed hydrogen ions are replaced by metallic cations.

4:12. CATION EXCHANGE

With the fundamental colloidal concepts already presented as a background, the exchange of one cation for another on soil colloids may be explained very simply. The tendency of hydrogen ions to force such a transfer will be used to illustrate this phenomenon. Cation exchange is one of the most common and most important of soil reactions.

SIMPLE EXAMPLE. Consider a near-neutral soil rather high in adsorbed calcium and functioning under optimum conditions of moisture and temperature. A substantial amount of organic and mineral acids is formed as the organic matter decomposes. The hydrogen ions thus generated will tend to replace the exchangeable calcium of the colloidal complex. This occurs not only because of mass action effects but also because under comparable conditions the hydrogen ion is adsorbed more strongly than is the calcium ion. The reaction may be shown simply as follows, where only one ion of the adsorbed calcium is represented as being displaced:

$$\text{Ca}\boxed{\text{Micelle}} + 2\text{H}^+ \rightleftharpoons \begin{smallmatrix}\text{H}\\\text{H}\end{smallmatrix}\boxed{\text{Micelle}} + \text{Ca}^{++}$$

The reaction takes place fairly rapidly and the interchange of calcium and hydrogen is chemically equivalent. Moreover, if the hydrogen ion concentration were decreased for any reason or the calcium ions are increased by adding a calcium-containing limestone for instance, the adjustment would be to the left in response to mass action. Conversely, if the hydrogen ions are increased or if the calcium ions are removed by leaching, the adjustment would be to the right. The soil is so dynamic that the equilization is constantly changing, oscillating back and forth as conditions dictate.

CATION EXCHANGE UNDER NATURAL CONDITIONS. With these principles in mind, suppose we write the reaction more completely and as it usually occurs in humid region surface soils. We shall assume, for the sake of simplicity, that the number of calcium, aluminum, hydrogen, and other metallic cations (M) are in the ratio 40, 20, 20, and 20 per micelle, respectively. The metallic cations are considered monovalent in this case:

$$\begin{matrix} Ca_{40} \\ Al_{20} \\ H_{20} \\ M_{20} \end{matrix} \boxed{Micelle} + 5H_2CO_3 \rightleftharpoons \begin{matrix} Ca_{38} \\ Al_{20} \\ H_{25} \\ M_{19} \end{matrix} \boxed{Micelle} + \begin{matrix} 2Ca(HCO_3)_2 \\ + M(HCO_3) \end{matrix}$$

Subject to loss in drainage

It is important to note that where sufficient rainfall is available to leach the calcium the reaction tends to go toward the right—that is, hydrogen ions are entering and calcium and other bases (M) are being forced out of the exchange complex into the soil solution. This occurs because the establishment of equilibrium is prevented by the loss of carbonates and bicarbonates in drainage. Obviously, so long as bases are removed and the hydrogen ion concentration is not correspondingly diminished, the adjustment will continue as pictured.

Note that aluminum is not shown as being replaced by the hydrogen. This is because of the tenacity with which the aluminum ion is held. Also, an increase in acidity would likely bring more aluminum into solution, which might even increase the adsorption of this element. This situation is discussed in greater detail on page 373.

It is to be noted also that one micelle is taken to represent the whole colloidal fraction and that the loss of metallic cations (calcium and base-forming cations) is just balanced by the gain in hydrogen ions. That is, the exchange is in *chemically equivalent amounts.*

In regions of low rainfall the calcium and other salts are not easily leached from the soil, as indicated. This does not permit the reaction to go to completion. That is, it prevents the removal of bases from the exchange sites, thereby keeping the soil neutral or above in reaction. The interaction of climate, biological processes, and cation exchange thus helps determine the properties of soils.

LOSS OF METALLIC ELEMENTS. The above reaction is an excellent illustration of cation exchange and also is one of the most important and far-reaching transfers that occurs in humid region surface soils. By this mechanism, calcium, and to a lesser extent magnesium, potassium, and sodium, are lost from the soil by leaching. At the same time, aluminum and hydrogen ions, both of which enhance the acidic properties of the soil, are increased. Thus, cation exchange accounts for the great loss of calcium suffered by humid region surface soils. At the same time it indicates why such soils tend to become acid so quickly (see pp. 379–80).

INFLUENCE OF LIME AND FERTILIZER. Cation exchange reactions are reversible. Hence, if some form of limestone or other basic calcium compound is applied to an acid soil, the reverse of the replacement just cited occurs. The active calcium ions replace the hydrogen and other cations by mass action. As a result, the clay becomes higher in exchangeable calcium and lower in adsorbed hydrogen and aluminum. And, as the soil solution adjusts to this altered proportion of bases and hydrogen, its pH is raised and its chemical makeup is modified.

One more illustration of cation exchange will be offered. If a soil is treated with a liberal application of a fertilizer containing potassium chloride, an exchange such as follows may occur (again M is considered to be monovalent):

$$
\begin{array}{cccc}
& & K_7 & 2CaCl_2 \\
Ca_{40} & & Ca_{38} & + \\
Al_{20}\;\boxed{Micelle} & +\;7KCl \;\rightleftharpoons\; & Al_{20}\;\boxed{Micelle} & HCl \\
H_{40} & & H_{39} & + \\
M_{20} & & M_{18} & 2MCl
\end{array}
$$

Some of the added potassium pushes its way into the colloidal complex and forces out equivalent quantities of calcium, hydrogen, and other elements which appear in the soil solution. The adsorption of the added potassium is considered to be advantageous because a nutrient so held remains largely in an available condition but is less subject to leaching than are most fertilizer salts. Hence, cation exchange is an important consideration not only for nutrients already present in soils but also for those applied in commercial fertilizers and in other ways.

It should be noted at this point that anionic fixation may take place in soils and in some cases assumes considerable importance. This is especially true with respect to the adsorption of phosphate ions. This phase is discussed on page 92.

4:13. CATION EXCHANGE CAPACITY

The cation exchange capacity (also termed "cation adsorption capacity") of soil colloids can be determined rather easily. In commonly used methods, the original adsorbed nutrients are replaced by barium, potassium, or ammonium ions and then the amount of adsorbed barium, potassium, or ammonium is determined.

EFFECT OF SOIL pH. It should be pointed out that the cation exchange in most soils increases with pH (see Fig. 4:11). At a very low pH value, only the "permanent" charges of the clays (see p. 86) and a small portion of the charges of organic colloids hold ions that can be replaced by cation exchange. On the majority of the organic colloid exchange sites and on some

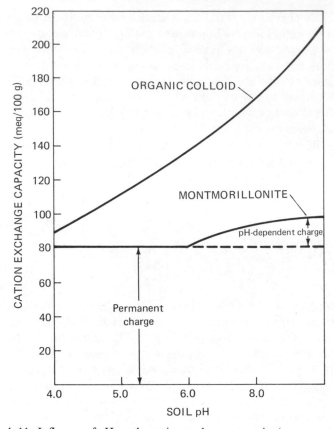

FIGURE 4:11. Influence of pH on the cation exchange capacity (a measure of the negative charge) of montmorillonite and humus. Note that below pH 6.0 the charge for the clay mineral is relatively constant. This charge is considered permanent and is due to ionic substitution in the crystal unit. Above pH 6.0 the charge on the mineral colloid increases because of ionization of hydrogen from exposed O—H groups at crystal edges. In contrast to the clay, essentially all of the charges on the organic colloid are considered pH dependent. [*Montmorillonite data from Coleman and Mehlich* (1); *organic colloid data from Helling et al.* (2).]

of those of the inorganic fraction, hydrogen and perhaps aluminum hydroxy ions are held so tightly as to resist replacement. Therefore, the cation exchange capacity is relatively lower than at high pH.

As the pH is raised, the hydrogen held by the remainder of the organic and inorganic colloids becomes ionized and is replaceable. Also, the adsorbed aluminum hydroxy ions (see p. 374) are removed, forming $Al(OH)_3$, thereby releasing additional exchange sites on the mineral colloids. The net result is an increase in the cation exchange capacity.

In most cases the cation exchange capacity is determined at a pH of 7.0 or above. This means that it includes most of those charges dependent on pH as well as the more or less permanent ones.

MILLIEQUIVALENTS. The cation exchange capacity is expressed in terms of *equivalents* or, more specifically, as *milliequivalents* per 100 grams. The term "equivalent" is defined as *1 gram atomic weight of hydrogen or the amount of any other ion that will combine with or displace this amount of hydrogen.* For monovalent ions such as Na^+, K^+, NH_4^+, and Cl^-, the equivalent weight and atomic weight are the same since they can replace or react with one H^+ ion. Divalent cations such as Ca^{++} and Mg^{++} can take the place of two H^+ ions. Consequently, their atomic weight must be divided by 2 to obtain the equivalent weight.

The milliequivalent weight of a substance is one thousandth of its atomic weight. Since the equivalent weight of hydrogen is about 1 gram, the term *milliequivalent* (meq) may be defined as *1 milligram of hydrogen or the amount of any other ion that will combine with or displace it.*

Thus, if a clay has a cation exchange capacity of 1 milliequivalent (1 meq per 100 grams), it is capable of exchanging 1 mg of hydrogen or its equivalent for every 100 grams of clay. This is 1 mg to 100,000 mg of clay, or 10 parts per million. Therefore, an acre–furrow slice of such a clay weighing 2 million pounds could adsorb 20 pounds of exchangeable hydrogen or its equivalent. A comparable figure for a hectare to a plow depth of 15 centimeters is 22 kilograms.

It is well to note this term—*equivalent*. It indicates that other ions also may be expressed in terms of milliequivalents. Consider calcium, for example. This element has an atomic weight of 40, compared to 1 for hydrogen. Each Ca^{++} ion has two charges and is thus equivalent to two H^+ ions. Therefore, the amount of calcium required to displace 1 mg of hydrogen is 40/2, or 20 mg. This, then, is the weight of 1 meq of calcium. If 100 grams of a certain clay is capable of exchanging a total of 250 mg of calcium, the cation exchange capacity is 250/20, or 12.5 meq per 100 grams. The milliequivalent method of expression is so convenient and so commonly used that a person dealing with the literature of soil science must be familiar with it.

The milliequivalent method of expression can be converted easily to practical field terms. For instance, 1 meq of hydrogen can be replaced on the colloids by 1 meq of $CaCO_3$ contained in ordinary limestone. The molecular weight of $CaCO_3$ (100) contains 2 equivalent weights. Since the amount of $CaCO_3$ needed is only 1 meq weight, 100/2 = 50 mg will be needed to replace 1 mg of hydrogen. In view of the fact that 1 meq of H^+ per 100 grams can be expressed as 20 pounds of hydrogen per acre–furrow slice, 1 meq of $CaCO_3$ per 100 grams is the same as $20 \times 50 = 1000$ pounds of $CaCO_3$ per acre–furrow slice. Thus, $\frac{1}{2}$ ton of $CaCO_3$ per acre–furrow slice has the potential of replacing 1 meq of H^+ per 100 grams of soil. Expressed in the metric system this figure is 1100 kilograms per hectare. These are practical relationships worth remembering.

CATION EXCHANGE CAPACITIES OF SOIL COLLOIDS. As might be expected, the cation exchange capacity of the colloidal fraction of soils exhibits a wide range because humus and several minerals may be present in varying amounts. The cation exchange capacities of humus, vermiculite, montmorillonite, hydrous mica and chlorites, kaolinite, and hydrous oxides are more or less in the order of 200, 150, 100, 30, 8, and 4 meq per 100 grams, respectively (see Table 4:3).

It is easy to see why the clay complex of Southern soils, when dominated by kaolinite, should have a low exchange capacity, ranging perhaps between 5 and 20 meq per 100 grams of soil. On the other hand, the clay mixtures functioning in the soils of the Midwest, where illite and montmorillonite are more likely to be prominent, have a much higher adsorptive capacity, ranging from 50 to possibly 100 meq, depending on conditions. Colloids of soils very high in humus (such as mucks) will have cation exchanges in excess of 180 meq per 100 grams.

TABLE 4:6. *Cation Exchange Capacity of a Wide Variety of Surface Soils from Various Parts of the United States* [a]

Milliequivalents per 100 grams of dry soil

Soil Type	Exchange Capacity	Soil Type	Exchange Capacity
Sand		*Silt Loam*	
Sassafras (N.J.)	2.0	Delta (Miss.)	9.4
Plainfield (Wis.)	3.5	Fayette (Minn.)	12.6
		Spencer (Wis.)	14.0
Sandy loam		Dawes (Nebr.)	18.4
Greenville (Ala.)	2.3	Carrington (Minn.)	18.4
Sassafras (N.J.)	2.7	Penn (N.J.)	19.8
Norfolk (Ala.)	3.0	Miami (Wis.)	23.2
Cecil (S.C.)	5.5	Grundy (Ill.)	26.3
Coltneck (N.J.)	9.9		
Colma (Calif.)	17.1	*Clay and clay loam*	
		Cecil clay loam (Ala.)	4.0
Loam		Cecil clay (Ala.)	4.8
Sassafras (N.J.)	7.5	Coyuco sandy clay (Calif.)	20.2
Hoosic (N.J.)	11.4	Gleason clay loam (Calif.)	31.5
Dover (N.J)	14.0	Susquehanna clay (Ala.)	34.2
Collington (N.J.)	15.9	Sweeney clay (Calif.)	57.5

[a] Data compiled from Lyon et al. (4).

4:14. CATION EXCHANGE CAPACITY OF WHOLE SOILS

VARIABILITY OF THE CATION EXCHANGE CAPACITY OF SOILS. Comparative cation exchange capacities (CEC) of a number of soils are shown in Table 4:6. Note the great range in the figures presented, the highest approaching 60 meq per 100 grams. This is to be expected because soils vary so markedly in humus content and in the amounts and kinds of clay present. It is one thing to have kaolinite as the major clay and quite another if montmorillonite is dominant, so different are their individual capacities to adsorb cations. Moreover, humic residues developed under different climates or from diverse plant tissues do not always possess the same adsorptive power. All these factors contribute to the exceptional variability and wide range in the data presented.

FACTORS AFFECTING THE CATION EXCHANGE CAPACITY. A more careful examination of the table suggests several factors affecting the cation exchange capacity. For example, finer textured soils tend to have higher cation exchange capacities than sandy soils. Further, within a given textural group, organic matter content and the amount and kind of clay probably influence the CEC. The marked difference between the CEC of the Cecil and Susquehanna clay soils in Alabama is due to the dominance of 1:1-type clays in the Cecil and of 2:1-type clays in the Susquehanna.

REPRESENTATIVE FIGURES. From the preceding discussion, average figures for the cation exchange capacity of silicate clay (0.5 meq for each 1 percent) and for well-humified organic matter (2.0 for each 1 percent) can be ascertained. By using these figures, it is possible to calculate in a rough way the cation exchange capacity of a humid temperate region surface soil from the percentages of clay and organic matter present.

4:15. PERCENTAGE BASE SATURATION OF SOILS

Two groups of adsorbed cations have opposing effects on soil acidity and alkalinity. Hydrogen and aluminum tend to dominate acid soils, both contributing to the concentration of H^+ ions in the soil solution. Adsorbed hydrogen contributes directly to the H^+ ion concentration in the soil solution. Al^{3+} ions do so indirectly through hydrolysis. This may be illustrated as follows:

$$Al^{3+} + H_2O \rightarrow Al(OH)^{++} + H^+$$

$$Al(OH)^{++} + H_2O \rightarrow Al(OH)_2^+ + H^+$$

Most of the other cations, called *exchangeable bases*, neutralize soil acidity. The proportion of the cation exchange capacity occupied by these bases is called the percentage *base saturation*. Thus, if the percentage base saturation is 80, four-fifths of the exchange capacity is satisfied by bases, the other by hydrogen and aluminum. The example in Fig. 4:12 should be helpful in showing this relationship.

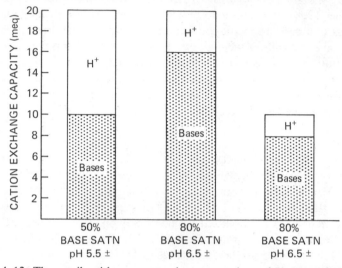

FIGURE 4:12. Three soils with percentage base saturations of 50, 80, and 80, respectively. The first is a clay loam; the second, the same soil satisfactorily limed; and the third, a sandy loam with a cation exchange capacity of only 10 me. Note especially that soil pH is correlated more or less closely with percentage base saturation. Also note that the sandy loam (*right*) has a higher pH than the acid clay loam (*left*), even though the latter contains more exchangeable bases.

PERCENTAGE BASE SATURATION AND pH. A rather definite correlation exists between the percentage base saturation of a soil and its pH. As the base saturation is reduced as a result of the loss in drainage of calcium and other metallic constituents, the pH also is lowered in a more or less definite proportion. This is in line with the common knowledge that leaching tends ordinarily to increase the acidity of humid region soils. The mechanism by which colloids exert a control on soil pH will be considered later (p. 378).

Within the range pH 5 to 6, the ratio for humid temperate region mineral soils is roughly at 5 percent base saturation change for every 0.10 change in pH. Thus, if the percentage base saturation is 50 percent at pH 5.5, it should be 25 and 75 percent at pH 5.0 and 6.0. This relationship is worth remembering.

HUMID VERSUS ARID REGION SOILS. Cation exchange data for representative surface soils from two temperate regions, one humid and the other semi-arid, are shown in Table 4:7. The differences are striking. The humid region soil is distinctly acid and about two-thirds base saturated. In contrast, the soil from a semiarid region has a pH near 7, and its percentage base saturation is nearly 100. The hydrogen and aluminum ions are prominent in the exchange complex of the humid region soil; calcium tends to dominate the soil from the semiarid region. Table 4:7 merits diligent study.

TABLE 4:7. *Cation Exchange Data for Representative Mineral Surface Soils in Humid Temperate and Semiarid Temperate Regions*

Characteristics	Humid Region Soil	Semiarid Region Soil
Exchangeable calcium (meq/100 g)	6–9	13–16
Other exchangeable bases (meq/100 g)	2–3	6–8
Exchangeable hydrogen and/or aluminum (meq/100 g)	4–6	1–2
Cation exchange capacity (meq/100 g)	12–18	20–26
Base saturation (%)	66.6	95 and 92
Probable pH	5.6–5.8	~7

4:16. CATION EXCHANGE AND THE AVAILABILITY OF NUTRIENTS

Exchangeable cations are generally available to both higher plants and microorganisms. By cation exchange, hydrogen ions from the root hairs and soil microorganisms replace nutrient cations from the exchange complex. They are forced into the soil solution, where they can be assimilated by the adsorptive surfaces of roots and soil organisms, or they may be removed by drainage water.

CATION SATURATION AND NUTRIENT ABSORPTION BY PLANTS. Several factors operate to expedite or retard the release of nutrients to plants. First, there is the proportion of the cation exchange capacity of the soil occupied by the nutrient cation in question. For example, if the percentage calcium saturation of a soil is high, the displacement of this cation is comparatively easy and rapid. Thus, 6 meq of exchangeable calcium in a soil whose exchange capacity is 8 probably would mean ready availability. But 6 meq when the total exchange capacity of a soil is 30 present quite the opposite condition. This is one reason why for a crop which requires abundant calcium, such as alfalfa, the base saturation of at least part of the soil should approach or even exceed 90 percent.

INFLUENCE OF ASSOCIATED IONS. A second important factor influencing the plant uptake of a given cation is the effect of the ions held in association with it. For example, potassium availability to plants has been shown to be limited by excessive quantities of exchangeable calcium. Likewise, in some cases high-exchangeable-potassium contents have depressed the availability of magnesium.

EFFECT OF TYPE OF COLLOID. Third, the several types of colloidal micelles differ in the tenacity with which they hold specific cations. This undoubtedly will affect the ease of cation exchange. At a given percentage base saturation, the tenacity with which calcium is held by montmorillonite is much greater than that by kaolinite. As a result, montmorillonite clay must be raised to about 70 percent base saturation before calcium will exhibit an ease and rapidity of exchange that will satisfy growing plants. A kaolinite clay, on the other hand, seems to liberate calcium much more readily, serving as a satisfactory source of this constituent at a much lower percentage base saturation. Obviously the need to add limestone to the two soils will be somewhat different, partly because of the factor under discussion.

4:17. OTHER PROPERTIES OF COLLOIDS—PLASTICITY, COHESION, SWELLING, SHRINKAGE, DISPERSION, AND FLOCCULATION

Without a doubt, the cation exchange property of colloids is outstanding. Yet, from the practical as well as the technical standpoint, certain other characteristics also assume considerable importance. We shall discuss plasticity, cohesion, swelling, shrinkage, dispersion, and flocculation. As might be expected, they are all surface phenomena, and their intensities depend upon the amount and nature of the interfaces presented by the colloids.

PLASTICITY. Soils containing more than about 15 percent clay exhibit *plasticity*—that is, pliability and the capacity to be molded. This property is probably due to the platelike nature of the clay particles and the lubricating though binding influence of the adsorbed water. Thus, the particles easily slide over each other, much like panes of glass with films of water between them.

Plasticity is exhibited only when soils are moist or wet. The range of moisture contents in which plasticity is evident is set by two *plastic limits*. The lower plastic limit is the moisture content below which the soil no longer can be molded. At moisture contents above the upper plastic limit the soil ceases to be plastic but instead tends to flow much as a liquid. While these limits have some usefulness for agricultural purposes, they have special meaning in the classification of soils for engineering purposes such as the bearing strength for a building or a highway bed.

A comparison of the upper and lower plastic limits of three types of clay is given in Table 4:8. Note that the limits and the range between them are much higher for montmorillonite than for kaolinite; illite is intermediate in its properties. Also, the saturating cation affects these limits; sodium gives mugh higher limits and range for the montmorillonite clay than does calcium. In general, soils with wide ranges between these two limits are difficult to handle in the field.

TABLE 4:8. *Plastic Limits of Three Types of Silicate Clays Saturated with Calcium or Sodium* [a]

Type of Clay	Calcium Saturated (%)		Sodium Saturated (%)	
	Lower Limit	Upper Limit	Lower Limit	Upper Limit
Montmorillonite	63	177	97	700
Illite	40	90	34	61
Kaolinite	36	73	26	52

[a] From White (5).

In a practical way, plasticity is extremely important because it encourages such a ready change in soil structure. This must be considered in tillage operations. As everyone knows, the cultivation of a fine-textured soil when it is too wet will result in a puddled condition detrimental to suitable aeration and drainage. With clayey soils, especially those of the montmorillonite type, plasticity presents a real problem. Suitable granulation therein is often difficult to establish and maintain.

COHESION. A second characteristic, somewhat related to plasticity, is *cohesion*. As the water of a clay gel is reduced, there is a seeming increase in the attraction of the colloidal particles for each other. This tendency of the clay particles to stick together probably is due primarily to the attraction of the clay particles for water molecules held in between them. Hydrogen bonding between clay surfaces and water and among the water molecules is the attractive force responsible for the cohesion. It results in cloddy field conditions which are not conducive to easy tillage.

As one might expect, montmorillonite and illite exhibit cohesion to a much more noticeable degree than does kaolinite or hydrous oxides. Humus, by contrast, tends to reduce the attraction of individual clay particles for each other.

SWELLING AND SHRINKAGE. The third and fourth characteristics of silicate clays to be considered are *swelling* and *shrinkage*. If the clay in question has an expanding crystal lattice, as is the case with montmorillonite, extreme swelling may occur upon wetting. Kaolinite and most hydrous oxides with a static lattice do not exhibit the phenomenon to any extent; illite and vermiculite are intermediate in this respect. After a prolonged dry spell, soils high in montmorillonite often are criss-crossed by wide, deep cracks which at first allow rain to penetrate rapidly (see Fig. 4:13). But later, because of swelling, such a soil is likely to close up and become much more impervious than one dominated by kaolinite.

Apparently, swelling, shrinkage, cohesion, and plasticity are closely related. They are dependent not only upon the clay mixture present in a soil and the dominant adsorbed cation but also upon the nature and amount of humus that accompanies the inorganic colloids. These properties of soils are responsible to no small degree for the development of soil structure, as has been previously stressed (see pp. 58–62).

FIGURE 4:13. A field scene showing the cracks that result when a soil high in clay dries out. The type of clay in this case was probably montmorillonitic. [*Photo courtesy U. S. Soil Conservation Service.*]

DISPERSION AND FLOCCULATION. A typical condition of a dilute colloidal suspension in water is that of complete *dispersion*; that is, the particles tend to repel each other, permitting each particle to act completely independent of the others. This condition in soil colloids is encouraged by the smallness of the colloidal particles. It is due even more, however, to their negative charge and hydration, the latter being enhanced by the swarm of hydrated cations around each micelle. Dispersion is encouraged by higher pH values, where the micelle electronegativity is at a maximum (see Fig. 4:11). Also, highly hydrated monovalent ions such as Na^+ which are not very tightly held by the micelles help to stabilize dispersed colloids. Apparently, these loosely held ions do not effectively reduce the electronegativity or *zeta potential* of the micelle, permitting the individual micelles to repel each other and to stay in dispersion. The zeta potential of a colloid–cation system can be determined by measuring the movement of colloidal particles in an electrical field. The rate of movement of the silicate clay particles toward the positive pole is directly proportional to the zeta potential. Thus, sodium-saturated colloidal particles will move more rapidly in an electrical field than will those saturated with calcium.

The zeta potential can be reduced in several ways. First, the pH can be lowered. This will have the effect of reducing the negative charge on the micelle (see p. 84) and of replacing at least some of the sodium or other monovalent ion with hydrogen. Second, a divalent or trivalent ion such as Ca^{++}, Mg^{++}, or Al^{3+} can be introduced to exchange with the monovalent sodium. These multicharged ions are more tightly adsorbed by the micelle, thereby reducing the zeta potential and enhancing the opportunities for micelle collisions and coagulation. Finally, adding simple salts will increase the concentration of cations around the micelle, thereby reducing the zeta potential and discouraging dispersion.

Each of these techniques will tend to encourage the opposite of dispersion, *flocculation*. This is a condition that is generally beneficial from the standpoint of agriculture since it is a first step in the formation of stable aggregates or granules. The ability of common cations to flocculate soil colloids is in the general order of Al > Ca and H > Mg > K > Na. This is fortunate since the colloidal complexes of humid and subhumid region soils are dominated by aluminum, hydrogen, and calcium; those of semiarid regions are high in calcium. These ions all encourage flocculation.

In limited areas of arid regions, sodium ions have become prominent on the exchange complex. This results in a dispersed condition of the soil colloids, making the soils largely impervious to water penetration. Most plants will not grow under these conditions. The sodium must be replaced before normal growth will occur.

It is clear that the six colloidal properties under discussion are of great importance in the practical management of arable soils. The field control of soil structure must definitely take them into account. With the colloidal viewpoint now provided, it might be worthwhile to review the discussion already offered (pp. 62–63) relating to structural management of cultivated lands.

4:18. CONCLUSION

No attempt will be made to summarize this chapter except to re-emphasize three things: (a) the unique and somewhat complicated physical and chemical organization of soil colloids, (b) their capacity to expedite certain phenomena vital to plant and animal life, and (c) the bearing of these phenomena on soil management and crop production.

Yet an understanding of these phases, even in detail, leaves the picture incomplete. The concept must be *bio*colloidal, suggesting that our knowledge of the colloidal state of matter should be kept in mind as we consider soil organisms, soil organic matter, and the genesis of humus.

REFERENCES

(1) Coleman, N. T., and Mehlich, A., "The Chemistry of Soil pH," *The Yearbook of Agriculture* (*Soil*) (Washington, D.C., U. S. Department of Agriculture, 1957).

(2) Helling, C. S., et. al., "Contribution of Organic Matter and Clay to Soil Cation Exchange Capacity as Affected by the pH of the Saturated Solution," *Soil Sci. Soc. Amer. Proc.,* **28**:517–20, 1964.

(3) Jackson, M. L., "Chemical Composition of Soils," in Bear, F. E., Ed., *Chemistry of the Soil,* 2nd ed. (New York: Reinhold, 1955).

(4) Lyon, T. L., Buckman, H. O., and Brady, N. C., *The Nature and Properties of Soils* (New York: Macmillan, Inc., 1952).

(5) White, W. A., "Atterberg Plastic Limits of Clay Minerals," *Amer. Mineralogist,* **34**:508–12, 1949.

Chapter 5

ORGANISMS OF THE SOIL[1]

HUMUS, like clay, is a product of degradation and synthesis. And the agencies responsible are the living organisms in the soil, both the animals (fauna) and the plants (flora). These organisms engineer a myriad of biochemical changes as decay takes place. They also physically churn the soil and help stabilize soil structure. Our concern for these organisms is in terms of their activity rather than their scientific classification. Consequently, the grouping employed (Table 5:1) is very broad and simple.

A vast number of organisms live in the soil. By far the greater proportion of these belong to plant life. Yet animals are not to be minimized, especially in the early stages of organic decomposition. Most soil organisms, both plant and animal, are so minute as to be seen only with the aid of the microscope. The number of the larger organisms ranging in size to that of the larger rodents is comparatively small. All are of significance in the intricate biological processes operative in soils.

5:1. ORGANISMS IN ACTION

The activities of soil flora and fauna are so interrelated as to make it rather difficult to study them independently. To illustrate this point, consider how various soil organisms are involved in the degradation of higher plant tissue. Even while the plants are growing they are subject to attack by soil organisms known as *herbivores*. Examples are parasitic nematodes, snails, slugs, and the larvae of some insects which attack plant roots. Likewise, soilborne termites and beetle larvae devour above-ground woody materials as do some of the larger mammals, such as woodchucks and mice.

PRIMARY CONSUMERS. As soon as a leaf or a stalk or a piece of bark drops to the ground, it is subject to a coordinated attack by microflora and by *detritivores*, animals which live on dead and decaying plant tissues (see Fig. 5:1). If a little moisture is present, bacteria and fungi initiate the attack. They are joined by mites, snails, beetles, millipedes, woodlice, springtails, earthworms, and enchytraeid worms. These animals chew or tear holes in the tissue, opening it up to more rapid attack by the microflora. Together with

---

[1] For reviews of soil organisms, see (1) and (3).

TABLE 5:1. *The More Important Groups of Organisms Commonly Present in Soils*
The grouping is very broad and general as the emphasis is to be placed not on classification but
upon biochemical activity.

Animals	Macro	Subsisting largely on plant materials	Small mammals—squirrels, gophers, woodchucks, mice, shrew Insects—springtails, ants, beetles, grubs, etc. Millipedes Sowbugs (woodlice) Mites Slugs and snails Earthworms
		Largely predatory	Moles Insects—many ants, beetles, etc. Mites, in some cases Centipedes Spiders
	Micro	Predatory or parasitic or subsisting on plant residues	Nematodes Protozoa Rotifers
Plants	Roots of higher plants		
	Algae	Green Blue-green Diatoms	
	Fungi	Mushroom fungi Yeasts Molds	
	Actinomycetes of many kinds		
	Bacteria	Aerobic Anaerobic and	Autotrophic Heterotrophic

microflora, they utilize the energy stored in the plant residues and are termed *primary consumers*.

While the action of the microflora is mostly chemical, that of the fauna is both physical and chemical. The animals chew the plant parts and move them from one place to another on the soil surface and even into the soil. Earthworms incorporate the plant residues into the mineral soil by literally eating their way through the soil. Larger animals, such as gophers, moles, voles, prairie dogs, and rats, also burrow into the soil and bring about considerable soil mixing and granulation. Detritivores, such as springtails, mites, and enchytraeid worms, edge their way into cracks and crevices in the soil.

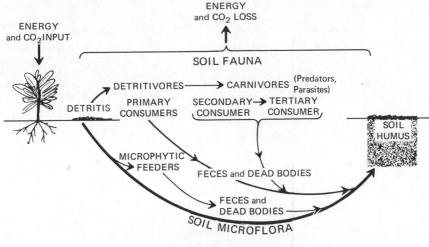

FIGURE 5:1. Diagram of the general pathway for the breakdown of higher plant tissue. Because they capture energy and CO_2, the higher plants are known as *primary producers*. When the debris from dead plants (detritus) falls to the soil surface, it is attacked by the soil fauna and microflora, the primary, secondary, or tertiary *consumers*. These organisms release energy and CO_2 and produce humus. Note that 80–90 percent of the total soil metabolism is due to the microflora.

SECONDARY AND TERTIARY CONSUMERS. The primary consumers are themselves food sources for predators and parasites existing in the soil. These are *secondary consumers* and include carnivores (animal consumers) and microphytic feeders which consume small plants (bacteria, fungi, algae, and lichens). Examples of carnivores are centipedes which consume small insects, springtails, spiders, nematodes, slugs and snails, and the European mole, which utilizes the earthworm as its primary diet source. Examples of microphytic feeders are some mites, springtails, termites, and protozoa which utilize the microflora as sources of food (see Fig. 5:1 and Table 5:2).

Moving farther up the food chain, the secondary consumers are prey for still other carnivores, called *tertiary consumers*. For example, ants consume centipedes, spiders, mites, and scorpions, which themselves can prey upon primary or other secondary consumers. Again, the microflora are intimately involved in decomposition of organic material associated with the fauna. In addition to their direct attack on plant tissue, they are active within the digestive tract of some of the animals. They also attack the finely shredded organic material in animal feces and later decompose the bodies of dead animals. For this reason they are referred to in Fig. 5:1 as the *ultimate decomposers*.

TABLE 5:2. *Examples of Microphytic Feeders and of Carnivores Which Act as Secondary and Tertiary Consumers Within or on Top of Soil*

Note that some animals (e.g., centipedes) are both secondary and tertiary consumers.

| Microphytic Feeders | | Carnivores | | | |
| | | Secondary Consumers | | Tertiary Consumers | |
Organism	Microflora Consumed	Predator	Prey	Predator	Prey
Springtails	Algae, Bacteria, Fungi	Mites	Springtails, Nematodes, Enchytraeids	Ants	Spider, Centipedes, Mites, Scorpions
Mites	Fungi, Algae, Lichens	Centipedes	Springtails, Nematodes, Snails, Slugs, Aphids, Flies	Centipedes	Spiders, Mites, Centipedes (other)
Protozoa	Bacteria and other microflora	Mole	Earthworm, Insects	Beetles	Spiders, Mites, Beetles (other)
Nematodes	Bacteria, Fungi				
Termites	Fungi				

5:2. ORGANISM NUMBERS, BIOMASS, AND METABOLIC ACTIVITY

The specific flora and fauna inhabiting soils are dependent upon many factors. The climate and in turn the vegetation greatly influence which organisms are dominant. Species composition in an arid desert will certainly be different from that in a humid forest area, which in turn will be quite different from that in a cultivated field. Soil temperature, acidity, and moisture relations are also factors which govern the activity of both the flora and fauna. For these reasons, it is not easy to predict the number, kinds, and activities of organisms which one might expect to find in a given soil. But there are a few generalizations that might be made. For example, vegetation under forests generally supports a more diverse fauna than do grasslands. However, the grassland fauna are metabolically more active and their total weight per acre is greater (see Table 5:3).

Compared to virgin areas, cultivated fields are generally lower in numbers and weight of soil organisms, especially the soil fauna. Exceptions may be soils which in the virgin state were originally very acid and which have been

TABLE 5:3. *Biomass of Groups of Soil Animals under Grassland and Forest Cover* [a]

The mass and in turn the metabolism is greatest under grasslands. The spruce with low-base-containing leaves encourages acid conditions and slow organic breakdown.

Group of Organisms	Biomass (g/m^2)		
		Forest	
	Grassland Meadow	Oak	Spruce
Herbivores	17.4	11.2	11.3
Detritivores			
Large	137.5	66.0	1.0
Small	25.0	1.8	1.6
Predators	9.6	0.9	1.2
Total	189.5	79.9	15.1

[a] Data from Macfadyen (1, pp. 3–17).

well limed and fertilized upon being tilled. Such cultivated soils may contain a higher microflora population than their untilled counterparts.

COMPARATIVE ORGANISM ACTIVITY. The activities of specific groups of soil organisms are commonly identified by (a) their numbers in the soil, (b) their weight per unit volume or area of soil (biomass), and (c) their metabolic activity. The numbers and biomass of groups of organisms which commonly occur in soils are shown in Table 5:4. Although the relative metabolic activities are not shown, they are generally related to biomass of the organisms.

As might be expected, the numbers are highest among the microorganisms, both plant and animal. So great are microfloral numbers that they dominate the biomass in spite of the minute size of each individual organism. Together with the earthworms, the microflora monopolize the metabolic activity in soils. It is estimated that 60–80 percent of the total soil metabolism is due to the microflora. Not only do they destroy plant residues but they function in the digestive tracts of animals and eventually decompose the dead bodies of all organisms. Furthermore, soil humus is one of the significant end products of their activities. For these reasons more attention will be given to the microflora along with earthworms, protozoa, and other micro animals.

Before leaving the other soil fauna, recognition should be given to their importance in soil formation and management. Rodents pulverize, mix and granulate soil, and incorporate organic materials into lower horizons. The medium-sized detritivores translocate and partially digest organic residues and leave their excrement for microfloral degradation. By living in the soil,

TABLE 5:4. *Relative Number and Biomass of Soil Flora and Fauna Commonly Found in Soils*

Since the metabolic activity is generally related to the biomass, it is obvious that the microflora and earthworms dominate the life of soils.

| Organisms | Values Common in Surface Soils | | |
	Number per square meter [a]	Number per gram	Biomass [b] (lb/AFS)
Microflora			
Bacteria	10^{13}–10^{14}	10^8–10^9	400–4,000
Actinomycetes	10^{12}–10^{13}	10^7–10^8	400–4,000
Fungi	10^{10}–10^{11}	10^5–10^6	500–5,000
Algae	10^9 –10^{10}	10^4–10^5	50–500
Microfauna			
Protozoa	10^9–10^{10}	10^4–10^5	15–150
Nematoda	10^6–10^7	10–10^2	10–100
Other fauna	10^3–10^5		15–150
Earthworms	30–300		100–1,000

[a] Generally considered 15 cm (6 in.) deep, but in some cases (e.g., earthworms) a greater depth is used.
[b] The biomass values are on a live weight basis. Dry weights are about 20–25% of these values.

many animals favorably affect its physical condition. Others utilize the soil as a habitat for destructive action against higher plants. In any case, all these organisms are part of perhaps the most intricate biological cycle in the world.

5:3. EARTHWORMS

The ordinary earthworm is, probably, the most important soil macro-animal. Of the more than 200 species known, the *Lumbricus terrestris*, a reddish organism, and *Allolobophora caliginosa*, pale pink in color, are the two most common, both in Europe and in eastern and central United States (see Fig. 5:2). In other parts of this country somewhat different species are dominant, depending on the nature of the habitat. In the tropics and semi-tropics still other types are prevalent, some small and others surprisingly large. In respect to species, it is rather interesting that *Lumbricus terrestris* is not native to America. As the forests and prairies were put under cultivation, this European worm rapidly replaced the native types, which could not withstand the change in soil environment. Virgin lands, however, still retain at least part of their native populations.

INFLUENCE ON SOIL FERTILITY AND PRODUCTIVITY. Earthworms are important in many ways. The amount of soil these creatures pass through their bodies annually may amount to as much as 15 tons of dry earth per acre, a startling figure. During the passage through the worms, not only the organic

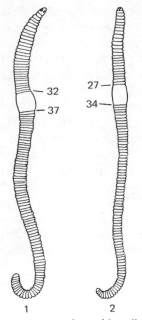

FIGURE 5:2. Two earthworms common in arable soils: (1) *Lumbricus terrestris* and (2) *Allolobophora caliginosa*. While of about the same size, the former is darker and redder, the latter being pale and somewhat pinkish. As the drawings also show, the girdles of the organisms are located somewhat differently in respect to the segments.

matter which serves the earthworms as food, but also the mineral constituents are subjected to digestive enzymes and to a grinding action within the animals. Earthworm casts on a cultivated field may weigh as much as 16,000 pounds per acre. Compared to the soil itself, the casts are definitely higher in bacteria and organic matter, total and nitrate nitrogen, exchangeable calcium and magnesium, available phosphorus and potassium, pH and percentage base saturation, and cation exchange capacity. The rank growth of grass around earthworm casts suggests an increased availability of plant nutrients therein. Earthworms are noted for their favorable effect on soil productivity.

OTHER EFFECTS.   Earthworms are important in other ways. The holes left in the soil serve to increase aeration and drainage, an important consideration in soil development. Moreover, the worms bring about a notable transportation of the lower soil to the surface. They also mix and granulate the soil by dragging into their burrows quantities of undecomposed organic matter such as leaves and grass which they use as food. In some cases, the accumulation is surprisingly large. In uncultivated soils, this is more important than in plowed land where organic matter is normally turned under in quantity.

Without a doubt, earthworms have definitely increased both the size and stability of the soil aggregates, especially in virgin soils.

FACTORS AFFECTING EARTHWORM ACTIVITY. Earthworms prefer a moist habitat that is reasonably well aerated. For this reason, they are found mostly in medium-textured upland soils where the moisture capacity is high, rather than in droughty sands or poorly drained lowlands. They must have organic matter as a source of food. Consequently, they thrive where farm manure or plant residues have been added to the soil. A few species are reasonably tolerant to low soil pH, but most earthworms thrive best where the soil is not too acid. This may be because the nutrition of some species depends upon certain lime-secreting glands. These species require a high level of exchangeable calcium for optimum activity.

Soil temperature affects earthworm numbers and their distribution in the soil profile. For example, a temperature of about 50°F (10°C) appears optimum for *Lumbricus terrestris*, earthworm numbers declining above or below this temperature. This relationship, as well as soil moisture requirements, probably accounts for the maximum earthworm activity noted in spring and autumn in temperate regions.

Earthworm counts must take into consideration the ability of these organisms to burrow deeply into the profile, thereby avoiding unfavorable moisture and temperature conditions. Although species differ markedly in the depth of soil they normally penetrate, cold weather or dry upper soil conditions will drive them into a more favorable environment deeper in the profile. Penetration as deep as 3 to 6 feet is not uncommon in temperate regions. Unfortunately, in barren soils a sudden heavy frost in the fall may kill the organisms before they can move lower in the profile. Soil cover is important in maintaining a high earthworm population under such circumstances.

Because of their sensitivity to soil and other environmental factors, there is a wide variation in the numbers of earthworms in different soils. In very acid soils under conifers, an average of fewer than one organism per square meter is common. In contrast, more than 500 per square meter have been found on rich grassland soils. The numbers commonly found in arable soils is from 30 to 300 per square meter (see Table 5:4), equivalent to from 120,000 to 1,200,000 per acre–furrow slice. The biomass or wet weight for this number would range from perhaps 100 to 1,000 pounds per acre.

5:4. SOIL MICROANIMALS

Of the abundant microscopic animal life in soils, two groups are particularly important—nematodes and protozoa. A third—the rotifers—deserves mention. The three groups will be considered in order (see Table 5:1).

NEMATODES. Nematodes—commonly called threadworms or eelworms—are found in almost all soils, often in surprisingly large numbers (see Table 5:4). These organisms are round and spindle-shaped, the caudal end usually being acutely pointed. In size, they are almost wholly microscopic, seldom being large enough to be seen at all readily with the naked eye (see Fig. 5:3).

Three groups of nematodes may be distinguished on the basis of their food demands: (a) those that live on decaying organic matter; (b) those that are predatory on other nematodes, bacteria, algae, protozoa, and the like; and (c) those that are parasitic, attacking the roots of higher plants to pass at least a part of their life cycle embedded in such tissue. The first and second groups are by far the most numerous in the average soil and most varied. They are important primarily because of their interrelationships with the soil microflora.

The last group, especially those of the genus *Heterodera*, is the most important to the plant specialist. Because of their pointed form and adaptable mouth parts, they find it easy to penetrate plant tissue. The roots of practically all plants are infested even in cool temperature regions and the damage done is often very great, especially to vegetable crops. Even in greenhouses, nematodes may become a serious pest unless care is taken to avoid infestation. Because of the difficulties encountered in control, an appreciable nematode infection is a serious matter.

PROTOZOA. Protozoa probably are the simplest form of animal life. Although one-celled organisms, they are considerably larger than bacteria and of a distinctly higher organization. Some are merely masses of naked protoplasm—amoeba—whereas others exhibit a much higher development and are even protected with siliceous or chitinous coverings. For convenience of discussion, soil protozoa are divided into groups: (a) amoeba, (b) ciliates or infusoria, and (c) flagellates. The presence of numerous cilia or hairs and of flagella, long whiplike appendages of protoplasm, are the bases for the two latter subdivisions (see Fig. 5:4). The flagellates are usually most numerous in soil, followed in order by the amoeba and the ciliates.

The protozoa are the most varied and numerous in the microanimal population of soils. More than 250 species have been isolated, sometimes as many as 40 or 50 of such groups occurring in a single sample of soil. Many of these organisms are highly adaptable and occur in habitats other than that of the soil. A considerable number of serious animal and human diseases are attributed to protozoan infections.

The numbers of protozoa in the soil are subject to extreme fluctuation even when conditions apparently are continuously favorable. Aeration, as well as the available food supply, is probably a very important factor. Most of the organisms are therefore confined to the surface horizons. Usually, the numbers are highest in the spring and autumn. Populations of a few thousand to several hundred thousand per gram of the upper inch or so of soil have

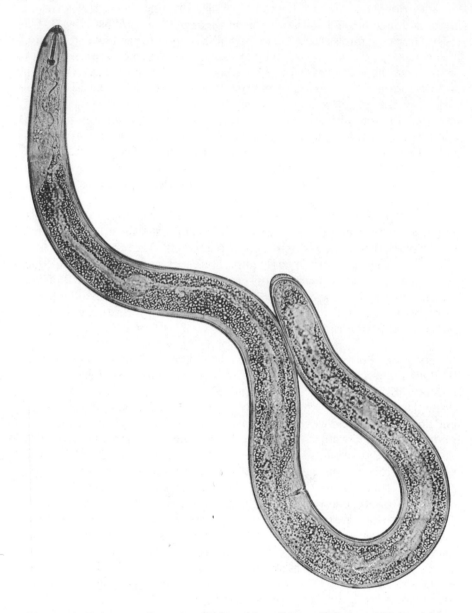

FIGURE 5:3. A nematode commonly found in soil (magnified about 235 times). More than 1,000 species of soil nematodes are known. They can cause serious damage to the roots of higher plants on which many of them feed. [*Photo courtesy William F. Mai, Cornell University.*]

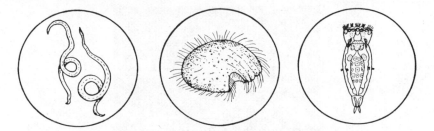

FIGURE 5:4. Parasitic nematodes (*left*), a ciliated protozoan (*center*), and a common rotifer (*right*). [*First two after Waksman.*]

been observed. Perhaps 1–10 billion protozoa of all kinds per square meter 15 centimeters deep[2] (10,000 to 100,000 per gram) might be considered a normal range. This might amount to a live weight of as much as 150 or even 200 pounds to the acre–furrow slice.

As a source of food, protozoa ingest bacteria and to a lesser extent other microflora as well. Although at an earlier time this was considered to be harmful, subsequent research suggests no such ill effect. In fact, there is some evidence that protozoan predation on bacteria may hasten the turnover of readily available nutrients.

ROTIFERS. This third group of soil microanimals thrives under moist conditions, especially in swampy land where their numbers may be great. As many as fifty different species have been found under such environs. These animals are mostly microscopic in size. Their anterior is modified into a retractile disk bearing circles of cilia which, in motion, give the appearance of moving wheels—hence, the name. These hairs sweep floating food materials into the animal. The posterior end of the rotifer tapers to a foot by which the rotifer can attach itself to convenient objects (see Fig. 5:4).

Just how important rotifers are in soils is unknown. No doubt they enter into the cycle of organic dissolution in a more or less important way, especially in peat bogs and in wet places occurring in mineral soils.

5:5. ROOTS OF HIGHER PLANTS

Higher plants are the primary producers of organic matter and storer of the sun's energy (see Fig. 5:1). Their roots grow and die in the soil and in so doing supply the soil fauna and microflora with food and energy. At the same time the living roots physically modify the soil as they push through cracks and make new openings of their own (see Fig. 5:5). Tiny initial channels are increased in size as the root swells and grows. By removing

---

[2] A common areal unit for comparative organism numbers is the square meter. The depth of surface soil is normally 15 cm or 6 inches.

FIGURE 5:5. Rootlets elongate rapidly and thereby are continually establishing new soil contacts. Were it not for this ready extension, crops growing on well-drained arable soils would soon suffer from lack of moisture as well as nutrients. Since the life of these shoots is very short, they continually contribute to the supply of readily decomposable tissue and greatly stimulate microbial activity. Note the excellent granulation of this soil. [*Photo courtesy N.Y. State College of Agriculture and Life Sciences.*]

moisture from the soil, plant roots bring about further physical stresses which help soil aggregate initiation. The roots also provide a mass of living organic matter which stabilizes these aggregates. As they later decompose, the roots provide building material for humus not only in the top few inches but to greater soil depths as well.

AMOUNTS OF ORGANIC TISSUE ADDED. A good crop of oats will produce perhaps 5,000 pounds of dry matter to the acre in its above-ground parts, whereas the figure for corn on the same basis would be about 8,000 pounds, and for sugar cane perhaps 15,000 pounds. If the roots left in the soil when such crops are harvested amount to even half these weights, the added organic residues are considerable. Many farmers do not appreciate the fact that the maintenance of a satisfactory supply of organic matter in arable soils is possible only because of root residues added in this automatic way. For a photo of elongating as well as dying rootlets, see Fig. 5:5.

RHIZOSPHERE. The roots of higher plants also function in a more intimate manner than as a source of dead tissue for the nutrition of soil microbes. When alive, they not only influence the equilibrium of the soil solution by the withdrawal of soluble nutrients, but they also have something to do in a direct way with nutrient availability. Organic acids are formed at the root surfaces and hence become effective solvents. The excretion of readily decomposable compounds such as amino acids as well as the sloughing of root tissue stimulate the microflora to a high intensity of action. The numbers of organisms in the immediate root zone, the *rhizosphere*, may be as many as 100 times as great as elsewhere in the soil, although a value of 10 times is probably more normal. This means that the adsorptive surfaces of the root hairs lie within the zone of unusual nutrient availability. This has already been examined (p. 37) and indicates in part why the roots of higher plants are classified as soil organisms. They not only force a transfer of nutrients but also promote an availability that, under other conditions, might be very slow.

5:6. SOIL ALGAE

Most algae are chlorophyll-bearing organisms and, like higher plants, are capable of performing photosynthesis. To do this, they must live at or very near the surface of the soil. However, a few forms obtain their energy largely from organic matter and live within and below the surface horizon. Several hundred species of algae have been isolated from soils, those most prominent being the same the world over. Soil algae are divided into three general groups: (a) blue-green, (b) green, and (c) diatoms.

Soil algae in vegetative form are most numerous in the surface layers, especially in the upper few inches of arable soils. In subsoils, most algae are present as resting spores, or cysts, or in vegetative forms that do not depend upon chlorophyll. Grassland seems especially favorable for the blue-green forms, whereas in old gardens, diatoms are often numerous. All of the ordinary types of algae are greatly stimulated by the application of farm manure.

Both the green and the blue-green algae outnumber the diatoms. A common range in algal population is from 1 to 10 billion per square meter 15 cm deep (10,000 to 100,000 per gram). Thus, algae undoubtedly contribute some to the organic content of the soil. Blue-green algae are especially numerous in rice soils, and when such lands are flooded and exposed to the sun, appreciable amounts of atmospheric nitrogen are fixed or changed to a combined form by these organisms. Moreover, algae fixation of nitrogen is enhanced when the rice crop occupies the paddies. Apparently, this stimulation is due to carbon dioxide, which is at an especially high level when the rice plants are growing vigorously.

5:7. SOIL FUNGI

Although the influence of fungi is by no means entirely understood, it is known that they play a very important part in the transformations of the soil constituents. Over 690 species have been identified, representing 170 genera. Like the bacteria and actinomycetes, fungi contain no chlorophyll and must depend for their energy and carbon on the organic matter of the soil.

The superficial characteristic that distinguishes fungi is the filamentous nature of their vegetative forms. Their mycelial threads may be simple and restricted or profusely branched. The special spore-forming or fruiting bodies of some groups attain macroscopic sizes. Fungal organisms may thus vary from the simple microscopic yeasts to mushroom and bracket fungi of extraordinary dimensions.

For convenience of discussion, fungi may be divided into three groups: (a) yeasts, (b) molds, and (c) mushroom fungi. Of these, only the last two are considered important in soils; yeasts occur in a very limited extent in such a habitat.

MOLDS. The distinctly filamentous, microscopic, or semimacroscopic fungi are commonly spoken of as molds (see Fig. 5:7). In soils, they play a role infinitely more important than the mushroom fungi, approaching or even excelling at times the influence of bacteria. They respond especially to soil aeration, their numbers and activities diminishing as soil–air content is reduced.

Molds will develop vigorously in acid, neutral, or alkaline soils, some being favored, rather than harmed, by lowered pH (see Fig. 14:8). Consequently, they are noticeably abundant in acid soils, where bacteria and actinomycetes offer only mild competition. This is especially important in decomposing the organic residues in acid forest soils.

The greatest numbers of molds are found in the surface layers, where organic matter is ample and aeration adequate. Many genera are represented, four of the more evident being *Penicillium, Mucor, Fusarium,* and *Aspergillus.* All of the common species occur in most soils, conditions determining which shall dominate. Their numbers fluctuate greatly with soil conditions, perhaps 0.1 to 1 million individuals to 1 gram of dry soil (10 to 100 billion/meter^2), representing a more or less normal range in population. This would amount to a range in biomass of 500 to 5,000 pounds to the acre. Molds are an important part of the general purpose, heterotrophic group of soil organisms that fluctuate so greatly in most soils. The fungal population is constantly changing not only in numbers but in respect to the species dominant. The complexity of the organic compounds being attacked seems to determine the particular mold or molds prevalent.

ACTIVITIES OF FUNGI. In their ability to decompose organic residues, fungi are the most versatile and perhaps the most persistent of any group. Cellulose, starch, gums, lignin, as well as the more easily affected proteins and sugars, readily succumb to their attack. In affecting the processes of humus formation and aggregate stabilization, molds are more important than bacteria. They are especially active in acid forest soils but play a more significant role than is generally recognized in all soils.

Moreover, fungi function more economically than bacteria in that they transform into their tissues a larger proportion of the carbon and nitrogen of the compounds attacked and give off as by-products less carbon dioxide and ammonium. Up to 50 percent of the substances decomposed by molds may become fungal tissue. However, they apparently cannot oxidize ammonium compounds to nitrates as do certain bacteria, nor can they *fix* or bind elemental nitrogen into combined forms. Nevertheless, soil fertility depends in no small degree on molds since they keep the decomposition process going after bacteria and actinomycetes have essentially ceased to function.

MYCORRHIZAE. An interesting and economically important association exists between numerous varieties of mushroom fungi and the roots of higher plants. The mycelia of the fungus infest the plant roots, giving an association called *mycorrhizae*, a term meaning "fungus root." This association has been known since the latter part of the nineteenth century, when it was first noted on certain forest tree species. It was only with the advent of specialized microbiological procedures and modern radiotracer techniques that the widespread occurrence and practical significance of mycorrhizae was firmly established.

Mycorrhizae are divided into two general classes based on the inter-relation of the threadlike fungus hyphae and the root cells. In the *ectotrophic* group the hyphae penetrate between cortex cells of the root but do not enter the cells. In contrast, the hyphae of the *endotropic* fungi actually penetrate the epidermal and cortex cells of the root. In each case, the fungi secrete appropriate enzymes which permit the penetration of the plant to which the fungus is adapted.

The ectotrophic organisms are easily identified and cultured and consequently have been studied longer than endotrophic fungi. They are common on the pine (Pinacae), birch (Betulaceae), and beech and oak (Fagaclae) families and are known to be essential for the establishment of some of the species.

The endotrophic mycorrhizae occur on most genera of seed plants not susceptible to the ectotrophic group and in some rain-forest tree species. Many cultivated crops, such as corn, onions, red clover, and strawberries, as well as shrubs and some trees, including maple, yellow poplar, sweet gum, redwood, and apple, are subject to the endotrophic association.

The root–fungus association is of mutual benefit to the host plant and the fungi. The growth of mycorrhizae-infected plants is superior to that of uninfected plants, especially on low-fertility soils (Fig. 5:6). The uptake of nutrients, particularly phosphorus, is increased markedly as a result of the

FIGURE 5:6. Invasion of plant roots by mycorrhizae is of considerable practical importance. (*Upper left*) The pine seedling on the left was innoculated with mycorrhizal fungi at 4 months; that on the right was untreated. (*Upper right*) The corn plant on the right was invaded by mycorrhizae; that on the left was not. (*Lower*) A section through a red maple root is shown. Note that the hyphae appear as coils. [*Tree root photos courtesy Edward Hacskaylo, U. S. Forest Service; corn photo courtesy J. W. Gerdeman, University of Illinois.*]

association. In return, the fungi apparently absorb carbohydrates from the plant roots.

The widespread occurrence of mycorrhizae is of great practical importance. It makes possible the growth of certain forest species in areas nearly devoid of plant nutrients—growth which would not occur in the absence of the mycorrhizae. In cultivated fields, the extent of the benefit is yet to be established. However, the known increased efficiency of nutrient uptake by mycorrhizae and their widespread occurrence suggests benefits of considerable magnitude.

5:8. SOIL ACTINOMYCETES

Actinomycetes resemble molds in that they are filamentous, often profusely branched, and produce fruiting bodies in much the same way. Their mycelial threads are smaller, however, than those of fungi. Actinomycetes are similar to bacteria in that they are unicellular and of about the same diameter. When they break up into spores, they closely resemble bacteria. On the basis of organization, actinomycetes occupy a position between true molds and bacteria. Although they often are classified with the fungi, they are sometimes called *thread bacteria* (see Fig. 5:7).

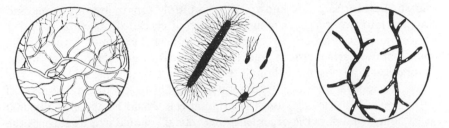

FIGURE 5:7. The three most important plant microorganisms of the soil. Fungal mycelium (*left*), various types of bacteria cells (*center*) and antinomycetes threads (*right*). The bacteria and antinomycetes are much more highly magnified than the fungus.

Actinomycetes develop best in moist, well-aerated soil. But in times of drought, they remain active to a degree not generally exhibited by either bacteria or molds. They are in general rather sensitive to acid soil conditions, their growth being practically prohibited in mineral soils at a pH of 5.0 or below. Their optimum development occurs at pH values between 6.0 and 7.5 (see Fig. 14:8). This marked relationship to soil reaction is sometimes taken advantage of in practice, especially to control potato scab, an actinomycete disease of wide distribution. It is possible by the use of sulfur to effect a reduction in pH sufficient to keep the disease under control (see p. 396).

NUMBERS AND ACTIVITIES OF ACTINOMYCETES. Except for bacteria, no other microorganisms are so numerous in the soil as the actinomycetes, their numbers sometimes reaching hundreds of millions, one tenth the number of bacteria. In actual live weight, they often exceed bacteria. Under especially favorable conditions, a biomass of more than 4,000 pounds of actinomycetes threads and spores might be present in an acre–furrow slice. These organisms are especially numerous in soils high in humus, such as old meadows or pastures, when the acidity is not too great. There they sometimes exceed in numbers all other microscopic forms of life. The addition of farm manure markedly stimulates their activities. The aroma of freshly plowed land that is so noticeable at certain times of the year is probably due to actinomycetes as well as to certain molds.

Actinomycetes undoubtedly are of great importance in the decomposition of soil organic matter and the liberation of nutrients therefrom. Apparently, they reduce to simpler forms even the more resistant compounds, such as cellulose, chitin, and phospholipids. The presence of actinomycetes in abundance in soils long under sod is an indication of this capacity to attack complex compounds.

5:9. SOIL BACTERIA

CHARACTERISTICS. Bacteria are single-cell organisms, one of the simplest and smallest forms of life known. They multiply rapidly by elongating and dividing into two parts. They are therefore often called *fission fungi*. Their almost unlimited capacity to increase in numbers is extremely important in soils. It allows certain groups quickly to assume their normal functions under favorable conditions, even though their numbers were originally small. If a single bacterium and every subsequent organism produced subdivided every hour, the offspring from the original cell would be about 17,000,000 in 24 hours. In 6 days the organisms theoretically would greatly surpass the earth in volume.

Bacteria may thus be considered a force of tremendous magnitude in the soil. They, with the fungi and actinomycetes, largely make up that general purpose group of 'microorganisms already mentioned that fluctuates so markedly in response to soil conditions.

Bacteria are very small; the larger individuals seldom exceed 4 or 5 microns (0.004 to 0.005 mm) in length, and the smaller ones approach the size of an average clay particle. The shape of bacteria is varied in that they may be nearly round, rodlike, or spiral. In the soil, the rod-shaped organisms seem to predominate (see Fig. 5:7). The surface exposed by soil bacteria is remarkable, amounting to about 10 square feet to the pound of dry soil, or about 460 acres to the acre–furrow slice.

BACTERIAL POPULATION IN SOILS. The numbers of bacteria present in soil are variable, as many conditions decidedly affect their growth. In general, the greatest population is in the surface horizons since conditions of temperature, moisture, aeration, and food are more favorable. The numbers of bacteria are high, normally ranging from a few hundred million to 3 billion in each gram of soil. A biomass of from a few hundred pounds to 2 tons live weight to the acre–furrow slice of fertile soils is commonly encountered. The bacterial flora, as well as the other soil organisms, fluctuates sharply with season, the numbers in humid temperate regions usually being the greatest in early summer and in the autumn.

In the soil, bacteria exist as mats, clumps, and filaments, called *colonies*, on and around the soil particles wherever food and other conditions are favorable. The jellylike mixture of mineral and organic colloidal matter makes an almost ideal medium for their development. Their existence depends on soil conditions, particularly food supply. Thus, there is a constant and rapid fluctuation—multiplication and death, often by starvation.

Many of the soil bacteria are able to produce spores or similar resistant bodies, thus presenting both a vegetative and a resting stage. This latter capacity is important as it allows the organisms more readily to survive unfavorable conditions.

SOURCE OF ENERGY. Soil bacteria are commonly classified under two heads, *autotrophic* and *heterotrophic*. The former obtain their energy from the oxidation of mineral constituents such as ammonium, sulfur, and iron, and most of their carbon from carbon dioxide. In numbers they are comparatively insignificant, but since they include the organisms that support nitrification and sulfur oxidation, they are tremendously important in the sustenance of higher plants.

Most soil bacteria are heterotrophic, that is, their energy and carbon both come directly from the soil organic matter. The general purpose decay and ammonifying bacteria, as well as the fungi and actinomycetes, are all heterotrophic in character.

IMPORTANCE OF BACTERIA. Bacteria as a group, almost without exception, participate vigorously in all of the organic transactions so vital if a soil is to support higher plants. They not only rival but often excel both fungi and actinomycetes in this regard. Also, they hold monopolies on three basic enzymic transformations: (a) *nitrification*, (b) *sulfur oxidation*, and (c) *nitrogen fixation*. If these were to fail, life for higher plants and for animals would be endangered. Observed from this angle, bacteria, the simplest and most numerous of all life forms, are perhaps basically the most consequential.

5:10. CONDITIONS AFFECTING THE GROWTH OF SOIL BACTERIA

Many conditions of the soil affect the growth of bacteria. Among the most important are the supplies of oxygen and moisture, the temperature, the amount and nature of the soil organic matter, the pH, and the amount of exchangeable calcium present. These effects are briefly outlined below:

1. Oxygen requirements:
   a. Some bacteria use mostly oxygen gas (aerobic).
   b. Some use mostly combined oxygen (anaerobic).
   c. Some use either of the above forms (facultative).
   d. All three of the above types usually function in a soil at one time.

2. Moisture relationships:
   a. Optimum moisture level for higher plants usually best for most bacteria.
   b. The moisture content affects oxygen supply (see 1 above).

3. Suitable temperature range:
   a. Bacterial activity generally the greatest at 70 to 100°F.
   b. Ordinary soil temperature extremes seldom kill bacteria.

4. Organic matter requirements:
   a. Used as energy source for majority of bacteria (heterotrophic).
   b. Organic matter not required as energy source for others (autotrophic).

5. Exchangeable calcium and pH relationships:
   a. High calcium, pH from 6 to 8 generally best for most bacteria.
   b. Calcium and pH values determine the specific bacteria present.
   c. Certain bacteria function at very low pH (± 3.0) and others at high pH values.
   d. Exchangeable calcium seems to be more important than pH.

5:11. Injurious Effects of Soil Organisms on Higher Plants

SOIL FAUNA. It has already been suggested that certain of the soil fauna are injurious to higher plants. For instance, rodents and moles may, in certain cases, greatly damage crops. Snails and slugs in some climates are exceedingly important pests, especially of vegetables, while the activity of ants, especially as to their transfer and care of aphids on certain plants, must be diligently combated by gardeners. Also, most plant roots are infested with nematodes, sometimes so seriously as to make the successful growth of certain crops both difficult and expensive. Tobacco, potatoes, and cotton are examples of commercial crops adversely affected by nematodes. Crop rotation and the development of resistant varieties are the most acceptable means of combating these pests.

MICROFLORA AND PLANT DISEASES. In general, it is the plant forms of soil life that exert the most devastating effects on higher plants. All three groups— bacteria, fungi, and actinomycetes—contribute their quota of plant diseases. However, fungi are responsible for most of the common soil-borne diseases of crop plants. Some of the more common diseases produced by soil flora are wilts, damping off, root rots, clubroot of cabbage and similar crops, and the actinomycetes scab of potatoes. In short, disease infestations occur in great variety, induced by many different organisms (see Fig. 5:8).

Injurious organisms live for variable periods in the soil. Some of them will disappear within a few years if their host plants are not grown, but others are able to maintain existence on almost any organic substance. Once a soil is infested, it is likely to remain so for a long time. Infection usually occurs easily. Organisms from infested fields may be carried on implements, plants, or rubbish of any kind. Even manure from animals having been fed infected

FIGURE 5:8. Some soil organisms are plant pathogens which cause considerable damage to crop plants. "Damping-off" organisms present in the flat on the right have attacked and killed the pepper plants. [*Photo courtesy Department of Plant Pathology, Cornell University.*]

plants may act as disease carriers. Erosion, if soil is washed from one field to another, may be a means of transfer.

DISEASE CONTROL BY SOIL MANAGEMENT. Prevention is one of the best defenses against diseases produced by such soil organisms, hence the strict quarantines often attempted. Once a disease has procured a foothold, it is often very difficult to eradicate it. Rotation of crops is adequate for some diseases but the absence of the host plant is often necessary.

The regulation of the pH is effective to a certain extent with potato scab and the clubroot of cabbage. If the pH is held somewhat below 5.3 or even 5.5, the former disease, which is due to an actinomycete, is much retarded. With clubroot, a fungus disease, the addition of calcium hydroxide until the pH of the soil is definitely above 7.0 seems fairly effective.

Wet, cold soils favor some seed rots and seedling diseases known as damping off. Good drainage and ridging help control these diseases. Steam or chemical sterilization is a practical method of treating greenhouse soils for a number of diseases. The breeding of plants immune to particular diseases has been successful in the case of a number of other crops.

COMPETITION FOR NUTRIENTS. Another way in which soil organisms may detrimentally affect higher plants, at least temporarily, is by competition for available nutrients. Nitrogen is the element usually most vigorously contested for, although organisms may utilize appreciable quantities of phosphorus, potassium, and calcium to the exclusion of the crops growing on the land. Competition for trace elements may even be serious. Soil organisms usually exact their nutrient quota first and higher plants must subsist on what remains available. This subject will be considered in greater detail in Chapter 6.

OTHER DETRIMENTAL EFFECTS. Under conditions of somewhat restricted drainage, active soil microflora may deplete the already limited oxygen supply of the soil. This may affect plants adversely in two ways. First, the plant roots require a certain minimum amount of O_2 for normal growth and nutrient uptake. Second, oxidized forms of several elements, including nitrogen, sulfur, iron, and manganese, will be chemically reduced by further microbial action. In the cases of nitrogen and sulfur, some of the reduced forms are gaseous and these elements may be lost to the atmosphere. Iron and manganese reduction may result in soluble forms of these elements being present in toxic quantities, especially if the soil is quite acid. Thus, nutrient deficiencies and toxicities, both microbiologically induced, can result from the same basic set of conditions.

5:12. COMPETITION AMONG SOIL MICROORGANISMS

In addition to competition between microorganisms and higher plants, there exists in soils an intense intermicrobial rivalry for food. When fresh organic matter is added, the vigorous heterotrophic soil organisms (bacteria, fungi, and actinomycetes) entirely supersede the less numerous autotrophic bacteria. Only after the rapid decay processes have spent their forces and humification is well advanced may the autotrophic bacteria function with any degree of vigor. Undoubtedly, such food competition is the rule and not the exception in soils (see Fig. 5:9).

ANTIBIOTICS PRODUCED IN SOILS. Besides the food competition just stressed, there is another type of microbial rivalry just as intense and even more deadly. Certain bacteria, fungi, and actinomycetes have the capacity to produce substances that will inhibit, or even actually kill, other microbes. Not only are organisms alien to the soil thus affected but also many of those that normally flourish vigorously therein.

The discovery that many soil organisms can produce *antibiotics*, as they are called, has in some respects revolutionized the treatment of certain human and animal diseases, greatly minimizing their seriousness. Many preparations carrying specific bactericidal ingredients are now on the market,

FIGURE 5:9. Organisms compete with each other in the soil. The growth of a fungus (fusarium) was rapid when this organism was grown alone in a soil (*left*), but when a certain bacteria (agrobacterium) was also introduced, the fungal growth did not appear. [*Photo courtesy M. A. Alexander, Cornell University.*]

such as penicillin, streptomycin, and aureomycin. Undoubtedly, many new and valuable substances of this type are yet to be discovered. Although the soil harbors numerous types of disease organisms, at the same time it supports others that are the source of life-saving drugs, the discovery of which marks an epochal advance in medical science.

5:13. EFFECTS OF AGRICULTURAL PRACTICE ON SOIL ORGANISMS

Changes in environment affect not only the number but the kinds of organisms in the soil. Placing either forested or grassland areas under cultivation brings about a drastic change in the soil environment. In the first place, the amount of plant residues (food for the organisms) is drastically reduced. Also, the species of higher plants are changed and generally less numerous. Monoculture or even rotations in the absence of weeds provides a much narrower range of original plant materials than is generally encountered in nature.

Tillage of the soil and applications of lime and fertilizer help to present a completely different environment to the soil inhabitants. Drainage or irrigation likewise can drastically affect soil moisture and aeration relations with their concomitant effects on soil organisms. Also, commercial agriculture transports organism species from one location to another, making possible their introduction in new areas.

It is obvious that agricultural practices have different effects on different organisms. There are a few generalizations that can be made, however. Figure 5:10, intended originally to show the ecological effects of agricultural practices on soil microarthropods, illustrates some principles relating to the total soil organism population. For example, agricultural practices generally

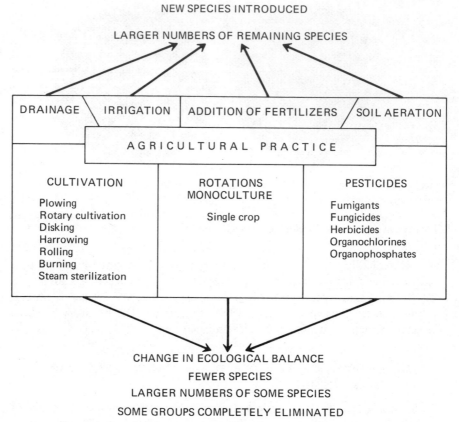

FIGURE 5:10. Ecological effects of agricultural practices on soil microarthropod population. The principles illustrated probably hold also for most other soil organisms. [*From Edwards and Lofty* (2).]

tend to reduce the species diversity and, especially with the soil fauna, the total organism population. At the same time, they can greatly increase or reduce the numbers of a given species, depending on the situation. Adding lime, fertilizers, and manures to an infertile soil will definitely increase the activities of certain bacteria and actinomycetes. Pesticides, on the other hand (especially the fumigants), can sharply reduce organism numbers, at least on a temporary basis. Likewise, the use of monoculture reduces species numbers but may actually increase the organism count of the remaining species.

Most changes in agricultural technology have ecological effects on soil organisms that can affect higher plants and animals, including man. For this reason, ecological side effects of modern technology must be carefully scrutinized. The effects of pesticides, both positive and negative, provide evidence of this fact.

5:14. ACTIVITIES OF SOIL ORGANISMS BENEFICIAL TO HIGHER PLANTS

In their influence on crop production, the soil fauna and flora are indispensable. Of their many beneficial effects on higher plants, only the most important can be emphasized here.

ORGANIC MATTER DECOMPOSITION. Perhaps the most significant contribution of the soil fauna and flora to higher plants is that of organic matter decomposition. By this process, plant residues are broken down, thereby preventing an unwanted accumulation. Furthermore, nutrients held in organic combinations within these residues are released for use by plants. Nitrogen is a prime example. At the same time, the stability of soil aggregates is enhanced not only by the slimy intermediate products of decay but by the more resistant portion, humus. Plants naturally profit from these beneficial chemical and physical effects.

INORGANIC TRANSFORMATIONS. The appearance in the soil of ammonium compounds and nitrates is the result of a long series of biochemical transfers beginning with proteins and related compounds (see p. 426). These successive changes are of vital importance to higher plants since the plants absorb most of their nitrogen in ammoniacal and nitrate forms.

The production of sulfates is roughly analogous to the biological simplification of nitrogen (see pp. 449–50). Here again a complicated chain of enzymic activities culminates in a simple soluble product—in this case the sulfate—the only important form utilized by higher plants.

Other biologically instigated inorganic changes that may be helpful to plants are those relating to mineral elements such as iron and manganese. In well-drained soils these elements are oxidized by autotrophic organisms to their higher valent states, in which forms their solubilities are very low at intermediate pH values. This keeps the greater portion of iron and manganese, even under fairly acid conditions, in insoluble and nontoxic forms. If such oxidation did not occur, plant growth would be jeopardized because of toxic quantities of these elements in solution.

NITROGEN FIXATION. Microbial activity of tremendous practical import is the fixation of elemental nitrogen. Nitrogen gas, so plentiful in the atmospheric air, cannot be used directly by higher plants. It must be in combined form before it can satisfy their nutritional needs. Two groups of bacteria participate in the capture of gaseous nitrogen: the *nodule organisms*, especially those of legumes, and *free-fixing bacteria* of several kinds.

The legume bacteria, as they are often called, use the carbohydrates of their hosts as an energy source, fix the nitrogen, and pass part of it on to the infected host. The free-fixing soil bacteria acquire their energy from the soil organic matter, fix the free nitrogen, and make it a part of their own tissue. When they die and decay, part of this nitrogen is available to higher plants.

It is obvious that the organisms of the soil must have energy and nutrients if they are to function efficiently. In obtaining them they break down organic matter, aid in the production of humus, and leave behind compounds that are useful to higher plants. These biotic features and their practical significance are considered in the next chapter, which deals with soil organic matter.

REFERENCES

(1) Burges, A., and Raw, F., Eds., *Soil Biology* (New York: Academic Press, 1967).

(2) Edwards, C. A., and Lofty, J. R., "Agricultural Practice and Soil Micro-arthropods," in Sheals, J. G., Ed., *The Soil Ecosystem.* Publ. 8 (London: The Systematics Association, 1969).

(3) Wallwork, J. A., *Ecology of Soil Animals* (London: MacGraw-Hill, 1970).

Chapter 6

ORGANIC MATTER OF MINERAL SOILS

ORGANIC matter influences physical and chemical properties of soils far out of proportion to the small quantities present. It commonly accounts for at least half the cation exchange capacity of soils and is responsible perhaps more than any other single factor for the stability of soil aggregates. Furthermore, it supplies energy and body-building constituents for the microorganisms whose general activities have just been considered.

6:1. SOURCES OF SOIL ORGANIC MATTER

The original source of the soil organic matter is plant tissue. Under natural conditions, the tops and roots of trees, shrubs, grasses, and other native plants annually supply large quantities of organic residues. A good portion of crop plants is commonly removed from cropped soils, but some of the tops and all of the roots are left in the soil. As these materials are decomposed and digested by soil organisms of many kinds, they become part of the underlying horizons by infiltration or by actual physical incorporation. Thus, higher plant tissue is the primary source not only of food for the various soil organisms but of organic matter, which is so essential for soil formation (see pp. 310–11).

Animals are usually considered secondary sources of organic matter. As they attack the original plant tissues, they contribute waste products and leave their own bodies as their life cycles are consummated. Certain forms of animal life, especially the earthworms, centipedes, and ants, also play an important rôle in the translocation of plant residues.

6:2. COMPOSITION OF PLANT RESIDUES

The moisture content of plant residues varies from 60 to 90 percent, 75 percent being a representative figure (see Fig. 6:1). On a weight basis, the dry matter is mostly carbon and oxygen, with less than 10 percent each of hydrogen and inorganic elements (ash). However, on an elemental basis (number of atoms of the elements), hydrogen predominates. Thus, there are

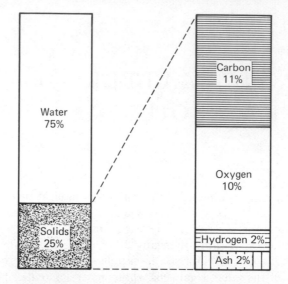

FIGURE 6:1. Composition of representative green plant tissue. Plant tissue solids are made up largely of carbon, hydrogen, and oxygen. The nitrogen is included in the ash.

8 hydrogen atoms for every 3.7 carbon atoms and 2.5 oxygen atoms. These three elements can truly be said to dominate the bulk of organic tissue in the soil.

Even though more than 90 percent of the dry matter is carbon, hydrogen, and oxygen, the other elements play a vital rôle in plant nutrition and in meeting microorganism body requirements. Nitrogen, sulfur, phosphorus, potassium, calcium, and magnesium are particularly significant, as are the micronutrients contained in plant materials. These will be discussed later in more detail.

The actual compounds in plant tissue are many and varied. However, they can be grouped into a small number of classes, as shown in Fig. 6:2. Representative percentages as well as ranges common in plant material are shown.

GENERAL COMPOSITION OF COMPOUNDS. The carbohydrates, which are made up of carbon, hydrogen, and oxygen, range in complexity from simple sugars to the celluloses. The fats and oils are glycerides of fatty acids such as butyric, stearic, and oleic. These are associated with resins of many kinds and are somewhat more complex than most of the carbohydrates. They, too, are made up mostly of carbon, hydrogen, and oxygen.

Lignins occur in older plant tissue such as stems and other woody tissues. They are complex compounds, some of which may have "ring" structures.

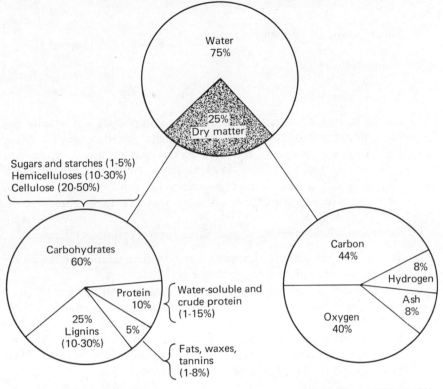

TYPES OF COMPOUNDS   ELEMENTAL COMPOSITION

FIGURE 6:2. Composition of representative green plant materials added to soils. All inorganic elements, including nitrogen, are represented in the ash. Common ranges in the percentage of compounds present are shown in parentheses. [*Data from Waksman* (7).]

The major components of lignins are carbon, hydrogen, and oxygen. They are very resistant to decomposition.

Of the various groups, the crude proteins are probably the most complicated. They carry not only carbon, hydrogen, and oxygen but also nitrogen and, in lesser amounts, such elements as sulfur, iron, and phosphorus. They are compounds of high molecular weight and many are of uncertain constitution. As found in fresh plant tissue, they are present in the colloidal state, which further complicates their study.

6:3. DECOMPOSITION OF ORGANIC COMPOUNDS

RATE OF DECOMPOSITION. Organic compounds vary greatly in their rate of decomposition. They may be listed in terms of their ease of decomposition as follows:

1. Sugars, starches, and simple proteins ⎫
2. Crude proteins   ⎬ rapidly decomposed
3. Hemicelluloses   ⎭
4. Cellulose   ⎫
5. Lignins, fats, waxes, etc.   ⎬ very slowly decomposed
                                      ⎭

It should be remembered that all of these compounds are usually de-composing simultaneously when fresh plant tissue is added to a soil. The rate at which decomposition occurs, however, decreases as we move from the top to the bottom of the list. Thus, sugars and water-soluble proteins are examples of readily available energy sources for soil organisms. Lignins are a very resistant source of food, although they eventually supply much total energy.

When organic tissue is added to soil, three general reactions take place:

1. The bulk of the material undergoes enzymatic oxidation with carbon dioxide, water, and heat as the major products.
2. The essential elements, nitrogen, phosphorus, and sulfur are released and/or immobilized by a series of specific reactions relatively unique for each element.
3. Compounds resistant to microbial action are formed either from compounds in the original plant tissues or by microbial synthesis. Each kind of reaction has great practical significance.

DECOMPOSITION A BURNING PROCESS. In spite of the differences in composi-tion of the various organic compounds, the similarity of the ultimate end products of decay is quite striking, especially if aerobic organisms are involved. Under such conditions the major portion of all of these compounds undergoes essentially a "burning" or oxidation process. The oxidizable fractions of organic materials are composed largely of carbon and hydrogen, which make up more than one half of the dry weight (see Fig. 6:1). Conse-quently, the complete oxidation of most of the organic compounds in the soil may be expressed as follows:

$$-[C, 4H] + 2O_2 \xrightarrow[\text{oxidation}]{\text{enzymic}} CO_2 + 2H_2O + \text{energy}$$

Carbon- and hydrogen-
containing compounds

It is recognized, of course, that many intermediate steps are involved in this over-all reaction. Also, important side reactions which involve elements other than carbon and hydrogen are occurring simultaneously. Neither of these facts, however, detracts from the importance of this basic reaction in accounting for most of the organic matter decomposition in the soil.

BREAKDOWN OF PROTEINS. The plant proteins and related compounds yield other very important products upon decomposition in addition to the carbon dioxide and water previously mentioned. For example, they break down into amides and amino acids[1] of various kinds, the rate of breakdown depending on conditions. Once these compounds are formed, they may be hydrolyzed readily to carbon dioxide, ammonium compounds, and other products. The ammonium compounds may be changed to nitrates, the form in which higher plants take up much of their nitrogen.

EXAMPLE OF ORGANIC DECAY. The process of organic decay in cyclic sequence is illustrated in Fig. 6:3. First, assume a situation where no readily decomposable materials are present in a soil. The microbial numbers and activity are low. Next, under favorable conditions, introduce an abundance of fresh, decomposable tissue. A marked change occurs immediately as the number of soil microorganisms suddenly increases many-fold. Soon microbial activity is at its peak, at which point energy is being liberated rapidly and carbon dioxide is being formed in large quantities. General purpose decay bacteria, fungi, and actinomycetes are soon fully active and are decomposing and synthesizing at the same time.

The soil organic matter at this stage contains a great variety of substances—intermediate products of all kinds, ranging from the more stable compounds, such as lignins, to microbial cells, both living and dead. The microbial tissue may even at times account for as much as one half of the organic fraction of a soil. Dead microbial cells soon decay, and the compounds present are devoured by living microbes, a process that is accompanied by the profuse evolution of carbon dioxide. As the readily available energy is used up and food supplies diminish, microbial activity gradually lessens and the general purpose soil organisms again sink back into comparative quiescence. This is associated with a release of simple products such as nitrates and sulfates. The organic matter now remaining is a dark, incoherent, and heterogeneous colloidal mass usually referred to as *humus*. The decomposition of both plant residues and soil organic matter is nothing more than a process of enzymic digestion. It is just as truly a digestion as though the plant materials entered the stomach of a domestic animal. The products of these enzymic activities, although numerous and tremendously varied, may be listed for convenience of discussion under three headings: (a) energy appropriated by the microorganisms or liberated as heat, (b) simple end products, and (c) humus. They will be considered in order.

---

[1] Amino acids are produced by replacing one of the alkyl hydrogens in an organic acid with NH_2. Acetic acid (CH_3COOH) thereby becomes amino acetic acid or glycine (CH_2NH_2COOH). Amides are formed from organic acids by replacing the hydroxyl of the carboxyl group with NH_2. Acetic acid (CH_3COOH) thus becomes acetamide (CH_3CONH_2).

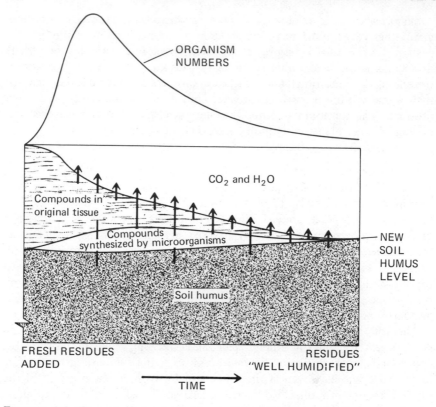

FIGURE 6:3. General changes in the chemical form of carbon- and hydrogen-containing compounds when plant residues are added to soil. Initially, organisms attack easily decomposable compounds such as sugars and celluloses, releasing CO_2 and H_2O and rapidly increasing their own numbers as well as the quantities of new compounds they synthesize. Note that as the initial organism buildup takes place, even the original soil organic matter is subject to some breakdown, CO_2 and H_2O probably being the decomposition products. As soon as the easily decomposed food is exhausted, microorganism numbers decline. The remaining microbes attack the more resistant compounds, both those in the original plant material and those that have been synthesized. In time, modified compounds from the original plant materials and new synthesized compounds, all of which are very slowly decomposed, become indistinguishable from the original soil humus. The time required for the process will depend upon the nature of both the residue and the soil.

6:4. ENERGY OF SOIL ORGANIC MATTER

The microorganisms of the soil must not only have substance for their tissue synthesis but energy as well. For most of the microorganisms, both of these are obtained from the soil organic matter. All manner of compounds are utilized as energy sources, some freely, others slowly and indifferently.

POTENTIAL ENERGY IN ORGANIC MATTER. Organic matter contains considerable potential energy, a large proportion of which is readily transferrable to other latent forms or is liberated as heat. Plant tissue, such as that entering the soil, has a heat value approximating 4 or 5 kilocalories per gram of air-dry substance. For example, the application of 10 tons of farm manure containing 5,000 pounds of dry matter would mean an addition in round numbers of 11 million kilocalories of latent energy. A soil containing 4 percent of organic matter carries from 150 to 180 million kilocalories of potential energy per acre–furrow slice. This is equivalent in heat value perhaps to 20 to 25 tons of anthracite coal.

Of this large amount of energy carried by the soil organic matter, only a part is used by soil organisms. The remainder is left in the residues or is dissipated as heat. This heat loss represents a large and continual removal of energy from the soil.

RATE OF ENERGY LOSS FROM SOILS. Certain estimates made at the Rothamsted Experiment Station, England, on two plots of the Broadbalk field give some idea concerning the rate of energy dissipation from soils (4). It was calculated that 1 million kilocalories per acre were lost annually from the untreated, low-producing soil, and about 15 million kilocalories were dissipated from the more productive soil, which was receiving liberal supplies of farm manure. The magnitude of such loss is surprising even for the poorer plot.

6:5. SIMPLE DECOMPOSITION PRODUCTS

As the enzymic changes of the soil organic matter proceed, simple products begin to manifest themselves. Some of these, especially carbon dioxide and water, appear immediately. Others, such as nitrate nitrogen, accumulate only after the peak of the vigorous decomposition is over and the general purpose decay organisms have diminished in numbers.

The more common simple products resulting from the activity of the soil microorganisms may be listed as follows:

Carbon	CO_2, CO_3^{--}, HCO_3^-, CH_4, elemental carbon
Nitrogen	NH_4^+, NO_2^-, NO_3^-, gaseous nitrogen
Sulfur	S, H_2S, SO_3^{--}, SO_4^{--}, CS_2
Phosphorus	$H_2PO_4^-$, HPO_4^{--}
Others	H_2O, O_2, H_2, H^+, OH^-, K^+, Ca^{++}, Mg^{++}, etc.

Some of the significant relationships of each of the five groups are presented in the following sections. They are considered in the order just listed.

6:6. THE CARBON CYCLE

Carbon is a common constituent of all organic matter. Consequently, its movements during the microbial digestion of plant tissue are extremely significant. Much of the energy acquired by the fauna and flora within the soil comes from the oxidation of carbon. As a result, its oxide is evolved continuously and in large amounts. The various changes this element undergoes within and without the soil are collectively designated the *carbon cycle* and are shown graphically in Fig. 6:4.

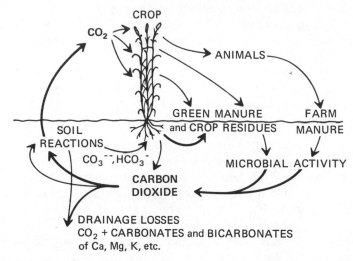

FIGURE 6:4. Transformations of carbon, commonly spoken of as the *carbon cycle*. Note the stress placed on carbon dioxide both within and without the soil.

RELEASE OF CO_2. As the compounds in plant residues are digested, carbon dioxide is given off. The soil is the main soil source of this gas, although small amounts are excreted by plant roots and are brought down in rain water. Under optimum conditions as much as 100 pounds of carbon dioxide per acre per day may be evolved, 20 to 30 pounds being more common. The carbon dioxide of the soil ultimately escapes in a large degree to the atmosphere, where it may again be used by plants, thus completing the cycle.

A lesser amount of carbon dioxide reacts in the soil, producing carbonic acid (H_2CO_3) and the carbonates and bicarbonates of calcium, potassium, magnesium, and other bases. The bicarbonates are readily soluble and may be lost in drainage or used by higher plants.

OTHER CARBON PRODUCTS OF DECAY. The simplification of organic matter results in other carbon products. Elemental carbon is found in soils to a certain extent; although not especially important, its presence is significant. Under certain conditions, methane (CH_4) and carbon bisulfide (CS_2) may be produced in small amounts. But of all the simple carbon products, carbon dioxide is by far the most abundant.

It is now obvious that the carbon cycle is all-inclusive since it involves not only the soil and its teeming fauna and flora and higher plants of every description, but also all animal life, including man himself. Its failure to function properly would mean disaster to all. It is an energy cycle of such vital import that, with its many ramifications, it might properly be designated the "cycle of life".

6:7. SIMPLE PRODUCTS CARRYING NITROGEN

Ammonium salts are the first inorganic nitrogen compounds produced by microbial digestion. Proteins split up into amino acids and similar nitrogenous materials which readily yield ammonium compounds by enzymic hydrolysis. These complex transformations are brought about by a large number of heterotrophic organisms—bacteria, fungi, and actinomycetes. The ammonium ion is readily available to microorganisms and most higher plants.

NITRIFICATION. If conditions are now favorable, ammonium ions are subject to ready oxidation, principally by two special purpose organisms, the nitrite and the nitrate bacteria. The process, nitrification, will be discussed in more detail later (see p. 428). However, it may be shown simply as follows:

$$2NH_4^+ + 3O_2 \xrightarrow[\text{oxidation}]{\text{enzymic}} 2NO_2^- + 2H_2O + 4H^+ + \text{energy}$$

$$2NO_2^- + O_2 \xrightarrow[\text{oxidation}]{\text{enzymic}} 2NO_3^- + \text{energy}$$

The autotrophic bacteria obtain energy by these reactions, which produce nitrates that are assimilated by plants and are also subject to leaching loss. The hydrogen ions formed show that nitrification tends to result in an increase in soil acidity. This effect is of even greater importance when dealing with commercial fertilizers containing ammonium salts such as $(NH_4)_2SO_4$. Extra increments of lime are often added to counteract this acidifying effect.

RELEASE OF GASEOUS NITROGEN. One more group of end products must be considered—gaseous nitrogen compounds. Under certain conditions, the reduction of nitrates and nitrites takes place in soils, and free nitrogen or oxides of nitrogen may be evolved. This transfer is most likely to occur in poorly drained soils or in acid soils containing nitrites. The processes are very complicated and as yet are not fully understood. They apparently may be chemical as well as biochemical. In any case, the reduction is serious since these nitrogen gases ordinarily are relatively inert and are not recombined into compounds useful to higher plants. Some authorities believe that soils lose considerable nitrogen in this way (see pp. 431–33).

6:8. SIMPLE PRODUCTS CARRYING SULFUR

Many organic compounds, especially those of a nitrogenous nature, carry sulfur, which appears in simple forms as decay progresses. General purpose heterotrophic types of organisms apparently simplify the complex organic compounds. The sulfur of these simplified by-products is then subjected to oxidation by special autotrophic bacteria. The final transformation carried to completion by the sulfur-oxidizing organisms may be shown as follows:

$$\underset{\substack{\text{Organic} \\ \text{combinations}}}{-[HS]} + 2O_2 \xrightarrow[\text{oxidation}]{\text{enzymic}} SO_4^{--} + H^+ + \text{energy}$$

The organisms involved obtain energy by the transfer and leave the sulfur as sulphate.

6:9. MINERALIZATION OF ORGANIC PHOSPHORUS

A large proportion of the soil phosphorus is carried in organic combinations (see p. 458). Upon attack by microorganisms the organic phosphorus compounds are mineralized; that is, they are changed to inorganic combinations. The particular forms, as already shown (p. 36), depend to a considerable degree upon soil pH. As the pH goes up from 5.5 to 7.5, the available phosphorus changes from $H_2PO_4^-$ to HPO_4^{--}. Both of these forms are available to higher plants. Since the small amount of phosphorus held in complex mineral combinations in soils usually is very slowly available, the organic sources mentioned above become especially important.

It must not be assumed, however, that the maintenance of soil organic matter at normal or even high levels will solve the phosphorus problem. Most field soils need liberal applications of phosphatic fertilizers. Yet strangely enough, the economic use of such phosphorus depends to a considerable degree upon the organic transformation previously described. Since microorganisms utilize phosphorus freely, some of that added commercially quickly becomes part of the soil organic matter. Thus, this phosphorus is held in an organic condition and is later mineralized by microbial activity.

6:10. HUMUS—GENESIS AND DEFINITION

The formation of humus, although an exceedingly complicated biochemical process, may be described in general terms rather simply. As organic tissue is incorporated into a moist warm soil it is immediately attacked by a host of different soil organisms. The easily decomposed compounds quickly succumb, first yielding intermediate substances and finally the simple, soluble products already enumerated.

HUMUS FORMATION. As the above decomposition occurs, two major kinds of organic compounds tend to remain in the soil: (a) resistant compounds of higher plant origin, such as oils, fats, waxes, and especially lignin[2]; and (b) new compounds, such as polysaccharides and polyuronides, which are synthesized by microorganisms and held as part of their tissue. The lignin and similar materials are at least partially oxidized during the decomposition, thereby increasing their reactivity. The compounds of microbial origin are not insignificant in quantity, studies having shown that up to one third of the organic carbon may be in this form. Apparently these two groups of compounds, one modified from the original plant material and one newly synthesized by the microorganisms, provide the basic framework for humus.

As these humic substances form, there are other side reactions of great practical import. These reactions permit nitrogen to become an integral part of the humus complex. Exactly how these reactions occur is not known, nor are all the specific forms in which the nitrogen is held. Nevertheless, the nitrogen compounds are thought to react with aromatic and quinone groups as well as polysaccharides. Among the reaction products are amino combinations, in which form about one half the nitrogen occurs.

Regardless of the mechanism by which nitrogen is bound, the important fact is that the resultant product, newly formed humus, is quite resistant to further microbial attack. Its nitrogen and other essential nutrients are thereby protected from ready solubility and dissipation.

PROTEIN–CLAY COMBINATIONS. Another means of stabilizing nitrogen in soil is through the reaction between certain clays and proteinaceous substances and other nitrogen compounds. Clays with expanding lattices, such as montmorillonite, seem to have such a faculty; the proteins and other nonionic molecules perhaps function as bases in satisfying the adsorption capacity of the inorganic colloids. The proteins seem to be protected against rapid decomposition. How important this is in the average soil is difficult to state. Nevertheless, it suggests that intermixed with the nitrogen-containing modified lignin and the polysaccharides, there may be clay–protein combinations all of which tend to protect nitrogen from microbial attack.

HUMUS DEFINED. From the previous discussion two facts are obvious: (a) humus is a mixture of complex compounds, not a single material; and (b) these compounds are either resistant materials which have been only modified from the original plant tissue, or compounds synthesized within microbial tissue which remain as the organisms die. These two facts lead to the following definition: *humus is a complex and rather resistant mixture*

---

[2] Lignin is a waxy resinous material that impregnates the cell walls of plants as they increase in age. It is exceedingly complex and varies with different plants and different tissues of the same plant. The chemical formulas of the various lignins are in doubt.

of brown or dark brown amorphous and colloidal substances modified from the original tissues or synthesized by the various soil organisms. It is a natural body, and although it is exceedingly variable and heterogenous, it possesses properties that distinguish it sharply from the original parent tissues and from the simple products that develop during its synthesis.

6:11. HUMUS—NATURE AND CHARACTERISTICS

Humus is highly colloidal, but unlike its mineral counterpart in the soil, it is amorphous and not crystalline. Moreover, its surface area and adsorptive capacity are far in excess of those exhibited by any of the clays. The cation exchange capacity of silicate clays commonly ranges from 8 to 150 meq per 100 grams. Comparable exchange capacities for well-developed humus ranges from 150 to 300.

In respect to adsorbed water, the contrast is of the same order. Humus will adsorb from a saturated atmosphere an amount of water equivalent to about 80 or 90 percent of its weight. Clay, on the other hand, may be able to acquire possibly only 15 to 20 percent. The significance of these figures regarding soil properties is obvious.

PHYSICAL PROPERTIES. The low plasticity and cohesion of humus is a significant practical feature. The maintenance of this constituent in fine-textured soils helps alleviate unfavorable structural characteristics induced by large quantities of clay. This is due to a considerable extent to granulation, which it so markedly encourages.

Another physical characteristic of outstanding interest is the color imparted to soils by humus. It is definitely significant that the development of a black pigment in humus varies with climate. In some Mollisol soils occurring in the northern semiarid regions with an annual rainfall of about 20 inches, the pigment is very dark and abundant. The pigmentation is less intense in humid-temperate zones, and the least coloration is found in the humus of the tropics and semitropics. Color thus in a very general way is an expression of climate. This is an important point because it indicates that organic pigmentation cannot always be used satisfactorily as a comparative measure of the amount of organic matter present in soils.

COLLOIDAL CONSTITUTION. Soil humus as a colloidal complex is organized in much the same way as clay. This relationship has already been discussed (pp. 94–95). The modified lignin, polyuronides, and other constituents function as complex micelles. Under ordinary conditions, these carry innumerable negative charges. But instead of being made up principally of silicon, oxygen, aluminum, and iron as are the silicate crystals, the humic micelles are composed mostly of carbon, hydrogen, and oxygen with minor quantities of nitrogen, sulfur, phosphorus, and other elements. The negative charges arise from exposed —COOH and —OH groups from which at least part of the hydrogen may be replaced by cation exchange.

The range in complexity, size, and molecular weight of the humic acids have already been discussed (see p. 95).

The humic micelles, like the particles of clay, carry a swarm of adsorbed cations (Ca^{++}, H^+, Mg^{++}, K^+, Na^+, etc.). Thus, humus colloidally may be represented by the same structural formula used for clay:

and the same reactions will serve to illustrate cation exchange in both (see p. 97). M represents other metallic cations, such as potassium, magnesium, and sodium.

EFFECT OF HUMUS ON NUTRIENT AVAILABILITY. One particular characteristic of humus merits attention—the capacity of this colloid when saturated with H^+ ions to increase the availability of certain nutrient bases such as calcium, potassium, and magnesium. It seems that an H-humus, as is the case with an H-clay, acts much like an ordinary acid and can react with soil minerals in such a way as to extract their bases. Acid humus has an unusual capacity to effect such a transfer since the organic acid is comparatively strong. Once the exchange is made, the bases so affected are held in a loosely adsorbed condition and are easily available to higher plants. A generalized reaction to illustrate this point follows, microcline being used as an example of the various minerals so affected:

$$KAlSi_3O_8 \; + \; H \boxed{Humus} \; \rightarrow \; HAlSi_3O_8 \; + \; K \boxed{Humus}$$
Microcline   Acid silicate   Adsorbed K

The potassium is changed from a molecular to an adsorbed status, in which condition it is rated as rather readily available to higher plants.

HUMUS–CLAY COMPLEX IN SOILS. In considering all the colloidal matter of a mineral soil, the fact that a mixture of many very different kinds of colloids is involved must be remembered. The crystalline clayey nucleus is markedly stable under ordinary conditions and is active mainly in respect to cation exchange. Conversely, the amorphous organic micelle is susceptible to slow but continuous microorganic attack and hence has a twofold activity—cation exchange and nutrient release. These contrasting details are as important practically as the colloidal similarities already noted.

6:12. DIRECT INFLUENCE OF ORGANIC COMPOUNDS ON HIGHER PLANTS

One of the early concepts regarding plant nutrition was that organic matter as such is directly absorbed by higher plants. Although this opinion was later discarded, there is some evidence that certain organic nitrogen compounds can be absorbed by higher plants, often rather readily. For example, some amino acids, such as alanine and glycine, can be absorbed directly. Such substances ordinarily do not satisfy plant needs for nitrogen as is indicated by the ready response that most plants make to an application of nitrates. The uptake of vanillic acid and other phenol carboxylic acids has been established by the use of radioactive carbon, but the significance of these acids in practical agriculture is not known.

The beneficial effects of an exceedingly small absorption of organic compounds might be explained by the presence of growth-promoting substances. In fact, it is quite possible that vitaminlike compounds are developed as organic decay progresses in soils. If this is true, the direct effect of humic substances upon higher plants might be much more important than has been suspected. Undoubtedly, hormones and vitamins are carried by the soil humus and may at times stimulate both higher plants and microorganisms.

On the other hand, some soil organic compounds may be harmful. As an example, dihydroxystearic acid, which is toxic to higher plants, was isolated from 20 soils of a group of 60 taken in 11 states. It may be, however, that such compounds are merely products of unfavorable soil conditions. And when such conditions are corrected, the "toxic matter" disappears. Apparently, good drainage and tillage, lime, and fertilizers reduce the probability of organic toxicity.

6:13. INFLUENCE OF SOIL ORGANIC MATTER ON SOIL PROPERTIES

Before discussing the practical maintenance of soil organic matter, a brief review of the most obvious influences of this all-important constituent follows:

1. Effect on soil color—brown to black.
2. Influence on physical properties:
   a. Granulation encouraged.
   b. Plasticity, cohesion, etc., reduced.
   c. Water-holding capacity increased.
3. High cation adsorption capacity:
   a. Two to thirty times as great as mineral colloids.
   b. Accounts for 30 to 90 percent of the adsorbing power of mineral soils.

4. Supply and availability of nutrients:
   a. Easily replaceable cations present.
   b. Nitrogen, phosphorus, and sulfur held in organic forms.
   c. Extraction of elements from minerals by acid humus.

6:14. CARBON–NITROGEN RATIO

Attention has been called several times to the close relationship existing between the organic matter and nitrogen contents of soils. Since carbon makes up a large and rather definite proportion of this organic matter, it is not surprising that the carbon to nitrogen ratio of soils is fairly constant. The importance of this fact in controlling the available nitrogen, total organic matter, and the rate of organic decay is recognized in developing sound soil management schemes.

RATIO IN SOILS. The ratio of carbon to nitrogen in the organic matter of the furrow slice of arable soils commonly ranges from 8:1 to 15:1, the median being between 10:1 and 12:1. In a given climatic region, little variation is found in this ratio, at least in similarly managed soils. The variations which do occur seem to be correlated in a general way with climatic conditions, especially temperature and the amount and distribution of rainfall. For instance, it is rather well established that the carbon–nitrogen ratio tend to be lower in soils of arid regions than in those of humid regions when annual temperatures are about the same. It is also lower in warmer regions than in cooler ones if the rainfalls are of about the same magnitude. Also the ratio is narrower for subsoils, in general, than for the corresponding surface layers.

RATIO IN PLANTS AND MICROBES. The carbon–nitrogen ratio in plant material is variable, ranging from 20:1 to 30:1 to legumes and farm manure to as high as 100:1 in certain strawy residues (see Fig. 6:6). Conversely, the carbon–nitrogen ratio of the bodies of microorganisms is not only more constant but much narrower, ordinarily falling between 4:1 and 9:1. Bacterial tissue in general is somewhat richer in protein than fungi and consequently has a narrower ratio.

It is apparent, therefore, that most organic residues entering the soil carry large amounts of carbon and comparatively small amounts of total nitrogen; that is, their carbon–nitrogen ratio is wide, and the C/N values for soils are in between those of higher plants and the microbes.

6:15. SIGNIFICANCE OF THE CARBON–NITROGEN RATIO

The carbon–nitrogen ratio in soil organic matter is important for two major reasons: (a) keen competition for available nitrogen results when residues having a high C/N ratio are added to soils, and (b) because this ratio is relatively constant in soils, the maintenance of carbon—and hence soil organic matter— is dependent to no small degree on the soil nitrogen level. The significance of the C/N ratio will become obvious as a practical example of the influence of highly carbonaceous material on the availability of nitrogen is considered.

PRACTICAL EXAMPLE. Assume that a representative cultivated soil in a condition favoring vigorous nitrification is examined. Nitrates are present in relatively large amounts and the C/N ratio is narrow. The general purpose decay organisms are at a low level of activity, as evidenced by low carbon dioxide production (see Fig. 6:5).

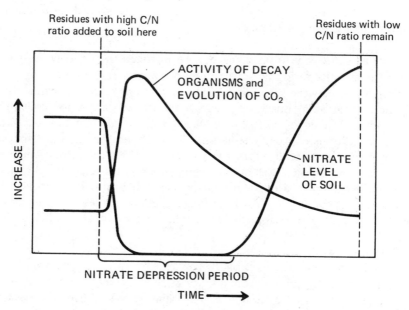

FIGURE 6:5. Cyclical relationship between the stage of decay of organic residues and the presence of nitrate nitrogen in soil. As long as the C/N ratio is wide, the general purpose decay organisms are dominant and the nitrifiers are more or less inactive. During the period of nitrate depression that results, higher plants can obtain but little nitrogen from the soil. The length of this period will depend upon a number of factors, of which the C/N ratio is of prime importance.

Now suppose that large quantities of organic residues with a wide C/N ratio (50:1) are incorporated in this soil under conditions supporting vigorous digestion. A change quickly occurs. The heterotrophic flora—bacteria, fungi, and actinomycetes—become active and multiply rapidly, yielding carbon dioxide in large quantities. Under these conditions, nitrate nitrogen practically disappears from the soil because of the insistent microbial demand for this element to build their tissues. And for the time being, little or no nitrogen, even ammoniacal, is in a form available to higher plants. As decay occurs, the C/N ratio of the plant material decreases since carbon is being lost and nitrogen conserved.

This condition persists until the activities of the decay organisms gradually subside due to a lack of easily oxidizable carbon. Their numbers decrease, carbon dioxide formation drops off, nitrogen ceases to be at a premium, and nitrification can proceed. Nitrates again appear in quantity and the original conditions again prevail except that, for the time being, the soil is somewhat richer both in nitrogen and humus. This sequence of events, an important phase of the carbon cycle, is shown in Fig. 6:5.

REASON FOR C/N CONSTANCY. As the decomposition processes continue, both carbon and nitrogen are now subject to loss—the carbon as carbon dioxide and the nitrogen as nitrates which are leached or absorbed by plants. It is only a question of time until their percentage rate of disappearance from the soil becomes approximately the same; that is, the percentage of the total nitrogen being removed equals the percentage of the total carbon being lost. At this point the carbon–nitrogen ratio, whatever it happens to be, becomes more or less constant, always being somewhat greater than the ratios characterizing microbial tissue. As already stated, the carbon–nitrogen ratio in humid temperate region soils, especially if under cultivation, usually stabilizes in the neighborhood of 10:1 to 12:1.

PERIOD OF NITRATE DEPRESSION. The time interval of nitrate depression (Fig. 6:5) may be long or short, depending on conditions. The rate of decay will lengthen or shorten the period as the case may be. And the greater the amount of residues applied, the longer will nitrification be blocked. Also, the narrower the C/N ratio of the residues applied, the more rapidly the cycle will run its course. Hence, alfalfa and clover residues should interfere least with nitrification and yield their nitrogen more quickly than if oats or wheat straw were plowed under. Also, mature residues, whether legume or nonlegume, have a much higher C/N ratio than do younger succulent materials (see Fig. 6:6). These facts are of much practical significance and should be considered when organic residues are added to a soil.

Moreover, cultivation by hastening oxidation should encourage nitrification while a vigorous sod crop such as bluegrass should discourage it. This is because the root and top residues maintain a wide C/N ratio, and the small quantities of nitrate and ammonical nitrogen which appear are immediately appropriated by the sod itself. All of this is illustrative of the influence exerted by the C/N ratio on the transfer of nitrogen in the soil and its availability to crop plants.

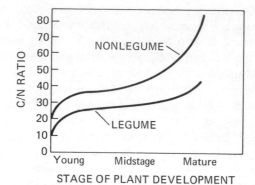

FIGURE 6:6. The C/N ratio of organic residues added to soil will depend upon the maturity of the plants turned under. The older the plants, the wider will be the C/N ratio and the longer will be the period of nitrate suppression. Obviously, leguminous tissue will have a distinct advantage over nonlegumes since the former will promote a more rapid organic turnover in soils.

C/N RATIO AND ORGANIC MATTER LEVEL. Since carbon and nitrogen are reduced to a more or less definite ratio (about 11:1), the amount of soil nitrogen largely determines the amount of organic carbon present when stabilization occurs. Thus, the greater the amount of nitrogen present in the original residue, the greater will be the possibility of an accumulation of organically combined carbon. And since a rather definite ratio (about 1:1.7) exists between the organic carbon and the soil humus, the amount of organic matter that can be maintained in any soil is largely contingent upon the amount of organic nitrogen present. The ratio between nitrogen and organic matter is thus rather constant. A value for the organic matter–nitrogen ratio of 20:1 is commonly used for average soils.

Practical deductions concerning the carbon–nitrogen ratio are thus clear cut. Apparently the ratio is related not only to the availability of soil nitrogen but also to the maintenance of soil organic matter. In practical handling of cultivated soils, both of these phases must receive due consideration.

6:16. AMOUNT OF ORGANIC MATTER[3] AND NITROGEN IN SOILS

The amounts of organic matter in mineral soils vary widely; mineral surface soils contain from a trace to 15 or 20 percent of organic matter.

---

[3] No figures are given as to the amount of humus present in mineral soils. This is partly because no very satisfactory method is available for its determination. Unfortunately there is no very satisfactory method of determining directly the exact quantity of total organic matter present in soil. The usual procedure is to find first the amount of organic carbon, which can be done quite accurately. This figure multiplied by the factor 1.7 will give the approximate amount of organic matter present.

The average organic matter and nitrogen contents of large numbers of soils from different areas of the United States are shown in Table 6:1. The relatively wide ranges in organic matter encountered in these soils even in comparatively localized areas is immediately apparent. Thus, the West Virginia soils have a range from 0.74 to more than 15 percent of organic matter. The several factors which may account for this wide variability as well as for the differences between averages in Table 6:1 will be considered in Section 6:17.

TABLE 6:1. *Average Nitrogen and Organic Matter Contents and Ranges of Mineral Surface Soils in Several Areas of the United States* [a]

Soils	Organic Matter (%)		Nitrogen (%)	
	Range	Av.	Range	Av.
240 West Va. soils	0.74–15.1	2.88	0.044–0.54	0.147
15 Pa. soils	1.70–9.9	3.60	—	—
117 Kansas soils	0.11–3.62	3.38	0.017–0.27	0.170
30 Nebraska soils	2.43–5.29	3.83	0.125–0.25	0.185
9 Minn. prairie soils	3.45–7.41	5.15	0.170–0.35	0.266
21 Southern Great Plains soils	1.16–2.16	1.55	0.071–0.14	0.096
21 Utah soils	1·54–4.93	2.69	0.088–0.26	0.146

[a] From Lyon et al. (3).

The data in the table are for surface soils only. The organic matter contents of the subsoils are generally much lower (see Fig. 6:7). This is readily explained by the fact that most of the organic residues in both cultivated and virgin soils are incorporated in or deposited on the surface. This increases the possibility of organic matter accumulation in the upper layers.

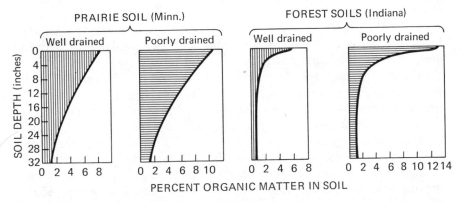

FIGURE 6:7. Distribution of organic matter in four soil profiles. Note that the prairie soils have a higher organic matter content in the profile as a whole compared to the corresponding soils developed under forest vegetation. Poor drainage results in a higher organic matter content, particularly in the surface horizon.

ORGANIC MATTER TO NITROGEN RATIO. One more significant feature of the data in Table 6:1 is the fact that there is about 20 times as much organic matter as nitrogen in the soils listed. This figure is fairly consistent as long as average data on a large number of soils are compared.

The explanation for this constancy is rather simple. It will be remembered that the C/N ratio of mineral soils is rather constant and that the organic matter content is about 1.7 times the carbon content. Thus, if a C/N ratio of 11.7:1 is assumed, the organic matter to nitrogen ratio is 11.7 × 1.7, or about 20:1. This figure is of considerable value in making rough calculations involving these two constituents.

6:17. Factors Affecting Soil Organic Matter and Nitrogen

INFLUENCE OF CLIMATE. Climatic conditions, especially *temperature* and *rainfall*, exert a dominant influence on the amounts of nitrogen and organic matter found in soils [see Jenny (2)]. As one moves from a warmer to a cooler climate, the organic matter and nitrogen of comparable soils tend to increase. At the same time, the C/N ratio widens somewhat. In general, the decomposition of organic matter is accelerated in warm climates; a lower loss is the rule in cool regions. Within belts of uniform moisture conditions and comparable vegetation, the average total organic matter and nitrogen increase from two to three times for each 10°C fall in mean annual temperature.

The situation is well illustrated by conditions in the Mississippi Valley region (see Fig. 6:8). Here, the northern Mollisols contain considerably greater amounts of total organic matter and nitrogen than those to the south. When the amount of original tissue annually added to the soil is considered, the tropics, where plants grow so luxuriantly, afford an even better example of the influence of temperature on the rapidity of decay and disappearance of organic materials.

Effective soil moisture also exerts a very positive control upon the accumulation of organic matter and nitrogen in soils. Ordinarily, under comparable conditions, the nitrogen and organic matter increase as the effective moisture becomes greater (see Fig. 6:8). At the same time, the C/N ratio becomes wider, especially in grasslands areas. The explanation lies not only in the rapidity of microbial action and hence a more complete humification in areas of moderate to low rainfall but also in the scantier vegetation of these regions. In arriving at this rainfall correlation, it must be remembered that the organic situation in any one soil is in large degree an expression of both temperature and precipitation plus other factors as well. Climatic influences never work singly.

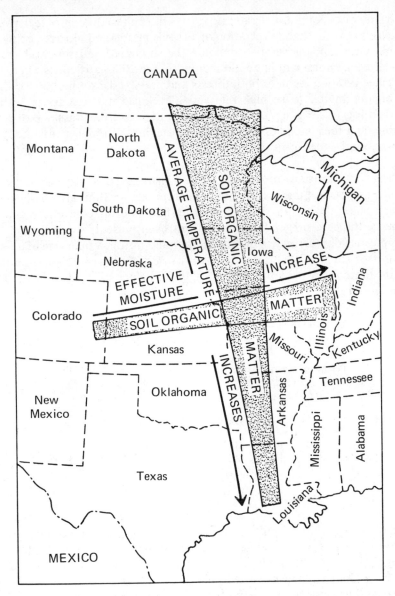

FIGURE 6:8. Influence of the average annual temperature and the effective moisture on the organic matter contents of grassland soils of the Midwest. Of course, the soils must be more or less comparable in all respects except for climatic differences. Note that the higher temperatures yield soils lower in organic matter. The effect of increasing moisture is exactly opposite, favoring a higher level of this constituent. These climatic influences affect forest soils in much the same way.

INFLUENCE OF NATURAL VEGETATION. It is difficult to differentiate between the effects of climate and vegetation on organic matter and nitrogen contents of soil. Grasslands generally dominate the subhumid and semiarid areas, while trees are dominant in humid regions. In those climatic zones where the natural vegetation includes both forests and grasslands, studies have shown the organic matter to be higher in soils developed under grasslands than under forests (see Fig. 6 :7). Apparently, the nature of the grassland organic residues and their mode of decomposition encourage a higher organic level than is found under forests.

EFFECT OF TEXTURE, DRAINAGE, AND OTHER FACTORS. Besides the two broader aspects just discussed, numerous local relationships are involved. In the first place, the *texture* of the soil, other factors being constant, seems to influence the percentage of humus and nitrogen present. A sandy soil, for example, usually carries less organic matter and nitrogen than one of a finer texture (see Table 6:2). This is probably because of the lower moisture content and the more ready oxidation occurring in the lighter soils. Also the natural addition of residues normally is less with the lighter soil.

TABLE 6:2. *Relationship Between Soil Texture and Approximate Organic Matter Contents of a Number of North Carolina Soils* [a]

Percent of organic matter calculated by multiplying %N × 20

Soil Type	Number of Soils	Organic Matter (%)	
		Topsoil	Subsoil
Cecil sands	15	0.80	0.50
Cecil clay loams	10	1.32	0.56
Cecil clays	27	1.46	0.64

[a] From Williams et al. (8).

Again, *poorly drained* soils, because of their high moisture relations and relatively poor aeration, are generally much higher in organic matter and nitrogen than their better drained equivalents (see Fig. 6:7). For instance, soils lying along streams are often quite high in organic matter. This is due in part to their poor drainage and the wash they receive from the uplands.

The *lime content* of a soil, its *erosion* status, and its *vegetative cover* are other factors that influence the accumulation and activity of the soil organic matter and its nitrogen.

INFLUENCE OF CROPPING. A very marked change in the soil organic matter content occurs when a virgin soil developed under either forest or prairie is brought under cultivation. This is illustrated in Fig. 6:9, which shows the

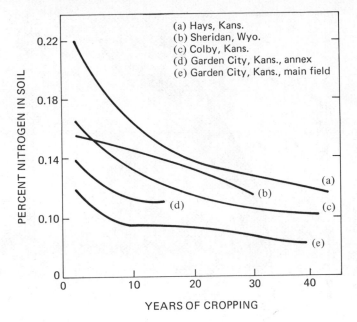

FIGURE 6:9. The effect of cropping on the soil nitrogen level at several locations in the mid-western part of the United States. [*Redrawn from Haas et al.* (1).]

decline in organic matter in several soils of the Midwest with time of cultivation. It is common to find cropped land much lower in both nitrogen and organic matter than comparable virgin areas. This is not too surprising since in nature all the organic matter produced by the vegetation is returned to the soil. In contrast in cultivated areas, much of the plant material is removed for human or animal food and relatively little finds it way back to the land. Also, soil tillage breaks up the organic residues and brings them into easy contact with soil organisms, thereby increasing the rate of decomposition. The depressing effect of cropping on soil organic matter goes far beyond the plow layer, however, as shown by the graph in Fig. 6:10. Subsoil layers are also depleted of their organic matter and nitrogen contents.

Soil organic matter and nitrogen contents are also influenced by level of crop production. Field trials in which no fertilizer was used have demonstrated dramatically the marked loss in soil organic matter and nitrogen contents with cropping. Crop rotation, especially if legumes are included, helps to maintain soil organic matter (see Table 6:3).

Soils that are kept highly productive by supplemental applications of fertilizers, lime, and manures, and by proper choice of disease-free, high-yielding varieties are apt to have a higher organic matter content than comparable less productive soils. The amounts of root and top residues to be returned to the soil are invariably dependent upon the level of soil

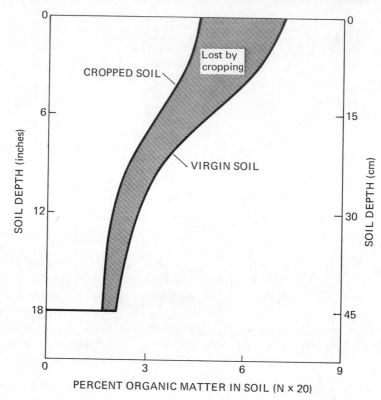

FIGURE 6:10. Average organic matter content of three North Dakota soils before and after an average of 43 years of cropping. About 25 percent of the organic matter was lost from the 0- to 6-inch layer as a result of cropping. [*From Haas et al. (1).*]

TABLE 6:3. *Organic Matter and Nitrogen Contents of Unfertilized Soil Plots at Wooster, Ohio, after Being Cropped in Various Ways for 32 Years* [a]

Cropping	Organic Matter (tons/acre)	Nitrogen (lb/acre)
Original crop land	17.5	2,176
Continuous corn	6.4	840
Continuous oats	11.4	1,425
Continuous wheat	11.0	1,315
Corn, oats, wheat, clover, timothy	13.4	1,546
Corn, wheat, clover [b]	14.8	1,780

[a] From Salter et al. (5).
[b] Continued for 29 years instead of 32.

productivity. It is well to remember, however, that even the most productive
tilled field soil will likely be considerably lower in organic matter than soil
in a nearly virgin area.

6:18. REGULATION OF SOIL ORGANIC MATTER

The preceding discussion has established two definite conclusions regarding the organic matter and nitrogen of cultivated soils. First, the inherent
capacity of soils to produce crops is closely and directly related to their
organic matter and nitrogen contents. Second, the satisfactory level of these
two constituents is difficult to maintain in the majority of farm soils.
Consequently, methods of organic matter additions and upkeep should
receive early consideration in all soil management programs.

SOURCE OF SUPPLY. Organic matter is added to cultivated soils in several
ways. One is the plowing under of crops when in an immature, succulent
stage. This is called *green-manuring*. Such crops as rye, buckwheat, oats,
peas, soybeans, and vetch, as well as others, lend themselves to this method
of soil improvement.

A second source of organic matter supply on many farms, especially in
dairy sections, is *farm manure*. When applied at the usual rates of 10 or 15
tons per acre during a five-year rotation, perhaps 1,000 to 1,500 pounds of
dry matter per acre go into the soil as a yearly average. Such additions greatly
aid in organic matter maintenance.

In the same category with farm manure but of much less concern are
artificial farm manures and *composts*. Composts are used mostly in the
management of nursery, garden, and greenhouse soils.

The third and probably the most important sources of organic residues
are the *current crops* themselves. Stubble, aftermath, and especially root
residues of various kinds left in the soil to decay make up the bulk of such
contributions. Few farmers realize how much residual root systems aid
in the conditioning of their soils. Without them, the practical maintenance
of the humus in most cases would be impossible.

In this connection it should also be remembered that the maintaining of
high crop yields through proper *liming* and *fertilizer practices* may be as
effective as adding manure or other organic residues. Apparently in some
cases organic matter can be maintained by simply raising bumper crops.
Optimum yields usually mean more residues to return to the soil and
certainly increase the amount of roots remaining after harvest.

CROPS AND CROP SEQUENCE. It has already been suggested that certain
sod crops, such as meadow and pasture grasses and legumes, tend to facilitate
humus accumulation. This is partly due to their liberal contributions of
organic residues, the slow decay of these materials, and their wide C/N ratios.

Under these conditions, little nitrogen can appear in the soil in the nitrate form and therefore rather slight losses of this constituent will occur. In legumes this may be partially offset by nitrogen fixation. Because the amount of humus depends to a considerable extent upon the amount of organic nitrogen, sod crops by their nitrogen economy promote the highest possible yields of humus.

Cultivated crops, on the other hand, remove as much or even more nitrogen at harvest than do sod crops. In addition, tillage and other features of their management encourage an extremely rapid rate of decay and dissipation of organic matter. Thus, intertilled crops are associated with humus reduction instead of humus accumulation. Small grains such as oats and wheat are in the same category—they are humus wasters but to a lesser degree.

In practice, crop sequences are so arranged as to offset depreciation with humus conservation. Much can be accomplished by using a suitable rotation (Table 6:3). Also, adequate liming and fertilization to provide bumper crops encourages residues which help maintain the organic level.

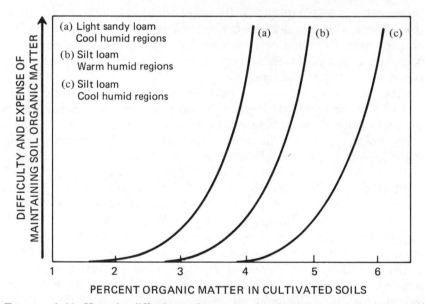

FIGURE 6:11. How the difficulty and expense of maintaining the organic matter of cultivated soils increase as the support level is raised. While the curves are quite similar, their position in relation to the percentage of organic matter possible varies with *texture*, (*a*) vs. (*c*), and *climate*, (*b*) vs. (*c*). Other factors also are involved, especially the type of crop rotation employed. In practice the average amount of organic matter maintained in a cultivated soil should be held at the maximum level economically feasible.

ECONOMY IN HUMUS MAINTENANCE. Since the rate at which carbon is lost from the soil increases very rapidly as the organic content is raised, the maintenance of the humus at a high level is not only difficult but also expensive (see Fig. 6:11). It is therefore unwise to hold the organic matter above a level consistent with crop yields that pay best. Just what this level should be will depend on climatic environments, soil conditions, and the particular crops grown and their sequence. Obviously, it should be higher in the Mollisol region of North Dakota than in central Kansas, where the temperature is higher, or in northern Montana, where the effective rainfall is lower. In any event, the soil organic matter and nitrogen should always be maintained at as high a level as economically feasible.

REFERENCES

(1) Haas, H. J. et al., *Nitrogen and Carbon Changes in Great Plains Soils as Influenced by Cropping and Soil Treatments,* Tech. Bull. 1167 (Washington, D.C.: U. S. Department of Agriculture, 1957).

(2) Jenny, H., *Factors of Soil Formation* (New York: McGraw-Hill, 1941).

(3) Lyon, T. L., Buckman, H. O., and Brady, N. C., *The Nature and Properties of Soils* (New York: Macmillan, Inc., 1952), p. 171.

(4) Russell, E. J., and Russell, E. W., *Soil Conditions and Plant Growth* (London: Longmans, Green, 1950), p. 194.

(5) Salter, R. M., Lewis, R. D., and Slipher, J. A., *Our Heritage—The Soil,* Extension Bull. 175, Ohio Agr. Exp. Sta., 1941, p. 9.

(6) Stevenson, F. J., "Origin and Distribution of Nitrogen in Soil," in *Agron 10: Soil Nitrogen,* Bartholomew, W. V., and Clark, F. E., Eds. (Madison, Wisc.: American Society of Agronomy, 1965).

(7) Waksman, S. A., *Humus* (Baltimore: Williams & Wilkins, 1948), p. 95.

(8) Williams, C. B., et al., *Report on the Piedmont Soils Particularly with Reference to Their Nature, Plant-Food Requirements and Adaptability for Different Crops,* Bull. 36, No. 2, North Carolina Dept. Agr., 1915.

Chapter 7

SOIL WATER: CHARACTERISTICS AND BEHAVIOR

WE are interested in soil–water relationships for several reasons. First, large quantities of water must be supplied to satisfy the evapo-transpiration requirements of growing plants. This water must be available when the plants need it, and most of it must come from the soil. Second, water acts as the solvent which, together with the dissolved nutrients, makes up the soil solution. The significance of this soil component has already been adequately stressed.

As we shall see in a succeeding chapter, soil moisture also helps control two other important components so essential to normal plant growth—soil air and soil temperature. And last but not least, the control of the disposition of water as it strikes the soil determines to a large extent the incidence of soil erosion—that devastating menace which constantly threatens the impairment or even destruction of our soils.

7:1. Structure and Related Properties of Water

Water participates directly in dozens of soil and plant reactions and indirectly affects many others. Its ability to do so is determined primarily by its structure. Water is a simple compound, its individual molecules containing one oxygen atom and two much smaller hydrogen atoms. The elements are bonded together covalently, each hydrogen proton sharing its single electron with the oxygen. The resulting molecule is not symmetrical, however. Instead of the atoms being arranged linearly (H—O—H) the hydrogen atoms are attached to the oxygen in sort of a V arrangement $\left(\begin{smallmatrix} H & & H \\ & \diagdown \diagup & \\ & O & \end{smallmatrix}\right)$ and are separated from each other by an angle of only 105 degrees. As shown in Fig. 7:1, this results in an asymmetric molecule with the shared electrons being closer to the oxygen than to the hydrogen. Consequently, the side on which the hydrogen atoms are located tends to be electropositive and the opposite side electronegative. This accounts for the

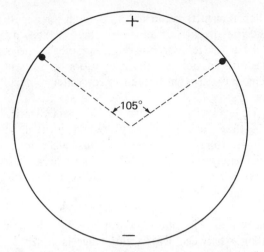

FIGURE 7:1. Two-dimensional representation of a water molecule, showing a large oxygen atom (large white circle) and two much smaller hydrogen atoms (black). The H—O—H angle is 105°, resulting in an asymmetrical arrangement. One side of the water molecule (the side with the two hydrogens) is electropositive and the other electronegative. This accounts for the dipolar nature of water.

polarity of water and in turn for many reactions so important in soil and plant science.

POLARITY. The property of polarity helps explain how water molecules relate to each other. Each water molecule does not act completely independently but rather is coupled with other neighboring molecules. The hydrogen or positive end of one molecule attracts the oxygen end of another resulting in a polymerlike grouping. The angle of association between the hydrogen atoms (105 degrees) encourages an open tetrahedral lattice structure similar to that of silicate minerals. This structural arrangement is essentially complete in ice and persists to a considerable degree in the liquid state of water. When ice melts, the lattice partially collapses, releasing some individual water molecules. These associate in smaller, less-ordered groups or become attached in the intermolecular spaces within the tetrahedral lattice. The net association is more tightly packed than in ice, thereby accounting for a higher density of liquid water as compared to ice.

As the temperature of water is increased, the lattice structure further disintegrates and a higher proportion of the water molecules act independently. Some of the molecules continue to be associated in polymers or clusters, however, even after the water has vaporized.

Polarity also accounts for a number of other important properties of water. For example, it explains why water molecules are attracted to electrostatically charged ions. Cations such as Na^+, K^+, and Ca^{++} become hydrated through their attraction to the oxygen, or negative, end of water molecules. Likewise, negatively charged clay surfaces attract water, this time through the hydrogen, or positive, end of the molecule. Polarity of water molecules also encourages the dissolution of salts in water since the ionic components have greater attraction for water molecules than for each other.

When water molecules become attracted to electrostatically charged ions or clay surfaces, they do so in closely packed clusters. In this state their free energy is lower than in pure water. Thus, when ions or clay particles become hydrated, energy must be released. The released energy is evidenced as *heat of solution* when ions hydrate, or as *heat of wetting* in the case of hydrating clay particles (see p. 43).

HYDROGEN BONDING. The phenomenon by which hydrogen atoms act as connecting linkages between water molecules is called *hydrogen bonding*. This is a relatively low energy coupling in which hydrogen atoms are bonded simultaneously to two different molecules. This bonding is responsible for the structural rigidity of kaolinite crystals. It also accounts for the polymerization and lattice structure of water and for the relatively high boiling point, specific heat, and viscosity of water compared to the same properties of other compounds having similar molecular weight but no hydrogen bonding.

Hydrogen bonding is responsible for many other couplings in nature. For example, certain protein configurations are thought to be held together by hydrogen bonds, as is the helix of deoxyribonucleic acid (DNA), the essential genetic material of all plant and animal cell nuclei. Certainly hydrogen bonding is of universal biological significance. It plays a most critical role in soil–water–plant relations.

COHESION VERSUS ADHESION. Hydrogen bonding suggests two basic forces which account for water retention and movement in soils. One is the attraction of molecules for each other (cohesion). The other is the attraction of water molecules for the solid surfaces (adhesion). By adhesion, solids hold water molecules rigidly at their soil–water interfaces. These molecules in turn hold by cohesion other water molecules further removed from the solid surfaces. Together, these forces make it possible for the soil solids to retain water and control its movement and utilization.

SURFACE TENSION. One other important property of water which influences markedly its behavior in soils is that of surface tension. This phenomena is commonly evidenced at liquid–air interfaces and results from the greater attraction of water molecules for each other than for the air above (see Fig. 7:2). The net effect is an inward force at the surface which causes water to behave as if its surface were covered with a stretched elastic mem-

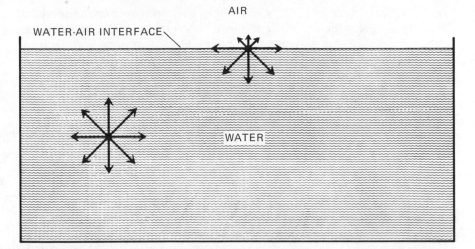

FIGURE 7:2. Comparative forces acting on water molecules at the surface and beneath the surface. Forces acting below the surface are equal in all directions since the water molecules are attracted equally by neighbouring water molecules. At the surface, however, the attraction of the air for the water molecules is much less than that of other molecules for each other. Consequently, there is a net downward force on the surface molecules, resulting in sort of a compressed film or membrane at the surface. This phenomenon is called *surface tension*.

brane. Because of the relatively high attraction of water molecules for each other, water has a high surface tension compared to that of most other compounds of similar molecular weight. As we shall see, surface tension is an important property, especially as a factor in the phenomenon of capillarity.

7:2. SOIL WATER ENERGY CONCEPTS

The retention and movement of water in soils, its uptake and translocation in plants, and its loss to the atmosphere are all energy-related phenomena. Different kinds of energy are involved, including potential, kinetic, and electrical. In the discussions which follow, *free energy* is the term we shall use to characterize the energy status of water. This is appropriate since free energy is sort of a summation of all other forms of energy. Also, its level in a substance is a general measure of the tendency of that substance to change. As we read the discussions on energy, we should keep in mind that all substances, including water, have a tendency to move or change from a higher to a lower free energy level. The significance of this truism to the soil–water–plant–atmosphere continuum will become obvious as we proceed.

As one might expect, there is great variability in the free energy levels of water in soils. However, the absolute level of free energy of water is not so critical as are *differences* in energy levels from one contiguous site to another. Thus, the tendency for soil water to move or to otherwise change its status is related primarily to differences in energy levels from one soil zone to another. The movement is from a zone where the free energy of the water is high to one where the free energy is low. Water will move readily from a standing water table (high free energy) to a dry soil (low free energy). Therefore, knowledge of the energy levels at various points in a soil makes possible predictions of the direction of water movement and gives some idea as to the forces to which the water is subjected.

FORCES AFFECTING FREE ENERGY. The discussion of properties of water in the previous section suggests two important forces affecting the free energy of soil water. The attraction of the soil solids (matrix) for water provides a *matric* force which markedly reduces the free energy of the adsorbed water molecules and even those held by cohesion. Likewise, the attraction of ions and other solutes for water resulting in *osmotic* forces tends to reduce the free energy of soil solution. Osmotic movement of pure water across a semipermiable membrane into a solution is evidence of the lower free energy state of the solution.

The third major force acting on soil water is *gravity*, which tends to pull the water downward. The free energy of soil water at a given elevation in the profile is thus higher than that of pure water at some lower elevation. It is this difference in free energy level which causes water to flow.

TOTAL SOIL WATER POTENTIAL. While the effects of each of the three major forces on changes in the free energy of soil water can be measured, it is the *total soil water potential* which ultimately determines soil water behavior. Technically, the total soil water potential is defined as "the amount of work that must be done per unit quantity of pure water in order to transport reversibly and isothermally an infinitesimal quantity of water from a pool of pure water at a specified elevation at atmospheric pressure to the soil water (at the point under consideration)." While it is impractical to make the measurements specified in this formal definition, the definition stresses that soil water potential is the *difference* between the energy state of soil water and that of pure free water.

Total soil water potential is in effect the sum of the contributions of the various forces acting on soil water:

$$P_t = P_g + P_m + P_o + \cdots$$

where P_t is the total soil water potential, P_g is the *gravitational potential*, P_m is the *matric potential*, and P_o is the *osmotic potential*. (Other less significant potentials are indicated by the dots.) Keep in mind that in each case the potential represents the difference in free energy levels of pure water and of soil water as the latter is affected by gravity and by the presence of the soil matric (solids) or of solutes. The relationship of soil water potential to free energy is shown in Fig. 7:3.

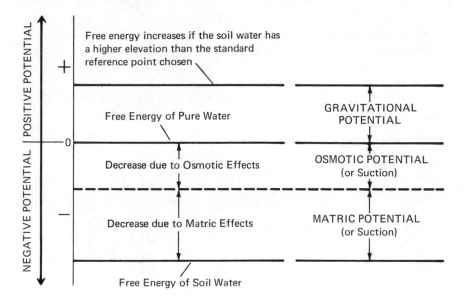

FIGURE 7:3. Relationship between the free energies of pure water and of soil water and the effect of elevation on free energy to illustrate the gravitational potential. Note that osmotic effects and the effects of attraction of the soil solids (matrix) for water both decrease the free energy of soil water. The extent of this decrease represents the osmotic and matric potentials, respectively. The effect of gravity is to increase the free energy if the standard reference point assigned to free water is at a lower elevation than the soil water in the profile. Note that both osmotic and matric potentials are negative, explaining why they are sometimes referred to as *suction* or *tension*. The gravitational potential is generally positive. The behavior of soil water at any one time will be affected by each of these three potentials.

GRAVITATIONAL POTENTIAL. The force of gravity acts on soil water the same as it does on any other body, the attraction being toward the earth's center. The gravitational potential (P_g) of soil water may be expressed mathematically as follows:

$$P_g = Gh$$

where G is the acceleration of gravity and h is the height of the soil water above a reference elevation. The reference elevation is usually chosen within the soil profile or at its lower edge, thus assuring that the gravitational potential of soil water above the reference point will always be positive.

Gravity plays an important role in removing excess water from the upper rooting zones following heavy precipitation or irrigation. It will be given further attention when the movement of soil water is discussed (pp. 182–84).

MATRIC AND OSMOTIC POTENTIALS. The relationship between these two components of total soil water potential is shown in Fig. 7:4. Figures 7:3 and 7:4 should be studied carefully to be certain the meaning of these two potentials is clear.

Matric potential is the result of two kinds of forces, adsorption and capillarity (see Fig. 7:5). The attraction of soil solids and their exchange-

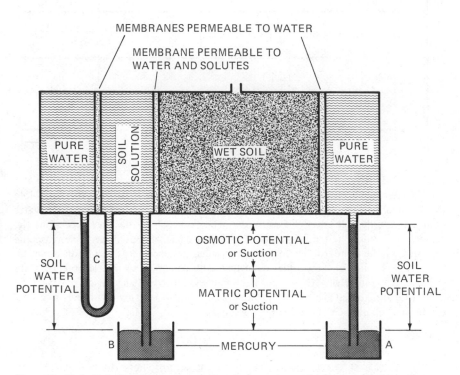

FIGURE 7:4. Relationship among osmotic, matric, and the combined soil water potentials. The system is assumed to be at equilibrium and at constant temperature. The combined potential from both osmotic and matric attractive forces (soil water potential) is evidenced by the attraction of pure water (*right*) through the membrane and is measured at equilibrium by the rise in mercury from vessel A. The osmotic potential is given by the difference in pressure between the pure water (*left*) and the soil solution (measured by manometer C). The matric potential is the difference between the combined and osmotic potentials and is measured by the height of rise of the mercury in vessel B. Since both osmotic and matric potentials are negative, they are sometimes referred to as suctions or tensions. The gravity potential is not shown in this diagram. [*Modified from Richards* (5).]

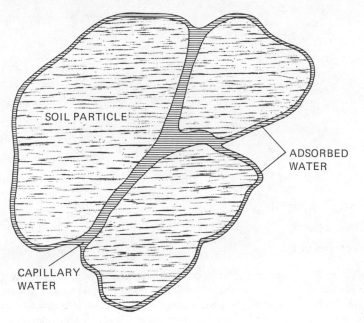

FIGURE 7:5. Two "forms" of water, which together give rise to matric potential. The soil solids tightly adsorb water, while capillary forces are responsible for water's being held in the capillary pores.

able ions for water (adsorption) was emphasized in a previous section, as was the loss of energy (heat of wetting) when the water is adsorbed. This attraction, along with the surface tension of water, also accounts for the capillary force (see pp. 178–81). The net effect of these two forces is to reduce the free energy of soil water as compared to that of unadsorbed or pure water. Consequently, matric potentials are always negative.

The matric potential exerts its effect not only on soil moisture retention but on mass soil water movement as well. The adsorptive and capillary forces tend to resist soil water movement except to adjust to free energy differences between water in adjoining zones of the soil profile. While such adjustments are slow, they are extremely important, especially in supplying water to plant roots.

The osmotic potential is attributable to the presence of solutes in the soil, or, in other words, to the soil solution. The solutes may be ionic or nonionic, but their net effect is to reduce the free energy of water. This is due primarily to the attraction of the solute ions or molecules for the water molecules. The process of osmosis is illustrated in Fig. 7:6. This figure should be studied carefully.

Unlike the matric potential, the osmotic potential has little effect on the mass movement of water in soils. Its major effect is on the uptake of water

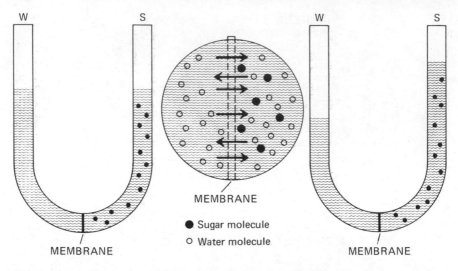

FIGURE 7:6. Illustration of the process of osmosis and of osmotic pressure. (*Left*) A U-tube containing water (W) in the left arm and a solution (S) of sugar in water in the right arm. These are separated by a membrane which is permeable to water molecules but not the dissolved sugar. (*Center*) Enlarged portion of the membrane with H_2O molecules moving freely from the water side to the solution side and vice versa. The sugar molecules, in contrast, are unable to penetrate the membrane. Since the effect of sugar is to decrease the free energy of the water on the solution side, more water passes from left to right than from right to left. At equilibrium (*right*) sufficient water has passed through the membrane to bring about significant differences in the heights of liquid in the two arms. The difference in the level in the W and S arms represents the osmotic potential. [*Modified from Keeton* (5)].

by plant roots. The root membrane, which transmits water more freely than solutes, permits the osmotic effects to be exerted, a matter of considerable importance if the solute content of soils is high. The osmotic potential also affects the movement of water vapor since water vapor pressure is lowered by the presence of solutes.

SUCTIONS AND TENSIONS. It should be remembered that both the matric and osmotic potentials are negative. This is due to the fact that both the osmotic and attractive forces responsible for these potentials reduce the free energy level of the soil water (see Fig. 7:3). Consequently, these negative potentials are sometimes referred to as *suctions* or *tensions*, indicating that they are responsible for the soil's ability to attract and adsorb pure water. The terms "suction" and "tension" have the advantage over "potential" in that they can be expressed in positive rather than negative units. Thus, soil solids are responsible for a *negative* potential which can be expressed as a *positive* tension or suction. For that reason the terms "suction" and "tension" will be used from here on to refer to the matric and osmotic negative potentials.

METHODS OF EXPRESSING ENERGY LEVELS. Several units have been used to express differences in energy levels of soil water. A common means of expressing suction (negative potential) is in terms of the height in centimeters of a unit water column whose weight just equals the suction under consideration. The greater the centimeter height, the greater is the suction measured. We may thus express the tenacity with which water is held in soils in centimeters of water or we may convert such readings into other units. For example, the logarithm of the centimeter height is often used. This unit, called "pF," has the advantage of being expressed in small numbers.

Another common means of expressing suction is that of *bars* or *atmospheres*—the standard atmosphere, which is the average air pressure at sea level, or 14.7 pounds per square inch. The term *millibar* (*mbar*) identifies $\frac{1}{1,000}$ atmosphere. The suction of water 10 cm high is about $\frac{1}{100}$ atmosphere (10 mbars), that of a column 100 cm high about $\frac{1}{10}$ atmosphere (100 mbars), that of a 1,000-cm column about 1 atmosphere (1,000 mbars, or 1 bar). Table 7:1 gives a comparison of these and other units used to measure soil potential and suction. The atmosphere or bar (and millibar) units are used in most cases in this text.

TABLE 7:1. *Approximate Equivalents of Common Means of Expressing Differences in Energy Levels of Soil Water*

Height of Unit Column of Water (cm)	Logarithm of Water Height (pF)	Atmosphere (bars)
10	1	0.01
100	2	0.01
346	2.53	$\frac{1}{3}$
1,000	3	1
10,000	4	10
15,849	4.18	15
31,623	4.5	31
100,000	5	100
1,000,000	6	1,000
10,000,000	7	10,000

7:3. SOIL MOISTURE CONTENT VERSUS SUCTION

The previous discussions suggest an inverse relation between the water content of soils and the suction or tension with which the water is held. Water is more apt to flow out of a wet soil than from one low in moisture. As one might expect, there are many factors which affect the relationship between soil water suction and moisture content. A few examples will illustrate this point.

SOIL MOISTURE–ENERGY CURVES. The relationship between soil water suction and moisture content of three soils of different texture is shown in Fig. 7:7. Note that there are no sharp breaks in the curves, indicating a

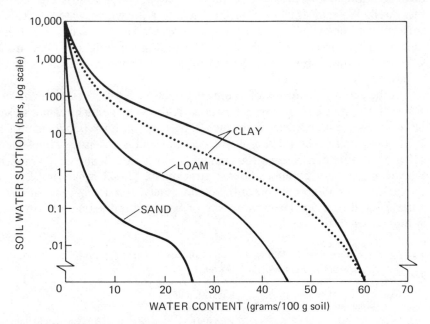

FIGURE 7:7. Soil moisture suction curves for three representative mineral soils. The solid lines show the relationship obtained by slowly drying completely saturated soils. The dotted line for the clay soil is the relationship expected when a dry soil is wetted. The difference between the two clay lines is due to *hysteresis*.

gradual decrease in suction with increased soil water and vice versa. As might be expected, the clay soil holds much more water at a given suction level than does loam or sand. Likewise, at a given moisture content the water is held much more tenaciously in the clay as compared to the other two soils. As we shall see, much of the water held by clay soils in the field is held so tightly that it cannot be removed by growing plants. In any case, the influence of texture on soil moisture retention is obvious.

The structure of a soil also influences its soil moisture–energy relationships. A well-granulated soil has more total pore space than has a similar soil where the granulation has been destroyed and the soil has become compacted. The reduced pore space may be reflected in a lower water-holding capacity. The compacted soil may also have a higher proportion of small- and medium-sized pores, which tend to hold the water with a greater suction than do the larger pores.

HYSTERESIS. An interesting phenomenon occurring when soils are alternately wetted and dried is illustrated by the upper two curves for the clay soil in Fig. 7:7. The upper (solid) curve is termed the *desorption* curve since it is obtained by slowly drying a completely saturated soil. The lower or *sorption* curve (dashed line) results from wetting an initially dry soil. The difference between the two curves is due to *hysteresis*. This phenomenon is caused by a number of factors, the most important of which is the entrapment of air as a soil is rewetted. This clogs some pores and prevents effective contact between others. Also, soil colloids shrink as the soil dries and upon rewetting do not always give the same kind of pore configuration. Hysteresis must be taken into consideration in research where great accuracy is required. It is generally of less significance in most applied work.

The soil moisture–suction curves in Fig. 7:7 have marked practical significance. They illustrate retention–energy relationships which influence important field processes, the two most important of which are the movement of water in soils and its uptake and utilization by plants. The curves should be referred to frequently as the more applied aspects of soil water behavior are considered.

7:4. MEASURING SOIL MOISTURE

Two general types of measurements relating to soil water are ordinarily used. First, by some methods the moisture content is measured directly or indirectly, and second, techniques are used to determine the soil moisture potential (for example, tension or suction).

MOISTURE CONTENT. The most common method of expressing soil moisture percentage is in terms of wet weight percentage or the grams of water associated with 100 grams of dry soil. Thus, if 100 grams of moist soil (soil and water) when dried loses 20 grams of water, the 80 grams of dry matter are used as a basis for the percentage calculation. Therefore, (20 ÷ 80) × 100 = 25 percent. The weight of the wet soil is undesirable as a basis for calculation since it changes with every moisture fluctuation. This unique method of expressing soil moisture percentage should be remembered as the following specific methods for its determination are described.

Moisture content is also sometimes expressed in *volume percentage*, that is, the volume of soil water as a percentage of the volume of the soil sample. This measure has the advantage of giving a better picture of the moisture available to roots in a given volume of soil.

The *gravimetric* method for measuring soil moisture is the one most commonly used to measure weight percentage. A known weight of a sample of moist soil, usually taken in cores from the field, is dried in an oven at a temperature of 100 to 110°C and weighed again. The moisture lost by heating represents the soil moisture in the moist sample.

The *resistance* method takes advantage of the fact that the electrical resistance of certain porous materials such as gypsum, nylon, and fiberglass is related to their water content. When these blocks with suitably embedded electrodes are placed in moist soil, they absorb soil moisture until equilibrium is reached. The electrical resistance in the blocks is determined by their moisture content and in turn by the tension or suction of water in the nearby soil. The relationship between the resistance reading and the soil moisture percentage can be determined by calibration. The blocks are used for measuring the moisture content in selected field locations over a period of time. They give reasonably accurate moisture readings over the range of 1 to 15 atmospheres suction.

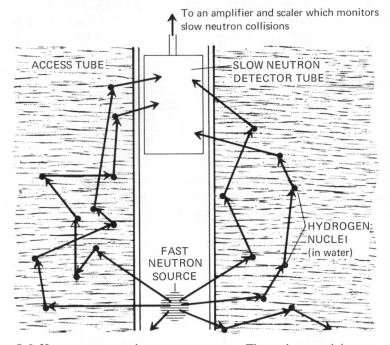

FIGURE 7:8. How a neutron moisture meter operates. The probe, containing a source of fast neutrons and a slow neutron detector, is lowered into the soil through an access tube. Neutrons are emitted by the source (for example, radium or americium–beryllium) at a very high speed. When these neutrons collide with a small atom such as hydrogen contained in soil water, their direction of movement is changed and they lose part of their energy. These "slowed" neutrons are measured by a detector tube and a scalar. The reading is related to the soil moisture content.

A unique method of determining soil moisture in the field involves *neutron scattering*. The neutron moisture meter is based upon the principle that hydrogen is relatively unique in its ability to drastically reduce the speed of fast-moving neutrons and to scatter them. The principle of a neutron moisture meter is illustrated in Fig. 7:8. Although expensive, these meters are versatile and give reasonably accurate measurements in mineral soils, where water is the primary source of combined hydrogen. This method has distinct limitations in organic soils, where much combined hydrogen is in forms other than water.

SUCTION METHODS. *Field tensiometers* such as the one shown in Fig. 7:9 measure the tension with which water is held in soils. Their effectiveness is based on the principle that water in the tensiometer equilibrates through a porous cup with adjacent soil water and that the suction in the soil is the same as the suction in the potentiometer. They are used successfully in determining the need for irrigation when the moisture is being kept near the field capacity. Their range of usefulness is between 0 and 0.8 bar suction.

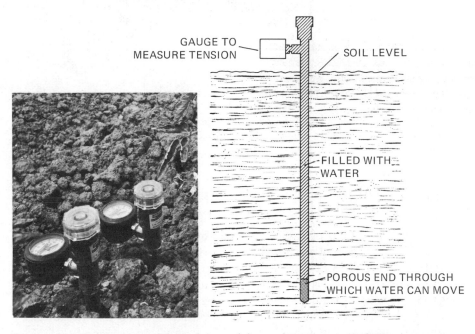

FIGURE 7:9. Tensiometer method of determining moisture stress. (*Left*) Tensiometers in place in the field. (*Right*) Cross section showing essential components of a tensiometer. Water will move through the porous end of the instrument in response to the pull of the soil. This creates a tension which is measured by the gauge. [*Photo courtesy T. W. Prosser Co., Arlington, California.*]

A *tension plate* apparatus is a form of tensiometer used under laboratory conditions. A core of soil is placed firmly on a porous plate to which a suction is applied. The soil core eventually reaches equilibrium with the porous plate. The sample is weighed and the relationship between suction and soil moisture content determined. The range of suitability of this apparatus is from 0 to 1 bar only.

A *pressure membrane* apparatus (see Fig. 7:10) is used to measure matric suction–moisture content relations at suction values as high as 100 bars. This important laboratory tool makes possible the accurate measurement of energy–soil moisture relations of a number of soil samples over a wide energy range in a relatively short time.

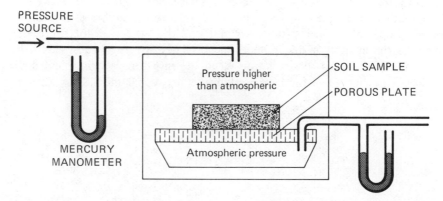

FIGURE 7:10. Pressure plate apparatus used to determine moisture content–matric suction relations in soils. An outside source of gas creates a pressure inside the cell. Water is forced out of the soil through a porous plate into a cell at atmospheric pressure. This apparatus will measure much higher soil suction values than will tensiometers or tension plates.

7:5. CAPILLARY FUNDAMENTALS AS THEY RELATE TO SOIL WATER

The phenomenon of capillarity is a common one, the classic example being the movement of water up a wick, the lower end of which is immersed in water. Capillarity is due to two forces: (a) the attractive force of water for the solids on the walls of channels through which it moves, and (b) the surface tension of water, which resists any form except that of a flat plane at the air–liquid interface. The basis for surface tension has already been discussed (see pp. 166–67).

CAPILLARY MECHANISM. Capillarity can be demonstrated by placing one end of a fine glass tube in water. The water rises in the tube, the smaller the tube bore the higher the height of rise (see Fig. 7:11). The water molecules

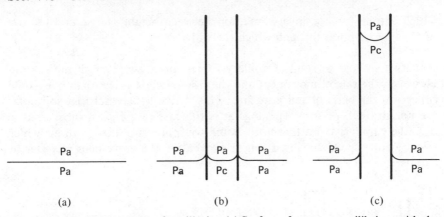

| | Pa |
| | Pc |

| Pa | Pa | Pa | Pa | Pa | Pa |
| Pa | Pa | Pc | Pa | Pa | Pa |

(a)   (b)   (c)

FIGURE 7:11. Phenomenon of capillarity. (a) Surface of water at equilibrium with the atmosphere. The pressure above and below the water surface is the same and is equal to the atmospheric pressure. When a fine glass tube is inserted in the liquid (b), water molecules are attracted to the surface of the tube on both the inside and outside. The curvature on the inside of the tube suggests that the pressure beneath the surface (Pc) is less than that of the atmosphere and less than that of the water outside the tube. Consequently, in time water is pushed up the tube (c) until the weight of the water in the tube provides a force equal to the difference in pressure below and above the curvature (meniscus) in the tube (Pa minus Pc).

are attracted to the sides of the tube giving a curved air–water interface. The pressure under this concave meniscus is less than atmospheric pressure,[1] causing the water in the vessel to push water up the capillary tube. When the downward gravitational force of the water in the tube equals the difference in force between atmospheric pressure and the pressure just under the meniscus, upward movement will cease. The height of rise in a capillary tube is inversely proportional to the tube diameter and is approximated as follows:

$$h = \frac{2T}{rdg}$$

where h is the height of capillary rise in the tube, T is the surface tension, r is the radius of the tube, d is the density of the liquid, and g is the force of gravity. For water, this equation reduces to the simple expression

$$h = \frac{0.15}{r}$$

---

[1] That this is true can be seen if we consider the analogy between the water surface film and a thin rubber pressure membrane placed.crosswise in a tube. If the pressure on one side of the membrane is higher than on the other, the membrane is deformed, the direction of the deformation (curvature) being toward the side with the lower pressure.

which emphasizes the inverse relation between height of rise and the size of the tube or pores through which the water rises.

HEIGHT OF RISE IN SOILS. Capillary forces are at work in all moist soils. However, the rate of movement and the rise in height is less than one would expect on the basis of soil pore size. This is due to the fact that soil pores are not straight, uniform openings as is the case of the glass tubes used to describe capillarity. Furthermore, some soil pores are filled with air which may be entrapped, slowing down or preventing the movement of water by capillarity (see Fig. 7:12).

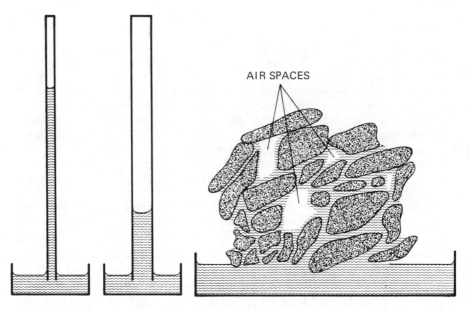

FIGURE 7:12. Upward movement by capillarity in glass tubes of different sizes as compared with capillarity in soils. While the mechanism is the same in the tubes and in the soil, adjustments are extremely irregular in soil because of the tortuous nature and variability in size of the soil pores and because of entrapped air.

The upward movement due to capillarity in soils is illustrated in Fig. 7:13. Usually the height of rise resulting from capillarity is greater with fine-textured soils if sufficient time is allowed and the pores are not too small. This is readily explained on the basis of the capillary size and the continuity of the pores. With sandy soils the adjustment is rapid, but so many of the pores are noncapillary that the height of rise cannot be great.

Although the principle of capillarity is traditionally illustrated as an upward adjustment, horizontal movement also takes place as response to capillarity. This is not unexpected since the same basic attractions between

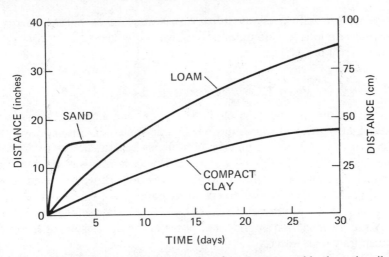

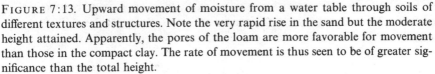

FIGURE 7:13. Upward movement of moisture from a water table through soils of different textures and structures. Note the very rapid rise in the sand but the moderate height attained. Apparently, the pores of the loam are more favorable for movement than those in the compact clay. The rate of movement is thus seen to be of greater significance than the total height.

soil pores and water are effective with horizontal pores as with the vertical ones.

The free energy concept is most certainly applicable to capillary movement. When such movement occurs, it does so from an area where the free energy of the soil water is high to one where it is lower. Thus, water movement will occur most easily from soil areas of high moisture level where low attractive forces of the soil matric results in high free energy levels of soil water.

7:6. TYPES OF SOIL WATER MOVEMENT

In discussing the characteristics of the different forms of moisture commonly recognized in soils, movement has been stressed again and again. And rightly so, as water is a notably dynamic soil constituent. Three types of movement within the soil are recognized—*unsaturated flow, saturated flow,* and *vapor equilizations.* Both saturated and unsaturated flow involve liquid water in contrast to vapor flow. We shall consider liquid water flow first.

The liquid flow of water takes place due to a *gradient* in soil water potential from one soil zone to another. The direction of flow is from a zone of higher to one of lower moisture potential. Saturated flow takes place when the soil pores in the wettest portion of the soil are completely filled (or saturated) with water. Unsaturated flow occurs when the pores in even the wettest soil zones are only partially filled with water. In each case energy–soil moisture relations are dominant. This will be evident as we consider the three types of movement.

7:7. SATURATED FLOW THROUGH SOILS

In most soils, at least part of the soil pores contain some air as well as water; that is, they are unsaturated. Under some conditions, however, at least part of a soil profile may be completely saturated; that is, all pores, large and small, are filled with water. The lower horizons of poorly drained soils are often saturated with water. Even portions of well-drained soils are sometimes saturated. Above stratified layers of clay, for example, the soil pores may all be saturated at times. During and immediately following a heavy rain or irrigation application, pores in the upper soil zones are often filled entirely with water.

The flow of water under saturated conditions is determined by two major factors, the hydraulic force driving the water through the soil and the ease with which the soil pores permit water movement. This can be expressed mathematically as

$$V = kf$$

where V is the total volume of water moved per unit time, f is the water moving force, and k is the hydraulic conductivity of the soil. It should be noted that the hydraulic conductivity of a saturated soil is essentially constant being dependent on the size and configuration of the soil pores. This is in contrast to the situation in an unsaturated soil, where hydraulic conductivity decreases with the moisture content.

An illustration of vertical saturated flow is shown in Fig. 7:14. The driving force, known as the *hydraulic gradient*, is the difference in height of water above and below the soil column. The volume of water moving down the column will depend upon this force as well as the hydraulic conductivity of the soils.

It should not be inferred from Fig. 7:14 that saturated flow occurs only vertically. Horizontal flow will occur due to the same hydraulic force. The rate of flow is not quite as rapid, however, since the force of gravity does not assist horizontal flow. An illustration of vertical and horizontal flow is shown in Fig. 7:15, which depicts the flow of irrigation water into two soils, a sandy loam and a clay loam. Most of the water movement was likely by saturated flow. The water moved much more rapidly in the sandy loam than in the clay loam. On the other hand, horizontal movement was much more evident in the clay loam.

FACTORS INFLUENCING THE HYDRAULIC CONDUCTIVITY OF SATURATED SOILS. Any factor influencing the size and configuration of soil pores will

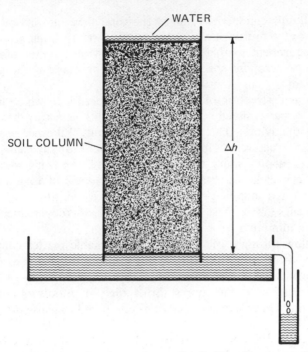

FIGURE 7:14. Saturated flow in a column of soil. All soil pores are filled with water. The force drawing the water through the soil is Δh, the difference in the height of water above and below the soil layer. This same force could be applied horizontally.

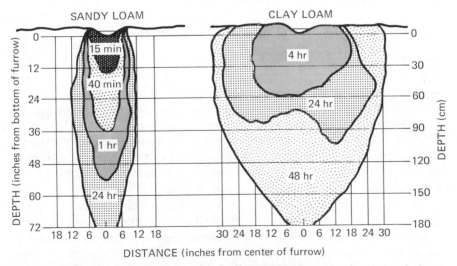

FIGURE 7:15. Comparative rates of irrigation water movement into a sandy loam (*left*) and a clay loam (*right*). Note the much more rapid rate of movement in the sand, especially in a downward direction. [*Redrawn from Cooney and Peterson* (2).]

influence hydraulic conductivity. Since the total flow rate in soil pores is proportional to the fourth power of the radius, flow through a pore 1 mm in radius is equivalent to that in 10,000 pores with a radius of 0.1 mm. Obviously the macropore spaces will account for most of the saturated water movement.

The texture and structure of soils are the properties to which hydraulic conductivity is most directly related. Sandy soils generally have higher saturated conductivities than finer-textured soils. Likewise, soils with stable granular structure conduct water much more rapidly than do those with unstable structural units, which break down upon being wetted. Fine clay and silt can clog the small connecting channels of even the larger pores. Fine-textured soils that crack during dry weather at first allow rapid water movement. Later, these cracks swell shut, thereby reducing water movement to a minimum.

Saturated flow through soils is affected by organic matter content and by the nature of the inorganic colloids. Organic matter helps to maintain a high proportion of macropores. In contrast, some types of clay are especially conducive to fine pores. Soils high in montmorillonite, for example, generally have low conductivities compared to soils with 1:1-type clays.

From a practical point of view saturated flow is very important, especially with poorly drained soils. We shall discuss these aspects in later chapters where percolation and soil drainage are considered.

7:8. UNSATURATED FLOW IN SOILS

Under field conditions most soil–water movement occurs where the soil pores are not completely saturated with water. The soil macropores are mostly filled with air, and the micropores (capillary pores) with water and some air. Furthermore, the irregularity of soil pores results in discontinuities between pockets of water not in contact with each other. Water movement under these conditions is very slow compared to that occurring when the soil is saturated. This fact is illustrated in Fig. 7:16 which shows the generalized relation between matric suction (tension) and conductivity. At zero or near zero suction, the tension at which saturated flow occurs, the hydraulic conductivity is orders of magnitude more rapid than at suctions of 0.1 bar and above, which characterize unsaturated flow.

At low suction levels hydraulic conductivity is higher in the sandy soil than in the clay. The opposite is true at higher suction values. This relationship is to be expected since the dominance of large pores in the coarse-textured soil encourages saturated flow. Likewise, the prominence of finer (capillary) pores in the clay soil encourages more unsaturated flow than in the sand.

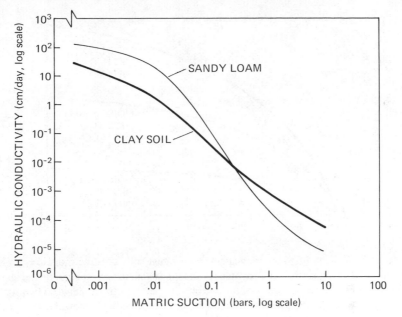

FIGURE 7:16. Generalized relationship between matric suction and hydraulic conductivity for a sandy soil and a clay soil. Saturation flow takes place at or near zero suction, while most of the unsaturated flow occurs at a suction of 0.1 bar or above.

FACTORS AFFECTING UNSATURATED FLOW.   Unsaturated flow is governed by the same general principles affecting saturated flow; that is, its direction and rate are related to the hydraulic conductivity and to a driving force, which in this case is *moisture suction gradient*. This gradient is the difference in suction between one soil zone and an adjoining one. Movement will be from a zone of low suction to one of high suction or from a zone of thick moisture films to one where the films are thin (see Fig. 7:21). The two forces responsible for this suction are the attraction of soil solids for water, and capillarity.

The influence of suction gradient is illustrated by the moisture curves shown in Fig. 7:17, where the rate of water movement from a moist soil into a drier one is shown. The higher the percentage of water in the moist soil, the greater is the suction gradient and the more rapid is the delivery. In this case the rate of movement obviously is a function of the suction gradient.

7:9. WATER MOVEMENTS IN STRATIFIED SOILS

The discussion up to now has dealt almost entirely with soils which are assumed to be quite uniform in texture and structure. In the field, layers

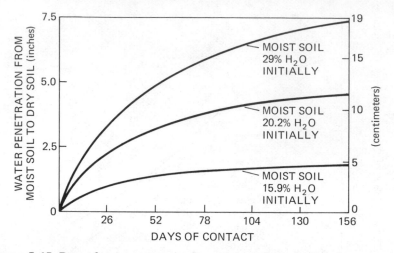

FIGURE 7:17. Rate of water movement from moist soils at three moisture levels to a drier one. The higher the water content of the moist soil, the greater will be the tension gradient and the more rapid will be the delivery. Water adjustment between two slightly moist soils at about the same water content will be exceedingly slow. [*After Gardner and Widtsoe* (4).]

differing in physical makeup from the overlying horizons are common. These have a profound influence on water movement and deserve specific attention.

Various kinds of stratification are found in many soils. Impervious silt or clay pans are common, as are sand and gravel lenses or other subsurface layers. In all these cases, the effect on water movement is similar—that is, the downward movement is impeded. The influence of layering can be seen by referring to Fig. 7:18. Apparently, the change in texture from that of the overlying material results in conductivity differences which prevent rapid downward movement.

The significance of this effect of stratification is obvious. For example, it definitely influences the amount of water the upper part of the soil holds in the field. The layer acts as a moisture barrier until a relatively high moisture level is built up. This gives a much higher field capacity (see p. 189) than that normally encountered in freely drained soils. It also illustrates a well-known weakness of the field-capacity concept, especially if it is to be related to a definite moisture tension value.

7:10. WATER VAPOR MOVEMENT

Water vaporization as it relates to soils may be distinguished for convenience of discussion as *internal* and *external*. In the one case, the change from the liquid to the vapor state takes place within the soil, that is, in the

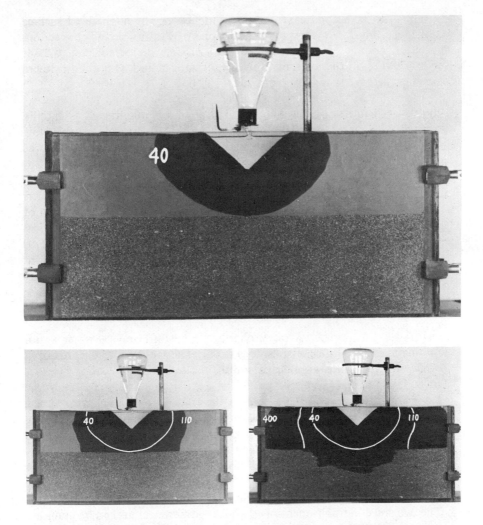

FIGURE 7:18. Downward water movement in soils having a stratified layer of coarse material. (*Top*) Water applied to the surface of a medium-textured top soil. Note that at the end of 40 minutes, downward movement is no greater than movement to the sides, indicating that in this case the gravitational force is insignificant compared to the tension gradient between dry and wet soil. (*Lower left*) The *downward* movement stops when a coarse textured layer is encountered. After 110 minutes, no movement into the sandy layer has occurred. After 400 minutes (*lower right*), the moisture content of the overlying layer becomes sufficiently high to give a moisture tension of 0.5 atmosphere or less, and downward movement into the coarse material takes place. Thus, sandy layers, as well as compact silt and clay, influence downward moisture movement in soils. [*Photo courtesy W. H. Gardner, Washington State University.*]

soil pores. In the second case, the phenomenon occurs at the land surface and the resulting vapor is lost to the atmosphere by diffusion and convection. The latter, commonly called *surface evaporation*, will be considered later (see p. 208). For the present, only vaporization and vapor adjustment tendencies *within* the soil are pertinent.

RELATIVE HUMIDITY OF THE SOIL AIR. The soil air is maintained essentially saturated with water vapor so long as the moisture suction is not below about 31 atmospheres. At this suction and less, water seems to be free enough to maintain the air at nearly 100 percent relative humidity. But when the moisture is held with a greater tenacity, water vaporizes with greater and greater difficulty and its vapor pressure becomes lower and lower.

This maintenance of the soil air at or very near a relative humidity of 100 percent is of tremendous importance, especially in respect to biological activities. This is perhaps the most important single feature in respect to the vapor–liquid interrelations within soils. Nevertheless, the actual amount of water present in vapor form in a soil at optimum moisture is surprisingly small, being at any one time perhaps not over 10 pounds in the upper 6 inches (15 cm) of an acre of soil.

MECHANICS OF WATER VAPOR MOVEMENT. The diffusion of water vapor from one area to another in soils does occur. The moving force is the *vapor pressure gradient*. This gradient is simply the difference in vapor pressure of two points a unit distance apart. The greater this difference, the more rapid is the diffusion and the greater is the transfer of vapor water during a unit period. Thus, if a moist soil where the vapor pressure is high is in contact with an air-dry layer where the vapor pressure is lower, a diffusion of water vapor into the drier area will tend to occur. Likewise, if the temperature of one part of a uniform moist soil mass is lowered, the vapor pressure of the air would be decreased and water vapor will tend to move in this direction. Heating will have the opposite effect.

The two soil conditions mentioned above—differences in relative humidity and in temperature—seem to set the stage for the movement of water vapor under ordinary field conditions. However, they may work at cross purposes and reduce vapor transfer tendencies to a minimum, or they may be so coordinated as to raise them to a maximum. The possible situation is set forth in Fig. 7:19.

Undoubtedly, some vapor transfer does occur within soils. The extent of the movement by this means, however, even from one continuous macropore to another, probably is not great if the soil water is within the range optimum for higher plants. In dry soils some moisture movement may take place in the vapor form. Such movement may be of some significance in supplying moisture to drought-resistant desert plants, many of which can exist at extremely low soil moisture levels.

SOIL HORIZONS

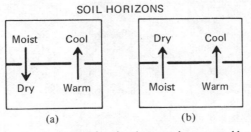

(a)   (b)

FIGURE 7:19. Vapor movement tendencies that may be expected between soil horizons differing as to temperature and moisture. In (a) the tendencies more or less negate each other, but in (b) they are coordinated and considerable vapor transfer might be possible if the liquid water in the soil capillaries does not interfere.

7:11. RETENTION OF SOIL MOISTURE IN THE FIELD

With the energy–soil moisture relations covered in previous sections in mind, we now turn to some more practical considerations. We shall start by following the moisture and energy relations of a soil during and following a very heavy rain or the application of irrigation water.

MAXIMUM RETENTION CAPACITY. Assume that water is applied to the surface of a well-granulated loam soil which is relatively uniform in texture and structure. The water might come from a heavy, steady rain or from irrigation. As the water enters the soil, air is displaced and the surface soil "wets up"— that is, the soil pores, large and small, are filled with water. Continued application will result in further downward movement and air replacement. At this point, all the pores in the upper part of the soil will be filled with water. The soil is said to be saturated with respect to water and is at its *maximum retentive capacity* (Fig. 7:20). The matric suction is essentially zero.

FIELD CAPACITY. If we now cut off the supply of water to the soil surface— that is, it stops raining or we shut off the irrigation water—there will be a continued relatively rapid downward movement of some of the water which is responding to the hydraulic gradient. After a day or so, this rapid downward movement will become negligible. The soil is then said to be at its *field capacity*. At this time an examination of the soil will show that water has moved out of the larger or *macropores* and that its place has been taken by air. The *micropores* or *capillary pores* are still filled with water and it is from this source that the plants will absorb moisture for their use. The matrix suction will vary slightly from soil to soil but is generally in the range of 0.1 to 0.2 bar. Moisture movement will continue to take place but the rate of movement is quite slow since it is now due primarily to capillary forces which are effective only in the micropores (Fig. 7:20).

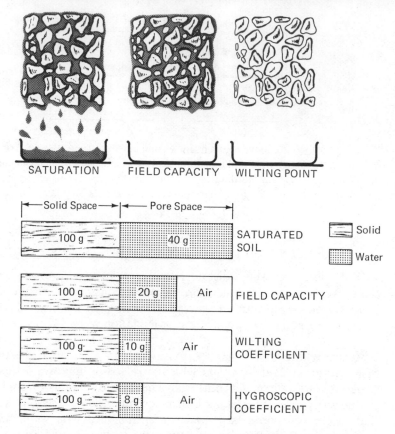

FIGURE 7:20. Volumes of solids, water, and air in a well-granulated silt loam at different moisture levels. The top bar shows the situation when a representative soil is completely saturated with moisture. This situation will usually occur for short periods of time during a rain or when the soil is being irrigated. Water will soon drain out of the larger or *macropores*. The soil is then said to be at the *field capacity*. Plants will remove moisture from the soil quite rapidly until the *wilting coefficient* is approached. Permanent wilting of the plants occurs at this point even though there is still considerable moisture in the soil (wilting coefficient). A further reduction in moisture content to the hygroscopic coefficient is illustrated in the bottom bar. At this point, the water is held very tightly, mostly by the soil colloids. [*Upper drawings modified from Irrigation on Western Farms published by the U.S. Departments of Agriculture and Interior.*]

WILTING COEFFICIENT. Plants growing in our soil will absorb water and will reduce the quantity of moisture remaining in the soil. Some of the water will be translocated from the roots to the leaves, where most of it will be lost by evapo-transpiration at the leaf surfaces. A second important avenue of loss is evaporation directly from the soil surface which will aid materially in the removal of soil moisture. Both of these losses are taking place simultaneously and are responsible for a markedly rapid rate of water dissipation from soils.

As the soil dries out, plants will begin to show the effects of reduced soil moisture. During the daytime they will tend to wilt, especially if temperatures are high and if there is some wind movement. At first this daytime wilting will be associated with renewed nighttime turgor or plant vigor. Ultimately, the rate of the supply of water to the plants will be so slow that the plant will remain wilted night and day. Although not dead, the plants are now existing in a permanently wilted condition and will die if water is not added. A measure of soil matric suction would show a value of about 15 bars.

An examination of the soil at this point will show a considerable amount of moisture remaining. The soil moisture content at this stage is called the *wilting coefficient* or the *critical moisture*. The water remaining in the soil is found in the smallest of the micropores and around individual soil particles (see Fig. 7:20). Obviously, a considerable amount of water present in soils is not available to higher plants. The soil moisture must be maintained considerably above the wilting coefficient if plants are to grow and function normally.

HYGROSCOPIC COEFFICIENT. To obtain a more complete picture of the soil moisture relations let us take a sample into the laboratory and allow the soil to dry out further. If it is kept in an atmosphere which is essentially completely saturated with water vapor, it will lose the liquid water held in even the smallest of the micropores. The remaining water will be associated with the surfaces of the soil particles, particularly the colloids, as adsorbed moisture. It is held so tightly that much of it is considered nonliquid and can move only in the vapor phase. The matric suction is about 31 bars. The moisture content of the soil at this point is termed the *hygroscopic coefficient*. As might be expected, soils high in colloidal materials will hold more water under these conditions than will sandy soils and those low in clay and humus (see Fig. 7:20 and Table 7:2).

SUCTION AND MOISTURE CONTENT. Figure 7:21 helps explain some of the observations as a wet soil is allowed to dry. The inverse relationship between moisture content (expressed in thickness of water films) and the tension (suction) at the outer edge of the film is shown. High moisture contents (thick films) are associated with low suctions. In contrast, when the moisture is low, the water is held at high suction values.

The simple case of water retention described in this section is also related directly to the suction–moisture content curves discussed in Section 7:3. This relationship is illustrated in Fig. 7:22, which shows the moisture content–matric suction relationship for a loam soil. The diagram at the upper part of this figure suggests physical and biological classification

TABLE 7:2. *Hygroscopic and Capillary Water Capacities of Various Soils* [a]

Soils	1 Organic Matter (%)	2 Hygroscopic Coefficient (%)	3 Field Capacity (moisture equiv.) (%)	4 Capillary Water (col. 3—2) (%)	5 Maximum Retentive Capacity (tension near 0) (%)
Western soils					
Sandy soil (Nebr.)	1.22	3.3	7.9	4.6	34.2
Red loam (N. Mex.)	1.07	10.0	19.2	9.2	49.0
Silt loam (Nebr.)	4.93	10.2	27.8	17.8	60.9
Black adobe (Ariz.)	2.22	12.9	25.8	12.9	60.3
Iowa soils					
Dickinson fine sand	2.13	3.4	7.6	4.2	44.5
Clarion sandy loam	3.01	6.9	15.5	8.6	58.0
Marshall silt loam	3.58	10.4	24.0	13.6	76.5
Wabash silty clay	5.91	16.1	30.4	14.3	87.0

[a] Data on Western soils from Alway and McDole (1), on Iowa soils from Russell (6).

schemes for soil water. However, this diagram is coupled intentionally with the moisture–suction curve to emphasize the fact that there are no clearly identifiable "forms" of water in soil. There is only a gradual change in suction with moisture content. This should be kept in mind in the next section as we discuss some commonly used soil moisture classification schemes.

7:12. CONVENTIONAL SOIL MOISTURE CLASSIFICATION SCHEMES

On the basis of observations of the drying of wet soils and of plants growing on these soils, two types of soil water classification schemes have been developed: (a) physical, and (b) biological. These schemes are useful in a practical way even though they lack the scientific bases which characterize the preceding moisture–energy discussions.

PHYSICAL CLASSIFICATION. From a physical point of view the terms *gravitational, capillary*, and *hygroscopic* waters are identified in Fig. 7:22. Water in excess of the field capacity (0.1 to 0.2 bar suction) is termed gravitational. Even though energy of retention is low, gravitational water is of little use to plants because it occupies the larger pores, thereby reducing soil aeration. Its removal from the soil in drainage is generally a requisite for optimum plant growth. It occupies the larger soil pores and moves readily under the force of gravity.

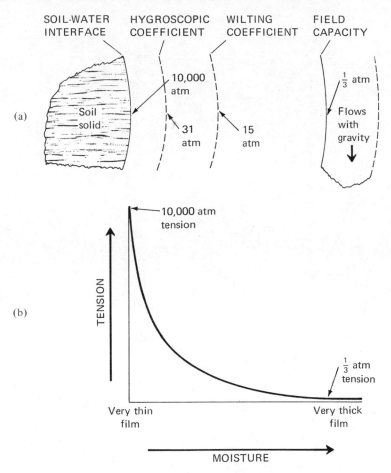

FIGURE 7:21. Relationship between thickness of water films and the tension with which the water is held at the liquid–air interface. The tension is shown in atmospheres. (*Upper*) Water film thickness at several moisture levels. (*Lower*) Logarithmic change in tension with increase in thickness of moisture film.

As the name suggests, capillary water is held in the pores of capillary size and behaves according to laws governing capillarity. Such water includes most of the water taken up by growing plants and exerts suctions between 0.1 and 31 bars.

Hygroscopic water is that bound tightly by the soil solids at suction values greater than 31 atmospheres. It is essentially nonliquid and moves primarily in the vapor form. Higher plants cannot absorb hygroscopic water, but some microbial activity has been found to take place in soils containing only hygroscopic water.

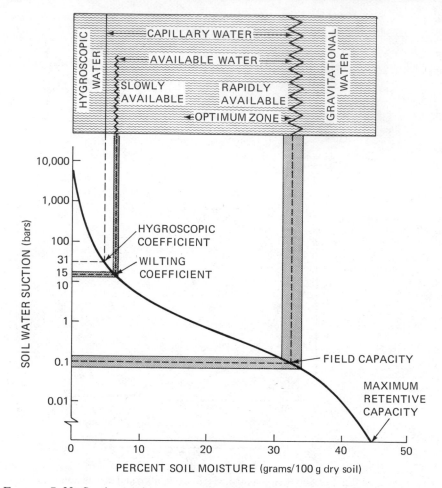

FIGURE 7:22. Suction–moisture curve of a loam soil as related to different terms used to describe water in soils. The wavy lines in the upper diagram suggests that measurements such as field capacity are not very quantitative. The gradual change in suction with soil moisture change discourages the concepts of different "forms" of water in soil. At the same time, such terms as "gravitational," "availability," etc., assist in the qualitative description of moisture utilization in soils.

BIOLOGICAL CLASSIFICATION. As one would expect, there is a definite relationship between moisture retention and its utilization by plants. Gravitational water is obviously of little use to plants and may be harmful. In contrast, moisture retained in the soil between the field capacity (0.1 to 0.2 bar) and the permanent wilting coefficient (15 bars) is said to be usable by plants and as such is *available water*. Water held at tensions greater than 15 bars is said to be *unavailable* to plants (Fig. 7:22).

In most soils, optimum growth of plants takes place when the soil moisture content is kept near the field capacity or at least does not approach the permanent wilting point. Thus, the moisture zone for optimum plant growth does not extend over the complete range of moisture availability.

While the various terms employed to describe soil water physically and biologically are useful in a practical way, at best they are only semi-quantitative. For example, measurement of field capacity tends to be rather arbitrary since the value obtained is affected by such factors as the initial soil moisture in the profile before wetting and the removal of water by plants and surface evaporation during the period of downward flow. Also, the determination as to when the downward movement of water due to gravity has "essentially ceased" is rather arbitrary. These facts stress once again that there is no clear line of demarcation between different "forms" of soil water.

7:13. FACTORS AFFECTING AMOUNT AND USE OF AVAILABLE SOIL MOISTURE

The amount of water plants absorb from soils is determined by a number of plant, climatic, and soil variables. Rooting habits, basic drought toler-ance, and stage and rate of growth are important plant factors. Significant climatic variables include air temperature and humidity and wind velocity and turbulence. These will be considered further in Chapter 8, which is concerned with vapor losses of moisture.

Among the important soil characteristics influencing water uptake by plants are (a) moisture suction relations (matric and osmotic), (b) soil depth, and (c) soil stratification or layering. Each will be discussed briefly.

MATRIC SUCTION. The effect of matric suction on the amount of available moisture in a soil should be obvious. These factors which affect the amount of water in a soil at the field capacity, and in turn at the wilting coefficient, will influence the available water. The texture, structure, and organic matter content all influence the quantity of water a given soil can supply to growing plants. The general influence of texture is shown in Fig. 7:23. Note that as fineness of texture increases, there is a general increase in available moisture storage, although clays frequently have a smaller capacity than do well-granulated silt loams. The comparative available water-holding capacities in terms of inches of water per foot of soil are also shown by this graph.

The influence of organic matter deserves special attention. A well-drained mineral soil containing 5 percent organic matter will probably have a higher available moisture capacity than a comparable soil with 3 percent organic matter. One might erroneously assume that this favorable effect is all due directly to the moisture-holding capacity of the organic matter. Such is not

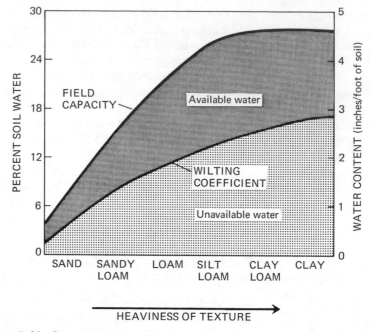

FIGURE 7:23. General relationship between soil moisture characteristics and soil texture. Note that the wilting coefficient increases as the texture becomes heavier. The field capacity increases until we reach the silt loams, then levels off. Remember, these are representative curves. Individual soils would probably have values different from these shown.

the case. Most of the benefit of organic matter in this case is attributable to its favorable influence on soil structure and in turn on soil porosity. Although humus does have a high field capacity, its wilting coefficient is proportionately high. Thus, the net direct contribution toward available moisture is less than one would suppose.

OSMOTIC SUCTION. The presence of salts in soils, either from applied fertilizers or as naturally occurring compounds, can influence soil water uptake. Osmotic suction effects in the soil solution will tend to reduce the range of available moisture in such soils by increasing the wilting coefficient. The total moisture stress in such soils at this point is the matric suction plus the osmotic suction of the soil solution. Although in most humid region soils this osmotic suction effect is insignificant, it becomes of considerable importance in some saline soils of arid and semiarid regions.

SOIL DEPTH AND LAYERING. All other factors being equal, deep soils will have greater available moisture-holding capacities than will shallow ones. For deep-rooted plants this is of practical significance, especially in those subhumid and semiarid regions where supplemental irrigation is not possible. Soil moisture measurements to depths as great as 5 to 6 feet are sometimes used as bases for predicting wheat yields in the Great Plains area of the United States. Shallow soils are obviously not well suited to these climatic conditions.

Soil stratification or layering will influence markedly the available water and its movement in the soil. Hardpans or impervious layers, for example, slow down drastically the rate of movement of water and also influence unfavorably the penetration of plant roots. They sometimes restrict root growth and effectively reduce the soil depth from which moisture is drawn. Sandy layers also act as barriers to soil moisture movement from the finer-textured layers above. Movement through a sandy layer is very slow at intermediate and high tensions. The moisture tension in the overlying layers must be less than about 0.5 atmosphere before movement into the sand will take place. The explanation for this unusual situation was apparent as we considered Section 7:9 dealing with moisture movement in soils (see Fig. 7:18).

The available moisture storage capacity of soils determines to a great extent their usefulness in practical agriculture. This capacity is often the buffer between an adverse climate and crop production. It becomes more significant as the utilization of water for all purposes—industrial, domestic, as well as agricultural—begins to tax the supply of this all-important natural resource.

7:14. How Plants are Supplied with Water— Capillarity and Root Extension

At any one time, only a small proportion of the soil water lies in the immediate neighborhood of the absorptive surfaces of plant root systems. Consequently, a question arises as to how the immense amount of water (see pp. 208–209) needed to offset transpiration is so readily and steadily acquired by vigorously growing crops. Two phenomena seem to account for this acquisition: (a) the capillary movement of the soil water to plant roots, and (b) the growth of the roots into moist soil.

RATE OF CAPILLARY MOVEMENT. When plant rootlets begin to absorb water at any particular point or locality in a moist soil, the thick water films in the soil pores are thinned and their energy of retention is increased. The pull of moisture in this direction is intensified and water tends to move toward the points of plant absorption. The rate of movement depends on the magnitude of the suction gradients developed and the conductivity of the soil pores. A more complete explanation of this mechanism of unsaturated flow is offered on pages 184–85.

With some soils, the above adjustment may be comparatively rapid and the flow appreciable; in others, especially fine-textured and poorly granulated clays, the movement will be sluggish and the amount of water delivered meager. Thus, a root hair, by absorbing some of the moisture with which it is in contact, automatically creates a suction gradient, and a flow of water is initiated toward its active surface.

How effective the above flow may be under field conditions is questionable. Many of the early investigators greatly overestimated the distance through which capillarity may be effective in satisfactorily supplying plants with moisture. They did not realize that the rate of water supply is the essential factor and that capillary delivery over appreciable distances is very slow. Plants must have large amounts of water delivered rapidly and regularly. The influence of capillarity is exerted through only a few centimeters so far as the hour-by-hour needs of plants are concerned.

The above statement must not be taken to mean that capillary adjustments in the aggregate are not important. It is not always necessary for capillary water to move great distances in the soil to be of significance to plants. As roots absorb moisture, capillary movement of no more than a few millimeters (if occurring throughout the soil volume) may be of practical importance. Capillary adjustment along with vapor movement is a major factor in supplying water for plants, especially these growing at low moisture contents. Since there is little root extension at moisture tensions approaching the wilting coefficient, the water must move to the plants.

RATE OF ROOT EXTENSION. The limited water-supplying capacity of capillarity over long distances directs our attention even more forcibly to the rate of root extension, and here early researchers made an underestimate. They failed to recognize the rapidity with which root systems expand and the extent to which new contacts are constantly established (see Fig. 5:5). During favorable growing periods, roots often elongate so rapidly that satisfactory moisture contacts are maintained even with a lessening water supply, especially with the aid of capillarity. The mat of roots, rootlets, and root hairs in a meadow, between corn or potato rows, or under oats or wheat is ample evidence of the enormous root system of plants. Table 7:3 provides information on roots of several common crop plants.

The rate of root extension is surprising even to those engaged in plant production. On the basis of the data available, the elongation may be rapid enough to take care of much of the water needs of a plant growing in a soil at optimum moisture. It is well to remember, however, that at any one time roots are in contact with no more than about 1 percent of the soil surface area. Consequently, to be absorbed much of the water must move from the soil to the root. The movement may be only a few millimeters, but in a soil well permeated with roots, that is all the movement needed. Thus, while capillary movement of water over distances of inches or feet is now known to be of little significance in the daily supply of water to

TABLE 7:3. *Observations of Roots and Root Hairs of Three Plants in One Cubic Inch of Soil* [a]

Plant	Number of Roots	Number of Root Hairs (thousands)	Combined length (ft)	Combined surface (in.^{2})
Oats	110	150	630	15
Rye	150	300	1,300	30
Kentucky bluegrass	2,000	1,000	4,000	65

[a] From Ditmer (3).

plants, capillarity operating over short distances throughout the rooting zone is a major factor in keeping water supplied to growing plants.

7:15. CONCLUSION

The characteristics and behavior of soil water appear to be complex. As we have gained more knowledge about them, however, it has become apparent that they are governed by relatively simple basic physical principles. Furthermore, researchers are discovering the similarity between these principles and those governing the uptake and utilization of soil moisture by plants. These principles are the subject of the next chapter.

REFERENCES

(1) Alway, F. J., and McDole, G. R., "The Relation of Movement of Water in a Soil to Its Hygroscopicity and Initial Moisture," *Jour. Agr. Res.*, **10**: 391–428, 1917.

(2) Cooney, J. J., and Peterson, J. E., *Avocado Irrigation*, Leaflet 50, California Agr. Ext. Sta., 1955.

(3) Ditmer, H. J., "A Comparative Study of the Subterranean Members of Three Field Grasses," *Science*, **88**: 482, 1938.

(4) Gardner, W., and Widtsoe, J. A., "The Movement of Soil Moisture," *Soil Science*, **11**: 230, 1921.

(5) Keeton, W. T., *Biological Science*, 2nd ed. (New York: W. W. Norton, 1972).

(6) Richards, L. A., "Physical Condition of Water in Soil," in *Agron 9: Methods of Soil Analysis*, Part I (Madison, Wisc.: American Society of Agronomy, 1965).

(7) Russell, M. B., "Soil Moisture Sorption Curves for Four Iowa Soils," *Proc. Soil Sci. Soc. Amer.*, **4**: 51–54, 1939.

Chapter 8

VAPOR LOSSES OF SOIL
MOISTURE AND THEIR
REGULATION[1]

WISE use and management of water are perhaps the most critical factor in schemes to increase food supplies, especially in the population-heavy developing nations. At the same time, industrial and domestic requirements for water have expanded dramatically, creating strong competition for traditional agricultural uses. This competition forces increased emphasis on efforts to manage and conserve water and to prevent its misuse.

The central role of soils and plants in the hydrological cycle is illustrated in Fig. 8:1. This figure also shows the interrelationship among the various uses of water. Further, it emphasizes the need to minimize the loss of water from plant and soil surfaces in either the vapor or liquid form. Liquid losses and soil erosion are covered in Chapter 9. Vapor losses from both soil and plant systems will now be discussed.

8:1. INTERCEPTION OF RAIN WATER BY PLANTS

Rainfall data are often dealt with as though all precipitation reaches the soil. Such is far from the case, especially where the vegetative cover is dense. Part of the rain or snow is intercepted by the plant leaves and stems and is evaporated directly into the atmosphere without reaching the soil.

The extent of plant interception of precipitation is determined by a number of factors. In areas with perennial standing cover, such as range and natural grasslands and especially forests, the annual interception is much more than where annual field crops are grown. Forested areas, chiefly those with year-round foliage (coniferous) are particularly effective in intercepting precipitation. One third to one half of the annual precipitation in such forests commonly does not reach the soil, being intercepted by the plant parts and evaporated directly into the atmosphere.

Field and vegetable crops which occupy the land for only part of the year intercept much less of the annual precipitation. Even so, it is not at all uncommon to find that 10–20 percent of the seasonal rainfall falling on

[1] For excellent reviews of soil–water–plant relations see Kozlowski (9).

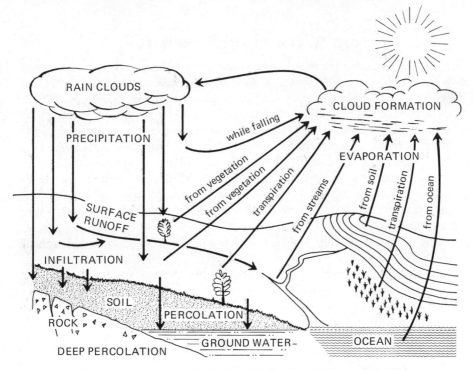

FIGURE 8:1. Diagram of the hydrologic cycle showing the interrelationships among soil, plants, and the atmosphere. Only water is lost from soils in evaporation and transpiration from plant surfaces. Solid particles are often included in surface runoff water, and nutrients in percolation water.

these crops does not reach the soil (Table 8:1). During periods of light rainfall a high proportion of the rain may be intercepted in a dense stand of a crop like alfalfa. While such losses are unavoidable, they must be taken into account when considering soil–water–plant relations.

TABLE 8:1. *Seasonal Interception of Rainfall by Crops at Bethany, Missouri, and Sussex, New Jersey*[a]

Average of 3 years' records for alfalfa and corn and 1 year for soybeans. Values in percent of total seasonal rainfall for each crop.

Fate of Rainfall	Alfalfa	Corn	Soybean
Direct to soil	64.7	70.3	65.0
Ran down stem	13.7	22.8	20.4
Total to soil	78.4	93.1	85.4
Remainder to atmosphere	21.6	6.9	14.6

[a] From Haynes (6).

8:2. THE SOIL–WATER–PLANT CONTINUUM

The sciences dealing with the behavior of soil water and those concerned with plant–water relationships developed rather independently. Consequently, different concepts and terminology emerged making it difficult to understand the processes of water absorption from the soil, its transport upward in the plant, and its loss to the atmosphere. The emergence of the concept of water potential brought with it the realization that the same basic principles applied for water in soil, plants, and the atmosphere. Further came the recognition of a unified and dynamic soil–plant–atmosphere system in which water flows from areas of high to those of low potential energy. The system has been termed the *soil–plant–atmosphere continuum*, or *SPAC* (11).

Figure 8:2 may help you to visualize in a general way the function of the SPAC. Water moves from the soil to the plant to the atmosphere and back to the soil. It seems reasonable that in this cycle or continuum the same basic principles govern the behavior of water whether it be in the soil, the plant, or the atmosphere. Thus, the movement of water in this continuum is determined by the moisture potential and by the resistance to flow set up by features such as the size and configuration of pores and the permeability of membranes.

SPAC ENERGY RELATIONS. In Chapter 7, the free energy level of water was seen to be a major controlling factor in determining soil–water behavior. The same can be said for soil–plant and plant–atmosphere relations. As water moves through the soil to plant roots, into the roots, across cells into stems, up the plant xylem to the leaves, and is evaporated from the leaf surfaces, its tendency to move is determined by differences in free energy levels of the water, or by the moisture potential.

The moisture potential must be higher in the soil than in the plant roots if water is to be absorbed from the soil. Likewise, movement to the xylem, up the stem through the xylem, and to the leaf cells is related to differences in moisture tension (see Fig. 8:3). The negative pressures found in stem xylem of plants suggest this relationship. Water potentials in the xylem may reach −20 bars at the wilting point for field crops and are commonly −20 to −25 bars in the upper limbs of tall trees growing in soils amply supplied with water. Desert plants will exist with potentials of −20 to −80 bars, although growth appears to stop at tensions below about −20 bars.

TWO POINTS OF RESISTANCE. Changes in water potential illustrated in Fig. 8:3 suggest two points of major resistance as the water moves through the SPAC: the movement of water into the plant from soil, and its evaporation

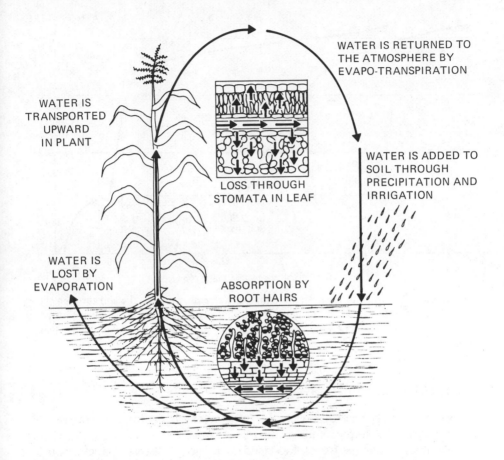

WATER IS RETURNED TO
THE ATMOSPHERE BY
EVAPO-TRANSPIRATION

WATER IS
TRANSPORTED
UPWARD
IN PLANT

WATER IS ADDED TO
SOIL THROUGH
PRECIPITATION AND
IRRIGATION

LOSS THROUGH
STOMATA IN LEAF

WATER IS
LOST BY
EVAPORATION

ABSORPTION BY
ROOT HAIRS

FIGURE 8:2. Soil–plant–atmosphere continuum. In recent years scientists have come to realize that a common set of basic principles governs relationships among water, soil, plants, and the atmosphere. This continuum is best illustrated by following water as it is added to soils through precipitation or irrigation, as it behaves in soils, is lost directly to the atmosphere from soils, or is absorbed by plants, transported upward, and subsequently evaporated into the atmosphere. Water behavior in all cases is subject to the same basic physical and chemical laws.

into the atmosphere from the surface of cells in the leaf. This is in accord with field observations that two primary factors determine whether plants are well supplied with water: (a) the rate at which water is supplied by the soil to the absorbing roots; and (b) the rate at which water is evaporated from the plant leaves. Factors affecting the soil's ability to supply water were discussed in Chapter 7. Attention will now be given to the loss of water by evaporation and factors affecting this loss.

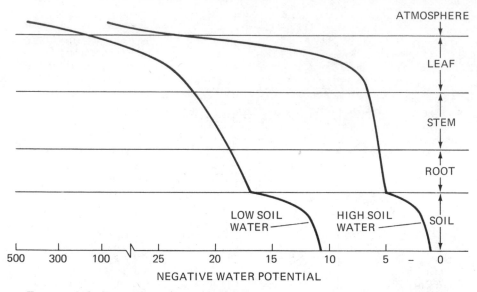

FIGURE 8:3. Change in moisture potential as water moves from the soil through the root, stem, and leaf to the atmosphere. Note that negative water potential is plotted, indicating a reduction in water potential as the water moves through the system. [*Adapted from Hillel* (7).]

8:3. EVAPO-TRANSPIRATION

Vapor losses from soils occur in two ways: (a) by the *evaporation* of water at the soil surface, and (b) by *transpiration* from the leaf surfaces of water absorbed by the plants and translocated to the leaves. The combined loss resulting from these two processes, termed *evapo-transpiration*, is responsible for most of the water removal from soils under normal field conditions. On irrigated soils located in arid regions, for example, it commonly accounts for the loss of 30 to 40 inches of water during the growing season of a crop such as alfalfa. Obviously, the phenomenon is of special significance to growing plants.

The rate of water loss by either evaporation from the soil or by transpiration is determined basically by differences in moisture potential identified as the *vapor pressure gradient*—the difference between the vapor pressure at the leaf or soil surface and that of the atmosphere. The vapor pressure gradient in turn is related to a number of other common climatic and soil factors, some of which will now be considered (see Fig. 8:4).

RADIANT ENERGY. For every gram of water vaporized, 540 calories of energy are required, whether the evaporation is from the soil directly or from leaf surfaces. The primary source of this energy is the sun (see Fig.

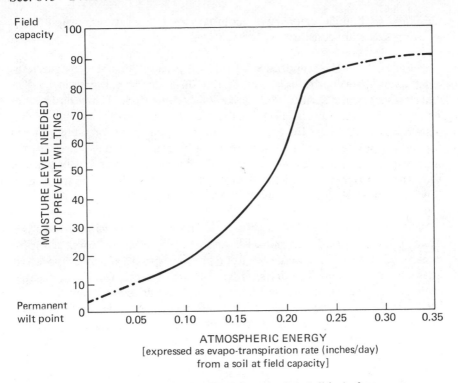

FIGURE 8:4. Moisture level needed to prevent turgor loss (wilting) of corn grown on a Colo silty clay loam under different conditions of atmospheric energy. The moisture level ranges from the permanent wilting point (0) to the field capacity (100). The atmospheric energy, which includes factors such as temperature, wind velocity, etc., is expressed in terms of the evapo-transpiration that it causes. The broken-line portions are extrapolations from actual measurements. [*Data adapted from Denmead and Shaw* (3).]

10:10). So long as ample soil moisture is available, there is a close relationship between evaporation and the absorption of this radiant energy. In arid regions the sparse cloud cover permits a high proportion of the solar radiation to strike the earth, maximizing opportunities for evapo-transpiration. In contrast, in regions characterized by cloudy days a smaller proportion of the solar radiation strikes the soil and plants, and the evaporative potential is not so great.

ATMOSPHERIC VAPOR PRESSURE. The vapor pressure of the atmosphere helps control evaporation from soils and plants. If it is low compared to the vapor pressure at the plant and soil surfaces, evaporation takes place rapidly. If it is high, as is the case of "humid" days, evaporation is slow. That atmospheric vapor pressure markedly influences evapo-transpiration is shown by the relatively high vapor losses from irrigated soils in arid climates. In humid regions with comparable temperatures, evapo-trans-piration losses are considerably less.

TEMPERATURE. The evaporation of water is greatly influenced by tempera-ture. Consequently, during warm or hot days, the vapor pressure at the leaf surfaces or the surface of a moist soil is quite high. Temperature does not have a similar direct effect on the vapor pressure of the atmosphere. As a result, on hot days there is a sharp difference in vapor pressure between leaf or soil surfaces and the atmosphere (a higher vapor pressure gradient) and evaporation proceeds rapidly. Plants and especially soils may be warmer than the atmosphere on bright, clear days, further emphasizing the impor-tance of temperature in controlling evapo-transpiration.

WIND. A dry wind will continually sweep away moisture vapor from a wet surface. The moist air thus moved is replaced by air with a lower moisture content, thereby maintaining the vapor pressure gradient, and evaporation is greatly encouraged. The drying effect of even a gentle wind is noticeable even though the air in motion may not be at a particularly low humidity level. Hence, the capacity of a high wind operating under a steep vapor pressure gradient to intensify evaporation both from soils and plants is tremendous. Farmers of the Great Plains dread the *hot winds* characteristic of that region.

SOIL MOISTURE SUPPLY. In discussing the influence of the other factors on evapo-transpiration, the assumption has been made that the soil and plant surfaces are plentifully supplied with moisture. Under these conditions, the climatic factors already considered will largely control vapor losses (see Fig. 8:4). At lower moisture contents, however, soil moisture tension will limit the rate of supply of water to the soil and plant surfaces, and evapo-transpiration losses will decrease accordingly (see Table 8:2).

TABLE 8:2. *Effect of Soil Moisture Level on Evapo-transpiration Losses*[a]
Where the surface moisture content was kept high, total evapo-transpiration losses were greater than when medium level of moisture was maintained.

Moisture Condition of Soil[b]	Evapo-transpiration (in.)	
	Corn	Alfalfa
High	17.7	24.4
Medium	12.7	20.5

[a] From Kelly (8).
[b] High moisture—irrigated when upper soil layers were 50 percent depleted of available water. Medium moisture—irrigated when upper layers were 85 percent depleted of available water.

As moisture is depleted from the soil surface and the root zone, evapo-ration losses decrease. The plant responds to this moisture deficiency by the closing of leaf stomata and ultimately by wilting. In the soil, capillarity will at first partially replenish moisture lost by evaporation at the surface. With time, however, the rate of loss by evaporation and plant uptake will deplete the surface soil. Capillarity to the upper layers will then be too slow to be of much practical importance. Under these conditions, some movement of moisture in the vapor phase from lower horizons will then take place.

Soil physicists are agreed that the depth to which soils may be depleted by capillarity and evaporation from the soil surface is far short of the 4, 5, or even more feet sometimes postulated. Some investigators think that a 20- to 24-inch depth is probably a maximum range. In most cases, only the water of the furrow slice sustains appreciable diminution by surface evapora-tion, the lower layers being depleted primarily by root absorption (see Table 8:3).

TABLE 8:3. *In a Humid Region Soil, More Than Half the Water Lost by Evapo-transpiration Came from the Surface Layer*[a]

In drier areas the lower zones furnish much more water, especially for deep-rooted crops.

Soil Depth (in.)	Evapo-transpiration[b] (in.)		
	Corn	Pasture	Woods
0–7	9.66	9.38	9.31
7–72	8.30	8.47	8.89

[a] Calculated from Dreibelbis and Amerman (5).
[b] Periods of measurement: corn, May 23–Sept. 25; pasture, Apr. 15–Aug. 23; woodland, May 25–Sept. 28.

8:4. MAGNITUDE OF EVAPORATION LOSSES

The combined losses by evaporation from the soil surface and by transpira-tion account for the *consumptive* use of water, which is a measure of the total water lost by evapo-transpiration in producing crops. This is an important practical figure, especially in areas where irrigation must be employed to meet crop needs.

CONSUMPTIVE USE. As might be expected, there is a marked variation in the consumptive use of water to produce different crops in different areas. All of the factors previously considered as influencing evapo-transpiration are operative. In addition, such plant characteristics as depth of rooting and length of growing season become important.

Consumptive use may vary from as little as 12 inches to as much as 85 inches or more. The low extreme might be encountered in cool mountain valleys where the growing seasons are short; the higher figure has been found in irrigated desert areas. The ranges commonly encountered are 15 to perhaps 30 inches in unirrigated, humid to semiarid areas and 20 to 50 inches in hot, dry regions where irrigation is used.

From a practical standpoint, daily consumption figures are in some cases more significant than those for the growing season. During the hot dry periods in the summer, daily consumptive use rates for corn, for example, may be as high as 0.4 to 0.5 inch. Even with deep soils having reasonably high capacities for storing available water, this rapid rate of moisture removal soon depletes the plant root zone of easily absorbed moisture. Sandy soils, of course, may lose most of their available moisture in a matter of a few days under these conditions. The importance of these vaporization losses is obvious.

SURFACE EVAPORATION VERSUS TRANSPIRATION. It is interesting to consider the comparative water losses from the soil surface and from transpiration. A number of factors determine these relative losses, including (a) plant cover in relation to soil surface; (b) efficiency of use of water by different plants; (c) proportion of time the crop is on the land, especially during the summer months; and (d) climatic conditions.

In humid regions, there is some evidence that vapor losses may be divided about equally between evaporation from the soil and transpiration. Naturally, this generalization would not hold in all cases since there are so many factors influencing vapor losses from both crops and soils.

Loss by evaporation from the soil is thought to be proportionately higher in drier regions than in humid areas. Such vapor loss is about 70 to 75 percent of the total rainfall for the Great Plains area of the United States. Losses by transpiration account for 20 to 25 percent, leaving about 5 percent for runoff.

8:5. EFFICIENCY OF WATER USE

The crop production obtained from the use of a given amount of water is an important figure, especially in areas where moisture is scarce. This efficiency may be expressed in terms of (a) consumptive use in pounds per pound of plant tissue produced, or (b) transpiration in pounds per pound of plant tissue produced. The second figure, called *transpiration ratio*, emphasizes the fact that large quantities of water are required to produce 1 pound of dry matter.

TRANSPIRATION RATIO. The transpiration ratio ranges from 200 to 500 for crops in humid regions, and almost twice as much for those of arid climates. The data in Table 8:4 give some idea of the water transpired by different crops.

TABLE 8:4. *Transpiration Ratios of Plants as Determined by Different Investigators*[a]

Crop	Harpenden, England	Munich, Germany	Dahme, Germany	Madison, Wisc.	Pusa, India	Akron, Colo.
Barley	258	774	310	464	468	534
Beans	209	—	282	—	—	736
Buckwheat	—	646	363	—	—	578
Clover	269	—	310	576	—	797
Maize	—	233	—	271	337	368
Millett	—	447	—	—	—	310
Oats	—	665	376	503	469	597
Peas	259	416	273	477	563	788
Potatoes	—	—	—	385	—	636
Rape	—	912	—	—	—	441
Rye	—	—	353	—	—	685
Wheat	247	—	338	—	544	513

[a] Data compiled by Lyon et al. (10).

Much of the variation observed in the ratios arises from differences in climatic conditions. Thus, in areas of more intense sunshine, the temperature is higher, the humidity is lower, and the wind velocity is frequently greater. All this tends to raise the transpiration ratio. This is illustrated in Fig. 8:5, which shows the transpiration ratio of four common crops as influenced by dryness of climate.

It is obvious from the transpiration ratios quoted in Table 8:4 that the amount of water necessary to mature the average crop is very large. For example, a representative crop of wheat containing 4,000 pounds of dry matter per acre and having a transpiration ratio of 500 will withdraw water from the soil during the growing season equivalent to almost 9 inches of rain. The corresponding figure for corn, assuming the dry matter as 8,000 pounds and the transpiration ratio as 350, would be over 12 inches. These amounts of water in addition to that evaporated from the surface must be supplied during the growing season. The possibility of moisture being the most critical factor in crop production is thus obvious.

FACTORS INFLUENCING EFFICIENCY. The efficiency of water use in crop production is influenced by climatic, soil, and nutrient factors. The climatic factors have already received adequate attention (pp. 204–205). Within a given climatic zone, the other effects can be seen.

In general, highest efficiency is found where optimum crop yields are being obtained. Conversely, where yields are limited by some factor or combination of factors, efficiency of water uptake is low and more water is required to produce a pound of plant tissue.

Maintaining the moisture content of a soil too low or too high for optimum growth will likely increase the water required to produce 1 pound of dry

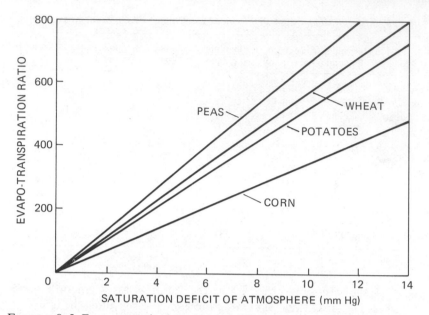

FIGURE 8:5. Evapo-transpiration ratios of different crop plants grown at locations differing in the saturation deficit of the atmosphere. In the drier climates (high saturation deficit) the corn uses much less water per unit of dry matter produced; peas use the most. [*Data from several sources as plotted by Bierhuizen and Slayter* (2).]

matter. Similarly, the amount of available nutrients and their balance condition are also concerned in the economic utilization of water. The more productive the soil, the lower the transpiration ratio, provided the water supply is held at optimum (see Fig. 8:6). Therefore, a farmer in raising the productivity of his soil by drainage, lime, good tillage, farm manure, and fertilizers provides at the same time for a greater amount of plant production for every unit of water utilized. The total quantity of water taken from the soil, however, will probably be larger.

8:6. EVAPORATION CONTROL: MULCHES AND CULTIVATION

Any material used at the surface of a soil primarily to prevent the loss of water by evaporation or to keep down weeds may be designated as a mulch. Sawdust, manure, straw, leaves, and other litter may be used successfully. Such mulches are highly effective in checking evaporation and are most practical for home garden use and for high-valued crops, including strawberries, blackberries, fruit trees, and such other crops as require infrequent, if any, cultivation.

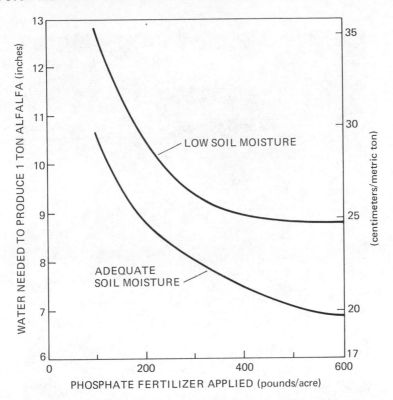

FIGURE 8:6. Effect of phosphate applications and soil moisture level on the amount of water required to produce one ton of alfalfa hay. Most efficient moisture utilization was obtained when the fertility level was high and the moisture supply adequate. [*From Kelly* (8).]

PAPER AND PLASTIC MULCHES. Specially prepared paper and plastics have been used as mulches (see Fig. 8:7). This cover is spread and fastened down either between the rows or over the rows, the plants in the latter case growing through suitable openings. This mulch can be used only with crops planted in rows or in hills. So long as the ground is covered, evaporation and weeds are checked and, in some cases, remarkable crop increases have been reported. Unless the rainfall is torrential, the paper or plastic does not seriously interfere with the infiltration of the rain water.

The paper and plastic mulches have been employed with considerable success in the culture of pineapples in Hawaii where the idea originated, and with other crops elsewhere. This type of mulch is also being used to some extent in trucking and vegetable gardening in the United States. The cost of the cover and the difficulty of keeping it in place limit the use of these materials to high-valued crops.

FIGURE 8:7. For crops with high cash value, plastic mulches are being used. The plastic is installed by machine (*above*) and at the same time the plants are transplanted (*below*). The plastics help control weeds, conserve moisture, encourage rapid early growth, and eliminate need for cultivation. Their high costs make them practical only with the highest-value crops. [*Photos courtesy K. Q. Stephenson, Pennsylvania State University.*]

STUBBLE MULCH. A mulch somewhat different from those already described, especially as to mode of establishment, is *stubble* or *trash mulch*, used mostly in subhumid and semiarid regions. In this case the mulching materials have been grown in place and consist of the refuse of the previous crop. Oat stubble, straw, cornstalks, and similar residues are good examples. By the use of suitable implements, a seedbed can be prepared and the crop planted, leaving the organic trash on the surface to act as a mulch (see Fig. 8:8). Modified tillage implements make the cultivation of row crops

FIGURE 8:8. (*Upper*) Preparation of a trash mulch by incomplete incorporation of surface residues. (*Lower*) In another field, the stubble mulch traps snow (right) in comparison with fallowed soil in the left. [*Photos courtesy U.S. Soil Conservation Service.*]

possible without greatly disturbing the surface layer. The use of such a mulch not only conserves moisture but also helps control wind erosion.

SOIL MULCH. One of the early misconceptions about the control of evaporation was that the formation of a *natural* or *soil mulch* was a desirable moisture-conserving practice. Years of experimentation and practice have shown that cultivating the surface soil to form a natural mulch does not necessarily conserve moisture. In fact, in some cases it may encourage moisture loss.

Three reasons are commonly advanced for the ineffectiveness of cultivation in moisture control. *First*, a large amount of moisture always escapes by evaporation long before the soil is dry enough for cultivation. *Second*, the precultivation loss has often already established a thin protective surface layer of air-dry soil which may successfully reduce evaporation. *Third*, the crop itself, if its root spread is wide and comparatively dense, intercepts upward-moving moisture that might otherwise reach the soil surface.

CULTIVATION AND WEED CONTROL. Perhaps the most important reason for cultivation of the soil is to control weeds. Evapo-transpiration of these unwanted plants can extract soil moisture far in excess of that used by the crop itself, especially a row crop (see Fig. 8:9). The widespread use of chemical herbicides, however, provides weed control without cultivation of the soil. Herbicides thus have become one of the most important tools in controlling evaporation. The necessity for tillage is thereby limited to situations where this practice is needed to maintain the proper physical conditions for plant growth.

FIGURE 8:9. Deep cultivation of intertilled crops such as corn (*right*) may result in serious root pruning. Hence cultivation for the killing of weeds should be shallow (*left*), even early in the season. If weeds are under control and the structure of the soil is satisfactory, further intertillage usually is unnecessary. [*From Donald* (4).]

8:7. VAPORIZATION CONTROL IN HUMID REGIONS

From what has been said, it seems that little in general can be done in humid regions to control evapo-transpiration from farm lands. Losses due to crop transpiration and interception are conceded to be beyond control. The more vigorous the growth of the crops, the larger will be these two removals, especially transpiration. The ineffectiveness of a soil mulch and the waste of water incidental to its establishment have been indicated. Whereas stubble and trash mulches are effective as well as feasible, they have not met with any great favor in humid regions. Paper and plastic mulches, while quite satisfactory in gardens and nurseries, are too expensive to be used with field crops.

This leaves the control of weeds as about the only effective and practicable means in humid regions of reducing the loss of water vapor from soils. While weed control may not result in a reduction of soil moisture loss *in toto*, it does ensure a greater amount of water available for crop transpiration. This, of course, is the ultimate objective in the moisture management of field soils. The attention now being given to improved methods of weed control is well placed.

8:8. VAPORIZATION CONTROL IN SEMIARID AND SUBHUMID REGIONS

When the average annual rainfall of a region is very much below 25 inches, it would seem that crop production would be impossible without irrigation. Yet between the limits of 15 to 20 inches, certain crop plants may be grown successfully, a feat made easier if much of the rainfall comes in the spring and early summer. The system adopted is often called *dry land farming*. It consists essentially of (a) using special varieties of such crops as wheat, corn, sorghums, and rye; (b) reducing the rate of seeding to correspond to the limited amounts of moisture that probably will be available; and (c) employing tillage practices that conserve and economize water. Only the latter phase will be discussed here.

TILLAGE OPERATIONS. The tillage operations practiced in dry land agriculture should keep weeds under control and at the same time reduce wind erosion to a minimum. The methods employed include making rough furrows at right angles to the prevailing high winds, listing, strip cropping, and the use of stubble mulch when practicable. At times of rain, the land should be highly permeable and retentive and should be protected from water erosion if the rainfall is torrential. As to vapor losses of water from the soil, the situation in general is much the same as that already described for humid regions except that in arid regions the humidity is generally low and the wind velocity high.

SUMMER FALLOW. Dry land farming often includes a *summer fallow*, the idea being to catch and hold as much of one season's rain as possible and carry it over for use during the next. The soil profile is used as a large reservoir.

A common procedure for summer fallowing is to allow the previous year's stubble to stand on the land until the spring of the year. The soil is then tilled before appreciable weed growth has taken place and the land is kept from weeds during the summer by occasional cultivation or by the use of chemical weed killers. In this way, evaporation loss is limited to that taking place at the surface of the soil.

The effectiveness of summer fallow in carrying over moisture from one year to another is somewhat variable. A conservation of perhaps one fourth of the fallow season rainfall would be expected under most conditions. Even though this efficiency of moisture storage is lower than would be desirable, yields of succeeding crops have been augmented significantly by summer fallow (see data in Table 8:5).

TABLE 8:5. *Influence of Summer Fallow in Alternate Years on the Moisture at Wheat Seeding Time and the Yield of Wheat Fallowing*[a]

Treatment	Available Water at Seeding Time (in.)	Wheat Yields
Mandan, N.D. (av. 20 years)		
Wheat after fallow	7.08	4,120 lb[b]
Wheat after wheat	2.48	2,300 lb
Hays, Kan. (av. 23 years)		
Wheat after fallow	7.96	27.3 bu
Wheat after wheat	2.90	17.4 bu
Garden City, Kan. (av. 13 years)		
Wheat after fallow	4.67	15.5 bu
Wheat after wheat	1.08	8.2 bu

[a] From Thysell (12) and Throckmorton and Meyers (13).
[b] Total plant yields.

Here, then, is a situation where an appreciable amount of moisture may be conserved by the proper handling of the land. But it is a case specific for dry land areas and should not be confused with the situation in humid regions. In the latter, summer fallowing is detrimental and unnecessary.

8:9. EVAPORATION CONTROL OF IRRIGATED LANDS

In the areas of low rainfall, or when precipitation is so distributed as to be insufficient for crops at critical times, *irrigation*, the artificial application of water to land, is common (see Fig. 8:10). In the United States, most of the irrigated land lies west of the hundredth meridian and in total embraces approximately 44 million acres. Moreover, irrigation in the more humid areas to the eastward is decidedly on the increase.

FIGURE 8:10. Typical irrigation scene. The use of easily installed siphons reduces the labor of irrigation and not only speeds up the rate of application of the water but also makes it easier to control. Note the upward capillary movement of water along the sides of the rows. [*Photo courtesy U.S. Soil Conservation Service.*]

CONTROL PRINCIPLES. The principles already discussed relating to the movement and distribution of moisture through the soil, its losses, and its plant relationships, apply just as rigidly for irrigation water as for that reaching the soil in natural ways. Although the problems inherent to liquid water, such as runoff and erosion, infiltration, percolation, and drainage, are subject to better control under irrigation, they are met and solved in much the same way as for humid region soils.

The situation regarding vapor losses is much more critical than even with contiguous soils under dry land cropping. In irrigated areas, conditions are ideal for evaporation and transpiration loss. The climate is dry, the sunshine intense, and wind velocities are often high. Furthermore, the soils are kept as moist as they are in well-drained humid region soils. These conditions are ideal for evaporation loss.

The only practical control of evaporation under these conditions is through irrigation practices. The surface of the soil should be kept only as moist as is needed for good crop production. The irrigation schedule should be such as to keep more than just the surface soil wet. Deep penetration of roots should be encouraged.

SALT ACCUMULATION. A phenomenon closely correlated with evaporation and often encountered on arid region lands under irrigation is that of the concentration of soluble salts at or near the soil surface. *Saline* and *alkali* soils may contain soluble salts in the upper horizons in sufficient quantities to inhibit the growth of many cultivated plants. Irrigation practices on these soils can improve or impair their usefulness as crop soils. By flood irrigation, some of the soluble salts can be temporarily washed down from the immediate surface. Subsequent upward movement of the water and evaporation at the soil surface may result in salt accumulation and concentration which are harmful to young seedlings (see Fig. 8:11). Although little can be done to prevent evaporation and upward capillary movement of the salt-containing water, timing and method of irrigation may prevent the salts from being concentrated in localized areas.

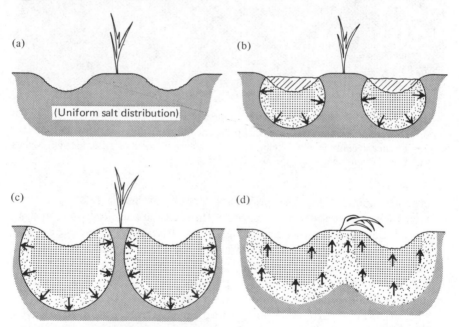

FIGURE 8:11. Effect of irrigation on salt movement in a saline soil. (a) Soil before irrigation. Notice the relatively uniform salt distribution. (b) Water running in the furrow. Some has moved downward in the soil carrying soluble salts with it. (c) Soon after irrigation. Water continues to move downward and toward center of row. (d) A week or so after irrigation. Upward movement of water by capillarity carries salts toward the surface. Salts tend to concentration in the center of the row and in some cases cause injury to plants. [*Concept obtained from Bernstein et al. (1).*]

REFERENCES

(1) Bernstein, L., et al., "The Interaction of Salinity and Planting Practice on the Germination of Irrigated Row Crops," *Proc. Soil Sci. Soc. Amer.*, **19**:240–243, 1955.

(2) Bierhuizen, J. F., and Slatyer, R. O., "Effect of Atmospheric Concentration of Water Vapor and CO_2 in Determining Transpiration–Photosynthesis Relationships of Cotton Leaves," *Agr. Meteor*, **2**:259, 1965.

(3) Denmead, O. T., and Shaw, R. H., "Availability of Soil Water to Plants as Affected by Soil Moisture Content and Meteorological Conditions," *Agron. Jour.*, **45**:385–390, 1962.

(4) Donald, L., "Don't Blame the Weather," *Comm. Fert.*, pp. 19–24, Feb. 1946.

(5) Dreibelbis, F. R., and Amerman, C. R., "How Much Topsoil Moisture Is Available to Your Crops," *Crops and Soils*, **17**:8–9, April–May 1965.

(6) Haynes, J. L., "Ground Rainfall Under Vegetative Canopy of Crops," *Jour. Amer. Soc. Agron.*, **6**:67–94, 1954.

(7) Hillel, D., *Soil and Water* (New York: Academic Press, 1971).

(8) Kelly, O. J., "Requirement and Availability of Soil Water," *Advan. in Agron.*, **6**:67–94, 1954.

(9) Kozlowski, T. T., Ed., *Water Deficits and Plant Growth,* a 3-Volume Treatise (New York: Academic Press, 1968 and 1972).

(10) Lyon, T. L., Buckman, H. O., and Brady, N. C., *The Nature and Properties of Soils* (New York: Macmillan (Inc.), 1952), p. 221.

(11) Philip, J. R., "Plant Water Relations: Some Physical Aspects," *Ann. Rev. Plant Physiol.*, **17**:245–268, 1966.

(12) Thysell, J. C., *Conservation and Use of Soil Moisture at Mandau, North Dakota,* Tech. Bull. 617 (Washington, D.C.: U. S. Dept of Agric., 1938).

(13) Throckmorton, R. I., and Meyers, H. E., *Summer Fallow in Kansas,* Bull. 293, Kansas Agr. Exp. Sta., 1941.

Chapter 9

LIQUID LOSSES OF SOIL WATER AND THEIR CONTROL

Two types of liquid losses of water from soils are recognized: (a) the downward movement of free water (percolation), which frees the surface soil and upper subsoil of superfluous moisture; and (b) the runoff of excess water over the soil surface (see Fig. 8:1). Percolation results in the loss of soluble salts (leaching), thus depleting soils of certain nutrients. Runoff losses generally include not only water but also appreciable amounts of soil (erosion). These two liquid losses of soil water with their concomitant effects will be considered in order.

9:1. PERCOLATION AND LEACHING—METHODS OF STUDY

Two general methods are available for the study of percolation and leaching losses—the use of an effective system of *tile drains* specially installed for the purpose, and the employment of *lysimeters* (from the Greek word *lysi*, meaning loosening, and *meter*, to measure). For the first method, an area should be chosen where the tile drain receives only the water from the land under study and where the drainage is efficient. The advantage of the tile method is that water and nutrient losses can be determined from relatively large areas of soil under normal field conditions.

The lysimeter method has been used most frequently to determine leaching losses. This method involves the measurement of percolation and nutrient losses under somewhat more controlled conditions. Soil is removed from the field and placed in concrete or metal tanks, or a small volume of field soil is isolated from surrounding areas by concrete or metal dividers (see Fig. 9:1). In either case water percolating through the soil is collected and measured. The advantages of lysimeters over a tile drain system are that the variations in a large field are avoided, the work of conducting the study is not so great, and the experiment is more easily controlled.

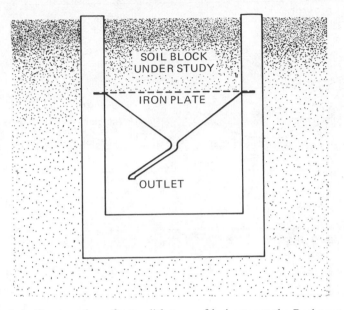

FIGURE 9:1. Cross section of monolith type of lysimeter at the Rothamsted Experiment Station, England.

9:2. PERCOLATION LOSSES OF WATER

When the amount of rainfall entering a soil becomes greater than its water-holding capacity, losses by percolation will occur. Percolation losses are influenced by the amount of rainfall and its distribution by evaporation, by the character of the soil, and by the presence of a crop.

PERCOLATION–EVAPORATION BALANCE. The relationship among precipitation, runoff, soil storage, and percolation for representative humid and subarid regions and for an irrigated arid region area is illustrated in Fig. 9:2. In the humid region, the rate of water infiltration into soils (precipitation less runoff) is commonly greater at least at some time of year than the rate of evapo-transpiration. As soon as the soil field capacity is reached, percolation into the substrata occurs.

In the example shown in Fig. 9:2, maximum percolation occurs during the winter and early spring, when evaporation is lowest. In contrast, during the summer little percolation occurs. Evapo-transpiration which exceeds the infiltration during these months results in a depletion of soil water. Normal plant growth is possible only because of moisture stored in the soil the previous winter and early spring.

The general trends in the semiarid region are the same as for the humid region. Soil moisture is stored during the winter months and utilized to

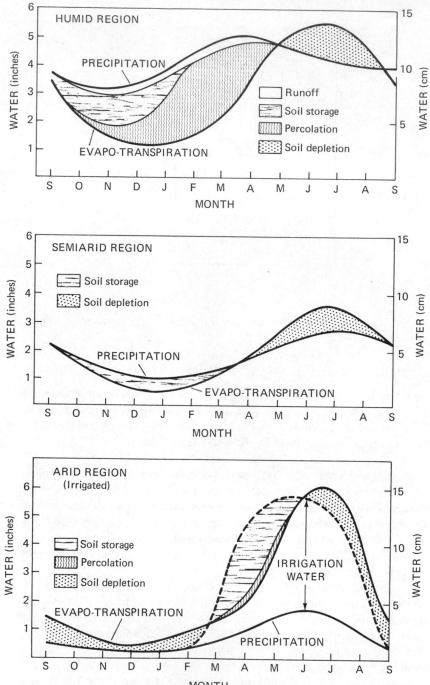

meet the moisture deficit in the summer. But because of the low rainfall, essentially no percolation out of the profile occurs. Water may move to the lower horizons but is absorbed by plant roots and ultimately lost by transpiration.

The irrigated soil in the arid region shows a unique pattern. Irrigation in the early spring along with a little rainfall provides more water than is being lost by evapo-transpiration. The soil is charged with water and some percolation may occur. During the summer, fall, and winter months this stored water is depleted since the amount being added is less than the evapo-transpiration.

BARE VERSUS CROPPED SOIL. The loss of water by percolation in cropped soil areas is generally less than that in bare areas unless surface runoff is excessively high from the bare soil. Data from lysimeter experiments in New York illustrate this point (see Fig. 9:3). The percentage of the 33 inches of precipitation lost by percolation was greater in the bare plots for both the Dunkirk and Volusia soils. In contrast, water lost by evaporation was greater in the cropped plots on which a rotation of corn, oats, wheat, and hay was grown. Percolation loss was higher on the moderately well drained Dunkirk soil than on the poorly drained Volusia. In both soils the percolation loss was probably higher than would have been expected under field conditions since in this experiment no surface runoff was permitted.

9:3. LEACHING LOSSES OF NUTRIENTS

The loss of nutrients through leaching is determined by climatic factors and soil–nutrient interactions. In regions where water percolation is high, the potential for leaching is also high (see Fig. 9:4). Such conditions exist in the United States in the humid East and in the heavily irrigated sections of the West. In these areas percolation of excess water is the rule, providing opportunities for nutrient removal. In the semiarid Great Plains, little nutrient leaching occurs since there is little percolation. Some nutrient leaching takes place in the Corn Belt, although much less than in the states to the east. In all cases, the growing of crops reduces the loss of nutrients by leaching.

FIGURE 9:2. Generalized curves for precipitation and evapo-transpiration for a humid region (*upper*), a semiarid region (*middle*), and an irrigated arid region (*lower*). Note the absence of percolation through the soil in the semiarid region. In each case water is stored in the soil. This moisture is released later when evapo-transpiration demands exceed the precipitation. In the semiarid region evapo-transpiration would likely be much higher if ample soil moisture were available. In the irrigated arid region soil the very high evapo-transpiration needs are supplied by irrigation. Soil moisture stored in the spring is utilized by later summer growth and by evaporation during the late fall and winter.

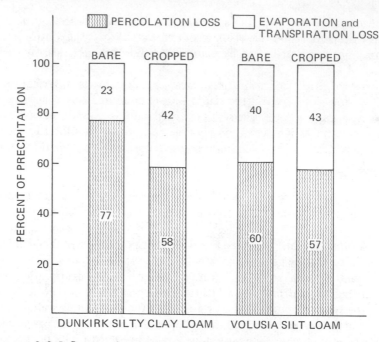

FIGURE 9:3. Influence of cropping on the percentage of annual precipitation lost by percolation through soil lysimeters and by evaporation from the soil and from leaf surface. No runoff was allowed in either soil. The Volusia silt loam is poorly drained; the Dunkirk is moderately well drained. [*Data calculated from Lyon et al.* (5).]

NUTRIENT–SOIL INTERACTION. Soil properties have a definite effect on nutrient leaching losses. Sandy soils generally permit greater nutrient loss than do clays, not only because of the lower rate of percolation in the finer textured soils but because of the nutrient-adsorbing power of these soils. For example, soluble phosphorus is quickly bound chemically in soils with appreciable amounts of iron and aluminum oxides, and 1:1-type silicate clays. Consequently, this element is lost very sparingly from medium- and fine-textured soils. Sulfates and nitrates react by anion exchange (p. 453) with iron and aluminum hydrous oxides. For this reason, sulfates and nitrates are less prone to leach from soils of the southeast which have red subsoils than from soils in similar climates when red-colored iron oxides are not so prominent.

The leaching of cations added in chemical fertilizers is affected by the soil's cation exchange capacity, those with a high capacity tending to hold the added nutrients and prevent their leaching. At the same time, such soils are often naturally high in exchangeable cations thus providing a large reservoir of nutrients, a small portion of which are continually subject to leaching.

FIGURE 9:4. Eastern part of the United States, showing regions varying in their susceptibility to leaching during the winter months. In region 1 the losses are insignificant; in region 4 they are high. Regions 2 and 3 are intermediate. [*Modified from Nelson and Uhland* (7).]

Some of the factors affecting the loss of nutrients by leaching are illustrated in Table 9:1, which includes data from lysimeter experiments at four locations. The loss of phosphorus is negligible in all cases. The leaching of nitrogen is dependent upon the crops grown and the amount of percolation occurring. The loss of cations is in rough proportion to their probable content in exchangeable form on the colloidal complex, calcium being lost in largest amounts and potassium least.

NUTRIENT LOSS AND FERTILIZATION. The rates of fertilizer additions in the lysimeter experiments reported in Table 9:1 are modest compared to those common today. For this reason, the nutrient losses given in Table 9:1 are probably considerably lower, at least for some of the nutrients, than would be expected from a heavily fertilized area, especially if water percolation were high. For example, nitrogen applications at rates of 150 pounds per acre and greater are common for corn fields. Likewise, crops such as vegetables and sugar cane are fertilized at rates which assure residual nutrients in the soil. These nutrients must react with the soil, be lost by

TABLE 9:1. *Average Annual Loss of Nutrients by Percolation Through Soils from Four Different Areas Using the Monolith Type of Lysimeters*

Runoff allowed from Illinois and Wisconsin lysimeters only.[a]

Condition	Pounds/Acre/Year					
	N	P	K	Ca	Mg	S
Rotation on a Scottish Soil (av. 6 years)						
No treatment	6.7	Trace	8.8	49.7	15.3	—
Manure and fertilizers	6.3	Trace	8.2	56.0	15.9	—
Manure, fertilizers, and lime	7.9	Trace	7.6	79.5	18.7	—
Muscatine (well drained)	76.6	—	1.2	89.8	46.3	10.5
Cowden (poorly drained)	6.1	—	0.6	10.9	3.4	1.4
Fayette Silt Loam, Wisconsin (av. 3 years)						
Fallow	—	—	1.2	37.8	18.5	2.9
Cropped to corn	—	—	0.4	14.7	5.8	0.8
New York Soils						
Dunkirk (bare)	69.0	Trace	72	398	63	53
Dunkirk (rotation)	7.8	Trace	57	230	44	43
Dunkirk (grass)	2.5	Trace	62	260	50	44
Volusia (bare)	43.0	Trace	64	323	41	35
Volusia (rotation)	6.6	Trace	57	250	27	33

[a] Data compiled by Buckman and Brady (3).

microbial action (for example, volatilization of nitrogen), or be lost through leaching.

The practice of fall application of nitrogen for crops to be planted the next spring offers considerable potential for leaching loss, especially in the humid East. Percolation losses are generally greatest in the spring prior to rapid growth. Generally nitrate losses would be expected to parallel the water lost by percolation. Similarly, some irrigated soils of the arid West which receive heavy nitrogen application prior to planting may suffer considerable loss of nitrates through leaching during the intensive irrigation period.

Because of the marked variability in climates, soils, cropping patterns, and rates and times of fertilizer application, it is not surprising to find a similar variability in the quantity of nutrients lost by leaching. In humid regions, however, the losses of at least the cations would be expected to be of the same order as the removal of these nutrients by crops (see Table 9:2). Growing concern over excessive nutrient enrichment of streams and lakes forces reconsideration of practices which encourage the leaching of nutrients from soils.

TABLE 9:2. *Comparison of the Average Annual Loss of Nutrients by Drainage from a Representative Humid Region Silt Loam, Cropped to a Standard Rotation, with the Nutrients Removed by an Average Well-Fertilized Rotation Crop*

Losses	Pounds/Acre, Annually					
	N	P	K	Ca	Mg	S
Leached from a representative silt loam	20	Trace	25	100	20	10
Removed by average rotation crop	120	22	100	40	30	20

CONTROL OF NUTRIENT LOSSES. There are two prime reasons for concern over the loss of essential elements by leaching. First is the obvious concern for keeping these nutrients in the soil, thereby supplying essential elements to crop plants. A second and equally significant reason is to keep the nutrients out of streams, rivers, and lakes. Such bodies, overly enriched with nutrients, promote the excessive growth of algae and other unwanted aquatic species.

Nutrient losses can be minimized by following a few simple rules. First, to the extent feasible, a crop should be kept on the land. Fall and winter cover crops can be grown following heavily fertilized cash crops such as corn, potatoes, and other vegetables. Second, fertilizer application rates should be no higher than can be clearly justified by scientific research trials. Third, in areas where water percolation from soils is common, the fertilizers should be added as close as feasible to the time of nutrient utilization by the crop plants. Even by following these suggestions, some leaching losses will occur. The aim should be to minimize these losses both for the sake of the farmer and for society generally.

9:4. LAND DRAINAGE

The effective utilization of soils with imperfect and poor natural drainage characteristics requires the removal of excess soil moisture. This is done by encouraging percolation through *land drainage*. The objective is to lower the moisture content of the upper layers of the soil, permitting oxygen penetration to the crop roots and the diffusion of carbon dioxide from these roots.

There is need for land drainage in select areas in most every climatic region. However, there are some regions where drainage is of more importance than others. For example, much of the flat coastal plain areas of the eastern United States require some form of drainage for best crop yields. Likewise, fine-textured soils of the Mississippi delta and of lake-laid soil areas in the glaciated regions of the United States commonly require some form of artificial drainage. Surprisingly enough, even irrigated lands of the arid West often require extensive drainage systems. This may be due to the need to remove excess salts or prevent upward seepage from upland areas.

Two general types of drainage systems are in use: *open* and *closed*. Ditch drainage represents the first group and tile drainage the second. Each has advantages which dictate their use in any given situation.

9:5. OPEN DITCH DRAINAGE

Drainage ditches are of several types. They may be deep and narrow such as those used in draining peat soils. They may be used merely as outlets for tile drains to transport the water to nearby streams or rivers. Or they may be relatively shallow and broad to permit *surface drainage* or the controlled removal of water from soils before it infiltrates the soil (see Fig. 9:5).

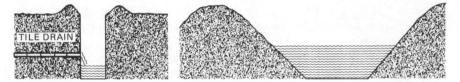

FIGURE 9:5. Two common types of drainage ditches. (*Left*) A small ditch with vertical sides such as is used to drain organic soil areas. Note the tile outlet from one side. (*Right*) A larger sloping-sided ditch of the type commonly used to transport drainage water to a nearby stream. The source of drainage water may be either tile or open drain systems.

Open drainage ditches have the advantage of large carrying capacities. For this reason they are an essential component of all drainage systems. In general, their cost per unit of water removed is relatively low. Disadvantages include cost of maintenance and interference with agricultural operations. Also, together with their banks, open ditches use valuable agricultural lands, which, if the area were underdrained, would be available for cropping.

LAND FORMING. The combination of surface drain ditches and *land forming* or smoothing is used increasingly as a means of rapid removal of surface water from soils. Depressions or ridges which prevent the movement of water to the drainage outlet are filled in or leveled off with precision using field leveling equipment (see Fig. 9:6). The resulting land configuration permits excess water to move slowly over the soil surface to the outlet ditch and thence to a natural drainage channel.

Land smoothing is a common practice in irrigated areas. Surface irrigation is made possible as is the removal of excess water by the outlet ditch. In humid areas the same methods are being put to use to remove excess

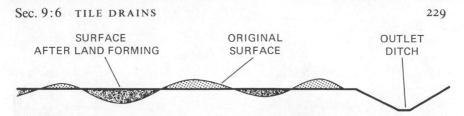

SURFACE
AFTER LAND FORMING

ORIGINAL
SURFACE

OUTLET
DITCH

FIGURE 9:6. Land surface before and after land forming or smoothing. Note that soil from the ridges fills in the depressions. Land forming makes possible the controlled movement of surface water to the outlet ditch, which transports it to a nearby natural waterway.

surface water. In combination with selectively placed tile drains, this method of orderly water removal shows great promise.

9:6. TILE DRAINS

Several means have been used to underdrain soils, the most important being the use of drainage tile. In some mineral soils, a "mole" drainage system is employed. A mole drain is simply an unwalled cylindrical channel, 3 or 4 inches in diameter, formed by pulling a pointed cylindrical metal plug through the soil at the desired depth. This leaves a compressed-wall channel through which the water can drain. While inexpensive to install, mole drains are relatively short lived and are generally not as suitable as tile drains.

NATURE AND OPERATION OF TILE DRAIN. Drainage tile are generally 12 or more inches long with a diameter varying with the amount of water to be carried. They are laid end to end in strings on the bottom of a trench of sufficient slope (see Fig. 9:7). The tile are then covered with earth, straw, or surface soil to facilitate the entrance of the water. The drainage water enters the tile through the joints, mostly from the sides and bottom (see Fig. 9:8). The tops of the joints may be covered with paper or other porous material to prevent the entrance of soil. A carefully protected outlet should be provided.

The function of a tile drain system is twofold: (a) to facilitate the removal of the superfluous water, and (b) to discharge it quickly from the land. It is most effective when the soil pores are large and numerous enough and so connected as to permit the rapid downward movement of gravity water.

SYSTEMS OF TILE INSTALLATION. Where the land possesses considerable natural drainage, the strings of tile are laid only along the depressions. This is referred to as the *natural* system of drainage, in that the tiles facilitate the quick removal of the water from the places of natural accumulation (see Fig. 9:9). But where the land is level or gently rolling, it often needs more uniform drainage, which requires installation of a *regular* system.

FIGURE 9:7. Field installation of drainage tile. (*Left*) Machine for trench digging and tile laying in operation. Note that tile are being placed in the trench bottom. (*Right*) Another field after the tile was laid. Note gravel alongside and under tile to permit ease of movement of excess drainage water into the tile. [*Photos courtesy U.S. Soil Conservation Service.*]

FIGURE 9:8. Demonstration of the saturated flow patterns of water toward a drainage tile. The water, containing a colored dye, was added to the surface of the saturated soil and drainage was allowed through the simulated drainage tile shown on the extreme right. [*Photo courtesy G. S. Taylor, Ohio State University.*]

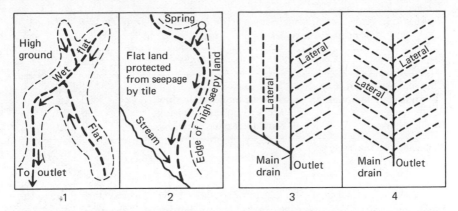

FIGURE 9:9. (*Left*) Natural (1) and interception (2) systems for laying tile drains. The object in both cases is to give efficient drainage at low cost. (*Right*) Gridiron (3) and fishbone (4) system for laying tile drains. Drainage on heavy level land can be effected only by a uniform layout system.

This may be of the *fishbone* or *gridiron* style or a modification or combination of the two. Where springs or seepage occur, a *cutoff* system must be devised.

Every regular system consists of two parts, the laterals and the main drain. The laterals are usually constructed of 4- or preferably 5-inch tile. The laterals should always enter the main at an angle of about 45 degrees. This causes a joining of the water currents with little loss of impetus and allows the more rapidly moving lateral streams to speed up the flow in the main drain. The size of the main depends on the amount and intensity of the rainfall, the acreage drained, and the slope. The main, of course, must be larger near the outlet than at any other point.

The grade, or fall, necessary for the satisfactory operation of a tile drain system varies with the system itself and the portion under consideration. The grade of the main drain may be very low, especially if the laterals deliver their water with a high velocity. In general, the grade will vary from 2 to 20 inches per 100 feet, 3 to 6 inches being most common.

The depth of the system beneath the surface of the land and the distance between laterals will vary with the nature of the soil. In most cases, however, a depth of about 3 feet is recommended. For sandy soils or poorly drained saline soils of arid regions, a depth of 4 feet or even more is commonly employed. On slowly permeable soils where the laterals are close together, a depth of $2\frac{1}{2}$ feet has been used. This is about the shallowest depth that can be used with safety on mineral soils because of the danger of tile breakage by heavy machinery.

The interval between tile lines is also reduced as the soil becomes more compact or finer in texture. On clay soils, the distance between the laterals is sometimes as low as 30 feet, although 50 to 75 feet is more common (see Table 9:3).

TABLE 9:3. *Suggested Spacing Between Tile Laterals for Different Soil and Permeability Conditions*[a]

Soil	Permeability	Spacing (ft)
Clay and clay loam	Very slow	30–70
Silt and silty clay loam	Slow to moderately slow	60–100
Sandy loam	Moderately slow to rapid	100–300
Muck and peat	Slow to rapid	50–200

[a] Modified from Beauchamp (1).

The maintenance cost of a properly installed tile drain system is low, the only special attention usually needed being the outlet. All outlets should be well protected so that the end of the tile is not loosened and the whole system endangered by clogging with sediment (see Fig. 9:10). It is well to embed the end tile in a masonry or concrete wall or block. The last 8 to 10 feet of tile may even be replaced by a galvanized iron pipe or with sewer tile, thus ensuring against damage by frost. The end tile of the system may be covered by a gate or by wire in such a way as to allow the water to flow out freely, preventing rodents from entering in dry weather.

9:7. BENEFITS OF LAND DRAINAGE

GRANULATION, HEAVING, AND ROOT ZONE. Draining the land promotes many conditions favorable to higher plants and soil organisms. Granulation is definitely encouraged. At the same time, heaving or the ill effects on perennial plants of alternate expansion and contraction due to freezing and thawing of soil water is alleviated. It is the heaving of small grain crops and the disruption of such taprooted plants as alfalfa and sweet clover that are especially feared (see Fig. 10:8). Also, by quickly lowering the water table at critical times, drainage maintains a sufficiently deep and effective root zone (see Fig. 9:11). By this means, the quantity of nutrients extractable by the plants is maintained at a higher level.

SOIL TEMPERATURE. The removal of excess water also lowers the specific heat of soil, thus reducing the energy necessary to raise the temperature of the layers thus drained (see p. 271). At the same time, surface evaporation, which has a cooling effect, may be reduced. The two effects together tend to make the warming of the soil easier. The converse of the old saying that in the spring "a wet soil is a cold soil" applies here. Good drainage is necessary if the land is to be satisfactorily cultivated and is imperative in the spring if the soil is to warm up rapidly. At this season of the year, wet soil may be 5 to 15°F cooler than a moist one.

FIGURE 9:10. This tile half-filled with sediment has lost much of its effectiveness. Such clogging can result from poorly protected outlets or inadequate slope of the tile line. In some areas of the western United States iron and manganese compounds may accumulate in the tile even when the system is properly designed. [*Photo courtesy I. B. Grass, U.S. Soil Conservation Service.*]

AERATION EFFECTS. Perhaps the greatest benefits of drainage, both direct and indirect, relate to aeration. Good drainage promotes the ready diffusion of oxygen to, and carbon dioxide from, plant roots. The activity of the aerobic organism is dependent upon soil aeration, which in turn influences the availability of nutrients, such as nitrogen and sulfur. Likewise, the change of toxicity from excess iron and manganese in acid soils is lessened if sufficient oxygen is present since the oxidized states of these elements are highly insoluble. Removal of excess water from soils is at times just as important for plant growth as is the provision of water when soil moisture is low.

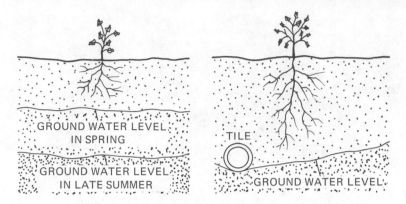

FIGURE 9:11. (*Left*) Root zone restriction that occurs when natural drainage is too slow. (*Right*) Lower water table and enlarged root zone that are developed by a properly installed tile drain.

9:8. RUNOFF AND SOIL EROSION

In most soils where the land is sloping or the soil is somewhat impermeable to water, a considerable amount of precipitation is likely to be lost by runoff. This loss has two serious consequences. First, crop plants are deprived of this water, which might otherwise have entered the soil; and second, the runoff water carries with it some of the valuable topsoil. This detachment and transfer of soil is called *erosion*.

EFFECTS OF RUNOFF. In some humid regions, loss by runoff may rise as high as 50 or 60 percent of the annual precipitation. In arid sections, it is usually lower unless the rainfall is of the torrential type, as it often is in the southwestern United States. While the loss of the water itself is deplorable, the soil erosion that accompanies it is usually even more serious. The surface soil is gradually taken away. This means not only a loss of the natural fertility but also of the nutrients that have been artificially added. Also, it is the finer portion of this soil that is always removed first, and this fraction, as already emphasized (p. 45), is highest in fertility.

Thus, the objectives of any scheme of soil management are seriously hindered. The furrow slice becomes comprised of subsoil, which usually is less fertile, and the maintenance of a satisfactory physical condition is made difficult. Many farmers, especially in the southern United States, are cultivating subsoils today unaware that the surface soils have been stolen from underfoot.

EXTENT OF EROSION. Extensive soil erosion damage occurred in the United States long before its seriousness was widely recognized. Researchers had identified areas where erosion was rampant, but public recognition of this problem came only in the 1930s. This was when the deterioration of many of our soils was emphasized by men such as H. H. Bennett, who later organized federal support for soil erosion control. His estimates indicated that nearly 50 million acres of cropland had been more or less ruined in the continental United States, another 50 million reduced to a marginal state of productivity, and erosion was making rapid progress on 200 million acres more. Excluding the 100 million acres classed as critically damaged, erosion is a serious threat to perhaps one third or even one half of the currently arable land in this country (see Fig. 9:12).

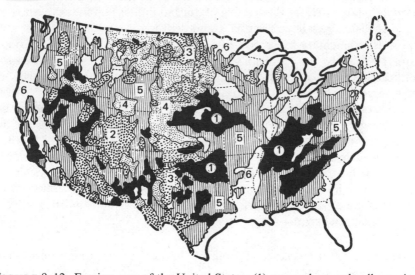

FIGURE 9:12. Erosion map of the United States: (1) severe sheet and gully erosion; (2) moderate to severe erosion of mesas and mountains; (3) moderate to severe wind erosion with some gullying; (4) moderate sheet and gully erosion with some wind action; (5) moderate sheet and gully erosion serious locally; (6) erosion rather unimportant. [*After U.S. Soil Conservation Service.*]

These estimates were made before the erosion situation in this country had been fully investigated and undoubtedly are too high. Nor was it realized that eroded lands, if not appreciably gullied, can in many cases be readily reclaimed. Nevertheless, soil erosion ravages should not be minimized. It is very serious in many sections of the United States, and special precaution must be taken to ensure its control. As late as 1967, a national survey by the U. S. Department of Agriculture showed soil erosion to be a prominent problem on about 221 million acres of cropland—more than half the total. To combat this problem, nearly 3,000 soil conservation districts have been established. Furthermore, public support is being provided to partially defray the expense of erosion-control methods.

The effects of erosion are no less serious in countries other than the United States—China, India, Syria, and Palestine, to name but a few. Moreover, erosion in ancient times evidently was a menace—in Greece, Italy, Syria, North Africa, and Persia. Perhaps the fall of empires such as of Rome was accelerated by the exhausting influence on agriculture of the washing away of fertile surface soils.

9:9. ACCELERATED EROSION—MECHANICS

Water erosion is one of the most common of geologic phenomena. It accounts in large part for the leveling of our mountains and the development of plains, plateaus, valleys, river flats, and deltas. The vast deposits that now appear as sedimentary rocks originated in this way. This is *normal* erosion. It operates slowly, yet inexorably. When erosion exceeds this normal rate and becomes unusually destructive, it is referred to as *accelerated*. This is the water action of special concern to agriculture.

Two steps are recognized in accelerated erosion—the *detachment* or loosening influence, which is a preparatory action; and *transportation* by floating, rolling, dragging, and splashing. Freezing and thawing, flowing water, and rain impact are the major detaching agencies. Raindrop splash and especially running water facilitate the carrying away of the loosened soil. In gullies, most of the loosening and cutting is due to water flow, but on comparatively smooth soil surfaces, the beating of the raindrops affects most of the detachment.

INFLUENCE OF RAINDROPS. Raindrop impact exerts three important influences: (a) it detaches soil; (b) its beating tends to destroy granulation; and (c) its splash, under certain conditions, affects an appreciable transportation of soil (see Fig. 9:13). So great is the force exerted by dashing rain that soil granules not only are loosened and detached but also they may be beaten to pieces. Under such hammering, the aggregation of a soil so exposed practically disappears. If the dispersed material is not removed by runoff, it may develop into a hard crust upon drying. Such a layer is unfavorable in many ways. For example, certain seedlings, such as beans, have difficulty in pushing through a soil crust.

The protective importance of a vegetative shield such as forest or bluegrass, or an artificial covering of straw or a stubble mulch is very great indeed. Such a protection, by absorbing the energy of rain impact, prevents the loss of both water and soil and reduces degranulation to a minimum.

TRANSPORTATION OF SOIL—SPLASH EFFECTS. In soil translocation, runoff water plays the major rôle. So familiar is the power of water to cut and carry that little more need be said regarding these capacities. In fact, so much

FIGURE 9:13. A raindrop (*left*) and the splash (*right*) that it creates when the drop strikes a wet bare soil. Such rainfall impact not only tends to destroy soil granulation and encourage sheet and rill erosion but it also effects considerable transportation by splashing. A ground cover, such as sod, will largely prevent this type of erosion. [*Photo courtesy U.S. Soil Conservation Service.*]

publicity has been given to runoff that the public generally ascribes to it all of the damage done by torrential rainfall.

However, splash transportation is under certain conditions of considerable importance (see Fig. 9:13). On a soil subject to easy detachment, a very heavy torrential rain may splash as much as 100 tons of soil an acre, some of the drops rising as high as 2 feet and moving horizontally perhaps 4 or 5 feet. On a slope or if the wind is blowing, splashing greatly aids and enhances runoff translocations of soil, the two together accounting for the total wash that finally occurs.

9:10. ACCELERATED EROSION—CAUSES AND RATE FACTORS

RATE OF EROSION. The "universal soil-loss equation" identifies the factors on which erosion losses depend (11). The equation is as follows:

$$A = RKLSCP$$

where A, the computed soil loss per unit area, is the product of the following factors:

R, rainfall
K, soil erodibility
L, slope length
S, slope gradient
C, crop management (vegetative cover)
P, erosion-control practice

Working together, these factors determine how much water enters the soil, how much runs off, and the manner and rate of its removal.

RAINFALL EFFECTS. Of the two phases, amount of *total rainfall* and its *intensity*, the latter is usually the more important. A heavy annual precipitation received in a number of gentle rains may cause little erosion, while a lower yearly rainfall descending in a few torrential downpours may result in severe damage. This accounts for the marked erosion often recorded in semiarid regions.

The *seasonal distribution* of the rainfall is also critical in determining soil erosion losses. For example, in the northern part of the United States, precipitation which runs off the land in the early spring when the soils are still frozen may bring about little erosion. The same amount of runoff a few months later, however, often carries considerable quantities of soil with it. In any climate, heavy precipitation occurring at a time of year when the soil is bare is likely to cause soil loss. Examples of such conditions are seedbed preparation time and the time after the harvesting of such crops as beans, sugar beets, and early potatoes.

NATURE OF THE SOIL. The two most significant soil characteristics influencing erosion are (a) infiltration capacity, and (b) structural stability. They are closely related. The infiltration capacity is influenced greatly by structural stability, especially in the upper soil horizons. In addition, soil texture, organic content (see Fig. 9:14), the kind and amount of swelling clays, soil depth, and the presence of impervious soil layers all influence the infiltration capacity.

The stability of soil aggregates affects the extent of erosion damage in another way. Resistance of surface granules to the beating action of rain saves soil even though runoff does occur. The marked granule stability of certain tropical clay soils high in hydrous oxides of iron and aluminum accounts for the resistance of these soils to the action of torrential rains. Downpours of a similar magnitude on temperate region clays would be disastrous.

SLOPE, TOPOGRAPHY, AND CHANNELS. The greater the *degree of slope*, other conditions remaining constant, the greater the erosion due to increased velocity of water flow. Also, more water is likely to run off. Theoretically, a doubling of the velocity enables water to move particles 64 times larger, allows it to carry 32 times more material in suspension, and makes the erosive power in total 4 times greater.

The *length* of the slope is of prime importance since the greater the extension of the inclined area, the greater is the concentration of the flooding water. For example, research in southwestern Iowa showed that doubling the length of a 9 percent slope increased the loss of soil by 2.6 times and the runoff water

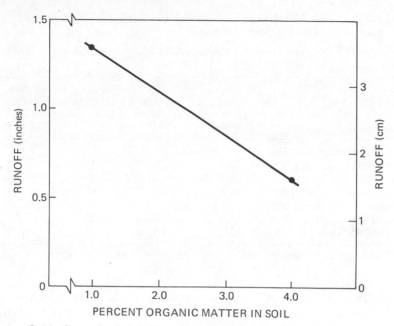
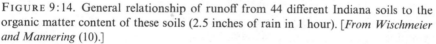

FIGURE 9:14. General relationship of runoff from 44 different Indiana soils to the organic matter content of these soils (2.5 inches of rain in 1 hour). [*From Wischmeier and Mannering* (10).]

by 1.8 times. This influence of slope is, of course, greatly modified by the size and general *topography* of the drainage area. Another modifying factor is the presence of *channels*, not only in the eroded area itself but in the water-shed. The development of such channels controls the intensity of water concentration.

EFFECT OF VEGETATIVE COVER. Forests and grass are the best natural soil protective agencies known and are about equal in their effectiveness. But their influence varies. For instance, a forest with a heavy ground cover of organic matter and a dense undergrowth is markedly superior to open woods with little organic accumulation. Also, the kind of grass, the thickness of its stand, and the vigor of its growth greatly affect erosiveness and are of great importance in control measures.

Field crops also vary in their influence. Some, such as wheat and oats, offer considerable obstruction to surface wash; others, especially intertilled crops, tend to encourage erosion (see Table 9:4). A trash or stubble mulch, where it can be utilized satisfactorily with field crops, would not only check erosion but also increase infiltration, reduce evaporation, and preserve granulation.

TABLE 9:4. *Losses from Erosion on a Shelley Silt Loam Soil at the Missouri Agricultural Experiment Station*[a]

The slope was 3.7 percent, the plots 90 feet long, and the average rainfall 40 inches. Average of 14 years.

Treatment	Runoff (% of rainfall)	Soil Lost (tons/acre/ year)	Relative Erosion (bluegrass = 1)	Time to Erode 7 in. of Soil (years)
No crop, plowed 4 in. deep and cultivated regularly	30.7	41.64	122	34
Corn grown continuously	29.4	19.72	58	50
Wheat grown continuously	23.3	10.10	30	100
Rotation; corn, wheat, and clover	13.8	2.78	8	368
Bluegrass sod continuously	12.0	0.34	1	3,043

[a] From Miller and Krusekopf (6).

EROSION-CONTROL PRACTICES. Sloping fields, especially those cultivated, erode easily if the rate of water runoff is not slowed down. Even sod and close-growing crops do not give sufficient protection under these conditions. Specific practices used to slow the water runoff include contour tillage, strip cropping, and terracing (pp. 243–45). The extent to which these practices are followed influences greatly the rate at which erosion occurs.

9:11. TYPES OF WATER EROSION

Three types of water erosion are generally recognized: *sheet, rill,* and *gully* (see Figs. 9:15 and 9:16). By *sheet* erosion, soil is removed more or less uniformly from every part of the slope. However, this type is often accompanied by tiny gullies irregularly dispersed, especially on bare land newly planted or in fallow. This is *rill* erosion. But where the volume of water is concentrated, the formation of large or small ravines by undermining and downward cutting occurs. This is called *gully* erosion. While all types may be serious, the losses due to sheet and rill erosion, although less noticeable, are undoubtedly the most important from the standpoint of field soil deterioration.

9:12. SHEET AND RILL EROSION—LOSSES UNDER REGULAR CROPPING

A number of different methods for the reduction and control of sheet and rill erosion may be utilized. For example, anything that will increase the absorptive capacity of the soil, such as surface tillage and more organic matter, will lessen the runoff over the surface. Trash and stubble mulches

FIGURE 9:15. The perched stones and pebbles shown in the picture are mute evidence of sheet erosion—the higher the pedestals, the greater the erosion. The soil under each rock has been protected from the beating action of the rain. The pedestal in the center is about 3 inches in height. [*Photo courtesy U.S. Soil Conservation Service.*]

are of great aid. The nature of the crops grown is an especially important factor.

INFLUENCE OF CROPS. Intertilled crops such as corn or potatoes may actually encourage erosion, especially if the rows possess much slope. Small grains tend to impede such loss, whereas grasses are perhaps the most effective in checking erosion. This is so well recognized that a sod is almost always recommended in places where serious erosion is likely to take place. A well-managed forest is of the same order of efficiency as grass.

The vegetative cover maintained by a given crop will vary depending upon its stage of growth. Runoff at different times of the year from plots at nine research stations in the Corn Belt illustrates this point (Table 9:5).

NUTRIENT LOSSES. The quantity of nutrients lost by erosion is high, as shown by the data in Table 9:6. These figures might be expected since the fine eroded material is higher in fertility than the whole soil. Experiments

FIGURE 9:16. Sheet and rill erosion on an unprotected slope. Serious gullying is imminent if protective measures are not instituted soon. Strip cropping or even terracing should be employed if cultivated crops are to be grown. Perhaps better, the field could be seeded and used as pasture or meadow. In any case a diversion ditch well up the slope would be advisable. [*Photo courtesy U.S. Soil Conservation Service.*]

TABLE 9:5. *Effect of Crop Stage (Increase in Ground Coverage) and Return of Crop Residues on Runoff on 678 Plot-Years of Corn at 9 Research Locations in the Corn Belt States*[a]

	Runoff (in.)	
Crop Stage	Residues Removed	Residues Returned
Rough fallow (plowing to seeding)	0.62	0.37
Seedbed (0–30 days after seeding)	0.47	0.26
Establishment (30–60 days after seeding)	0.51	0.24
Growing crop (establishment to harvest)	0.51	0.22
Residue (crop harvest to next plowing)	0.98	0.52
Total	3.09	1.61

[a] From Wischmeier and Mannering (9).

have shown organic matter and nitrogen to be up to 5 times as high in eroded material as in the original soil. Comparable figures for phosphorus and potassium are 3 and 2, respectively. Moreover, a higher proportion of the nutrients in the erodate is usable by crop plants.

Erosion losses, even on a 4 percent slope, may easily exceed the removal of nutrients by crops occupying the land. This seems to be true especially for calcium, magnesium, and potassium. The data in Table 9:6 emphasize once again the continual loss of valuable nutrients by the process of erosion. They also re-emphasize the need to keep the soil covered and the value of close-growing crops such as small grains or clovers in providing that cover. Removal of nutrients by crops is a necessary part of the production process; that by erosion loss is not.

TABLE 9:6. *Nutrients Removed Annually in the Missouri Erosion Experiment[a] Compared to the Yearly Draft of an Average Field Crop*

	Nutrients Removed (lb/acre)					
Condition	N	P	K	Ca	Mg	S
Erosion removal						
Corn grown continuously	66	18	605	220	87	17
Rotation: corn, wheat, and clover	26	8	214	85	29	6
Crop removal						
Average for standard rotation	120	22	100	40	30	20

[a] The erosion data are averages of 2 years only.

9:13. SHEET AND RILL EROSION—METHODS OF CONTROL

HIGH FERTILITY. One of the most important and at the same time the most underrated means of erosion control is the maintenance of high soil fertility and productivity. The growing of bumper crops not only gives a maximum of ground cover but supplies sufficient organic matter to aid in the main-tenance of this all-important soil constituent. The increased permeability of soils to water under such conditions is certainly a factor of major importance. Although not usually recognized as erosion-control features, the wise use of fertilizers, lime, and manure may do more to prevent soil erosion, especially sheet and rill, than some of the more obvious mechanical means of control.

MECHANICAL MEASURES. In the growing of corn and similar crops, it is important that the cultivation be across the slope rather than with it. This is called *contour tillage*. On long slopes subject to sheet and rill erosion, the fields may be laid out in narrow strips across the incline, alternating the tilled crops, such as corn and potatoes, with hay and grain. Water cannot achieve an undue velocity on the narrow strips of cultivated land, while the hay and grain definitely check the rate of runoff. Such a layout is called *strip cropping* and is the basis of much of the erosion control now advocated (see Fig. 9:17).

FIGURE 9:17. Aerial photograph of fields in Kentucky where strip cropping is being practiced. [*Photo courtesy U.S. Soil Conservation Service.*]

When the cross strips are laid out rather definitely on the contours, the system is called *contour strip cropping*. The width of the strips will depend primarily upon the degree of slope and the permeability of the land and its erodibility. Table 9:7 gives some idea of the practicable widths. Contour strip cropping often is guarded by diversion ditches and waterways between fields. When their grade is high, they should be sewn with grass to prevent underwashing.

TABLE 9:7. *Suggested Strip Widths for Contour Strip Cropping*[a]

Slope (%)	Width of Strip (ft)
2– 7	88–100
7–12	74– 88
12–18	60– 74
18–24	50– 60

[a] From Wischmeier and Smith (11).

When simpler methods of checking sheet erosion on cultivated lands are inadequate, it is advisable to resort to *terraces* constructed across the slope. These catch the water and conduct it away at a gentle grade. Opportunity is thus given for more water to soak in. The *Mangum* terrace and its modifications are now largely in use. Such a terrace is generally a broad bank of earth with gently sloping sides, contouring the field at a grade from 6 to 8 inches to 100 feet. It is usually formed by moving soil to the terrace ridge from both the lower and upper sides. The interval between the successive embankments depends on the slope and erodibility of the land.

Since the terrace is low and broad, it may be cropped without difficulty and offers no obstacle to cultivating and harvesting machinery. It wastes little or no land and is quite effective if properly maintained. Where necessary, waterways are sodded. Such terracing is really a more or less elaborate type of contour strip cropping.

9:14. Gully Erosion and Its Control

Small gullies, while at first insignificant, soon enlarge into unsightly ditches or ravines. They quickly eat into the land above, exposing its subsoil and increasing the sheet and rill erosion, already undesirably active (see Fig. 9:16). If small enough, such gullies may be plowed in and seeded to grass.

When the gully erosion is too active to be thus checked and the ditch is still small, dams of rotted manure or straw at intervals of 15 or 20 feet are very effective. Such dams may be made more secure by strips of wire netting staked below them. After a time, the ditch may be plowed in and the site of the gully seeded and kept in sod permanently. It therefore becomes a *grassed waterway*, an important feature of most successful erosion control systems.

With moderately sized gullies, larger dams of various kinds may be utilized. These are built at intervals along the channels. Piles of brush or woven wire tied to the soil by stakes driven into the ground make effective dams.

With very large gullies, dams of earth, concrete, or stone are often successfully used. Most of the sediment is deposited above the dam and the gully is slowly filled. The use of semipermanent check dams, flumes, and paved channels are also recommended on occasion. The only difficulty with engineering features at all extensive is that the cost may exceed the benefits derived or even the value of the land to be served. They thus may prove uneconomical.

9:15. Wind Erosion—Its Importance and Control

Wind erosion, although most common in arid and semiarid regions, occurs to some extent in humid climates as well. It is essentially a dry weather phenomenon and hence creates a moisture problem. All kinds of soils and soil materials are affected and at times their more finely divided portions are carried to great heights and for hundreds of miles.

In the great dust storm of May 1934, which originated in western Kansas, Texas, Oklahoma, and contiguous portions of Colorado and New Mexico, clouds of powdery debris were carried eastward to the Atlantic seaboard and even hundreds of miles out over the ocean. Such activity is not a new phenomenon but has been common in all geologic ages. The wind energy that gave rise to the wind-blown deposits now so important agriculturally in the United States and other countries belongs in this category (see Fig. 11:12).

The destructive effects of wind erosion are often very serious. Not only is the land robbed of its richest soils, but crops are either blown away or left to die with roots exposed or are covered up by the drifting debris. Even though the blowing may not be great, the cutting and abrasive effects upon tender crops is often disastrous. The area in the United States subject to greatest wind action is the Great Plains. Here the mismanagement of plowed lands and the lowered holding power of the range grasses due to overgrazing have greatly encouraged wind action. In dry years, as experience has shown, the results have been most deplorable. Figure 9:18 shows relative wind erosion forces at four locations in the United States.

AREAS OF EXCESSIVE WIND EROSION. Two great *dust bowls* (see Fig. 9:11) exist in the United States. One, the larger, occupies much of western Kansas, Oklahoma, and Texas, extending into southeastern Colorado and eastern New Mexico. The other, irregular and somewhat scattered, lies across the centers of three states—Nebraska and the two Dakotas. Eastern Montana is also affected. Curiously enough, areas of severe water erosion lie contiguous to these so-called dust bowls, while gully and sheet erosion, caused by torrential rainfall, are not at all uncommon within their borders.

Possibly 12 percent of the continental United States is somewhat affected by wind erosion, 8 percent moderately so, and perhaps 2 or 3 percent greatly. Although most of the damage is confined to regions of low rainfall, some serious wind erosion occurs in humid sections. Sand dune movement is a good example. More important agriculturally, sandy soils and peat soils cultivated for many years are often affected by the wind. The drying out of the finely divided surface layers of these soils leaves them both extremely susceptible to wind erosion.

MECHANICS OF WIND EROSION. As was the case for water erosion, the loss of soil by wind movement involves two processes: (a) detachment, and (b) transportation. The abrasive action of the wind results in some detachment of tiny soil grains from the granules or clods of which they are a part. When the wind is laden with soil particles, however, its abrasive action is greatly increased. The impact of these rapidly moving grains dislodges

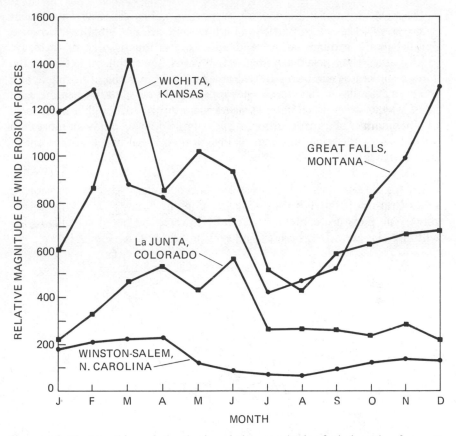

FIGURE 9:18. Monthly variation in the relative magnitude of wind erosion forces at four locations in the United States. One would expect little soil movement in Winston-Salem, N.C., not only because of the low wind erosion forces but because of the probable high soil moisture content. [*From Skidmore and Woodruff* (8).]

other particles from soil clods and aggregates. These particles are now ready for movement.

The transportation of the particles once they are dislodged takes place in several ways. The first and most important is that of *saltation*, or movement of soil by a short series of bounces along the surface of the ground. The particles remain fairly close to the ground as they bounce, seldom rising more than a foot or so. Depending on the conditions, this process may account for 50 to 75 percent of the total movement.

Saltation also encourages *soil creep*, or the rolling and sliding along the surface of the larger particles. The bouncing particles carried by saltation strike the large aggregates and speed up their movement along the surface. Soil creep may account for 5 to 25 percent of the total movement.

The most spectacular method of transporting soil particles is by movement in *suspension*. Here, dust particles of a fine sand size and smaller are moved parallel to the ground surface and upward. Although some of them are carried at a height no greater than a few yards, the turbulent action of the wind results in others being carried miles upward into the atmosphere and hundreds of miles horizontally. They return to the earth only when the wind subsides and when precipitation washes them down. Although it is the most obvious manner of transportation, suspension seldom accounts for more than 40 percent of the total and is generally no more than about 15 percent of the movement.

FACTORS AFFECTING WIND EROSION. Susceptibility to wind erosion is related rather definitely to the moisture content of soils. Wet soils do not blow. The moisture content is generally lowered by hot dry winds to the wilting point and lower before wind erosion takes place (Fig. 9:19).

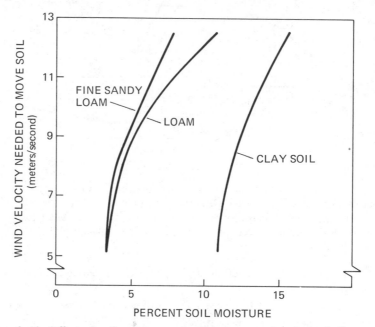

FIGURE 9:19. Effect of soil moisture and soil texture on the wind velocity required to move soils. Soil moisture and soil texture play a highly significant role in determining the susceptibility of soils to wind erosion. [*Redrawn from Bisal and Hsiek* (2).]

Other factors known to influence wind erosion are (a) wind velocity and turbulence, (b) soil surface condition, (c) soil characteristics, and (d) the nature and orientation of the vegetation. Obviously, the rate of wind movement, especially gusts having greater than average velocity, will influence erosion. Tests have shown that wind speeds of about 12 miles per hour are required to initiate soil movement. At higher wind speeds the soil movement is proportional to the wind velocity cubed. Thus, the quantity of soil carried by wind goes up very rapidly as speeds above 12 miles per hour are reached.

Wind turbulence will also influence the transporting capacity of the atmosphere. Although the wind itself has some direct influence in picking up fine soil, the impact of wind-carried particles on those not yet disengaged is probably more important.

Wind erosion is less severe where the soil surface is rough. This roughness can be obtained by proper tillage methods, which leave large clods or ridges on the soil surface. Leaving stubble mulch is perhaps an even more effective way of reducing wind-borne soil losses.

In addition to moisture content, several other soil characteristics influencing wind erosion are (a) mechanical stability of dry soil clods and aggregates, (b) the presence of a stable soil crust, and (c) bulk density and size of erodible soil fractions. The clods must be resistant to the abrasive action of wind-carried particles. If a soil crust resulting from a previous rain is present, it too must be able to withstand impact without deteriorating. The importance of cementing agents here is quite apparent.

Soil particles of a given size and bulk density are apparently more susceptible to erosion than are others. Apparently those particles about 0.1 mm in diameter are most erodible, those larger or smaller in size being less susceptible to movement. Particles about 0.1 mm in size apparently are responsible to a degree for that movement of larger or smaller particles which does occur. By saltation, the most erodible particles bounce against larger particles, causing surface creep, and against smaller dust particles, resulting in movement in suspension.

The presence on the land of vegetation or of a stubble mulch will reduce wind erosion hazards, especially if the rows run perpendicular to the prevailing wind direction. This effectively presents a barrier to wind movement. In addition, plant roots help bind the soil and make it less susceptible to wind damage.

CONTROL OF WIND EROSION. Factors just discussed give clues as to methods of reducing wind erosion. Obviously, if the soil can be kept moist there is little danger of wind erosion. A vegetative cover also discourages soil blowing, especially if the plant roots are well established. In dry-farming areas, however, sound moisture-conserving practices require summer fallow on some of the land, and hot, dry winds reduce the moisture in the soil surface. Consequently, other means must be employed on cultivated lands of these areas.

By roughing the soil surface, the wind velocity can be decreased and some of the moving particles trapped. Stubble mulch has proved to be effective

in this manner. Tillage to provide for a cloddy surface condition should be
at right angles to the prevailing winds. Likewise, strip cropping and alternate
strips of cropped and fallowed land should be perpendicular to the wind.
Barriers such as tree shelterbelts (see Fig. 9:20) are effective in reducing
wind velocities for short distances and for trapping drifting soil.

FIGURE 9:20. Shrubs and trees make good wind breaks and add beauty to a North
Dakota farm homestead. [*Photo courtesy U.S. Soil Conservation Science.*]

   In the case of the blowing of sands, sandy loams, and cultivated peat
soils in humid regions, various control devices are used. Windbreaks and
tenacious grasses and shrubs are especially effective (see Fig. 13:3). Picket
fences and burlap screens, while less efficient as windbreaks than such trees
as willows, are often preferred because they can be moved from place to
place as crops and cropping practices are varied. Rye, planted in narrow
strips across the field, is sometimes used on peat lands. All of these devices
for wind erosion control, whether applied in arid or humid regions and
whether vegetative or purely mechanical, are, after all, but phases of the
broader problem of soil moisture control.

9:16. CONSERVATION TREATMENT NEEDS IN THE UNITED STATES

The 1967 inventory of soil and water conservation needs (9) shows about 64 percent of the cropland in the United States (278 million acres) in need of some kind of conservation treatment (see Table 9:8). This treatment may vary from good management of crop residues or annual crops to more intensive management practices such as those involving close growing crops and the use of engineering features such as terraces and diversion ditches. About 4 percent of the cropland is not suited for cultivation and should be converted to forests or grasslands.

TABLE 9:8. *Estimate of the Adequacy of Conservation Treatments on Private Lands Used for Different Purposes in the United States in 1967*[a]

	Millions of Acres		
	Adequately Treated	Additional Treatment Needed	No Treatment Feasible
Cropland	160	278	—
Pastureland	28	71	1
Rangeland	112	251	19
Forestland	178	285	—
Forest Grazed	29	107	—
Other land	40	15	—

[a] From U. S. Department of Agriculture (9).

Conservation treatments are also needed for most of the pastures, rangeland, and forestland in the United States (see Table 9:8). Improvement of existing cover and the establishment of either grassland or forest species is most important on these lands. Treatment needs are related mostly to the conservation management and control of water, both that which enters the soil and that which is lost through surface runoff.

9:17. SUMMARY OF SOIL MOISTURE REGULATION

In the effective control of soil water, four more or less closely related phases must be considered. They are (a) weed transpiration, (b) surface evaporation, (c) percolation and leaching, and (d) runoff and erosion.

Weed transpiration and surface evaporation, by drawing on the available soil water, compete directly and seriously with the crop. Cultivation is the remedy commonly recommended for weeds, although herbicides are being utilized more and more in connection with a suitable rotation. Surface evaporation is subject to little control except through the use of a plastic organic mulch.

Percolation losses are serious because of the nutrients carried away and because of the waste of the water. In some cases, however, the removal of such water is necessary in order to promote soil aeration and to encourage other desirable conditions. Keeping crops on the land as much as possible is the only practicable means of reducing leaching losses of nutrients.

The detrimental influence of runoff over the surface is due to the bodily removal of soil as well as the loss of the water itself. Control measures involve organic matter maintenance to encourage infiltration, an effective vegetative cover, and mechanical practices such as contour strip cropping to allow an orderly removal of excess surface water.

REFERENCES

(1) Beauchamp, K. H., "The Drainage—Its Installation and Upkeep," *The Yearbook of Agriculture* (Water) (Washington, D.C.: U. S. Department of Agriculture, 1955), p. 513.

(2) Bisal, F., and Hsiek, J., "Influence of Moisture on Erodibility of Soil by Wind," *Soil Science*, **102**:143–46, 1966.

(3) Buckman, H. O., and Brady, N. C., *The Nature and Properties of Soils*, 7th ed. (New York: Macmillan, Inc., 1969).

(4) Grass, L. B., "Tile Clogging by Iron and Manganese in Imperial Valley, California," *Jour. Soil Water Conservation*, **24**:135–38, 1969.

(5) Lyon, T. L., et al., *Lysimeter Experiments III* (Memoir 134, 1930) and *IV* (Memoir 194, 1936), Cornell Univ. Agr. Exp. Sta.

(6) Miller, M. F., and Krusekopf, H. H., *The Influence of Systems of Cropping and Methods of Culture on Surface Runoff and Soil Erosion.* Res. Bull. 177, Missouri Agr. Exp. Sta., 1932.

(7) Nelson, L. B., and Uhland, R. E., "Factors that Influence Loss of Fall-Applied Fertilizers and Their Probable Importance in Different Sections of the United States." *Soil Sci. Soc. Amer. Proc.*, **19**:492–96, 1955.

(8) Skidmore, E. L., and Woodruff, N. P., *Wind Erosion Forces in the United States and Their Use in Predicting Soil Loss*, Agr. Handbook 346 (Washington, D.C.: U. S. Department of Agriculture, 1968).

(9) U. S. Department of Agriculture, *National Inventory of Soil and Water Conservation Needs* (*Basic Statistics*) (Washington, D.C.: USA Statistical Bulletin No. 461, 1971).

(10) Wischmeier, W. H., and Mannering, J. V., "Effect of Organic Matter Content of the Soil on Infiltration," *Jour. Soil Water Conservation*, **20**:150–52, 1965.

(11) Wischmeier, W. H., and Smith, D. W., *Predicting Rainfall—Erosion Losses from Cropland East of the Rocky Mountains*, Agr. Handbook 282 (Washington, D.C.: U. S. Department of Agriculture, 1965).

Chapter 10

SOIL AIR AND
SOIL TEMPERATURE

As indicated in Chapter 1, approximately one half of the total volume of a representative mineral surface soil is occupied by solid materials. The remaining nonsolid or pore space is occupied by water and gases. The previous three chapters (7, 8, and 9) have placed major emphasis on the soil moisture phase. They have, nevertheless, constantly emphasized the interrelationship of soil air and soil water; one cannot be affected without changing the other.

In spite of the fact that this reciprocal relationship already has been duly stressed, there are certain aspects of soil aeration which merit separate and special consideration. This is true particularly concerning certain plant processes.

It is logical also to include in this chapter another important physical property—the temperature of the soils. Although subject to little direct control under field conditions, soil temperature is closely related to soil moisture and soil air and exerts a tremendous influence upon microorganisms and higher plants.

10:1. SOIL AERATION DEFINED

In keeping with the edaphological viewpoint of this text, soil aeration will be considered in relation to *plant growth*. Thus, *a well-aerated soil is one in which gases are available to growing aerobic organisms (particularly higher plants) in sufficient quantities and in the proper proportions to encourage optimum rates of the essential metabolic processes of these organisms.*

A soil in which aeration is considered satisfactory must have at least two characteristics. First, sufficient space devoid of solids and water should be present. Second, there must be ample opportunity for the ready movement of essential gases into and out of these spaces. Water relations largely control the *amount of air space* available, but the problem of adequate *air exchange* is probably a much more complicated feature. The supply of oxygen, a gas constantly being used in biological reactions, must be continually renewed. At the same time, the concentration of CO_2, the major product of these or similar reactions, must not be allowed to build up excessively in the air spaces.

253

Two biological reactions largely account for these O_2 and CO_2 changes. They are (a) the respiration of higher-plant roots, and (b) the aerobic decomposition of incorporated organic residues by microorganisms. Although differing in many respects, these processes are surprisingly similar with regard to the gases involved. This similarity makes it possible to express both reactions in terms of the following very general equation:

$$(C) + O_2 \rightarrow CO_2$$

Both processes result in the oxidation of carbonaceous compounds, oxygen being used up and CO_2 being evolved. Either a deficiency of O_2 or an excess of CO_2 would thus tend to cut down the rate of these reactions. In fact, both inhibitions commonly operate concurrently.

10:2. SOIL AERATION PROBLEMS IN THE FIELD

Under actual field conditions there are generally two situations which may result in poor aeration in soils: (a) when the moisture content is excessively high, leaving little or no room for gases; and (b) when the exchange of gases with the atmosphere is not sufficiently rapid to keep the concentration of soil gases at desirable levels. The latter may often occur even when sufficient *total air space* is available.

EXCESS MOISTURE. In the first case, essentially a waterlogged condition is established. This may be quite temporary, but, nevertheless, often seriously affects plant growth. Such a situation is frequently found on poorly drained, fine-textured soils having a minimum of macropores through which water can move rapidly. It also occurs in soils normally well drained if the rate of water supply to the soil surface is sufficiently rapid. A low spot in a field or even a flat area in which water tends to stand for a short while is a good example of this condition.

Such complete saturation of the soil with water can be disastrous for certain plants in a short time, a matter of a few hours being critical in some cases. Plants previously growing under conditions of good soil aeration are actually more susceptible to damage from flooding than are plants growing on soils where poor aeration has prevailed from the very start.

Prevention of this type of poor aeration requires the rapid removal of excess water either by land drainage or by controlled runoff. Because of the large volume of small capillary pores in some soils, even these precautions leave only a small portion of the soil volume filled with air soon after a rain. This often makes the artificial drainage of heavy soils surprisingly effective.

GASEOUS INTERCHANGE. The seriousness of inadequate *interchanges* of gases between the soil and the free atmosphere above it is dependent primarily on two factors: (a) the rate of biochemical reactions influencing the soil gases, and (b) the actual rate at which each gas is moving into or out of the soil. Obviously, the more rapid the usage of O_2 and the corresponding release of CO_2, the greater will be the necessity for the exchange of gases. Factors markedly affecting these biological reactions, such as temperature, organic residues, etc., undoubtedly are of considerable importance in determining the air status of any particular soil (see pp. 256–58).

The exchange of gases between the soil and the atmosphere above it is facilitated by two mechanisms: (a) *mass flow*, and (b) *diffusion*. Mass flow of air is apparently due to pressure differences between the atmosphere and the soil air and is relatively unimportant in determining the total exchange that occurs. However, in the upper few inches of soil diurnal changes in soil temperature may result in mass flow of some significance. The extent of mass flow is determined by such factors as soil and air temperatures, barometric pressure, and wind movements.

Apparently most of the gaseous interchange in soils occurs by diffusion. Through this process each gas tends to move in a direction determined by its own partial pressure (see Fig. 10:1). The partial pressure of a gas in a mixture is simply the pressure this gas would exert if it alone were present in the volume occupied by the mixture. Thus, if the pressure of air is 1 atmosphere, the partial pressure of oxygen, which makes up about 21 percent of the air by volume, is approximately 0.21 atmosphere.

Diffusion allows extensive movement from one area to another even though there is no overall pressure gradient. Thus, even though the total

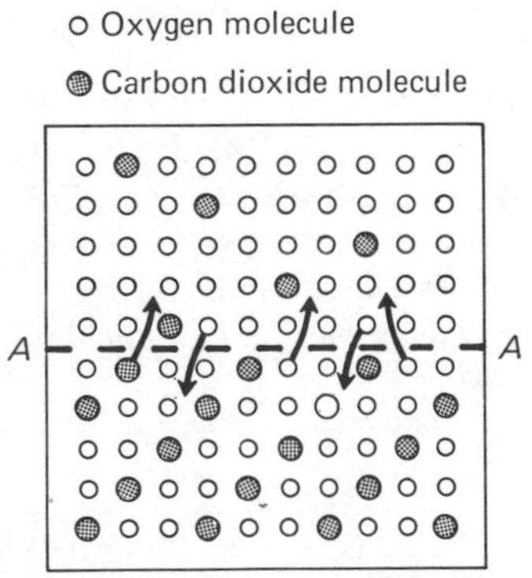

FIGURE 10:1. How the process of diffusion takes place. The total gas pressure is the same on both sides of boundary $A–A$. The partial pressure of oxygen is greater, however, in the top portion of the container. Therefore, this gas tends to diffuse into the lower portion, where fewer oxygen molecules are found. The carbon dioxide molecules, on the other hand, move in the opposite direction, owing to the higher partial pressure of this gas in the lower half. Eventually equilibrium will be established when the partial pressures of O_2 and CO_2 are the same on both sides of the boundary.

soil–air pressure and that of the atmosphere may be the same, a higher concentration of O_2 in the atmosphere will result in a net movement of this particular gas into the soil. An opposite movement of CO_2 and water vapor is simultaneously taking place since the partial pressures of these two gases are generally higher in the soil air than in the atmosphere. A representation of the principles involved in diffusion is given in Fig. 10:1.

In addition to being affected by partial pressure differences, diffusion seems to be directly related to the volume of pore spaces filled with air. On heavy-textured topsoils, especially if the structure is poor, and in compact subsoils, the rate of gaseous movement is seriously slow. Moreover, such soils allow very slow penetration of water into the surface layer. This prevents the rapid replacement of air high in CO_2 and the subsequent inward movement of atmospheric air after a heavy rain or after irrigation.

OXYGEN DIFFUSION RATES. Perhaps the best measurement of the aeration status of a soil is the *oxygen diffusion rate (ODR)*. This determines the rate at which oxygen can be replenished if it is used by respiring plant roots or replaced by water (10).

Figure 10:2 is a graph showing how ODR decreases with soil depth. Even when atmospheric air with 21 percent O_2 was used, the ORD rate at 38 inches was less than half that at 4.5 inches. When a lower oxygen concentration was used, the ODR decreased even more rapidly with depth. Note that root growth ceased when the ODR dropped to about $20 \text{ g} \times 10^{-8}/\text{cm}^2/\text{min}$.

10:3. COMPOSITION OF SOIL AIR

In Table 10:1 data are given showing the composition of soil air at various locations as compared to that of the atmosphere itself. These data are for topsoils only and consequently do not afford complete information

TABLE 10:1. *Composition of Soil Air at Various Locations Compared to That of the Atmosphere*[a]

Location	Percentage by Volume		
	O_2	CO_2	N_2
Soil air			
England	20.65	0.25	79.20
Iowa	20.40	0.20	79.40
New York	15.10	4.50	81.40
Atmospheric air			
England	20.97	0.03	79.0

[a] Compiled by Lyon et al. (5).

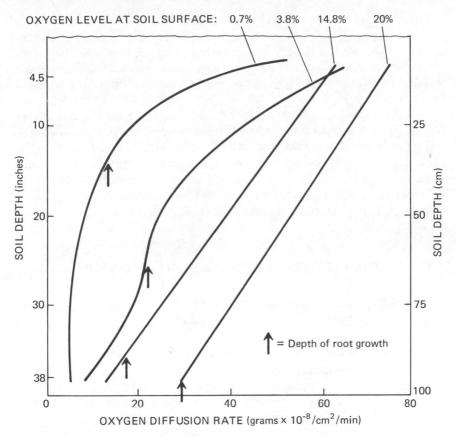

FIGURE 10:2. Effect of soil depth and oxygen concentration at the surface on the oxygen diffusion rate (ODR). Arrows indicate snapdragon root penetration depth. Even with a 20 percent O_2 level at the surface, the diffusion rate at 38 inches is less than half that at the surface. Note that when the ODR drops to about $20 \times 10^{-8}/cm^2/min$, root growth ceases. [*Redrawn from Stolzy et al. (9).*]

as to the variabilities encountered especially in the subsoil. For example, under conditions of extremely poor aeration O_2 contents of soil air as low as 1 percent have been found (1). They do illustrate, however, the higher CO_2 concentration and lower O_2 content of soil air as compared to that of atmosphere. Also, there is a general inverse relationship between oxygen and carbon dioxide contents, that of O_2 decreasing as the CO_2 increases.

Although the actual differences in CO_2 contents are not impressive, comparatively speaking they are significant. Thus, when the soil air contains only 0.25 percent CO_2, this gas is *more than 8 times* as concentrated as it is in the atmosphere. In cases where the CO_2 content becomes as high as 10 percent, there is more than 300 times as much present as is found in the air above.

Compared to the atmosphere, soil air usually is much higher in water vapor, being essentially saturated except at or very near the surface of the soil. This fact has already been stressed in connection with the movement of water. Also the concentration of gases such as methane and hydrogen sulfide, which are formed by organic matter decomposition, is somewhat higher in soil air.

In addition to the free soil air already discussed, the small quantities of certain gases dissolved in the soil moisture must be considered. Also, the soil colloids are thought to hold at their surfaces small quantities of the various gases by physical adsorption. Undoubtedly, gases held in these two ways are of some importance in the study of soil aeration. Thus, oxygen so held may be utilized in oxidation reactions, while dissolved carbon dioxide is of universal importance, especially concerning soil pH and the solubility of soil minerals.

10:4. FACTORS AFFECTING THE COMPOSITION OF SOIL AIR

The composition of soil air is largely dependent upon the amount of *air space available* together with the rates of *biochemical reactions* and *gaseous interchange*.

The total porosity of soils is determined chiefly by the bulk density (see p. 53). This, in turn, is related to factors such as soil texture and structure and soil organic matter. Not all the pore space in field soils is filled with air, some being occupied by water. In both poorly drained and well-drained soils, a high proportion of the pore space is filled with water immediately after a heavy rain or irrigation. If this situation persists for a period of time, only small amounts of oxygen will be available for plants.

The concentrations of both O_2 and CO_2 are definitely related to the biological activity in the soil. Microbial composition of organic residues apparently accounts for the major portion of the CO_2 evolved. Incorporation of large quantities of manure, especially if moisture and temperature are optimum, will alter the soil–air composition appreciably. Respiration by higher plants and the continuous contribution of their roots to the organic mass by sloughage are also significant processes. These influences are well illustrated by data from Russell and Appleyard (7) in Fig. 10:3. The effects of a growing crop and especially that of organic manures are clearly shown.

SUBSOIL VERSUS TOPSOIL. As might be expected, subsoils are usually more deficient in oxygen than are topsoils. The total pore space as well as the average size of the pores is generally much less in the deeper horizons. Boynton (1), working on orchard sites, determined the aeration status of several soils with depth. The average oxygen contents for the months of May and June in two of the soils studied are shown in Fig. 10:4.

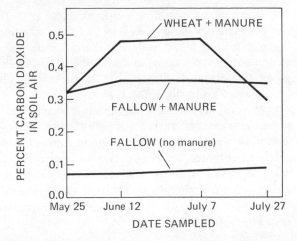

FIGURE 10:3. Carbon dioxide content of soil air at four different sampling dates, showing the effects of manure and cropping. Air samples were taken from the topsoil only. An unaccountable variation occurred in the data of June 10; hence this sampling data was omitted. [*From Russell and Appleyard* (7).]

In both cases the oxygen percentage in the soil air decreased with depth, the rate of decrease being much more rapid with the heavier soil. Throughout the depth sampled, the lighter-textured sandy loam contained a much higher oxygen percentage. In both cases the decrease in O_2 content could be largely, but not entirely, explained by a corresponding increase in CO_2 content. For example, CO_2 percentages as high as 14.6 were recorded for the lower horizons of the heavier-textured soil.

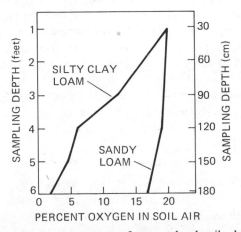

FIGURE 10:4. Average oxygen content of two orchard soils during the months of May and June 1938. Normal root growth and functions would occur at much deeper levels on the sandy soil. [*From Boynton* (1).]

SEASONAL DIFFERENCES. As would be expected, there is marked seasonal variation in the composition of soil air. Most of this variation can be accounted for by soil moisture and soil temperature differences. High soil moisture tends to favor a low oxygen and high CO_2 level in the soil air. In temperate regions, this situation often prevails in the winter and late spring when the soil moisture is generally highest.

Because soils are normally drier during the summer months, opportunity for gaseous exchange is greatest during this period. This results in relatively high O_2 and low CO_2 levels. Some exceptions to this rule may be found, however. Since high summer temperatures also encourage rapid microbiological release of CO_2, a given soil containing easily decomposable organic matter may have higher CO_2 levels in the summer than in the winter. The dependence of soil–air composition on soil moisture and soil temperature is of vital significance.

10:5. EFFECTS OF SOIL AERATION ON BIOLOGICAL ACTIVITIES

EFFECTS ON MICROORGANISMS. Probably the most apparent effect of poor soil aeration on microbiological processes is a decrease in the rate of organic matter oxidation. This decrease seems to be associated more with a lack of O_2 than with an excess of CO_2. The slow rate of decay of plant residues in swampy areas is a somewhat exaggerated example of how a lack of oxygen prevents rapid organic matter decomposition.

All aerobic organisms are unable to function properly in the absence of gaseous oxygen. For example, special purpose bacteria such as those responsible for the oxidation of nitrogen and sulfur are relatively ineffective in poorly aerated soils. This is also true for the symbiotic nitrogen fixers and for some of the nonsymbiotic (*Azotobacter*) groups (see p. 439).

The microorganism population, therefore, is drastically affected by soil aeration. Only the anaerobic and facultative organisms function properly under poor aeration conditions. This is possible because of their ability to utilize combined oxygen. Their activities consequently yield reduced forms of certain elements such as iron and manganese which are often toxic to higher plants. Organic toxins may also develop. Further attention will be given these effects later.

EFFECTS ON ACTIVITIES OF HIGHER PLANTS. Higher plants are adversely affected in at least four ways by conditions of poor aeration: (a) the growth of the plant, particularly the roots, is curtained; (b) the absorption of nutrients is decreased as is (c) the absorption of water; and (d) the formation of certain inorganic compounds toxic to plant growth are favored by poor aeration.

PLANT GROWTH. The ability of different plant species to grow in soils with low soil–air porosity varies greatly (see Fig. 10:5). Tomatoes, for example, require high air porosities (up to 20–30 percent) for best growth. In contrast, timothy, if well supplied with nitrogen, can grow with a very low air porosity, and rice grows normally submerged in water. Furthermore, the tolerance of a given plant to low porosity may be different for seedlings than for rapidly growing plants. The tolerance of red pine to restricted drainage during its early development and its poor growth or even death on the same site at later stages is a case in point.

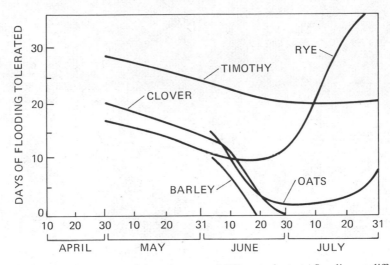

FIGURE 10:5. Comparison of the tolerance of different plants to flooding at different stages of plant growth under Finnish conditions. [*From Saukko* (8).]

In spite of these wide variations in soil–air porosity limitation, soil physicists generally consider that if the soil porosity is reduced below 10–12 percent, soil oxygen renewal is extremely slow and most plants are likely to suffer.

The abnormal effect of insufficient aeration on root development has been observed often. With root crops such as carrots and sugar beets the influence is most noticeable. Abnormally shaped roots of these plants are common on compact, poorly aerated soils. Even with sod crops the presence of an impervious soil layer generally results in restricted growth, particularly of the smaller roots. An example of this is shown in Fig. 10:6, in which is pictured the root system of rape plants growing on a compacted soil. The lack of extensive root penetration within the impervious layer is noteworthy. Such restricted root systems can hardly be expected to effectively absorb sufficient moisture and nutrients for normal plant growth.

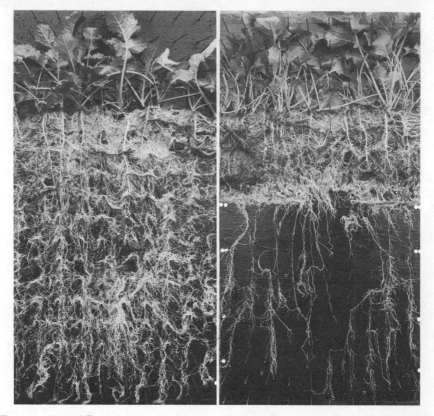

FIGURE 10:6. Effect of soil compaction on the development of the root system of the rape plant. (*Left*) Subsoil loosened before planting. (*Right*) Compacted layer at plow depth not loosened. [*From work of H. C. de Roo, Connecticut Agricultural Experiment Station.*]

Apparently, different levels of soil oxygen are required for the various functions of roots. Boynton and coworkers (2) found that apple tree roots required at least 3 percent oxygen in the soil air to subsist, while 5 to 10 percent was sufficient for the growth of existing roots. At least 12 percent O_2 was required for new root growth. Their results clearly emphasize the complexity of soil aeration problems.

Researchers have found that the *oxygen diffusion rate* (ODR) is of critical importance to growing plants. For example, the growth of roots of most plants cease when the ODR drops to about $20 \text{ g} \times 10^{-8}/\text{cm}^2/\text{min}$. Top growth is normally satisfactory so long as the ODR remain above $30\text{–}40 \text{ g} \times 10^{-8}/\text{cm}^2/\text{min}$. In Table 10:2 are found some field measurements of ODR along with comments about the condition of the plants. Note the sensitivity of sugar beets to low ODR and the general tendency for difficulty if the ODR gets below the critical level.

Naturally some plants are affected more by low ODR than are others. Grasses tend to be more tolerant of low diffusion rates than legumes. Sugar beets and alfalfa require higher rates than Ladino clover.

TABLE 10:2. *Relationship Between Oxygen Diffusion Rates (ODR) and the Condition of Different Plants*[a]

When the ODR drop below about 40 g $\times$ 10^{-8}/cm^2/min, the plants appear to suffer. Sugar beets require high ODR even at 30 cm depth. ODR values in g $\times$ 10^{-8}/cm^2/min.

Plant	Soil Type	ODR at Three Soil Depths			Remarks
		10 cm	20 cm	30 cm	
Broccoli	Loam	53	31	38	Very good growth
Lettuce	Silt loam	49	26	32	Good growth
Beans	Loam	27	27	25	Chlorotic plants
Sugar beets	Loam	58	60	16	Stunted tap root
Strawberries	Sandy loam	36	32	34	Chlorotic plants
Cotton	Clay loam	7	9	—	Chlorotic plants
Citrus	Sandy loam	64	45	39	Rapid root growth

[a] From Stolzy and Letey (10).

NUTRIENTS AND WATER. A deficiency of oxygen has been found to curtail nutrient and water absorption by plants. The exact reasons for the effects of aeration of these two processes are not well understood. However, these processes are known to be influenced quite markedly by the rate of root respiration. Apparently, the *energy* of respiration is utilized in bringing about at least part of the nutrient and water absorption. Since a supply of oxygen must be available if roots are to respire normally, a deficiency of this gas results in sluggish nutrient and water uptake. It is surprising as well as ironical that an oversupply of water in the soil tends to reduce the amount of water absorbed by plants.

The effect of aeration on nutrient absorption is of considerable practical significance. Under poor aeration conditions, for example, plants exhibit nutrient-deficiency symptoms on soils fairly well supplied with available nutrient elements. Also, on certain soils, improper tillage may destroy the granulation, leaving conditions which lead to *inefficient nutrient* utilization. The desirability of the frequent cultivation of heavy, poorly granulated soils when planted to row crops is undoubtedly related to this problem of nutrient uptake.

Soil aeration conditions in the field may vary drastically from day to day. Heavy rainfall, excessive irrigation, or flooding may bring about a temporary but complete absence of soil air. Plants will suffer, depending on their stage of growth and genetic tolerance (see Fig. 10:5).

SOIL COMPACTION AND AERATION. All negative effects of soil compaction are not due to poor aeration. Soil layers can become so dense as to impede the growth of roots even if an adequate O_2 supply is available. For example, some compacted soil layers adversely affect cotton more by preventing root penetration than by lowering available oxygen content.

10:6. OTHER EFFECTS OF SOIL AERATION

As previously indicated, anaerobic decomposition of organic materials is much slower than that occurring when ample gaseous oxygen is available. Moreover, the products of decomposition are entirely different. For example, the complete anaerobic decomposition of sugar occurs as follows:

$$C_6H_{12}O_6 \rightarrow 3CO_2 + 3CH_4$$

Sugar   Methane

Less complete decomposition yields other products, such as organic acids, which under extreme conditions may accumulate in toxic quantities. Examples of acids occurring by anaerobic decay are lactic, butyric, and citric acids.

When decomposed anaerobically, organic nitrogen compounds usually yield amines and gaseous nitrogen compounds in addition to ammonia. Of course, these nitrogen compounds remain in reduced forms, not being subject to nitrification under anaerobic conditions. The absence of oxygen thus completely changes the nature of the decay processes and the rates at which they occur.

TABLE 10:3. *Oxidized and Reduced Forms of Several Important Elements*

Element	Normal Form in Well-Oxidized Soils	Reduced Form Found in Waterlogged Soils
Carbon	CO_2	CH_4
Nitrogen	NO_3^-	N_2, NH_4^+
Sulfur	SO_4^{--}	H_2S, S^{--}
Iron	Fe^{3+} (ferric oxides)	Fe^{++} (ferrous oxides)
Manganese	Mn^{4+} (manganic oxides)	Mn^{++} (manganous oxides)

The chemical forms of certain elements found in soils in the oxidized and reduced states are shown in Table 10:3. In general, the oxidized forms are much more desirable for most of our common crops in humid regions. The advantages of the oxidized states of the nitrogen and sulfur compounds have already been discussed (see p. 36). The desirability of the higher-valent forms of iron and manganese is associated with the solubility of these elements at least in humid region soils. The reduced states, being more soluble than the oxidized forms, are often present in such quantities as to be toxic. This is true especially if the poor aeration occurs in an acid soil. Lack of lime thus intensifies the adverse effects of poor aeration on iron and manganese toxicity.

The insolubility of the oxidized forms of iron and manganese under alkaline conditions may result in a deficiency of these elements. Consequently, the interrelation of soil pH and aeration is very important in determining whether an excess or a deficiency of iron and manganese is likely to occur.

In addition to the chemical aspect of these differences, *soil color* is markedly influenced by the oxidation status of iron and manganese. Colors such as red, yellow, and reddish brown are encouraged by well-oxidized conditions. More subdued shades such as grays and blues predominate if insufficient O_2 is present. These facts are utilized in field methods of determining the need for drainage. Imperfectly drained soils usually show a condition wherein alternate streaks of oxidized and reduced materials occur. The *mottled* condition indicates a zone of alternate good and poor aeration, a condition not conducive to proper plant growth.

10:7. AERATION IN RELATION TO SOIL AND CROP MANAGEMENT

In view of the overwhelming importance of soil oxygen to the growth of most of the common crop plants, the question naturally arises as to the practical means of facilitating its supply. Interestingly enough, almost all methods employed for aeration control are involved directly or indirectly with the management of soil water.

Measures encouraging soil aeration logically fall into two categories: (a) those designed to remove excess soil moisture, and (b) those concerned with the aggregation and cultivation of the soil.

Both surface drainage and underdrainage are essential if an aerobic soil environment is expected. Since the nonsolid spaces in the soil are shared by air and water, the removal of excess quantities of water must take place if sufficient oxygen is to be supplied. The importance of surface runoff and tile drainage in this respect have already been indicated. (For the influence of drainage, see pp. 232–33).

The maintenance of a stable soil structure is an important means of augmenting good aeration. Pores of macrosize, usually greatly encouraged by large stable aggregates, are soon freed of water following a rain, thus allowing gases to move into the soil from the atmosphere. Organic matter maintenance by addition of farm manure and growth of legumes is perhaps the most practical means of encouraging aggregate stability, which in turn encourages good drainage and better aeration. These features received attention previously (see pp. 60–62).

In heavy soils, however, it is often impossible to maintain optimum aeration without resorting to the mechanical stirring of the soil by some type of cultivation. Thus, in addition to controlling weeds, cultivation in many cases has a second very important function—that of aiding soil aeration. Yields of row crops, especially those with large tap roots such as sugar beets and rutabagas, are often increased by frequent light cultivations that do not injure the fibrous roots. Part of this increase, undoubtedly, is due to aeration.

CROP–SOIL ADAPTATION. In addition to the direct methods of controlling soil aeration, one more phase of soil and crop management is of practical importance—that of crop–soil adaptation. The seriousness of oxygen deficiency depends to a large degree on the crop to be grown. Alfalfa, fruit and forest trees, and other deep-rooted plants require deep, well-aerated soils and are quite sensitive to a deficiency of oxygen, even in the lower soil horizons. Shallow-rooted plants such as grasses and alsike and ladino clovers, conversely, do very well on soils which tend to be poorly aerated, especially in the subsoil. These facts are significant in deciding what crops should be grown and how they are to be managed in areas where aeration problems are acute.

The dominating influence of moisture on the aeration of soils is universally apparent. And the control of the moisture means at least a partial control of the aeration. This leads to a second important physical property at times influenced by soil water—that of soil temperature.

10:8. SOIL TEMPERATURE

The temperature of the soil affects markedly its usefulness to man (6). In cold soils chemical and biological rates are slow. Biological decomposition is at a near standstill, thereby limiting the rate at which nutrients such as nitrogen, phosphorus, sulfur, and calcium are made available. For example, nitrification does not begin in the spring until the soil temperature reaches about 40°F, the most favorable limits being 80–90°F. Also, plant processes such as seed germination and root growth occur only above certain critical soil temperatures. Likewise, the absorption and transport of water and nutrient ions by higher plants is adversely affected by low temperatures.

Plants vary widely in the soil temperature at which they grow best. There is variability in the optimum temperature for different plant processes. For example, corn germination requires a soil temperature of 45–50°F and is optimum near 100°F. Dry matter production for corn is optimum when soil temperature is 80–85°F, although this apparently varies under different conditions of air temperature and soil moisture. Potato tubers develop best when the soil temperature is between 60 and 70°F. Oats also grow best at about 70°F, although the roots of this plant apparently do better when the soil temperature is about 60°F (see Fig. 10:7).

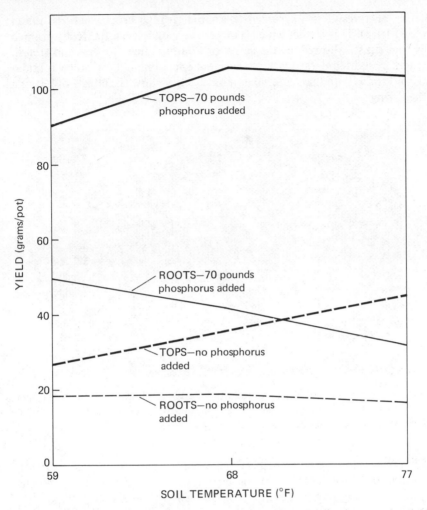

FIGURE 10:7. Effect of soil temperature and phosphorus applications on the yield of oat roots and tops in a greenhouse experiment. Top yields were increased by raising the soil temperature. The opposite was true for the roots when phosphorus was added at the rate of 70 pounds per 2 million pounds of soil. [*From Case et al.* (3).]

In addition to the direct influence of temperature of plant and animal life the effect of freezing and thawing must be considered. Frost action along with later thawing is responsible for some of the physical weathering which takes place in soils. As ice forms in rock crevices it forces the rocks apart, causing them to disintegrate. Similarly, alternate freezing and thawing subjects the soil aggregates and clumps and rocks to pressures and thus alters the physical setup in the soil. Freezing and thawing of the upper layers of soil can also result in *heaving* of perennial forage crops such as alfalfa (see Fig. 10:8). This action, which is most severe on bare, imperfectly drained soils, can drastically reduce the stand of alfalfa, some clovers, and trefoil. Changes in soil temperature have the same effect on shallow house foundations and roads with fine material as a base. They show the effects of heaving in the spring.

FIGURE 10:8. Alternate freezing and thawing of soils result in the "heaving" of perennial plants out of the ground. The above alfalfa crowns were exposed as a result of this type of action during the winter and spring months.

The temperature of soils in the fields is dependent directly or indirectly upon at least three factors: (a) the net amount of heat the soil absorbs, (b) the heat energy required to bring about a given change in the temperature of a soil, and (c) the energy required for changes such as evaporation, which are constantly occurring at or near the surface of soils. These phases will now be considered.

10:9. ABSORPTION AND LOSS OF SOLAR ENERGY

The amount of heat absorbed by soils is determined primarily by the quantity of effective solar radiation reaching the earth. This so-called global radiation represents only part of the total solar radiation. The remainder, before it reaches the earth, is reflected back into the atmosphere by clouds, is absorbed by atmospheric gases, or is scattered into the atmosphere (see Fig. 10:9). In relatively cloud-free arid regions as much as 75 percent of the solar radiation reaches the ground. In contrast, only 35 to 40 percent may

SOLAR BEAM RADIATION

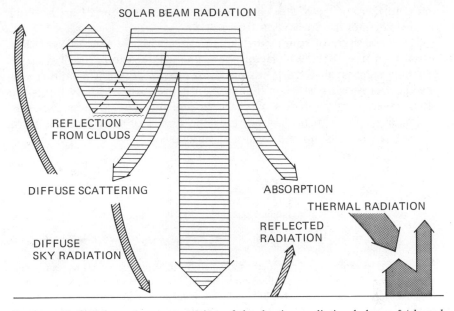

REFLECTION
FROM CLOUDS

DIFFUSE SCATTERING   ABSORPTION

                                        THERMAL RADIATION

                                   REFLECTED
DIFFUSE   RADIATION
SKY RADIATION

FIGURE 10:9. Schematic representation of the daytime radiation balance. [*Adapted from Tanner* (11).]

reach the ground in cloudy humid areas. An average global figure is about 50 percent.

Some 30 to 45 percent of the radiation energy reaching the earth is reflected back to the atmosphere or is lost by thermal radiation. Of that which is not returned to the atmosphere (termed net radiation) up to about 5 percent is used to energize photosynthesis and metabolic reactions in plants. Most of the remainder is utilized to evaporate water from soil and plant surfaces. Unless the soil is dry and little energy for vaporization is required, only about 5 to 15 percent of the net radiation is commonly stored as heat in the soil and plant cover.

Solar radiation in any particular locality depends fundamentally upon climate. But the amount of energy entering the soil is in addition affected by other factors such as (a) the color, (b) the slope, and (c) the vegetative cover of the site under consideration. It is well known that dark soils will absorb more energy than light-colored ones and that red and yellow soils will show a more rapid temperature rise than will those that are white. This does not imply that dark soils are always warmer. Actually the opposite may be true since dark soils usually are high in organic matter and consequently hold large amounts of water, which must also be warmed and evaporated.

Observation has shown that the nearer the angle of incidence of the sun's rays approaches the perpendicular, the greater will be the absorption (see Fig. 10:10). As an example, a southerly slope of 20 degrees, a level soil, and a northerly slope of 20 degrees receive energy on June 21 at the 42nd parallel north in the proportion 106:100:81.

The temperature of southward slopes varies with the time of year. For instance in the Northern Hemisphere the southeasterly inclination is generally warmest in the early season, the southerly slope during midseason, and the southwesterly slope in the fall. Southern or southeasterly slopes are often preferred by gardeners. Orchardists and foresters consider exposure an important factor regarding not only the species or variety of tree to be grown but also sunscald and certain plant diseases.

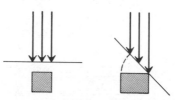

FIGURE 10:10. Area of soil warmed as affected by the angle at which the sun's rays strike the soil. If a given amount of radiation from the sun strikes the soil at right angles (*left*), the radiation is concentrated in a relatively small area and the soil warms up quite rapidly. If the same amount of radiation strikes the soil at a 45° angle (*right*), the area affected is larger, the radiation is not so concentrated, and the soil warms up more slowly. This is one of the reasons why north slopes tend to have cooler soils than south slopes. It also accounts for the colder soils in winter as compared to summers.

Whether the soil is bare or covered with vegetation is another factor markedly influencing the amount of insulation received. The effect of a forest is universally recognized. Even an ordinary field crop such as bluegrass has a very noticeable influence, especially upon temperature fluctuations. Bare soils warm up more quickly and cool off more rapidly than those covered with vegetation or with artificial mulches. Frost penetration during the winter is considerably greater in bare noninsulated land.

10:10. SPECIFIC HEAT OF SOILS

Another major factor affecting the temperature relations of a soil is its specific heat or its thermal capacity compared with that of water. Specific heat may be expressed as a ratio of the quantity of heat required to raise the temperature of a given substance from 15 to 16°C compared to that required for the same temperature rise of an equal weight of water. The importance of this property in soil temperature control undoubtedly is great. The mere absorption of a given amount of heat by a soil does not necessarily assure a rapid rise in temperature. Everything else being equal, a soil with a high specific heat exhibits much less rapid temperature change than does one having a low specific heat.

Under actual field conditions, the soil moisture content determines more than any other factor the energy required to raise the temperature of soils. For instance, the dry weight specific heat of mineral soils, in spite of variations in texture and organic matter, is about 0.20. But if the moisture is advanced to 20 percent, the specific heat of the wet mass becomes 0.33, while an increase of 30 percent of moisture raises the wet weight specific heat to 0.38.[1] Obviously, therefore, since moisture is one of the major factors determining the heat capacity of a soil, it has much to do with the rates of warming up and cooling off of soils.

10:11. HEAT OF VAPORIZATION

Soil moisture is also of major importance in determining the amount of heat used in the process of evaporation of soil water. Vaporization results from an increased activity of the soil–water molecules, which is made possible by the absorption of heat from the surrounding environment. This results in a cooling effect, especially at the surface, where most of the evaporation occurs. To evaporate 1 gram of water at 68°F requires about 540 gram calories. To evaporate 1 pound of water at the same temperature requires 245 kilogram calories, which is sufficient to lower the temperature of 1 cubic foot of a representative mineral soil at optimum moisture about 28°F if all of the energy of evaporation comes from the soil and its water. Such a figure is hypothetical because only a part of the heat of vaporization comes from the soil itself. Nevertheless, it indicates the tremendous cooling influence of evaporation.

The low temperature of a wet soil is due partially to evaporation and partially to high specific heat. The temperature of the upper few inches of a wet soil is 6–12°F lower than that of a moist or dry soil. This is a significant factor in the spring, when a few degrees will make the difference between the germination or lack of germination of crop seeds. Stand failures are commonly due to wet cold soils.

---

[1] These figures can be easily verified. For example, consider the above soil at 20 percent moisture content. Since soil moisture is expressed as grams of water per 100 grams of soil solids, there are in this case 20 grams of H_2O with each 100 grams of soil. The number of calories required to raise the temperature of 20 grams of water by 1°C is

$$20 \text{ g} \times 1 \text{ cal/g} = 20 \text{ cal}$$

The corresponding figure for the 100 grams of soil solids is

$$100 \text{ g} \times 0.2 \text{ cal/g} = 20 \text{ cal}$$

Thus, a total of 40 calories is required to raise the temperature of 120 grams of the moist soil by 1°C. Since the *specific heat* is the number of calories required to raise the temperature of 1 gram of wet soil by 1°C, in this case it is

$$\frac{40}{120} = 0.33 \text{ cal/g}$$

SOIL COLOR AND TEMPERATURE. At this point it is appropriate again to emphasize the interrelationship of soil color and soil temperature. Although dark-colored soils absorb heat readily, such soils, because of their usually high organic matter status, often have high moisture contents. Under such soil conditions, the evaporation and specific heat relationships become especially important. Consequently, a dark-colored soil, particularly if it happens to be somewhat poorly drained, may not warm up as quickly in the spring as will a well-drained light-colored soil nearby.

10:12. MOVEMENT OF HEAT IN SOILS

Before dealing with actual temperature data, the movement of thermal energy in the soil will be considered. As already mentioned, much of the solar radiation is dissipated into the atmosphere. Some, however, slowly penetrates the profile largely by *conduction*. While this type of movement is influenced by a number of factors, the most important is probably the moisture content of the soil layers. Heat passes from soil to water about 150 times more easily than from soil to air. As the water increases in a soil, the air decreases and the transfer resistance is lowered decidedly. When sufficient water is present to join most of the soil particles, further additions will have little effect on heat conduction. Here again the major rôle of soil moisture is obvious.

The significance of conduction in respect to field temperatures is not difficult to comprehend. It provides a means of temperature adjustments but, because it is slow, the subsoil changes tend to lag behind those to which the surface layers are subjected. Moreover, the changes occurring are always less in the subsoil. In temperate regions, surface soils in general are expected to be warmer in summer and cooler in winter than the subsoil, especially the lower horizons of the subsoil.

10:13. SOIL TEMPERATURE DATA

The temperature of the soil at any time depends on the ratio of the energy absorbed and that being lost. The constant change in this relationship is reflected in the *seasonal*, *monthly*, and *daily* temperatures. The accompanying data (Fig. 10:11) from College Station, Texas, are representative of average seasonal temperatures in relation to soil depth.

It is apparent from these figures that the seasonal variations of soil temperature are considerable even at the lower depths. The surface layers vary more or less according to air temperature and therefore exhibit a greater fluctuation than the subsoil. On the average, the surface 6-inch layer of soil is warmer than the air at every season of the year, while the subsoil is warmer in autumn and winter but cooler in the spring and summer because of its protected position and the lag in conduction.

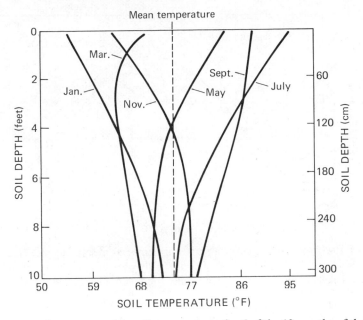

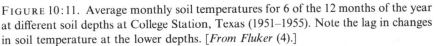

FIGURE 10:11. Average monthly soil temperatures for 6 of the 12 months of the year at different soil depths at College Station, Texas (1951–1955). Note the lag in changes in soil temperature at the lower depths. [*From Fluker* (4).]

This lag in temperature change in the subsoil is especially noticeable when the monthly movement of soil temperature at Lincoln, Nebraska, is considered. The monthly average temperature plotted in Fig. 10:12 shows a greater annual range in temperature for the surface soil than for the air. Daily data would, of course, show the greatest range for the soil air. Changes in soil temperature except at the very surface are gradual, whereas the air may vary many degrees in a very short time.

The daily and hourly temperatures of the atmospheric air and the soil in temperate zones may show considerable agreement or marked divergence according to conditions. Fluctuations are naturally more rapid in the case of air temperatures and usually are greater in temperate regions. However, the maximum temperature of a dry surface soil may definitely exceed that of the air, possibly approaching in some cases 125 or 130°F. But in winter even surface soils do not fall greatly below freezing.

With a solar control and a clear sky, the air temperature in temperate regions rises from morning to a maximum at about two o'clock. The surface soil, however, does not reach its maximum until later in the afternoon, owing to the usual lag (Fig. 10:13). This retardation is greater and the temperature change is less as the depth increases. The lower subsoil shows little daily or weekly fluctuation, the variation, as already emphasized, being a slow monthly or seasonal change.

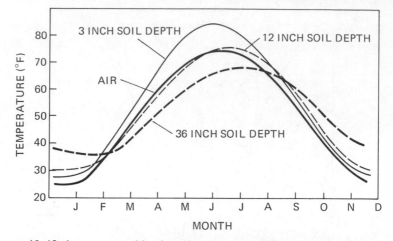

FIGURE 10:12. Average monthly air and soil temperatures at Lincoln, Nebraska (12 years). Note that the 3 inch soil layer is consistently warmer than the air above and that the 36 inch soil horizon is cooler in spring and summer but warmer in the fall and winter than surface soil.

10:14. SOIL TEMPERATURE CONTROL

The temperature of field soils is subject to no radical human regulation. The use of organic mulches will give lower and more uniform surface–soil temperatures (Fig. 10:13). However, such a practice has limited application over large acreages and is used mostly for gardens and flower beds.

Soil management methods, especially those influencing soil moisture, do provide for small but biologically vital modifications. The temperature regions of a well-drained and a poorly drained soil in a humid, temperate

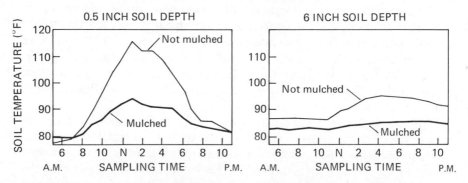

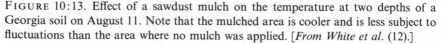

FIGURE 10:13. Effect of a sawdust mulch on the temperature at two depths of a Georgia soil on August 11. Note that the mulched area is cooler and is less subject to fluctuations than the area where no mulch was applied. [*From White et al.* (12).]

region will be compared briefly. The poorly drained soil has a high specific heat. Therefore, large amounts of radiant energy must be absorbed to raise the soil temperature by those few extra degrees so badly needed in early spring. And since the excess water will not percolate through this soil, much of it must be removed by evaporation, another costly process in terms of energy utilization. The low temperatures of poorly drained soils, especially in the spring, are well known. A range from 6 to 12°F lower in the surface layer than in comparable well-drained areas is expected. Drainage is, of course, the only practicable measure.

In addition to the control relations already mentioned, there is the influence of downward-moving water on soil temperature. In general, it has a cooling effect, except possibly in the early spring. Precipitation is usually cooler than the soil in temperate regions, especially in the summer. But even if rainwater should be 10°F warmer than the soil—an improbable assumption—an average rain would raise the temperature of the surface 6 inches only slightly, and this would be quickly offset by the cooling effects of surface evaporation.

As was the case with soil air, the controlling influence of soil water on soil temperature is apparent everywhere. Whether a question of acquisition of insolation, loss of energy to the atmosphere, or the movement of heat back and forth within the soil, the percentage of water present is always important. Water regulation seems to be the key to what little practical temperature control it is possible to exert on field soils.

REFERENCES

(1) Boynton, D., *Soils in Relation to Fruit Growing in New York. Part XV. Seasonal and Soil Influences on Oxygen and Carbon Dioxide Levels of New York Orchard Soils*, Bull. 763, Cornell Univ. Agr. Exp. Sta., 1941.

(2) Boynton, D., et al., "Are There Different Critical Oxygen Concentrations for the Different Phases of Root Activity?" *Science*, **88**: 569–70, 1938.

(3) Case, V. W., et al., "The Influences of Soil Temperature and Phosphorus Fertilizers of Different Water-Solubilities on the Yield and Phosphorus Uptake by Oats," *Soil Sci. Soc. Amer. Proc.*, **28**: 409–12, 1964.

(4) Fluker, B. J., "Soil Temperature," *Soil Science*, **86**: 35–46, 1958.

(5) Lyon, T. L., Buckman, H. O., and Brady, N. C., *The Nature and Properties of Soils* (New York: Macmillan, 1952), p. 278.

(6) Nielson, K. F., and Humphries, E. C., "Effects of Root Temperature on Plant Growth," *Soils and Fertilizers*, **29**: 1–7, 1966.

(7) Russell, E. J., and Appleyard, A., "The Atmosphere of the Soil: Its Composition and the Causes of Variation," *Jour. Agr. Sci.*, **7**: 25–6, 1915.

(8) Saukko, P., "Flood Damage to Crops" (in Swedish), *Grundförbattring*, 4:26–34, 1950–51.

(9) Stolzy, L. H., et al., "Root Growth and Diffusion Rates as Functions of Oxygen Concentration," *Soil Sci. Amer. Proc.*, 2:463–67, 1961.

(10) Stolzy, L. H., and Letey, J., "Characterizing Soil Oxygen Conditions with a Platinum Microelectrode," *Advan. in Agron.*, 16:249–79, 1964.

(11) Tanner, C. B., "Evaporation of Water from Plants and Soil," in Kozlowski, T. T., Ed., *Water Deficits and Plant Growth* (New York: Academic Press, 1968).

(12) White, A. W., et al., "The Effect of Sawdust on Crop Growth and Physical and Biological Properties of Cecil Soil," *Soil Sci. Soc. Amer. Proc.*, 23:365–68, 1959.

Chapter 11

ORIGIN, NATURE, AND CLASSIFICATION OF PARENT MATERIALS

T HE influence of weathering is evident on all sides. Nothing escapes it. It breaks up the country rocks, modifies or destroys their physical and chemical characteristics, and carries away the soluble products and even some of the solids as well. The unconsolidated residues, the *regolith*, are left behind. But weathering goes further. It synthesizes a soil from the upper layers of this heterogeneous mass. Hence, the upper part of the regolith may be designated as *parent material* of soils. These parent materials are not always allowed to remain undisturbed on the site of their development. Climatic agencies may shift them from place to place until they are allowed to rest long enough for a soil profile to develop. A study of weathering and of the parent materials that result is a necessary introduction to soil formation and classification.

The consideration of parent materials will begin with a brief review of kinds of rocks from which the regolith and in turn the soils have formed.

11:1. CLASSIFICATION AND PROPERTIES OF ROCKS

The rocks found in the earth's crust are commonly classified as *igneous, sedimentary,* or *metamorphic.* Those of igneous origin are formed from molten lava and include such common rocks as *granite* and *diorite* (Fig. 11:1). They are composed of primary minerals such as (a) quartz, (b) the feldspars, and (c) the dark-colored minerals, including biotite, augite, and hornblende. In general, gabbro and basalt (Fig. 11:2), which are high in the dark-colored, iron- and magnesium-containing minerals, are more easily weathered than are the granites and other lighter-colored rocks.

Sedimentary rocks have resulted from the deposition and recementation of weathering products of other rocks. For example, quartz sand weathered from a granite and deposited in the bottom of a prehistoric sea may through geological changes have become cemented into a solid mass called a *sandstone.* Similarly, recemented clays are termed *shale.* Other important sedimentary

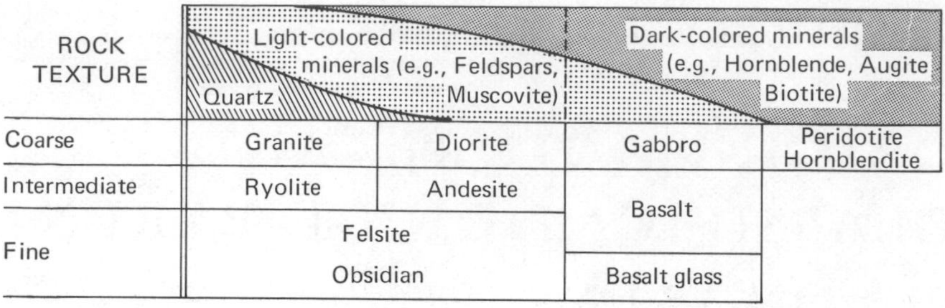

ROCK TEXTURE	Light-colored minerals (e.g., Feldspars, Quartz Muscovite)		Dark-colored minerals (e.g., Hornblende, Augite Biotite)	
Coarse	Granite	Diorite	Gabbro	Peridotite Hornblendite
Intermediate	Ryolite	Andesite	Basalt	
Fine	Felsite			
	Obsidian		Basalt glass	

FIGURE 11:1. Classification of some igneous rocks in relation to mineralogical composition and the size of mineral grains in the rock (rock texture). Light-colored minerals and quartz are generally more prominent than are the dark-colored minerals.

rocks are listed in Table 11:1 with their dominant minerals. The resistance of a given sedimentary rock to weathering is determined by the particular dominant minerals and by the cementing agent.

Metamorphic rocks are those which have formed by the metamorphism or change in form of other rocks. Igneous and sedimentary masses subjected to tremendous pressures and high temperature have succumbed to metamorphism. Igneous rocks are commonly modified to form *gneisses* and *schists*; those of sedimentary origin such as sandstone and shale may be changed to *quartzite* and *slate*. Some of the common metamorphic rocks are shown in Table 11:1. As with those of igneous and sedimentary origin, the particular mineral or minerals which dominate a given metamorphic rock will influence its resistance to weathering. (See Table 11:2 for a listing of the more common minerals.)

TABLE 11:1. *Some of the More Important Sedimentary and Metamorphic Rocks and the Minerals Commonly Dominant in Them*

Sedimentary Rocks	Dominant Mineral	Metamorphic Rocks	Dominant Mineral
Limestone	Calcite ($CaCO_3$)	Gneiss	(Varies)[a]
Dolomite	Dolomite [$CaMg(CO_3)_2$]	Schist	(Varies)[a]
Sandstone	Quartz (SiO_2)	Quartzite	Quartz (SiO_2)
Shale	Clays	Slate	Clays
Conglomerate	(Varies)[b]	Marble	Calcite ($CaCO_3$)

[a] The minerals present are determined by the original rock, which has been changed by metamorphism. Primary minerals present in the igneous rocks commonly dominate these rocks, although some secondary minerals are also present.

[b] Small stones of various mineralogical makeup will be cemented into conglomerate.

TABLE 11:2. *The More Important Original and Secondary Minerals Found in Soils*

The original minerals are also found abundantly in igneous and metamorphic rocks. Secondary minerals are commonly found in sedimentary rocks.

Name	Formula
Original Minerals	
Quartz	SiO_2
Microcline ⎱ Orthoclase ⎰	$KAlSi_3O_8$
Na plagioclase	$NaAlSi_3O_8$
Ca plagioclase	$CaAl_2Si_2O_8$
Muscovite	$KAl_3Si_3O_{10}(OH)_2$
Biotite	$KAl(Mg, Fe)_3Si_3O_{10}(OH)_2$
Hornblende[a]	$Ca_2Al_2Mg_2Fe_3, Si_6O_{22}(OH)_2$
Augite[a]	$Ca_2(Al, Fe)_4(Mg, Fe)_4Si_6O_{24}$
Secondary Minerals	
Calcite	$CaCO_3$
Dolomite	$CaMg(CO_3)_2$
Gypsum	$CaSO_4 \cdot 2H_2O$
Apatite	$Ca_5(PO_4)_3 \cdot (Cl, F)$
Limonite	$Fe_2O_3 \cdot 3H_2O$
Hematite	Fe_2O_3
Gibbsite	$Al_2O_3 \cdot 3H_2O$
Clay minerals	Al silicates

[a] These are approximate formulas only since these minerals are so variable in their composition.

11:2. WEATHERING—A GENERAL CASE

Before considering the various kinds of parent materials which develop from rocks, attention will be given to the general trends of weathering responsible for the formation of the regolith and the soils. Some of the changes which take place during weathering are illustrated in Fig. 11:2. Study it carefully, keeping in mind that it deals primarily with weathering as it occurs in temperate regions.

Weathering is basically a combination of destruction and synthesis. Rocks, the original starting point in the weathering process, are first broken down into smaller rocks and eventually into the individual minerals of which they are composed. Simultaneously, rock fragments and the minerals therein are attacked by weathering forces and are changed to new minerals either by minor modifications (alterations) or by complete chemical changes. These changes are accompanied by a continued decrease in particle size and by the release of soluble constituents, most of which are subject to loss in drainage waters.

The minerals which are synthesized are shown (Fig. 11:2) in two groups: (a) the silicate clays, and (b) the very resistant end products, including iron

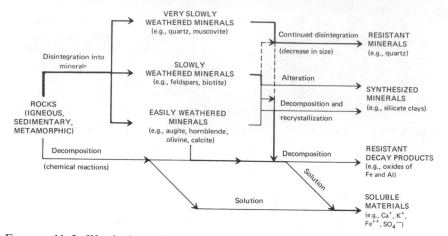

FIGURE 11:2. Weathering pathways which take place under moderately acid conditions common in humid temperate regions. Major paths of weathering are indicated by the heavier arrows, minor pathways by broken lines. As one would expect, climate modifies the exact relationships. In arid regions physical breakdown (disintegration) would dominate, and soluble ions would not be lost in large quantities. In humid regions decomposition becomes more important, especially under tropical conditions.

and aluminum oxides. Along with the very resistant primary minerals such as quartz, these two groups dominate temperate region soils.

THE PROCESSES. Two basic processes are involved in the changes indicated in Fig. 11:2. These are mechanical and chemical. The former is often designated as *disintegration*, the latter as *decomposition*. Both are operative, moving from left to right in the weathering diagram. Disintegration results in a decrease in size of rocks and minerals without appreciably affecting their composition. By decomposition, however, definite chemical changes take place, soluble materials are released, and new minerals are synthesized or are left as resistant end products. Mechanical and chemical processes may be outlined as follows:

1. Mechanical (disintegration):
   a. Temperature—differential expansion of minerals, frost action, and exfoliation.
   b. Erosion and deposition—by water, ice, and wind.
   c. Plant and animal influences.
2. Chemical (decomposition):
   a. Hydrolysis.
   b. Hydration.
   c. Carbonation and related acidity processes.
   d. Oxidation.
   e. Solution.

11:3. MECHANICAL FORCES OF WEATHERING

TEMPERATURE EFFECTS. Variations of temperature, especially if sudden or wide, greatly influence the disintegration of rocks. During the day rocks become heated and at night often cool much below the temperature of the air. This warming and cooling is particularly effective as a disintegrating agent. Rocks are aggregates of minerals which differ in their coefficients of expansion upon being heated. With every temperature change, therefore, differential stresses are set up which eventually must produce cracks and rifts, thus encouraging mechanical breakdown.

Because heat is conducted slowly, the outer surface of a rock is often at a markedly different temperature than the inner and more protected portions. This differential heating and cooling set up lateral stresses which, in time, may cause the surface layers to peel away from the parent mass. This phenomenon is referred to as *exfoliation* and at times is sharply accelerated by the freezing of included water.

The presence of water, if freezing occurs, greatly increases these mechanical effects. The force developed by the freezing of water is equivalent to about 150 tons to the square foot, an almost irresistible pressure. This is equivalent to a pressure of about 142 atmospheres or over 4 times the negative tension exerted upon the water molecules at the liquid–air interface of a soil when its moisture content is at the hygroscopic coefficient. It widens cracks in huge boulders and dislodges mineral grains from smaller fragments. This influence of temperature is not ended when rocks are reduced to fragments. It continues in the parent materials and finally in the resultant soil. Freezing and thawing are of great practical importance in altering the physical condition of medium- and fine-textured soils.

INFLUENCE OF WATER, ICE, AND WIND. Rain water beats down upon the land and then travels oceanward, continually shifting, sorting, and reworking unconsolidated materials of all kinds. When loaded with such sediments, water has a tremendous cutting power, as is amply demonstrated by the gorges, ravines, and valleys the country over. The rounding of sand grains on an ocean beach is further evidence of the abrasion which accompanies water movement.

Ice is an erosive and transporting agency of tremendous capacity, and next to water is perhaps the most important and spectacular physical agent of weathering. One must visit Greenland or Alaska to realize its power. The abrasive action of glaciers as they move under their own weight disintegrates rocks and minerals alike. Not only do glaciers affect the underlying solid rock, but also they grind and mix unconsolidated materials which have been picked up as they move over the countryside (see Fig. 11:8). Even though they are not so extensive at the present time, glaciers in ages past have been responsible for the transportation and deposition of parent materials over millions of acres. The importance of ice as a mechanical agent is not to be under-estimated (see pp. 293–95).

Wind has always been an important carrying agent and, when armed with fine debris, exerts an abrasive action also. Dust storms of almost con-tinental extent have occurred in the past with the result that tons of material have been filched from one section and transferred to another. As dust is transferred and deposited, abrasion of one particle against another occurs. The rounded rock remnants in some arid areas of the west are caused largely by wind action.

PLANTS. Simple plants, such as mosses and lichens, grow upon exposed rock, there to catch dust until a thin film of highly organic material accu-mulates. Higher plants sometimes exert a prying effect on rock which results in some disintegration. This is most obvious in the case of tree roots in rocky sections. Such influences, as well as those exerted by animals, are of little import in producing parent material when compared to the drastic physical effects of water, ice, wind, and temperature changes.

11:4. CHEMICAL PROCESSES OF WEATHERING

Scarcely has the disintegration of rock material begun when its decompo-sition usually is also apparent. This is especially noticeable in warm humid regions, where chemical and physical processes are particularly active and markedly accelerate each other.

HYDROLYSIS. The reaction of minerals with water is perhaps the most important way by which chemical breakdown occurs. Hydrolysis, which is a decomposition reaction, is important in the weathering of a wide range of minerals, including the feldspars and micas. The change may be indicated as follows, using microcline ($KAlSi_3O_8$) as the mineral undergoing hydro-lysis:

$$KAlSi_3O_8 + HOH \rightarrow HAlSi_3O_8 + KOH$$

$$2HAlSi_3O_8 + 8HOH \rightarrow Al_2O_3 \cdot 3H_2O + 6H_2SiO_3$$

The potassium released by this reaction is soluble and can be adsorbed by the soil colloids, used by plants, or removed in the drainage water. The aluminum and silicon compounds may recrystallize into a clay mineral such as kaolinite. If conditions are suitable, one or both may remain in the soil as their respective oxides or they may be removed in the drainage.

HYDRATION. Hydration involves the rigid attachment of H and OH ions to the compound being attacked. In most cases, these ions become an integral part of the mineral crystal lattice. As micas become hydrated, some hydrogen and hydroxyl ions move in between the platelike layers. In so doing, they tend to expand the crystal and make it more porous, thus hastening other decomposition processes.

A good example of hydration is the development of limonite from hematite, which may be shown as follows:

$$2Fe\,O_3 + 3H_2O \rightarrow 2Fe_2O_3{\cdot}3H_2O$$
Hematite   Limonite
(red)   (yellow)

When the products of hydration dry out because of varying weather conditions, dehydration may occur. Thus, limonite may readily be changed to hematite with a noticeable change in color. Variations in the color of subsoils in the Southeastern states are due to a considerable extent to the above reaction.

CARBONATION AND OTHER ACIDIC PROCESSES. The disintegration and decomposition of mineral matter are accelerated by the presence of the hydrogen ion in percolating waters. Even that contained in carbonic acid is effective in this respect. For example. this acid is known to result in the chemical solution of calcite in limestone; the following reaction illustrates what takes place:

$$CaCO_3 + H_2CO_3 \rightarrow Ca(HCO_3)_2$$
Calcite   Soluble
                     bicarbonate

Other acids much stronger than carbonic are also present in most humid region soils. They include very dilute inorganic acids such as HNO_3 and H_2SO_4 as well as some organic acids. Also present are the hydrogen ions associated with soil clays which are available for reaction with other soil minerals.

An example of the reaction of hydrogen ions with soil minerals is that of an acid clay with a feldspar such as anorthite:

$$CaAlSi_2O_8 + \begin{matrix} H \\ H \end{matrix} \boxed{Micelle} \rightarrow H_2AlSi_2O_8 + Ca \boxed{Micelle}$$
Anorthite   Acid
                            silicate

The mechanism is simple, the hydrogen ions of the acid clays replacing the bases of the fresh minerals which they may closely contact. An acid silicate results subject to recrystallization, thus producing clay. Thus, "clay begets clay" by a mechanism that depends upon the removal of bases from the weathering mass of minerals by insoluble inorganic acids.

OXIDATION. Of the various chemical changes caused by weathering, oxidation is usually one of the first to be noticed. It is particularly manifest in rocks carrying iron, an element which is easily oxidized. In some minerals iron is present in reduced ferrous (Fe^{++}) form. If oxidation to the ferric ion takes place while iron is still part of the crystal lattice, other ionic adjustments must be made since a three-valent ion is replacing a two-valent one. These adjustments result in a less stable crystal, which is then subject to both disintegration and decomposition.

In other cases, ferrous iron may be released from the crystal and is almost simultaneously oxidized to the ferric form. An example of this is the hydration of olivine and the release of ferrous oxide, which may be immediately oxidized to ferric oxide (hematite):

$$3MgFeSiO_4 + 2H_2O \rightarrow H_4Mg_3Si_2O_9 + SiO_2 + 3FeO$$

Olivine   Serpentine   Ferrous
                                                   oxide

$$4FeO + O_2 \rightarrow 2Fe_2O_3$$

Ferrous   Hematite
oxide

SOLUTION. The solvent action of water and the ions it carries as it moves through and around rocks and minerals further the weathering process. The solution effects of dissolved carbon dioxide and the hydrogen ions in water have already been mentioned, as has the release of soluble potassium by hydrolysis or orthoclase. In each case, ions of the alkali metals (Na, K, etc.) or the alkaline earths (Ca, Mg, etc.) were represented as being solubilized rather readily (see Table 11:3). Less spectacular is the slower dissolution of some iron, silicon, and aluminum. These elements are also subject to solution, the specific climatic conditions determining the extent to which it occurs.

TABLE 11:3. *Comparative Loss of Mineral Constituents as Weathering Takes Place in a Granite and in a Limestone*

The losses are compared to that of aluminum, which was considered in these cases to have remained constant during the weathering process.

Granite to Clay[a]		Limestone to Clay[b]	
Constituent	Comparative Loss (%)	Constituent	Comparative Loss (%)
CaO	100.0	CaO	99.8
Na_2O	95.0	MgO	99.4
K_2O	83.5	Na_2O	76.0
MgO	74.7	K_2O	57.5
SiO_2	52.5	SiO_2	27.3
Fe_2O_3	14.4	Fe_2O_3	24.9
Al_2O_3	0.0	Al_2O_3	0.0

[a] From Merrill (3).
[b] From Diller (2).

11:5. FACTORS AFFECTING WEATHERING OF MINERALS

It has been established that the following general factors have a marked effect on the weathering process: (a) climatic conditions, (b) physical properties, and (c) chemical characteristics of the rocks and minerals. These will each be discussed briefly.

CLIMATIC CONDITIONS. The climatic conditions, more than any other factor, will control the kind and rate of weathering which take place if sufficient time is allowed. Under conditions of low rainfall, for example, mechanical processes of weathering dominate, resulting in decreased particle size with little change in composition. The presence of more moisture encourages chemical as well as mechanical changes, resulting in new minerals and soluble products. In humid temperate regions, silicate clays are among the minerals synthesized. They account for much of the strong agricultural character of soils in these areas.

Weathering rates are generally more rapid in humid tropical areas, where the more resistant products of chemical weathering such as the hydrous oxides of iron and aluminum are prominent. This is because less resistant minerals have succumbed to the intense weathering common to these areas. Even quartz, the most resistant of the common macrograined primary minerals, disappears in time under these conditions.

Climate also largely controls the dominant types of vegetation present. In this way it indirectly influences the biochemical reactions in soils and in turn their effect upon mineral weathering. For example, soils developing under conifer trees whose needles are low in metallic cations are commonly more acid than soils developed under grassland or most deciduous trees.

PHYSICAL CHARACTERISTICS. Particle size, hardness, and degree of cementation are three physical characteristics that influence weathering. In rocks, large crystals of different minerals encourage disintegration since there is some variation in the amount of expansion and contraction occurring with each mineral as temperatures change. The resulting stress helps to develop cracks and break these rocks into their mineral components. Finer-grained materials are apparently more resistant to mechanical breakdown.

Particle size also influences the chemical breakdown of minerals. In general, a given mineral is more susceptible to decomposition when present in fine particles than in large grains. The much larger surface area of finely divided material presents greater opportunity for chemical attack. Apparently, the greater ease of weathering of small particles is more pronounced with some minerals than with others. For example, quartz particles of sand size are extremely resistant to chemical weathering. In contrast, clay-sized quartz, although not subject to ready decomposition, is not as resistant as many of the other minerals in similar-sized particles.

Hardness and cementation evidently influence weathering primarily by their effect on the rate of disintegration into particles small enough for decomposition. Thus, a dense quartzite or a sandstone cemented firmly by a slowly weathered mineral will resist mechanical breakdown and present a small amount of total surface area for chemical activity. Porous rocks such as volcanic ash or coarse limestones, on the other hand, are readily broken down into smaller particles. These have in total a larger surface area for chemical attack and are obviously more easily decomposed.

CHEMICAL AND STRUCTURAL CHARACTERISTICS. For minerals of a given particle size, chemical and crystalline characteristics determine the ease of decomposition. Minerals such as gypsum ($CaSO_4 \cdot 2H_2O$) which are slightly soluble in water are quickly removed if there is adequate rainfall. Water charged with carbonic acid likewise dissolves less soluble minerals such as calcite and dolomite (see p. 283). Consequently, these minerals are seldom found in the surface of the regolith in areas with even moderate rainfall.

The dark-colored primary minerals, sometimes called ferromagnesian because of their iron and magnesium contents, are more susceptible to chemical weathering than are the feldspars and quartz. The presence of iron (and perhaps less tightness of crystal packing) helps account for the more rapid breakdown of the ferromagnesian minerals. The ease with which iron-containing minerals become "stained" with oxides of this element is well known.

Tightness of packing of the ions in the crystal units of minerals is thought to influence their weathering rates. For example, olivine and biotite which are relatively easily weathered have crystal units less tightly packed than zircon and muscovite, comparable minerals quite resistant to weathering.

Climatic and biotic conditions will likely determine the relative stability of the various soil-forming minerals. Consequently, one cannot present a listing of minerals based on their resistance to weathering under all climatic conditions. However, studies (1) of minerals remaining in the soil and the regolith under various environmental conditions have led to the following general order of weathering resistance of the sand- and silt-sized particles of some common minerals: quartz (most resistant) > muscovite, potassium feldspars > sodium and calcium feldspars > biotite, hornblende, and augite > olivine > dolomite and calcite > gypsum. It is expected that this order would be changed slightly depending on the climate and other environmental conditions. This listing accounts for the absence of gypsum, calcite, and dolomite in soils of humid regions and for the predominance of quartz in the coarser fraction of most soils.

11:6. WEATHERING IN ACTION—GENESIS OF PARENT MATERIALS

A physical weakening, caused usually by temperature changes, initiates the weathering process, but it is accompanied and supplemented by certain chemical transformations. Such minerals as the feldspars, mica, hornblende. and the like suffer hydrolysis and hydration, while part of the combined iron is oxidized and hydrated. The minerals soften, lose their lustre, and become more porous. If hematite or limonite is formed, the decomposing mass becomes definitely red or yellow. Otherwise, the colors are subdued.

Coincident with these changes, such cations as calcium, magnesium, sodium, and potassium undergo carbonation and solution. These constituents are removed by leaching, leaving a residue deprived of its easily soluble bases. As the process goes on, all but the most resistant of the original minerals disappear and their places are occupied by (a) secondary hydrated silicates that recrystallize into highly colloidal clay, and (b) more resistant products, such as iron and aluminum oxides (see Fig. 11:2).

The intensity of the various forces will fluctuate with climate. Under arid conditions, the physical forces will dominate and the resultant soil material will contain the less-weathered minerals. Temperature changes, wind action, and water erosion will be accompanied by a minimum of chemical action.

In a humid region, however, these forces are more varied, and practically the full quota will be active. Vigorous chemical changes will accompany disintegration, resulting in greater fineness of the product. Clayey materials will be more evident. Moreover, the processes will be accelerated and intensified by decaying organic matter.

The forces of weathering lose their intensity as one goes downward from the soil surface. Both the physical and chemical transformations are in general somewhat less intense. The lower organic matter content is a factor, as is the decrease in porosity and aeration of the underlayers. This differentiation with depth is the forerunner of a profile development if and when soil genesis takes place.

11:7. GEOLOGICAL CLASSIFICATION OF PARENT MATERIALS

In discussing weathering, we assumed that parent material might lie in its original position above bedrock for centuries or it might be moved to new positions by the mechanical forces of nature. On this basis, two groups of inorganic parent materials, *sedentary* and *transported*, are usually recognized. The second may be subdivided according to the agencies of transportation and deposition as follows (see Fig. 11:3):

1. Sedentary   Still at original site   Residual

2. Transported

$\left\{\begin{array}{l} \text{Gravity} \\ \\ \text{Water} \\ \\ \text{Ice} \\ \text{Wind} \end{array}\right.$

Colluvial

$\left\{\begin{array}{l} \text{Alluvial} \\ \text{Marine} \\ \text{Lacustrine} \end{array}\right.$

Glacial

Aeolian

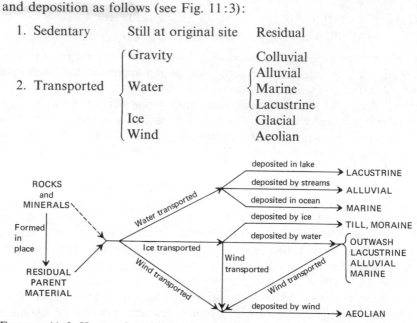

FIGURE 11:3. How various kinds of parent material are formed, transported, and deposited.

While these terms properly relate only to the placement of the parent materials, it has become customary to use them loosely in referring to the soils that have been developed by the weathering of these deposits—for example, glacial soils, alluvial soils, and residual soils. Such groupings are very general, however, since a wide diversity occurs within each soil group.

11:8. RESIDUAL PARENT MATERIAL

Residual parent material develops in place from the underlying rock below and is rarely transported to another site. If typically developed, it has usually experienced long and often intense weathering. In a warm, humid climate, it is likely to be thoroughly oxidized and well leached. Even though it may come from rocks such as limestone, it is often comparatively low in calcium because this constituent has in many cases been leached out. Red and yellow colors are characteristic when weathering has been intense as in the Piedmont Plateau of the eastern United States. In cooler and especially drier climates, residual weathering is much less drastic and the oxidation and hydration of the iron may be hardly noticeable. Also, the lime content is higher and the colors of the debris subdued. Tremendous areas of this type of debris are found on the Great Plains and in other regions of the western United States.

Residual materials are of wide distribution on all of the continents. In the United States, a glance at the physiographic map (Fig. 11:4) shows six great eastern and central provinces—the Piedmont Plateau, Appalachian

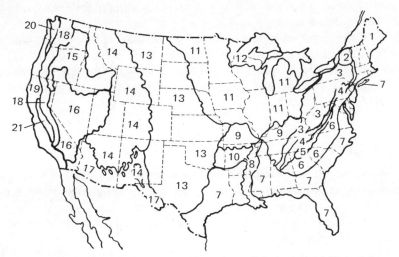

FIGURE 11:4. Generalized physiographic and regolith map of the United States. The regions located are as follows:

1. New England: mostly glaciated crystalline rocks.

2. Adirondacks: glaciated crystalline and sedimentary rocks.

3. Appalachian Mountains and plateaus: shales and sandstones.

4. Limestone valleys and ridges: mostly limestone.

5. Blue Ridge mountains: sandstones and shales.

6. Piedmont Plateau: crystalline rocks.

7. Atlantic and Gulf coastal plain: sands, clays, and limestones.

8. Mississippi flood plain and delta: alluvium.

9. Limestone uplands: mostly limestone and shale.

10. Sandstone uplands: mostly sandstone and shale.

11. Central lowlands: mostly glaciated sedimentary rocks of many kinds. Great areas are overlaid with loess, a wind deposit of great agricultural importance (see Fig. 11:12).

12. Superior uplands: glaciated crystalline and sedimentary.

13. Great Plains region: sedimentary rocks of many kinds.

14. Rocky Mountain region: mountains, uplands, and valleys.

15. Northwest intermountain: mostly igneous crystalline rock. Great areas in Columbia and Snake river basins are covered by loess (see Fig. 11:9).

16. Great Basin: gravels, sands, alluvial fans from various rocks.

17. Southwest arid region: gravel, sand and other debris of desert and mountain.

18. Sierra Nevada and Cascade mountains: mountains uplands, and valleys.

19. Pacific Coast province: mountains and valleys, mostly sedimentary rocks.

20. Puget Sound lowlands: glaciated sedimentary.

21. California central valley: alluvium and outwash.

Mountains and plateaus, the limestone valleys and ridges, the limestone uplands, the sandstone uplands, and the Great Plains region. The first three groups alone encompass about 10 percent of the area of the United States. In addition, great expanses of these sedentary accumulations are found west of the Rocky Mountains.

A great variety of soils occupy these regions covered by residual debris since climate and vegetation, two of the determining factors in soil characterization, vary so radically over this great area. It has already been emphasized that the profile of a mature soil is largely a reflection of climate and its accompanying vegetation.

11:9. COLLUVIAL DEBRIS

Colluvial debris is made up of the fragments of rock detached from the heights above and carried down the slopes mostly by gravity. Frost action has much to do with the development of such deposits. Talus slopes, cliff detritus, and similar heterogeneous materials are good examples. Avalanches are made up largely of such accumulations.

Parent material developed from colluvial accumulation is usually coarse and stony since physical rather than chemical weathering has been dominant. At the base of slopes in regions of medium- to fine-textured material such as loess, some superior soils develop. However, colluvial materials are generally not of great importance in the production of agricultural soils because of their small area, their inaccessibility, and their unfavorable physical and chemical characteristics.

11:10. ALLUVIAL STREAM DEPOSITS

There are three general classes of alluvial deposits: (a) flood plains, (b) alluvial fans, and (c) deltas. They will be considered in order.

FLOOD PLAINS. A stream on a gently inclined bed usually begins to swing from side to side in variable curves, depositing alluvial material on the inside of the curves and cutting on the opposite banks. This results in *oxbows* and *lagoons*, which are ideal for the further deposition of alluvial matter and development of swamps. This state of meander naturally increases the probability of overflow at high water, a time when the stream is carrying much suspended matter. Part of this sediment is deposited over the flooded areas, the coarser near the channel, building up natural levees, the finer farther away in the lagoons and slack water. Thus, there are two distinct types of first bottom deposits—*meander* and *flood*. As might be expected, flood plain deposits are variable, ranging texturally from gravel and sands to silt and clay.

If there is a change in grade, a stream may cut down through its already well-formed alluvial deposits, leaving *terraces* on one or both sides. Often two or more terraces of different heights may be detected along some valleys, marking a time when the stream was at these elevations.

Flood plain deposits are found to a certain extent beside every stream, the greatest development in the United States occurring along the Mississippi (see Fig. 11:5). This area varies from 20 to 75 miles in width, and from Cairo, Illinois, to the Gulf is over 500 miles long. The soils derived from such sediments usually are very rich but they may require drainage and protection from overflow.

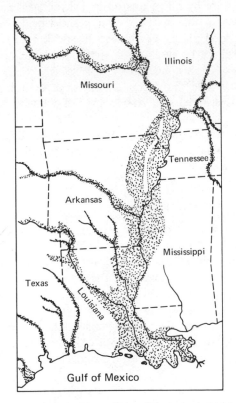

FIGURE 11:5. Flood plain and delta of the Mississippi River. This is the largest continuous area of alluvial soil in the United States.

ALLUVIAL FANS. Where streams descend from uplands, a sudden change in gradient sometimes occurs as the stream emerges at the lower level. A deposition of sediment is thereby forced, giving rise to *alluvial fans*. They differ from deltas in their location and in the character of their debris. Fan material often is gravelly and stony, somewhat porous, and, in general, well drained.

Alluvial fan debris is found over wide areas in arid and semiarid regions. The soils therefrom, when irrigated and properly handled, often prove very productive. In humid regions, especially in certain glaciated sections, such deposits also occur in large enough areas to be of considerable agricultural importance.

DELTA DEPOSITS. Much of the finer sediment carried by streams is not deposited in the flood plain but is discharged into the body of water to which the stream is tributary. Unless there is sufficient current and wave action, some of the suspended material accumulates, forming a delta. Such delta deposits are by no means universal, being found at the mouths of only a small proportion of the rivers of the world. A delta often is a continuation of a flood plain, its front so to speak, and is not only clayey in nature but is likely to be swampy as well. The deltas of the Nile and Po are good examples of these conditions.

Delta sediments, where they occur in any considerable acreage and are subject to flood control and drainage, become rather important agriculturally. The combination deltas and flood plains of the Mississippi, Ganges, Po, Tigris, and Euphrates rivers are striking examples. Egypt, for centuries the granary of Rome, bespeaks the fertility and productivity of soils originating from such parent material.

11:11. MARINE SEDIMENTS

Much of the sediment carried away by stream action is eventually deposited in the oceans, seas, and gulfs, the coarser fragments near the shore, the finer particles at a distance. Also, considerable debris is torn from the shoreline by the pounding of the waves and the undertow of the tides. If there have been changes in shoreline, the alternation of beds will show no regular sequence and considerable variation in topography, depth, and texture. These deposits have been extensively raised above sea level along the Atlantic and Gulf coasts of the United States and elsewhere, and have given origin to large areas of valuable soils (see Fig. 11:6).

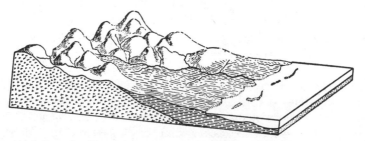

FIGURE 11:6. Location of marine sediments and their relation to the uplands. The emerged coastal plan has already suffered some dissection from stream action.

Marine deposits have been worn and weathered by a number of agencies. First, the weathering and erosion necessary to throw them into stream suspension were sustained. These were followed by the sorting and solvent action of the stream itself. Next, the sediment was swept into the ocean to be deposited and stratified after being pounded and eroded by the waves for years. At last came the emergence of the deposits above the sea and the final action of weathering. The latter effects are of great significance since they determine the topography and, to a considerable extent, the chemical and physical nature of the resultant parent material.

Marine sediments, although having been subjected to weathering a shorter time than some of the residual debris, are generally more worn and usually carry less of the mineral nutrient elements. Their silica content is high and they are often sandy, especially along the Atlantic seaboard of the United States. But in the Atlantic and Gulf coastal flatwoods and the interior pinelands of Alabama and Mississippi, clayey deposits are not uncommon. These marine clays may be dominated by kaolinite, illite, or montmorillonite.

In the continental United States, the marine deposits of the Atlantic and Gulf coastal provinces (see Fig. 11:4) occupy approximately 11 percent of the country and are very diversified, owing to source of material, age, and the climatic conditions under which they now exist. Severe leaching as well as serious erosion occurs in times of heavy rainfall. In spite of this, however, their soils support a great variety of crops when adequately supplied with organic matter properly cultivated and carefully fertilized.

11:12. The Pleistocene Ice Age

During the Pleistocene, northern North America, northern and central Europe, and parts of northern Asia were invaded by a succession of great ice sheets. Certain parts of South America and areas in New Zealand and Australia were similarly affected. Antarctica undoubtedly was capped with ice much as it is today.

It is estimated that the Pleistocene ice at its maximum extension covered perhaps 20 percent of the land area of the world. It is surprising to learn that present-day glaciers, which we consider as mere remnants of the Great Ice Age, occupy about one third as much land surface. The respective amounts of ice, however, are not in the same proportion, since our living glaciers are comparatively thin and are definitely on the wane.

In North America, the major centers of ice accumulation were in central Labrador and the western Hudson Bay region, with a minor concentration in the Canadian Rockies. From the major centers, great continental glaciers pushed outward in all directions but especially southward, covering from time to time most of what is now Canada and the northern part of the United States. The southernmost extension was down the Mississippi

Valley, where the least resistance was met because of the lower and smoother topography (see Fig. 11:7).

Central North America and Europe apparently sustained four distinct ice invasions. In the United States they are identified as Nebraskan, Kansan, Illinoian, and Wisconsin. The glacial debris of the eastern United States and Canada is practically all of Wisconsin age. These invasions were separated by long interglacial ice-free intervals which are estimated to

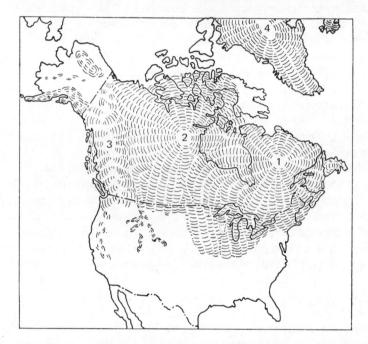

FIGURE 11:7. Maximum development of continental glaciation in North America. The four centers of ice accumulation are numbered. Apparently the eastern and central United States were invaded from the Labradorian (1) and Hudson Bay (2) centers. Note the marked southerly advance of the ice in the Mississippi Valley, where topography offered little resistance. To the east the Appalachian highlands more or less blocked the ice invasion.

have covered a total time period considerably longer than the periods of glaciation. Some of the interglacial intervals evidently were times of warm or ,semitropical climate in regions that are now definitely temperate. The total length of the Pleistocene ice age is estimated at 1 to 1.5 million years (see Table 11:4). According to recent studies using radioactive carbon (^{14}C), the glacial ice disappeared from northern Iowa and central New York possibly only about 12,000 years ago. We may be now enjoying the mildness of another interglacial period.

TABLE 11:4. *Nomenclature Used to Identify Periods of Glaciation and*
Interglaciation in North America During the Pleistocene Period

Period Name	Type of Period	Approximate Date of Starting
Modern period	Interglacial	10,000 B.C.
Wisconsin	Glacial	
Sangamonian	Interglacial	
Illinoian	Glacial	
Yarmouthian	Interglacial	
Kansan	Glacial	
Aftonian	Interglacial	
Nebraskan	Glacial	1–1.5 million B.C.

Of the major interglacial periods, the first (Aftonian) persisted perhaps 200,000 years, the second (Yarmouth) may have lasted for 300,000 years, and the third (Sangamon), which preceded the Wisconsin ice advance, possibly endured for 125,000 years. Assuming the Pleistocene age to have continued for not more than 1,000,000 years, it is obvious that glacial ice occupied what is now the north central and northeastern United States less than half of the period designated as the Great Ice Age.

As the glacial ice was pushed forward, it conformed to the unevenness of the areas invaded. It rose over hills and mountains with surprising ease. Not only was the existing regolith with its mantle of soil swept away but hills were rounded, valleys filled, and, in some cases, the underlying rocks were severely ground and gouged. Thus, the glacier became filled with rock wreckage, carrying much of its surface and pushing great masses ahead (see Fig. 11:8). Finally, when the ice melted and the region again was free, a mantle of glacial drift remained—a new regolith and fresh parent material for soil formation.

The area covered by glaciers in North America is estimated as 4,000,000 square miles, while perhaps 20 percent of the United States is influenced by the deposits. An examination of Figs. 11:7 and 11:9 will indicate the magnitude of the ice invasion at maximum glaciation in this country.

11:13. GLACIAL TILL AND ASSOCIATED DEPOSITS

The materials deposited directly by the ice are commonly called *glacial till*. Till is a mixture of rock debris of great diversity, especially as to size of particles. Boulder clay, so common in glaciated regions, is typical of the physical heterogeneity expected.

Glacial till is found mostly as irregular deposits called *moraines*, of which there are various kinds. *Terminal* moraines characterize the southernmost extension of the various glacial lobes when the ice margin was stationary

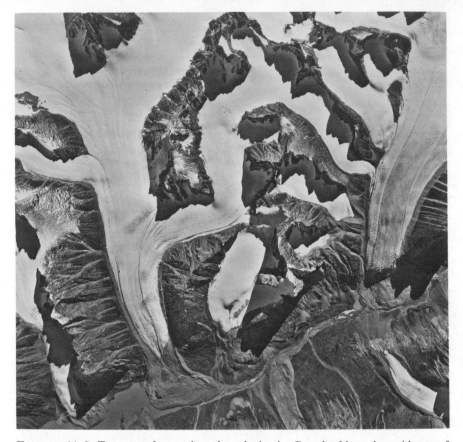

FIGURE 11:8. Tongues of a modern-day glacier in Canada. Note the evidence of transport of materials by the ice and the "glowing" appearance of the two major ice lobes. [*Photo* A–16817–102 *courtesy National Air Photo Library, Surveys and Mapping Branch, Canadian Department of Energy, Mines and Resources.*]

long enough to permit an accumulation of debris (see Fig. 11:10). Many other moraines of a *recessional* nature are found to the north, marking points where the ice front became stationary for a time as it receded by melting. While moraines of this type are generally outstanding topographic features, they give rise to soils that are rather unimportant due to their small area and unfavorable physiography.

The *ground moraine*, a thinner and more level deposition laid down as the ice front retreated rapidly, is of much more importance. It has the widest extent of all glacial deposits and usually possesses a favorable agricultural topography. Associated with the moraine in certain places are such special features as *kames, eskers*, and *drumlins*.

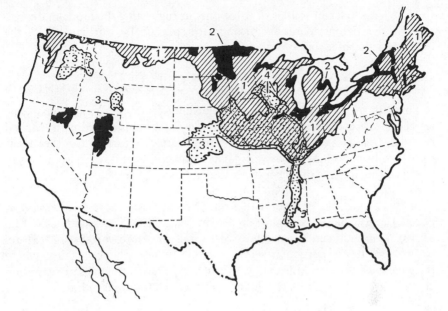

FIGURE 11:9. Areas in the United States covered by the continental ice sheet and the deposits either directly from, or associated with, the glacial ice. (1) till deposits of various kinds; (2) glacial-lacustrine deposits; (3) the loessial blanket—note that the loess overlies great areas of till in the Midwest; (4) an area, mostly in Wisconsin, that escaped glaciation. It is partially loess covered.

An outstanding feature of glacial till materials is their variability. This is because of the diverse ways by which the debris was laid down, of differences in the chemical composition of the original materials, and of fluctuations in the grinding action of the ice. As might be expected, the soils derived from such soil material are most heterogeneous. Such variations indicate

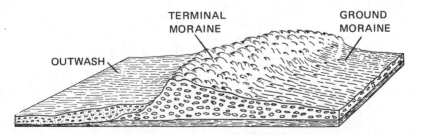

FIGURE 11:10. Relationship between a terminal moraine, its outwash, and its associated ground moraine. Note the differences in topography and in the general nature of the three deposits. The ground moraine is the most widely distributed. The direction of ice retreat is to the right.

that the term "glacial soil" is of value only in suggesting the mode of deposition of the parent materials. It indicates practically nothing about the characteristics of a soil so designated.

11:14. GLACIAL OUTWASH AND LACUSTRINE SEDIMENTS

Torrents of water were constantly gushing from the great Pleistocene ice sheet, especially during the summer. The vast loads of sediment carried by such streams were either dumped immediately or carried to other areas for deposition. So long as the water had ready egress, it flowed rapidly away to deposit its load as outwash of various kinds.

OUTWASH PLAINS. A type of deposit of great importance is the *outwash* plain, formed by streams heavily laden with glacial sediment (see Fig. 11:10). This sediment is usually assorted and therefore variable in texture. Such deposits are particularly important in valleys and on plains, where the glacial waters were able to flow away freely. These valley fills are common in the United States, both north and south of the terminal moraine.

GLACIAL LAKE DEPOSITS. In many cases the ice front came to a standstill where there was no such ready escape for the water, and ponding occurred as a result of damming action of the ice (see Fig. 11:11). Often, very large lakes were formed which existed for many years. Particularly prominent were those south of the Great Lakes in New York, Ohio, Indiana, Michigan, and in the Red River Valley (see Fig. 11:9). The latter lake, called Glacial

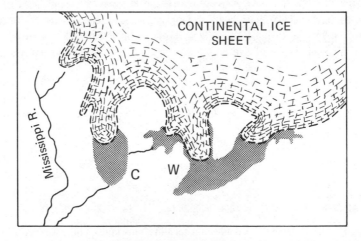

FIGURE 11:11. When conditions of topography were favorable, the Wisconsin ice sheet acted as a great waning dam. A stage in the development of glacial lakes in Chicago (C) and Warren (W) is represented. [*After Daly* (2).]

Lake Agassiz, was about 750 miles long and 250 miles wide at its maximum extension. The others, individually much smaller, covered great stretches of country. Large lakes also occurred in the intermountain regions west of the Rockies as well as in the Connecticut Valley in New England and elsewhere.

With the glacial ice melting rapidly in the hills and higher valleys, these lakes were constantly fed by torrents from above which were laden with sediment derived not only directly from the ice but also from the unconsolidated till sheet over which they flowed. Consequently, large deposits were made in these glacial lakes, ranging from coarse delta materials near the shore to fine silts and clay in the deeper and stiller waters. Such materials now cover large areas both in the United States and Canada, and their weathering has given rise to what are loosely designated as *glacial lake* soils.

The soils developed from these lake sediments are most heterogeneous. Because of climatic differences, weathering has been variable, and profile contrasts are great. Extending westward from New England along the Great Lakes to the broad expanse of the Red River Valley, these deposits have produced some of the most important soils of the Northern states. They also occur in the intermountain regions of the United States, where they have given rise to agriculturally important soils, especially when irrigated.

11:15. GLACIAL-AEOLIAN DEPOSITS

During the glaciation, much fine material was carried miles below the front of the ice sheets by streams that found their source within the glaciers. This sediment was deposited over wide areas by the overloaded rivers. The accumulations occurred below the ice front at all points; in the United States they reached their greatest development in what are now the Mississippi and Missouri valleys. Some of the debris found on the Great Plains probably had a similar origin, coming from glaciers debouching from the Rockies. All this, added to the great stretches of unconsolidated till in the glaciated regions and the residual material devoid of vegetation on the Great Plains, presented unusually favorable conditions for wind erosion in dry weather.

ORIGIN AND LOCATION OF LOESS. Wind erosion was responsible for the deposition of a very important soil material in the midwest and in parts of the northwest. In the Mississippi Valley, extensive deposits of glacial outwash left on river flood plains during the summer were subject to wind movement during the winter when melting had stopped and the flood plains were dry. This wind-blown material, called *loess*, was deposited in the uplands, the thickest deposits being found where the valleys were widest. The fine silty material covered existing soils and parent materials, both original and glacial in origin.

Loess is found over wide areas in the United States, concealing the unconsolidated material below. It covers eastern Nebraska and Kansas, southern and central Iowa and Illinois, northern Missouri, and parts of southern Ohio and Indiana, besides a wide band extending southward along the eastern border of the Mississippi River (see Figs. 11:9 and 11:12). Extensive loessial deposits also occur in the Palouse region in Washington and Idaho and other areas of northwestern United States. Along the Missouri and Mississippi rivers, its accumulation has formed great bluffs which impart a characteristic topography to the region. Farther from the rivers, the deposits, an upland type, are shallower and smoother in topography.

NATURE OF LOESS. Loess is usually silty in character and has a yellowish-buff color unless it is very markedly weathered or carries a large amount of humus. The larger particles are usually unweathered and angular. Quartz seems to predominate, but large quantities of feldspar, mica, hornblende, and augite are found. The vertical walls and ridges formed when this deposit is deeply eroded constitute one of its most striking physical characteristics.

Loess has given rise in the central United States and elsewhere to soils of considerable diversity. A comparison of the loessial area of the midwest (Fig. 11:12) with the map of soil orders (Plate 1, p. 304) will show the presence of loess in six distinct soil regions. This situation, while indicating the probability of great differences in the fertility and productivity of loess soils, especially emphasizes the influence of climate as the final determinant of soil characteristics.

OTHER AEOLIAN DEPOSITS. Aeolian deposits other than loess occur, such as volcanic ash and sand dunes. Soils from volcanic ash occur in Montana, Idaho, Nebraska, and Kansas. They are light and porous and not always of great agricultural value.

Sand dunes are of little value under any condition and become a menace to agriculture if they are moving. Examples of such deposits are found in the great Sahara and Arabian deserts. Smaller areas occur in this country in Nebraska, Colorado, New Mexico, and along the eastern seaboard.

11:16. AGRICULTURAL SIGNIFICANCE OF GLACIATION

The Pleistocene glaciation in most cases has been of some benefit, especially agriculturally. The leveling and filling actions when drift was abundant have given a smoother topography more suited to farming operations. The same can be said for the glacial lake sediments and loess deposits. Also, the parent materials thus supplied are geologically fresh, and the soils derived therefrom usually are not drastically leached. Young soils generally are higher in available nutrients and, under comparable conditions, superior to crop-producing power.

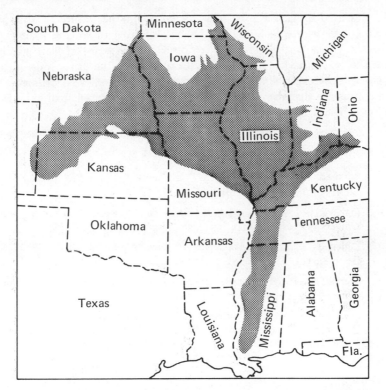

FIGURE 11:12. Approximate distribution of loess in central United States. The soil that has developed therefrom is generally a silt loam, often somewhat sandy. Note especially the extension down the eastern side of the Mississippi River and the irregularities of the northern extensions. Smaller areas of loess occur in Washington, Oregon, and Idaho (see Fig. 11:9).

TABLE 11:5. *Relationship Between Certain Parent Materials in Pennsylvania and Soil Drainage Class*[a]

		Percent of Profiles in Drainage Classes	
Parent Material Class	Number of Profiles	Well and Moderately Well Drained	Somewhat Poorly, Poorly, and Very Poorly Drained
Alluvium	23	81	19
Limestone	23	100	0
Till	52	65	35
Shale	77	74	26
Sandstone	21	100	0

[a] From Peterson et al. (5).

In glaciated areas there is great variation in the nature of soils which develop. The drainage characteristics of a number of soils in Pennsylvania shown in Table 11:5 illustrate this point. Soils developed from glacial till tended to be somewhat more poorly drained than the other soils, although soils developed from shale and alluvium were not all well drained. These data emphasize that the characteristics of parent material strongly influence the kinds of soil which develop.

REFERENCES

(1) Barshad, I., "Soil Development," in *Chemistry of the Soil,* Bear, F. E., Ed. (New York: Reinhold, 1955), pp. 1–52.

(2) Daly, R. A., *The Changing World of the Ice Age* (New Haven, Yale University Press, 1934).

(3) Diller, J. S., *Educational Series of Rock Specimens*, Bull. 150, U. S. Geol. Survey, 1898, p. 385.

(4) Merrill, G. P., *Weathering of Micaceous Gneiss*, Bull. 8, Geol. Soc. Amer., 1879, p. 160.

(5) Peterson, G. W., Cunningham, R. L., and Matelski, R. P., "Moisture Characteristics of Pennsylvania Soils: III. Parent Material and Drainage Relationships," *Soil Sci. Soc. Amer. Proc.,* **35**:115–19, 1971.

Chapter 12

SOIL FORMATION, CLASSIFICATION, AND SURVEY

To study satisfactorily any heterogeneous group in nature, some sort of classification is necessary. This is especially true of soils. The value of experimental work of any kind is seriously restricted and may even be misleading unless the relation of one soil to another is known. The crop requirements in any region depend to a marked degree on the soils in question and on their profile similarities and differences.

In arriving at such an understanding, three phases must be considered: (a) soil genesis, or the evolution of a soil from its parent material; (b) soil classification, in this case especially as it applies to the United States; and (c) soil survey, its interpretation and utilization. Soil genesis will receive attention first, starting with the five major factors influencing soil formation.

12:1. FACTORS INFLUENCING SOIL FORMATION

Studies of soils throughout the world have shown that the kinds of soil that develop are largely controlled by five major factors:

1. *Climate* (particularly temperature and precipitation).
2. *Living organisms* (especially the native vegetation).
3. *Nature of parent material* $\begin{cases} \text{Texture and structure.} \\ \text{Chemical and mineralogical composition.} \end{cases}$
4. *Topography of area.*
5. *Time* that parent materials are subjected to soil formation.

CLIMATE. From an over-all standpoint, climate is perhaps the most influential factor. It determines in no small degree the nature of the weathering that occurs. For example, temperature and precipitation exert profound influences on the rates of chemical and physical processes—the essential means by which profile development is affected. Consequently, if allowed ample opportunity, climatic influences eventually tend to dominate the soil formation picture.

PATTERNS OF SOIL ORDERS AND SUBORDERS OF THE UNITED STATES

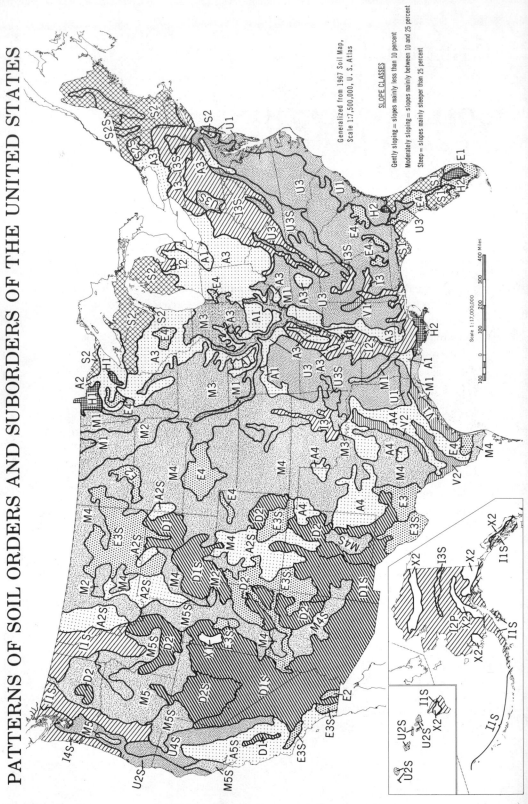

Generalized from 1967 Soil Map,
Scale 1:7,500,000, U. S. Atlas

SLOPE CLASSES

Gently sloping = slopes mainly less than 10 percent

Moderately sloping = slopes mainly between 10 and 25 percent

Steep = slopes mainly steeper than 25 percent

Scale 1:17,000,000

100   0   100   200   300   400 Miles

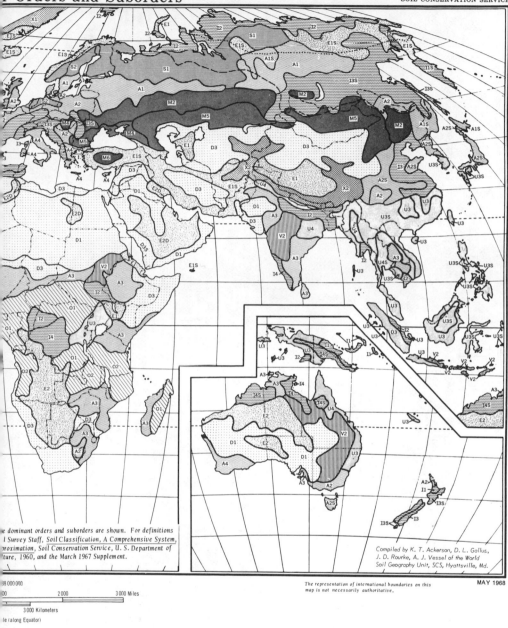

e dominant orders and suborders are shown. For definitions
l Survey Staff, Soil Classification, A Comprehensive System,
roximation, Soil Conservation Service, U. S. Department of
lture, 1960, and the March 1967 Supplement.

Compiled by K. T. Ackerson, D. L. Gallut,
J. D. Rourke, A. J. Vessel of the World
Soil Geography Unit, SCS, Hyattsville, Md.

88 000 000

| 00 | 2000 | 3 000 Miles |

3 000 Kilometers

le (along Equator)

The representation of international boundaries on this
map is not necessarily authoritative.

MAY 1968

borders according to the new comprehensive soil classification system. (Map courtesy Soil

PLATE 1. Patterns of soil orders and suborders of the United States using the new comprehensive soil classification system. (Map courtesy Soil Survey Division, U. S. Soil Conservation Service.)

U. S. DEPARTMENT OF AGRICULTURE

LEGEND

Only the dominant orders and suborders are shown. Each delineation has many inclusions of other kinds of soil. General definitions for the orders and suborders follow. For complete definitions see Soil Survey Staff, Soil Classification, A Comprehensive System, 7th Approximation, Soil Conservation Service, U. S. Department of Agriculture, 1960 (for sale by U. S. Government Printing Office) and the March 1967 supplement (available from Soil Conservation Service, U. S. Department of Agriculture). Approximate equivalents in the modified 1938 soil classification system are indicated for each suborder.

ALFISOLS . . . Soils with gray to brown surface horizons, medium to high base supply, and subsurface horizons of clay accumulation; usually moist but may be dry during warm season

A1   AQUALFS (seasonally saturated with water) gently sloping; general crops if drained, pasture and woodland if undrained (Some Low–Humic Gley soils and Planosols)

A2   BORALFS (cool or cold) gently sloping; mostly woodland, pasture, and some small grain (Gray Wooded soils)

A2S BORALFS steep; mostly woodland

A3   UDALFS (temperate or warm, and moist) gently or moderately sloping; mostly farmed, corn, soybeans, small grain, and pasture (Gray–Brown Podzolic soils)

HISTOSOLS . . . Organic soils

H1   FIBRISTS (fibrous or woody peats, largely undecomposed) mostly wooded or idle (Peats)

H2   SAPRISTS (decomposed mucks) truck crops if drained, idle if undrained (Mucks)

INCEPTISOLS . . . Soils that are usually moist, with pedogenic horizons of alteration of parent materials but not of accumulation

I1S ANDEPTS (with amorphous clay or vitric volcanic ash and pumice) gently sloping to steep; mostly woodland; in Hawaii mostly sugar cane, pineapple, and range (Ando soils, some Tundra soils)

I2   AQUEPTS (seasonally saturated with water) gently sloping; if drained, mostly row crops, corn, soybeans, and cotton; if undrained, mostly woodland or pasture (Some Low–Humic Gley soils and Alluvial soils)

I2P AQUEPTS (with continuous or sporadic permafrost) gently sloping to steep; woodland or idle (Tundra soils)

I3   OCHREPTS (with thin or light–colored surface horizons and little organic matter) gently to moderately sloping; mostly pasture, small grain, and hay (Sols Bruns Acides and some Alluvial soils)

I3S OCHREPTS gently sloping to steep; woodland, pasture, small grains

I4S UMBREPTS (with thick dark–colored surface horizons rich in organic matter) moderately sloping to steep; mostly woodland (Some Regosols)

ULTISOLS . . . Soils that are usually moist with horizon of clay accumulation and a low base supply

U1   AQUULTS (seasonally saturated with water) gently sloping; woodland and pasture if undrained, feed and truck crops if drained (Some Low–Humic Gley soils)

U2S HUMULTS (with high or very high organic–matter content) moderately sloping to steep; woodland and pasture if steep, sugar cane and pineapple in Hawaii, truck and seed crops in Western States (Some Reddish–Brown Lateritic soils)

U3   UDULTS (with low organic–matter content; temperate or warm, and moist) gently to moderately sloping; woodland, pasture, feed crops, tobacco, and cotton (Red–Yellow Podzolic soils, some Reddish–Brown Lateritic soils)

U3S UDULTS moderately sloping to steep; woodland, pasture

U4S XERULTS (with low to moderate organic–matter content, continuously dry for long periods in summer) range and woodland (Some Reddish–Brown Lateritic soils)

VERTISOLS . . . Soils with high content of swelling clays and wide deep cracks at some season

V1   UDERTS (cracks open for only short periods, less than 3 months in a year) gently sloping; cotton, corn, pasture, and some rice (Some Grumusols)

V2   USTERTS (cracks open and close twice a year and remain open more than 3 months); general crops, range, and some irrigated crops (Some Grumusols)

AREAS with little soil . . .

X1   Salt flats

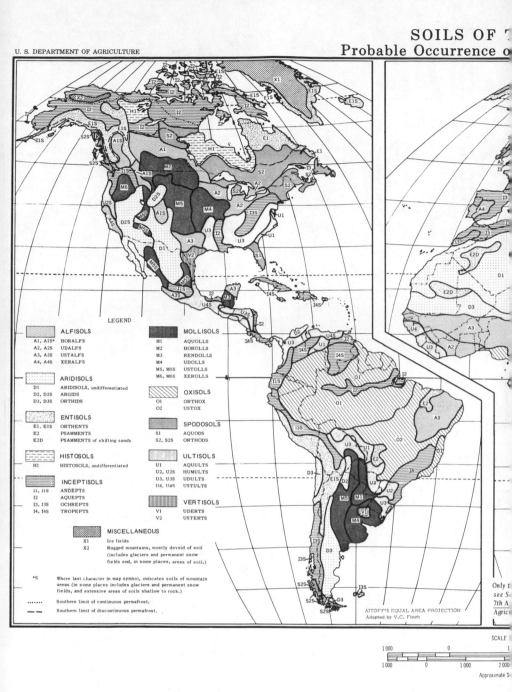

LEGEND

	ALFISOLS		MOLLISOLS
A1, A1S*	BORALFS	M1	AQUOLLS
A2, A2S	UDALFS	M2	BOROLLS
A3, A3S	USTALFS	M3	RENDOLLS
A4, A4S	XERALFS	M4	UDOLLS
		M5, M5S	USTOLLS
	ARIDISOLS	M6, M6S	XEROLLS
D1	ARIDISOLS, undifferentiated		OXISOLS
D2, D2S	ARGIDS	O1	ORTHOX
D3, D3S	ORTHIDS	O2	USTOX
	ENTISOLS		SPODOSOLS
E1, E1S	ORTHENTS	S1	AQUODS
E2	PSAMMENTS	S2, S2S	ORTHODS
E2D	PSAMMENTS of shifting sands		
	HISTOSOLS		ULTISOLS
H1	HISTOSOLS, undifferentiated	U1	AQUULTS
		U2, U2S	HUMULTS
	INCEPTISOLS	U3, U3S	UDULTS
I1, I1S	ANDEPTS	I14, I14S	USTULTE
I2	AQUEPTS		
I3, I3S	OCHREPTS		VERTISOLS
I4, I4S	TROPEPTS	V1	UDERTS
		V2	USTERTS
	MISCELLANEOUS		
X1	Ice fields		
X2	Rugged mountains, mostly devoid of soil (includes glaciers and permanent snow fields and, in some places, areas of soil.)		

*S   Where last character in map symbol, indicates soils of mountain
       areas (in some places includes glaciers and permanent snow
       fields, and extensive areas of soils shallow to rock.)
.... Southern limit of continuous permafrost.
— —   Southern limit of discontinuous permafrost.

AITOFF'S EQUAL AREA PROJECTION
Adapted by V.C. Finch

Only t
see S
7th A
Agric

SCALE

1 000   0   1
1 000   0   1 000   2 000
Approximate S

PLATE 2. Generalized world soil map showing the probable occurrence of orders and su
Survey Division, U. S. Soil Conservation Service.)

A4 USTALFS (warm and intermittently dry for long periods) gently or moderately sloping; range, small grain, and irrigated crops (Some Reddish Chestnut and Red–Yellow Podzolic soils)

A5S XERALFS (warm and continuously dry in summer for long periods, moist in winter) gently sloping to steep; mostly range, small grain, and irrigated crops (Noncalcic Brown soils)

ARIDISOLS . . . Soils with pedogenic horizons, low in organic matter, and dry more than 6 months of the year in all horizons

D1 ARGIDS (with horizon of clay accumulation) gently or moderately sloping; mostly range, some irrigated crops (Some Desert, Reddish Desert, Reddish–Brown, and Brown soils and associated Solonetz soils)

D1S ARGIDS gently sloping to steep

D2 ORTHIDS (without horizon of clay accumulation) gently or moderately sloping; mostly range and some irrigated crops (Some Desert, Reddish Desert, Sierozem, and Brown soils, and some Calcisols and Solonchak soils)

D2S ORTHIDS gently sloping to steep

ENTISOLS . . . Soils without pedogenic horizons

E1 AQUENTS (seasonally saturated with water) gently sloping; some grazing

E2 ORTHENTS (loamy or clayey textures) deep to hard rock; gently to moderately sloping; range or irrigated farming (Regosols)

E3 ORTHENTS shallow to hard rock; gently to moderately sloping; mostly range (Lithosols)

E3S ORTHENTS shallow to hard rock; steep; mostly range

E4 PSAMMENTS (sand and loamy sand textures) gently to moderately sloping; mostly range in dry climates, woodland or cropland in humid climates (Regosols)

MOLLISOLS . . . Soils with nearly black, organic-rich surface horizons and high base supply

M1 AQUOLLS (seasonally saturated with water) gently sloping; mostly drained and farmed (Humic Gley soils)

M2 BOROLLS (cool or cold) gently or moderately sloping, some steep slopes in Utah; mostly small grain in North Central States, range and woodland in Western States (Some Chernozems)

M3 UDOLLS (temperate or warm, and moist) gently or moderately sloping; mostly corn, soybeans, and small grains (Some Brunizems)

M4 USTOLLS (intermittently dry for long periods during summer) gently to moderately sloping; mostly wheat and range in western part, wheat and corn or sorghum in eastern part, some irrigated crops (Chestnut soils and some Chernozems and Brown soils)

M4S USTOLLS moderately sloping to steep; mostly range or woodland

M5 XEROLLS (continuously dry in summer for long periods, moist in winter) gently to moderately sloping; mostly wheat, range, and irrigated crops (Some Brunizems, Chestnut, and Brown soils)

M5S XEROLLS moderately sloping to steep; mostly range

SPODOSOLS . . . Soils with accumulations of amorphous materials in subsurface horizons

S1 AQUODS (seasonally saturated with water) gently sloping; mostly range or woodland; where drained in Florida, citrus and special crops (Ground-Water Podzols)

S2 ORTHODS (with subsurface accumulations of iron, aluminum, and organic matter) gently to moderately sloping; woodland, pasture, small grains, special crops (Podzols, Brown Podzolic soils)

S2S ORTHODS steep; mostly woodland

X2 Rockland, ice fields

NOMENCLATURE

The nomenclature is systematic. Names of soil orders end in sol (L. solum, soil), e. g., ALFISOL, and contain a formative element used as the final syllable in names of taxa in suborders, great groups, and sub-groups.

Names of suborders consist of two syllables, e. g., AQUALF. Formative elements in the legend for this map and their connotations are as follows:

and – Modified from Ando soils; soils from vitreous parent materials

aqu – L. aqua, water; soils that are wet for long periods

arg – Modified from L. argilla, clay; soils with a horizon of clay accumulation

bor – Gr. boreas, northern; cool

fibr – L. fibra, fiber; least decomposed

hum – L. humus, earth; presence of organic matter

ochr – Gr. base of ochros, pale; soils with little organic matter

orth – Gr. orthos, true; the common or typical

psamm – Gr. psammos, sand; sandy soils

sapr – Gr. sapros, rotten; most decomposed

ud – L. udus, humid; of humid climates

umbr – L. umbra, shade; dark colors reflecting much organic matter

ust – L. ustus, burnt; of dry climates with summer rains

xer – Gr. xeros, dry; of dry climates with winter rains

AUGUST 1967

For every 10°C rise in temperature, the rate of chemical reactions doubles. Furthermore, biochemical changes by soil organisms are most sensitive to temperature as well as moisture. The influence of temperature and effective moisture on the organic matter contents of soils (see Fig. 6:8) in the Great Plains of the United States is evidence of the strong influence of climate on soil characteristics. The very modest profile development characteristic of arid areas contrasted with the deep-weathered profiles of the humid tropics is further evidence of climatic control.

Figure 12:1 shows the distribution of major vegetative types in the United States. Note the general correlation between climate and vegetation. Comparison of Figs. 12:1 and 12:7 will indicate the general relationship between natural vegetation and the kinds of soil that develop.

Climatic influences are also expressed through or in combination with the other factors. Thus, much of the influence of climate is due to the measure of control which it exercises over natural vegetation. In humid regions, plentiful rainfall provides an environment favorable for the growth of

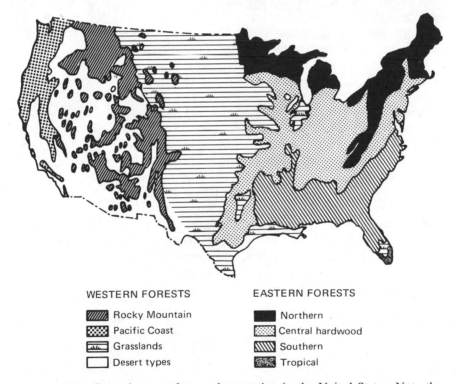

WESTERN FORESTS   EASTERN FORESTS

▨ Rocky Mountain   ■ Northern
▧ Pacific Coast   ▨ Central hardwood
▨ Grasslands   ▨ Southern
□ Desert types   ▨ Tropical

FIGURE 12:1. General types of natural vegetation in the United States. Note the relationship between the dominant vegetation and the kinds of soil which develop (see Figure 12:7 and Plate 1). [*Redrawn from a more detailed map by the U. S. Geological Survey.*]

trees. In contrast, grasslands are the dominant native vegetation in semiarid regions. Thus, climate exerts much of its influence through a second soil-forming factor, the living organisms.

LIVING ORGANISMS. The major role of living organisms in profile differentiation cannot be overemphasized. Organic matter accumulation, profile mixing, nutrient cycling, and structural stability are all made possible by the presence of organisms in the soil. Also, nitrogen is added to the soil system by microorganisms alone or in association with plants. And vegetative cover reduces natural erosion rates, thereby slowing down the rate of mineral surface removal. It is obvious that the nature and number of organisms growing in and on the soil will play a vital role in the kind of soil that develops.

The most striking evidence of the effect of vegetation on soil formation is seen in comparing properties of soils formed under grassland and forest vegetation (see Fig. 12:2). For example, the organic matter content of the "grassland" soils is generally much higher, especially in the subsurface horizons. This gives the soil darker color and higher moisture- and cation-holding capacity as compared to the "forest" soil. Also, structural stability tends to be encouraged by the grassland vegetation.

The mineral element content in the leaves, limbs, and stems of the natural vegetation strongly influences the characteristics of the soil which develop. Coniferous trees, for example, tend to be low in metallic cations such as calcium, magnesium, and potassium. The cycling of nutrients from the litter falling from these trees (see Fig. 12:3) will be low compared to that of some deciduous trees, the litter of which is much higher in bases. Consequently, soil acidity is much more likely to develop under pine vegetation with its low base status than under ash or elm trees. Also the removal of bases by leaching is encouraged by coniferous vegetation.

As might be expected, there is an interaction among the natural vegetation, other soil organisms, and the characteristics of soil which develops. Thus, in soils developed under native grasslands, nonsymbiotic nitrogen fixers (*Azotobacter*) tend to flourish. This provides an opportunity for an increase in both the nitrogen and organic matter contents of the soil. Vegetation is thus seen to be a critical factor in determining soil character.

PARENT MATERIAL. The nature of the parent material, even in humid regions where weathering intensity is high, may still profoundly influence the characteristics of even fully mature soils. For example, the texture of sandy soils is determined largely by parent materials. In turn, the downward movement of water is controlled largely by the texture of the parent material. The chemical and mineralogical compositions of parent material often not only determine the effectiveness of the weathering forces but in some instances partially control the natural vegetation. The presence of limestone

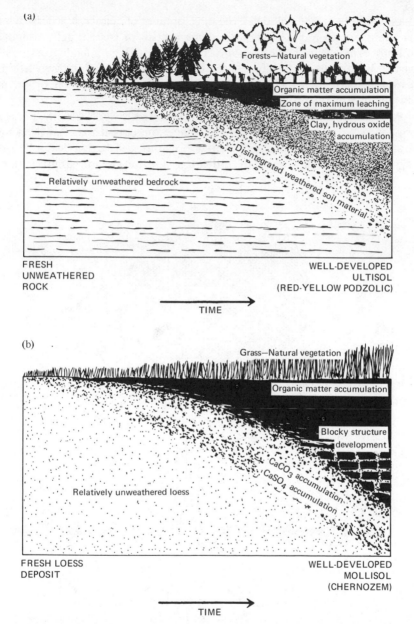

FIGURE 12:2. How two soil profiles may have developed (a) from the weathering in place of solid rock and (b) from the weathering of wind-deposited loess. Organic matter accumulation in the upper horizons occurs in time, the amount and distribution depending on the type of natural vegetation present. Clay and iron oxide accumulate and characteristic structures develop in the lower horizons. The end products differ markedly from the soil materials from which they form.

in a humid region soil will delay the development of acidity, a process which the climate encourages. In addition, the species of trees usually found on limestone materials are commonly relatively high in metallic cations or bases. By continually bringing these bases to the surface, the vegetation is responsible for a further delay in the process of acidification or, in this particular case, the progress of soil development.

Parent material has a marked influence on the type of clay minerals present in the soil profile. In the first place, the parent material itself may contain clay minerals, perhaps from a previous weathering cycle. Second, as our study of clay minerals indicated, the nature of parent material

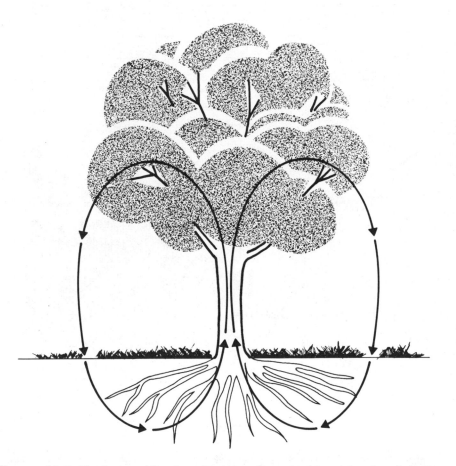

FIGURE 12:3. Nutrient recycling is an important factor in determining the relationship between vegetation and the soil that develops. If residues from the vegetation are low in bases, acid weathering conditions are favored. High-base-containing plant residues tend to neutralize acids, thereby favoring only slightly acid to neutral weathering conditions.

definitely influences the nature of the clay which develops. Illite, for example, tends to form from the mild weathering of potash-containing micas, while montmorillonite is favored by base-rich minerals high in calcium and magnesium. Certainly the kind of soil which develops would be markedly affected by the nature of the clay minerals present. These in turn are greatly influenced by parent material.

TOPOGRAPHY. The topography of the land may be such as to hasten or delay the work of climatic forces. In smooth flat country the rapidity with which excess water is removed is much less than if the landscape were rolling. The rolling topography encourages some natural erosion of the surface layers, which, if extensive enough, may eliminate the possibility of a deep soil. On the other hand, if water stands for part or all of the year on a given area, the climatic influences become relatively ineffective in regulating soil development.

There is a definite interaction among topography, vegetation, and soil formation. In the grassland–forest transition zones, trees commonly occupy the slight depressions in an otherwise prairie vegetation. This is apparently a moisture effect. As would be expected, however, the nature of the soil in the depressions is quite different from that in the uplands. Topography, therefore, is significant not only as a modifier of climatic and vegetative effects but often as a major control in local areas.

TIME. The actual length of time that materials have been subjected to weathering plays a significant role in soil formation. Perhaps the best evidence of the importance of time may be obtained by a comparison of the soils of a glaciated region with those in a comparable area untouched by the ice sheet. As a general rule, the influence of parent material is much more apparent in the soils of glaciated regions, where insufficient time has elapsed since the disappearance of the ice to permit the changes in the composition of the glacial deposit that are necessary for full development of soils.

Soils located on alluvial or lacustrine materials have generally not had as much time to develop as the surrounding upland soils. Some coastal plain parent materials have been uplifted only in recent geological time, and consequently the soils thereon have had relatively little time for weathering.

The interaction among the various factors affecting soil formation is obvious but must be emphasized. Seldom is a single factor solely responsible for the kind of soil that develops. Vegetation effects are definitely related to climate and to a lesser degree to parent material and topography. The time it takes for a horizon to develop will definitely be related to the parent material, the climate, and the vegetation. It is easy to visualize the interdependence of these factors in determining the kind of soil that develops.

12:2. WEATHERING AND SOIL PROFILE DEVELOPMENT

As shown in Chapter 11, mineral soils have originated from the uncon-solidated materials (the regolith) that cover the country rock. The weathering processes of disintegration and decomposition which have given rise to this regolith are in general destructive processes. Thus, rocks and minerals are destroyed or altered. And soluble nutrients are subject to loss by leaching. It seems incongruous that these same seemingly destructive processes can promote the genesis of natural bodies called soils, and yet this is the case.

There are no distinct stages in the development of soils. Even if stages are identified, they seem to overlap and blend together to give a continuum of genetic processes. Only a few of the obvious happenings can be identified as movement occurs from solid rock or recently deposited soil material to a well-developed soil profile.

DISINTEGRATION AND SYNTHESIS. The process of disintegration of solid rock makes possible a foothold for living organisms. Decomposing minerals release nutrients which nourish simple plant and animal forms. Alteration of primary minerals to silicate and other clays destroys one mineral while giving birth to another. These clays hold reserves of nutrients and water, permitting plants to gain a foothold. Residues from the plants return to the weathering mass and are altered to humus, which exceeds even clay in its nutrient- and water-holding capacity.

Thus, *silicate clays, humus*, and *living organisms* together with life-giving *water* become prime constituents as soil characteristics develop. They affect markedly the kind and extent of layering or horizon differentiation that occurs. A general description of how soil layers form will illustrate this point.

ORGANISMS AND ORGANIC MATTER. As soon as plants gain a foothold in a weathering rock or in recently deposited soil material, the development of a *soil profile* has begun. Residues of plants and animals remain on and in the soil material. As these decay and are mixed with mineral matter by living organisms, the first evidence of layering occurs. The upper part of the soil mass becomes slightly darked in color than the deeper layers. And it assumes structural stability due to the presence of the organic matter. Thereby, the surface horizon begins to appear in the young soil.

The presence of decaying organic matter accelerates soil layering in other ways. Acids released from organic decomposition enhance the break-down of base-containing minerals, yielding soluble nutrients and secondary minerals such as the silicate clays and the oxides of iron and aluminum. These products may simply enrich the upper layers in which they are formed, or they may be moved downward by percolating waters, eventually to accumulate as layers at a lower depth in the developing soil. This downward movement and accumulation again dictates layer or horizon formation. Upper horizons may be depleted of nutrients and clay-sized minerals, whereas deeper layers tend to be enriched in these same constituents (see Fig. 12:2).

As these chemical and physical changes occur, living organisms continue to play a vital role. They physically manipulate, move, and bind the soil particles and structural units, helping to provide stability of the horizons. At the same time, by burrowing or moving through the soil they help mix materials from adjacent horizons, thereby reversing the process of distinct horizon differentiation.

NUTRIENT RECYCLING. Living organisms team up with soil water to provide an important mechanism for stabilizing the acid–base ratio of the soil solution in given weathering situations. This is done by *recycling nutrients* through the soil (see Fig. 12:3). Soluble elements are absorbed by plants from the soil body, translocated to the upper plant parts, released again upon the death of the plant, and moved downward into the soil by percolating water ready to be recycled again. The nutrient content of the plant parts and the amount of rainfall thereby establish the ionic environment of the weathering soil. Plants high in mineral elements help to maintain a high metallic cation concentration in the soil solution. Plants of low nutrient content cannot maintain such high concentrations, especially if they grow, as they often do, in areas of high rainfall and high leaching potential. Nutrient recycling thus helps control the acid–base balance of the weathering solutions and the ultimate horizons which develop.

WATER'S ROLE. Soil water is active from the very beginning in helping to enhance the development of soil layers. In the first place, its presence is essential for plant growth and for most of the chemical reactions whereby mineral breakdown occurs. Its movement in the regolith is of no less significance to soil development. This is shown by the nutrient recycling just described and by the downward movement of silicate clays, oxides of iron and aluminum, and salts of various kinds. In regions of low rainfall, upward movement and evaporation of water result in the accumulation of salts at the soil surface or at some point below the surface. Water is seen to be the principal transport within the soil body.

Water affects the nature of soil horizons in other important ways. If water can freely drain from the weathering area, aerobic conditions generally exist. The resulting soil is generally weathered to a considerable depth and contains oxidized minerals and nutrient elements, and root penetration is uninhibited by excess water.

Contrasting this situation is that found in wet areas. Oxidative weathering is minimized and undecomposed organic matter tends to accumulate at or near the mineral surface. Reduced conditions characterize the soil horizons, which are exploited to only a limited extent by growing plants.

Horizon differentiation is indeed affected directly by presence of excess water.

ACQUIRED VERSUS INHERITED CHARACTERISTICS. Each of these examples illustrates how, over a period of time, significant changes take place in the parent material. Layers or horizons develop which may be leached of certain minerals or may be enhanced by the deposition or the formation in place of others. Thus, the weathered material has some characteristics which are *acquired*[1] in contrast to those *inherited* from the parent material. In the early stages of soil formation inherited soil properties dominate, but as the soil develops acquired characteristics become more prominent and eventually may become dominant.

12:3. THE SOIL PROFILE

The layering or horizon development described in the previous section eventually gives rise to natural bodies called soils. Each soil is characterized by a given sequence of these horizons. This sequence is termed a *soil profile*. Attention will now be given to the major horizons making up soil profiles and the terminology used to describe them.

For convenience in study and description, the layers resulting from soil-forming processes are grouped under four heads: O, A, B, and C. The subdivisions of these are called *horizons*. Their common sequence of occurrence within the profile is shown in Fig. 12:4.

ORGANIC HORIZONS. The O group are organic horizons which form above the mineral soil. They result from litter derived from dead plants and animals. Occurring commonly in forested areas, they are generally absent in grassland regions. Specific horizons and their description follow.

O1: organic horizon wherein the original forms of the plant and animal residues can be recognized by the naked eye.

O2: organic horizon wherein the original plant and animal forms cannot be so distinguished.

A (ELUVIAL) HORIZONS. The A group are mineral horizons which lie at or near the surface and are characterized as zones of maximum leaching or eluviation (eluvial from the Latin words *ex* or *e*, meaning out, and *luv*,

---

[1] The presence of quartz is perhaps the most common example of an *inherited* characteristic, although the presence of any mineral, such as feldspar or mica, carried over unchanged into a soil, is a good illustration. Color also is often inherited, as in the case of a red parent sandstone or shale. Even clay, if originally a part of the parent material, may be rated as inherited. Examples of *acquired* characteristics are as easily cited. Those due to organic matter, to clay formed as the soil developed, and to products of weathering such as red and yellow iron oxides are common. Certainly profile layering and the development of various structural forms within the various horizons are due largely to environmental influences.

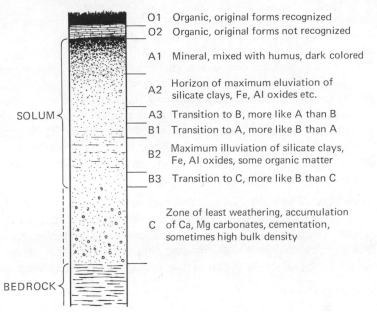

O1 Organic, original forms recognized

O2 Organic, original forms not recognized

A1 Mineral, mixed with humus, dark colored

A2 Horizon of maximum eluviation of silicate clays, Fe, Al oxides etc.

A3 Transition to B, more like A than B

B1 Transition to A, more like B than A

B2 Maximum illuviation of silicate clays, Fe, Al oxides, some organic matter

B3 Transition to C, more like B than C

C Zone of least weathering, accumulation of Ca, Mg carbonates, cementation, sometimes high bulk density

SOLUM

BEDROCK

FIGURE 12:4. Theoretical mineral soil profile, showing the major horizons that may be present. Any particular profile may exhibit only part of these illustrated. In addition, however, it may exhibit more detailed subhorizons than those indicated. The solum includes the A and B plus some fragipans and duripans of the C horizon.

meaning washed). Beginning with the surface, these horizons are designated as A1, A2, etc. (see Fig. 12:4).

A1: topmost mineral horizon, containing a strong admixture of humified organic matter which tends to impart a darker color than that of the lower horizons.

A2: horizon of maximum eluviation of clay, iron, and aluminum oxides and a corresponding concentration of resistant minerals, such as quartz, in the sand and quartz sizes. Generally lighter in color than the A1.

A3: transition layer between A and B with properties more nearly like those of A1 or A2 than of the underlying B. Sometimes absent.

B (ILLUVIAL) HORIZONS. The B horizons follow and include layers in which illuviation from above or even below has taken place (illuvial from the Latin words *il*, meaning in, and *luv*, meaning washed). It is the region of maximum accumulation of materials such as iron and aluminum oxides and silicate clays. In arid regions, calcium carbonate, calcium sulfate, and other salts may accumulate in the lower B.

The B horizons are sometimes referred to as the *subsoil*. This is incorrect, however, since the B horizons may be incorporated at least in part in the plow layer or they may be considerably below the plow layer in the case of soils with deep A horizons.

The specific B horizons are referred to as B1, B2, etc.

B1: a transition layer between A and B with properties more nearly like B than A. Sometimes absent.

B2: zone of maximum accumulation of clays and hydrous oxides. These may have moved down from upper horizons or may have formed in place. Organic matter content is generally higher than that of A2. Maximum development of blocky or prismatic structure or both.

B3: transition between B and C with properties more like those of B than those of C below.

C HORIZON. The C horizon is the unconsolidated material underlying the solum (A and B). It may or may not be the same as the parent material from which the solum formed. It is outside the zones of major biological activities and is little effected by solum-forming processes. Its upper layers may in time become a part of the solum as weathering and erosion continue.

HORIZONS IN A GIVEN PROFILE. It is at once evident that the profile of any one soil probably will not show all of the horizons that collectively are cited in Fig. 12:4. The profile may be immature or may be overly influenced by some local condition such as poor drainage, texture, or topography. Again some of the horizons are merely transitional and may at best be very indistinct. With so many factors involved in soil genesis, a slight change in their coordination may not encourage the development of some layers. As a result, only certain horizons are consistently found in well-drained and uneroded mature soils. The ones most commonly found are: O2, if the land is forested; A1 or A2, depending on circumstances; B2; and C. Conditions of soil genesis will determine which others are present and their clarity of definition.

When a virgin soil is put under cultivation, the upper horizons become the furrow slice. The cultivation, of course, destroys the original layered condition of this portion of the profile and the furrow slice becomes more or less homogeneous. In some soils the A horizons are of sufficient depth so that not all of the A is included in the furrow slice (Fig. 12:5). In other cases where the A is quite thin, the plowline is just at the top of or even down in the B.

Many times, especially on cultivated land, serious erosion has produced a *truncated* profile. As the surface soil was swept away, the plowline was gradually lowered in order to maintain a sufficiently thick furrow slice. Hence, the furrow, in many cases, is almost entirely within the B zone and the C horizon is correspondingly near the surface. Many farmers, especially in the southern states, are today cultivating the subsoil without realizing the ravages of erosion. In profile study and description, such a situation requires careful analysis.

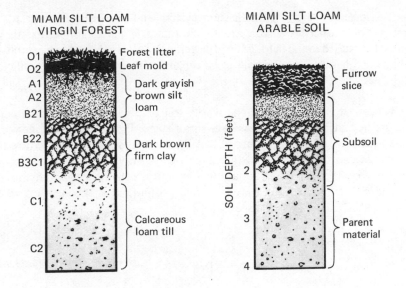

FIGURE 12:5. Generalized profile of the Miami silt loam, one of the Alfisols or Gray-Brown Podzolic soils of the eastern United States. A comparison of the profile of the virgin soil with its arable equivalent shows the changes that may occur as the land is plowed and cultivated. The surface layers are mixed by tillage. If erosion occurs, they may disappear, at least in part, and some of the B horizon will be included in the furrow slice.

12:4. CONCEPT OF INDIVIDUAL SOILS

Soils that have developed by the various processes discussed in this and the preceding chapter differ greatly from place to place. They vary in many profile characteristics, such as degree of horizon differentiation, depth, clay and organic matter contents, and wetness. These differences are noted not only from continent to continent or region to region but from one part of a given field to another. In fact, notable differences sometimes occur within a matter of a few feet. Furthermore, the changes in soil properties occur gradually as one moves from one location to another.

To study intelligently the soil at different places in the earth's surface and to communicate in an orderly manner information about it, man has developed systems of soil classification. These permit classification of *the soil* into a number of individual units or natural bodies which are called *soils*. This involves two basic steps. First, the kind and range of soil properties which are to characterize each soil unit is determined. Second, each unit is given a name, such as a Cecil clay, a Barnes loam, or a Miami silt loam. In this way, when a given soil unit is discussed by name, its identity and characteristics are known by the soil scientist just as is the case for plants when botanists discuss them by their species names.

To develop a useful classification system and to establish the kinds and ranges of properties which are to characterize given soil units, the soil must be studied in the field. Even after tentative classes have been established, they must be tested in the field to be certain of their utility in a classification scheme.

PEDON. It is obvious that the whole soil or even large areas of it cannot be studied at one time. Therefore, small three-dimensional samples of the soil are used. Such a sampling unit is large enough so that the nature of its horizons can be studied and the range of its properties identified. It varies in size from about 1 to 10 square meters and is called a *pedon* (rhymes with "head on" from the Greek word *pedon*, ground). It is the smallest volume that can be called "a soil" (see Fig. 12:6).

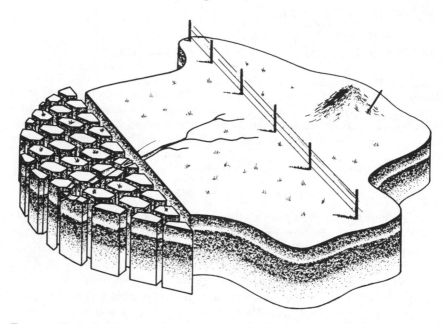

FIGURE 12:6. A schematic diagram to illustrate the concept of a pedon and how a group of closely related pedons are grouped together in a field soil mapping unit. While individual pedons may differ slightly from each other, their behavior under field conditions is sufficiently similar to justify their being included in the same field mapping unit.

Because of its very small size, a pedon obviously cannot be used as the basic unit for a workable field soil classification system. However, a group of pedons, termed a *polypedon*, closely associated in the field and similar in their properties are of sufficient size to serve as a basic classification unit. Such a grouping approximates what in the United States has been called a *soil series*. More than 7,000 soil series have been characterized in this country. They are the basic units used in the field classification of the nation's soils.

Two extremes in the concept of soils have now been identified. One extreme is that of a natural body called *a* soil characterized by a three-dimensional sampling unit (pedon), related groups of which are termed a soil series. At the other extreme is *the* soil, a collection of all these natural bodies which is distinct from water, solid rock, and other natural parts of the earth's crust. These two extremes represent opposite ends of elaborate soil classification schemes which man has used to organize knowledge of soils.

12:5. SOIL CLASSIFICATION IN THE UNITED STATES

The history of the classification of soils (5) suggests three kinds of concepts: (a) soils as a habitat for crops, (b) soils as a superficial mantle of weathered rock, and (c) soils as natural bodies. The first of these, because of its relation to practical agriculture, is the oldest. From the time man first began cultivating his own crops he noticed differences in the productive capacity of soils and classified them, if only in terms of "good" and "bad" for his purposes. The early Chinese, Egyptian, Greek, and Roman civilizations acknowledged differences in soils as media for plant growth. Such recognition is common even today as soils are described as being good "cotton," "soybean," "alfalfa," or "wheat" soils.

The concept of soils as the uppermost mantle of weathered rock may too have been in the minds of our ancient ancestors. However, it was not until geology became a science in the late eighteenth and early nineteenth centuries that this concept became prominent. While in no way replacing the "plant media" concept, it brought into being such terms of classification as "sandy" and "clayey" soils as well as "limestone" soils and "lake-laid" soils. These terms have a geological connotation.

The Russian soil scientist V. V. Dokuchaev and his associates were the first to develop the concept of soils as natural bodies which were subject to classification by man. Further, they noted a definite relationship among climate, vegetation, and soil characteristics. Dokuchaev's monograph in 1883 developed this thesis. Some of the terms used until recent times in the United States to describe soils were of Russian origin, examples being *Chernozem* (black earth) and *Podzol* (under ash).

The first organized attempts in the United States to classify soils and to map them in the field starting in 1898 were dominated by the crop habitat and geological concepts. The practical agricultural significance of soils and their classification seemed to have had highest priority, but the study of soils in the field forced a partial geological bias as well. Although the natural body concept as developed by the Russians was known to a few American scientists late in the nineteenth century, it was not until the early part of the nineteenth century that it received attention. C. F. Marbut of the U. S. Department of Agriculture promoted this concept along with the importance of the soil profile in identifying the natural bodies. His soil classification scheme, presented first in 1927 and in more detail in 1936, was based on the concept of soils as natural bodies.

With the background laid by Marbut, scientists of the U. S. Department of Agriculture developed in 1938 (and improved in 1949) a system of soil classification based on the natural body concept. This system also had a strong genetic bias, that is, the classification of a given soil was dependent to a considerable degree on the supposed processes by which it had formed. The difficulty of ascertaining for certain which genetic processes were responsible for a given soil was one of the weaknesses of this system, although it served very well for a period of about 25 years.

The classification system currently in use in the United States and in some other parts of the world as well is based on the properties of soils as they are found in the field. Published in 1960 it maintains the natural body concept but is less dependent upon genetic influences. This system will now be discussed in general terms. For comparative purposes an outline will also be given of the older (1949) system. This will provide at least general background for reading the literature published prior to 1960.

12:6. SOIL CLASSIFICATION—NEW COMPREHENSIVE SYSTEM

The classification system currently in use in the United States is that developed by the Soil Survey Staff of the U. S. Department of Agriculture (7, 8). It is relatively new, having been developed and evaluated through several stages or "approximations." The seventh approximation has been officially adopted as the *Comprehensive Soil Survey System*. It will be used in this text.

The new system has many features to recommend it. However, there are two which students of soil science may find most useful. *First* is the fact that the primary bases for identifying different classes in the system are the properties of soils as found in the field—properties which can be measured quantitatively. Furthermore, the measurements so obtained can be verified by others. This lessens the likelihood of controversy over the place of a given soil in the classification system. Such controversy is common when scientists deal with systems where genesis or presumed genesis is the basis for the classification.

The *second* significant feature of the new system is the nomenclature employed especially for the broader classification categories. The names give a definite connotation of the major characteristics of the soils in question—a connotation easily understood in many languages since Latin or

Greek root words are the bases for the names. Consideration will be given to the nomenclature used after brief reference is made to the major criteria for the system—soil properties.

BASES OF SOIL CLASSIFICATION. The new system is based on the properties of soils as they are found today. While one of the objectives of the system is to group soils similar in genesis, the specific criteria used to place soils in these groups are those of soil properties. The advantages of this system over that based primarily on soil genesis or presumed soil genesis (6) are as follows:

1. It permits classification of soils rather than soil-forming processes.
2. It focuses on the soil rather than related sciences such as geology and climatology.
3. It permits the classification of soils of unknown genesis—only the knowledge of their soil properties is needed.
4. It permits greater uniformity of classification as applied by a large number of soil scientists. Differences in interpretation of how a soil was formed do not influence its classification under this scheme.

It should not be assumed that the new comprehensive classification system ignores soil genesis. Since soil properties, the basis for the new system, are often related directly to soil genesis, it is difficult to emphasize soil properties without at least indirectly emphasizing soil genesis as well.

It is not possible to generalize with respect to the kinds of soil properties used as criteria for soil classification. All of the chemical, physical, and biological properties presented in this text are subject to use in this classification scheme. A few examples of the criteria used include moisture, temperature, color, texture, and structure of the soil. Chemical and mineral properties such as contents of organic matter, clay, iron and aluminum oxides, pH, percentage base saturation, and the presence of salts may also be mentioned. Soil depth is still another important criterion for classification.

Among the most significant of the properties used as a basis for classification is the presence or absence of certain diagnostic soil horizons. Because of their prominence in helping to determine the place of a soil in the classification system, they will be given somewhat more detailed attention.

DIAGNOSTIC HORIZONS. To illustrate that soil properties are the primary criteria for classifying soils under the new system, brief mention will be made of certain *diagnostic* surface and subsurface horizons. The diagnostic surface horizons are called epipedons (from the Greek words *epi*, over, and *pedon*, soil). The epipedon includes the upper part of the soil darkened by organic matter, the upper eluvial horizons, or both. It may include part of the B horizon (see p. 313) if the latter is significantly darkened by organic matter. Six epipedons are recognized but only four are of any importance in the soils of the United States. The other two, called *anthropic* and *plaggen*, are the result of man's intensive use of soils. They are found in parts of Europe and probably Asia. The major features of the four horizons are given in Table 12:1.

Many subsurface horizons characterize different soils in the system. Those that are considered diagnostic horizons are shown along with their major features in Table 12:1. In addition, there are a number of subsurface horizons in which materials have accumulated such as *gypsic* (gypsum), *calcic* (calcium and magnesium carbonates), and *sodic* (soluble salts). Also, *pans* (duripans and fragipans) are cemented layers which restrict water movement and root penetration. Each of these layers can be used as a distinctive property to help place a soil in its proper class.

TABLE 12:1. *Major Features of Diagnostic Horizons Used to Differentiate at the Higher Levels of the Comprehensive Classification Scheme*

Diagnostic Horizon	Major Feature
	Surface Horizons (Epipedons)
Mollic	Thick, dark colored, high base saturation, strong structure
Umbric	Same as Mollic except low base saturation
Ochric	Light colored, low organic content, may be hard and massive when dry
Histic	Very high in organic content, wet during some part of year
	Subsurface Horizons
Argillic	Silicate clay accumulation
Natric	Argillic, high in sodium, columnar or prismatic structure
Spodic	Organic matter, Fe and Al oxide accumulation
Cambic	Changed or altered by physical movement or by chemical reactions
Agric	Organic and clay accumulation just below plow layer
Oxic	Primarily mixture of Fe, Al oxides, and 1:1-type minerals

CATEGORIES OF THE SYSTEM. There are six categories of classification in the new system: (a) order (the broadest category), (b) suborder, (c) great group, (d) subgroup, (e) family, and (f) series (the most specific category). These categories may be compared with those used for the classification of plants. The comparison would be as shown in Table 12:2, where white clover (*Trifolium repens*) and Miami silt loam are the examples of plants and soils, respectively.

Just as *Trifolium repens* identifies a specific kind of plant, the Miami silt loam identifies a specific kind of soil. The similarity continues up the classification scale. Several soil types are grouped together in a single soil series, and a family generally consists of groups of soil series just as the

Leguminosae family of plants includes several genera in addition to the clovers. The same similarity between the soil and plant classification schemes can be followed to the highest categories—that of phylum for plants and of order for soils. With this general background, a brief description of the six soil categories and the nomenclature used in identifying them is presented.

TABLE 12:2. *Comparison of the Classification of a Common Cultivated Plant, White Clover* (Trifolium repens), *and a Soil, Miami Silt Loam*

Plant Classification		Soil Classification
Phylum—Pterophyta		Order—Alfisol
Class—Angiospermae		Suborder—Udalf
Subclass—Dicotyledoneae	Increasing specificity	Great Group—Hapludalf
Order—Rosales		Subgroup—Typic Hapludalf
Family—Leguminosae		Family—Fine loamy, mixed, mesic
Genus—*Trifolium*		Series—Miami
Species—*repens*		Phase—Miami, eroded phase

The *order* category is based largely on morphology, but soil genesis is an underlying factor. A given order includes soils whose properties suggest that they are not too dissimilar in their genesis. As an example, soils developed under grassland vegetation have the same general sequence of horizons and are characterized by a thick, dark, epipedon (surface horizon) high in bases. They are thought to have been formed by the same general genetic processes and are thereby usually included in the same order, Mollisol.

Suborders are subdivisions of orders which emphasize genetic homogeneity. Thus, wetness, climatic environment, and vegetation are characteristics which help determine the suborder in which a given soil is found.

Diagnostic horizons (p. 319) are used to differentiate the *great groups* in a given suborder. Soils in a given great group are thought to have the same kind and arrangement of these horizons.

Subgroups are subdivisions of the great groups. The typical or central concept of a great group makes up one subgroup (Typic). Other subgroups may have characteristics that are intergrades between those of the central concept and those of another great group (for example, *Aquic Hapludult*).

The *family* category is the least well defined. However, properties important to the growth of plants are used to differentiate families. This category permits the grouping of members of subgroups having in common similar characteristics, such as texture, mineral content, pH, soil temperature, and soil depth.

The *series* category is essentially the same as that which has been in use for years in the United States. It is defined as a "collection of soil individuals essentially uniform in differentiating characteristics and in arrangement of horizons" (7). This category is given further attention later (see p. 337).

NOMENCLATURE. An important feature of the system is the nomenclature used to identify different soil classes. The names of the classification units are combinations of syllables, most of which are derived from the Latin or Greek and are the root words in several modern languages. Since each part of a soil name conveys a concept of soil character or genesis, the name automatically describes the general kind of soil being classified. For example, soils of the order *Aridisol* (Latin *aridus*, dry, and *solum*, soil) are characteristic of arid or dry places. Those of the order *Inceptisols* (Latin *inceptum*, beginning, and *solum*, soil) are soils with only the beginnings of profile development. Thus, the names of orders are combinations of (a) formative elements, which generally define the characteristics of the soils, and (b) the ending *sol*.

The names of suborders automatically identify the order of which they are a part. For example, soils of the suborder *Aquolls* are the wetter soils (Latin *aqua*, water) of the Mollisol order. Likewise, the name of the great group identifies the suborder and order of which it is a part. *Argiaquolls* are Aquolls with clay or argillic (Latin *argilla*, white clay) horizons.

The nomenclature as it relates to the different categories in the classification system might be illustrated as follows:

$$M \; o \; l \; l \; i \; s \; o \; l \text{———— order}$$
$$A \; q \; u \; o \; l \; l \text{———————— suborder}$$
$$A \; r \; g \; i \; a \; q \; u \; o \; l \; l \text{——————— great group}$$
$$\text{Typic } A \; r \; g \; i \; a \; q \; u \; o \; l \; l \text{—————— subgroup}$$

Note that the three letters *oll* identify each of the lower categories as being in the Mollisol order. Likewise, the suborder name *Aquoll* is included as part of the great group and subgroup name. Given only the subgroup name, the great group, suborder, and order to which the soil belongs are automatically known.

Family names in general identify groups of soil series similar in texture, mineral composition, and in soil temperature at a depth of 20 inches. Thus the name *fine, mixed, mesic* applies to a family with a fine texture, mixed clay mineral content, and mesic (8–15°C) soil temperature.

Soil series names have local significance since they normally identify the particular locale in which the soil is found. Thus, names such as Fort Collins, Cecil, Miami, Norfolk, and Ontario are used to identify the soil series. When the texture name of the surface horizon is added to that of the series, the soil type name has been identified. Fort Collins loam and Cecil clay are examples.

With this brief explanation of the nomenclature of the new system, the order category of the system will now be considered.

12:7. SOIL ORDERS

Ten orders are recognized. With the exception of one order (Entisols), which roughly corresponds to the azonal soils of the 1949 system (pp. 340–45), the new orders show little resemblance to those formerly used. The names of these orders and their approximate equivalents in the old system are shown in Table 12:3. Note that all order names have a common ending *sol* from the Latin *solum*, meaning soil.

TABLE 12:3.   *Names of Soil Orders in the Comprehensive Soil Classification System, Their Derivation, and Approximate Equivalents in the Old System*

| Name[b] | Formative Element[a] | | Approximate Equivalents in the Old System |
	Derivation	Pronunciation	
Entisol	Nonsense symbol	re*cent*	Azonal, some Low-Humic Gley soils
Vertisol	L. *verto*, turn	in*vert*	Grumusols
Inceptisol	L. *inceptum*, beginning	in*cept*ion	Ando, Sol Brun Acide, some Brown Forest, Low-Humic Gley, and Humic Gley soils
Aridisol	L. *aridus*, dry	ar*id*	Desert, Reddish Desert, Sierozem, Solonchak, some Brown and Reddish Brown soils and associated Solonetz
Mollisol	L. *mollis*, soft	mo*ll*ify	Chestnut, Chernozem, Brunizem (Prairie), Rendzinas, some Brown, Brown Forest, and associated Solonetz, and Humic Gley soils
Spodosol	Gk. *Spodos*, wood ash	*Podz*ol; odd	Podzols, Brown Podzolic soils, and Groundwater Podzols
Alfisol	Nonsense symbol	Ped*alf*er	Gray-Brown Podzolic. Gray Wooded, and Non-Calcic Brown soils, Degraded Chernozems, and associated Planosols and some Half-Bog soils
Ultisol	L. *ultimus*, last	*ult*imate	Red-Yellow Podzolic soils, Reddish-Brown Lateritic soils of the U. S., and associated Planosols and Half-Bog soils
Oxisol	F. *oxide*, oxide	*ox*ide	Laterite soils, Latosols
Histosol	Gk. *histos*, tissue	*hist*ology	Bog soils

[a] The italicized letters in the pronunciation column are used in the suborder and great group categories to identify the order to which they belong.
[b] Note that all orders end in sol (Latin *solum*, soil).

ENTISOLS (RECENT SOILS).   These are mineral soils without natural genetic horizons or with only the beginnings of such horizons. The central concept of this order are soils in deep regolith with no horizons except a plow layer. Included are the extremes of highly productive soils on recent alluvium and infertile soils on barren sands. Shallow soils on bedrock are included also. The common characteristics of all Entisols is lack of significant profile development.

Soils of this order are found under a wide variety of climatic conditions. For example, in the Rocky Mountain region and in southwest Texas, shallow, medium-textured Entisols (Orthents) over hard rock are common (called Lithosols in the old classification system). They are used mostly as range land. Sandy Entisols (Psamments) are found in Florida, Alabama, and Georgia and typify the sand hill section of Nebraska (see Plate 1). Psamments are used mostly for grazing in the drier climates. They may be forested or used for cropland in humid areas. Some of the citrus-, vegetable-, and peanut-producing areas of the South are typified by Psamments (Regosols, old system).

Entisols are probably found under even more widely varied climatic conditions outside the United States (see Plate 2). Psamments are typical of the shifting sands of the Sahara Desert and Saudi Arabia. Large areas of Psamments dominate parts of southern Africa and of central and north-central Australia. Entisols having medium to fine textures (Orthents) are found in northern Quebec and parts of Alaska, Siberia, and Tibet. Orthents are typical of some mountain areas such as the Andes in South America and some of the uplands of an area extending from Turkey westward to Pakistan.

As might be expected, the agricultural productivity of the Entisols varies, greatly depending on their location and properties. When adequately fertilized and when their water supply is controlled, some of these soils are quite productive. However, restrictions on their depth, clay content, or water balance limit the intensive use of large areas of these soils.

VERTISOLS (Latin *verto*, turn). This order of mineral soils is characterized by high content of swelling-type clays, which in dry seasons cause the soils to develop deep, wide cracks. A significant amount of material from the upper part of the profile may slough off into the cracks, giving rise to a partial "inversion" of the soil. This accounts for the term *invert*, which is used to characterize this order in a general way.

In the past, most of these soils have been called *Grumusols*. Because of their excessive shrinking, cracking, and shearing they are generally unstable and present problems when used for building foundations, highway bases, and even for agricultural purposes.

There are several small but significant areas of Vertisols in the United States (see Plate 1). Two are located in humid areas, one in eastern Mississippi and western Alabama and the other along the southeast coast of Texas. These soils are of the Udert (Latin *udus*, humid) suborder because, owing to the moist climate, cracks do not persist for more than three months of the year.

Two Vertisol areas are found in east central and southern Texas, where the climate is drier. Since cracks persist for more than three months of the year, the soils belong to the Ustert (Latin *ustus*, burnt) suborder, characteristic of areas with hot, dry summers.

In India, the Sudan, and eastern Australia large areas of Vertisols are found (see Plate 2). These soils probably are of the Ustert suborder since dry weather persists long enough for the wide cracks to stay open for periods of three months or longer.

In each of the major Vertisol areas listed above some of the soils are used for crop production. Even so, their very fine texture and their marked shrinking and swelling characteristics make them less suitable for crop production than soils in the surrounding areas. They are sticky and plastic when wet and hard when dry. As they dry out following a rain, the period of time when they can be plowed or otherwise tilled is very short. This is a limiting factor even in the United States, where powerful tractors and other mechanical equipment make it possible to plow, prepare seedbeds, plant, and cultivate very quickly. The limitation becomes even more severe in India and the Sudan, where slow-moving animals are used to till the soil. Not only can the cultivators not perform tillage operations on time, but they are limited to the use of small, near primitive tillage implements because their animals cannot pull larger equipment through the "heavy" soil.

In spite of their limitations, Vertisols are widely tilled, especially in India and in the Sudan. Sorghum, corn, millet, and cotton are crops commonly grown. Unfortunately, the yields are generally low. Further research and timely soil management are essential if these large soil areas are to help produce the food crops these countries so badly need.

INCEPTISOLS (Latin *inceptum*, beginning). Inceptisols may be termed young soils since their profiles contain horizons that are thought to form rather quickly and result mostly from alteration of parent materials. The horizons do not represent extreme weathering. Horizons of marked accumulation of clay and iron and aluminum oxides are absent in this order. The profile development of soils in this order is more advanced than that of the Entisol order but less so than that of the other orders.

Soils formerly classified as Brown Forest, Ando, and Sols Brun Acide typify this order. Many agriculturally useful soils are included along with others whose productivity is limited by factors such as imperfect drainage (see Table 12:2).

Inceptisols are found in the United States and on each of the continents (see Plates 1 and 2). For example, some Inceptisols called Andepts (productive soils developed from volcanic ash) are found in a sizable area of Oregon, Washington, Idaho, and in Ecuador and Columbia in South America. Some called Ochrepts (Greek *ochros*, pale), with thin, light-colored surface horizons, extend from southern New York through central and western

Pennsylvania, West Virginia, and eastern Ohio. Ochrepts dominate an area extending from southern Spain through central France to central Germany. They are also present in Chile, north Africa, eastern China, and western Siberia.

Tropepts (Inceptisols of tropical regions) are found in northwestern Australia, central Africa, southwestern India, and in southwestern Brazil. Areas of wet Inceptisols or Aquepts (Latin *aqua*, water) are found along the Amazon and Ganges rivers.

As might be expected, there is considerable variability in the natural productivity of Inceptisols. For instance, those found in the Pacific Northwest are quite fertile and provide us with some of our best wheat lands. In contrast, some of the low-organic-containing Ochrepts in southern New York and northern Pennsylvania are not naturally productive. They have been allowed to reforest following earlier periods of crop production.

ARIDISOLS (Latin *aridus*, dry). These mineral soils are found mostly in dry climates. Except where there is ground water or irrigation, the soil layers are dry throughout most of the year. Consequently, they have not been subjected to intensive leaching. They have an ochric epipedon generally light in color and low in organic matter. They may have a horizon of accumulation of calcium carbonate (calcic), gypsum (gypsic), or even more soluble salts (salic). If groundwater is present, conductivity measurements show the presence of soluble salts.

Aridisols include most of the soils of the arid regions of the world, such as those formerly designated as Desert, Reddish Desert, Sierozem, Reddish Brown, and Solonchak (see Table 12:3).

A large area of Aridisols called Argids (Latin *argilla*, white clay), which have a horizon of clay accumulation, occupies much of the southern parts of California, Nevada, Arizona, and central New Mexico. The Argids also extend down into northern Mexico. Smaller areas of Orthids (Aridisols without clay accumulation) are found in several western states.

Vast areas of Aridisols are present in the Sahara desert in Africa, the Gobi and Taklamakan deserts in China, and the Turkestan desert of the Soviet Union. Most of the soils of southern and central Australia are Aridisols, as are those of southern Argentina, southwestern Africa, West Pakistan, and the Middle East countries.

Without irrigation, Aridisols are not suitable for growing cultivated crops. Some areas are used for sheep or goat grazing, but the production per unit area is low. Where irrigation water is available Aridisols can be made most productive. Irrigated valleys of the western United States are among the most productive in the country.

MOLLISOLS (Latin *mollis*, soft). This order includes some of the world's most important agricultural soils. They are characterized by a *mollic epipedon* or surface horizon, which is thick, dark, and dominated by divalent cations. They may have an argillic (clay), natric, albic, or cambic horizon but not an oxic or spodic one (see Table 12:1). The surface horizons generally have granular or crumb structures and are not hard when the soils are dry. This justifies the use of a name which implies softness.

Most of the Mollisols have developed under prairie vegetation (see Fig. 12:7). Grassland soils of the central part of the United States, formerly classified as Chernozem, Brunizem (Prairie), Chestnut, and Reddish Prairie along with associated Humic Gley and Planosols, make up the central core of this order. However, some soils developed under forest vegetation, such as certain soils formerly classified as Brown Forest, have a mollic epipedon and are included among the Mollisols. They, along with the grassland soils which dominate this order, are among the more productive cultivated soils in the world.

Mollisols are dominant in the Great Plains states (see Plate 1). Those in the eastern and more humid part of this region are called udolls (Latin *udus*, humid). A region extending from North Dakota to southern Texas is characterized by Ustolls (Latin *ustus*, burnt), which are intermittently dry during the summer. Further west in parts of Idaho, Utah, Washington, and Oregon are found sizable areas of Xerolls (Greek *Xeros*, dry), which are the driest of the Mollisols.

The largest area of Mollisols outside the United States is that stretching from east to west across the heartland of the Soviet Union (see Plate 2). Other sizable areas are found in Mongolia and northern China and in northern Argentina, Paraguay, and Uruguay.

The native fertility of Mollisols dictates that they be rated among the world's best soils. When first cleared for cultivation, their high native organic matter released sufficient nitrogen and other nutrients to produce bumper crops even without fertilization. Yields on these soils were unsurpassed by other unirrigated areas. Even today, when moderate to heavy fertilization gives a competitive advantage to less fertile soils in more humid areas, Mollisols still rate among the best. Photographs of profiles of two mollisols are shown in Fig. 12:8.

SPODOSOLS (Greek *spodos*, wood ash). Spodosols are mineral soils which have a spodic horizon, a subsurface horizon with an accumulation of organic matter, and oxides of aluminum with or without iron oxides. This illuvial horizon usually occurs under an eluvial horizon, normally an albic horizon (light in color, hence justifying reference to "wood ash") (see Table 12:1).

These soils form mostly on coarse-textured, acid parent materials subject to ready leaching. They occur only in humid climates and are most common where it is cold and temperate. Forests are the natural vegetation under which most of these soils have developed.

Species low in metallic ion contents, such as pine trees, seem to encourage

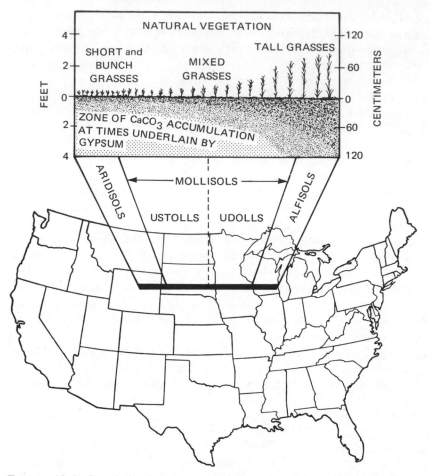

FIGURE 12:7. Correlation between natural vegetation and certain zonal soil groups is graphically shown for a strip of territory in north central United States. The control, of course, is climate. Note the greater organic content and deeper zone of calcium accumulation as one proceeds from the drier areas in the west toward the more humid region where prairie soils are found.

the development of Spodosols. As the litter from these low-base species decompose, strong acidity develops. Percolating water leaches acids down into the profile. The upper horizons succumb to this intense acid leaching. Most minerals except quartz are removed. In the lower horizons, oxides of aluminum and iron as well as organic matter precipitate, thus yielding the interesting Spodosol profiles.

Many of the soils in the northeastern United States, including those of northern Michigan and Wisconsin which were formerly classified as Podzol,

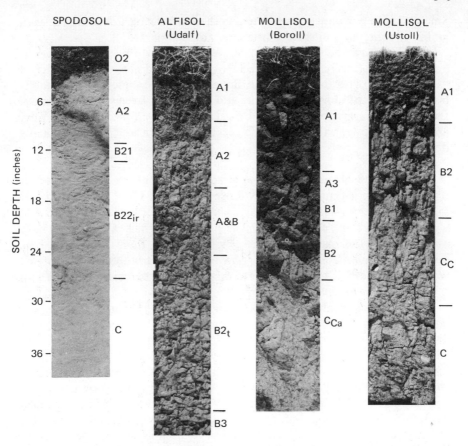

FIGURE 12:8. Monoliths of profiles representing four soil orders. The suborder names are also shown (in parentheses). Note the spodic horizons in the Spodosol characterized by humus (h) and iron (ir) accumulation. In the Alfisol is found the illuvial clay horizon $B2_t$. The thick dark surface horizon (mollic epipedon) characterizes both Mollisols. Note that the zone of calcium accumulation (C_{Ca}) is higher in the Ustoll, which has developed in a dry climate.

Brown Podzolic, and Ground Water Podzol, belong to this order (see Plate 1). Most of them are Orthods, the "common" Spodosols described above. Some, however, are Aquods since they are seasonally saturated with water and possess characteristics associated with this wetness such as very high surface organic accumulation, mottling in the albic horizon, and the development of a hard pan (duripan) in the albic horizon. Important areas of Aquods occur in Florida.

The area of Spodosols in the northeast extends up into Canada. Other large areas of this soil order are found in northern Europe and Siberia.

Smaller but important areas are found in the southern part of South America and in cool mountainous areas of temperate regions.

Spodosols are not naturally fertile. When properly fertilized, however, these soils can become quite productive. For example, the productive "Potato" soils of northern Maine are Spodosols, as are some of the vegetable-producing soils of Florida, Michigan, and Wisconsin. Even so, the low native fertility of most spodosols makes them uncompetitive for tilled crops. They are covered mostly with forests, the vegetation under which they originally developed.

ALFISOLS. Alfisols are moist mineral soils having no mollic epipedon or oxicor spodic horizons. Typified by those soils called Gray-Brown Podzolic in the old system, Alfisols have gray to brown surface horizons, medium to high base status, and contain an illuvial horizon in which silicate clays have accumulated. The clay horizon is generally more than 35 percent base saturated. This horizon is termed *argillic* if only silicate clays are present and *natric* if, in addition to the clay, it is more than 15 percent saturated with sodium and has prismatic or columnar structure.

The Alfisols appear to be more strongly weathered than the Inceptisols but less so than the Spodosols. They are formed mostly in humid region areas under native deciduous forests, although grass is the native vegetation in some cases. In addition to the Gray-Brown Podzolic soils, Alfisols include most of those formerly called Noncalcic Brown and Gray Wooded soils and some of those called Planosols, Half Bog, and Solodized Solonetz.

Some of our best agricultural soils are in the Alfisol order. It is typified by Udalfs (Latin *udus*, humid) in Ohio, Indiana, Michigan, Wisconsin, Minnesota, Pennsylvania, and New York. It includes sizable areas of Xeralfs (Greek *xeros*, dry) in central California; some cold-climate Boralfs (Greek *boreas*, northern) in the Rockies; Ustalfs (Latin *ustus*, burnt) in areas of hot summers, including Texas and New Mexico; and wet Alfisols, Aqualfs (Latin *aqua*, water), in parts of the midwest.

Alfisols occur in other countries having climatic environments similar to those of areas just mentioned (see Plate 2). A large area dominated by Boralfs is found in northern Europe stretching from the Baltic States through western Russia. A second large area is found in Siberia. Ustalfs are prominent in the southern half of Africa, eastern Brazil, eastern India, and in southeast Asia. Large areas of Udalfs are found in central China, in England, France and Central Europe, and southeast Australia. Xeralfs are prominent in southwest Australia, Italy, and central Spain.

In general, Alfisols are quite productive soils. Their medium to high base status, generally favorable texture, and location (except for some Xeralfs) in humid and subhumid regions all favor good crop yields. In the United States, these soils rank favorably with the Mollisols and Ultisols in their productive capacity.

ULTISOLS (Latin *ultimus*, last). This order contains most soils formerly called Red-Yellow Podzolic and Reddish-Brown Lateritic soils and Rubrozems along with some called Humic Gley, Low Humic Gley, and Ground Water Laterite soils. They are usually moist soils and develop under warm to tropical climates. Ultisols are more highly weathered and acidic than the Alfisols but generally are not so acid as the Spodosols. They have argillic (clay) horizons with base saturations lower then 35 percent. Except for the wetter members of the order, their subsurface horizons are commonly red or yellow in color, evidence of accumulation of free oxides of iron. They still have some weatherable minerals, however, in contrast to the Oxisols. Ultisols are formed on old land surfaces, normally under forest vegetation, although savannah or even swamp vegetation is common.

Most of the soils of the southeastern part of the United States fall in this order. Udults, the moist but not wet Ultisols, extend from the east coast (Maryland to Florida) to and beyond the Mississippi River Valley and are the most extensive of soils in the humid southeast. Humults (high in organic matter) are found in Hawaii, eastern California, Oregon, and Washington. Xerults (drier Ultisols) are common in southern Oregon and northern and western California.

Ultisols are prominent on the east and northeast coasts of Australia. Large areas of Udults are located in southeast Asia, including southern China. Important areas are also found in southern Brazil and Paraguay.

Although Ultisols are not naturally as fertile as Alfisols or Mollisols, they respond well to good management. They are located mostly in regions of long-growing seasons and of ample moisture for good crop production. Their clays are usually of the 1:1 type along with oxides of iron and aluminum, which assures ready workability. Where adequate chemical fertilizers are applied, these soils are quite productive. In the United States the better Ultisols compete well with Mollisols and Alfisols as first-class agricultural soils.

OXISOLS (French *oxide*, oxide). These are the most highly weathered soils in the classification system. Their most important diagnostic feature is the presence of a deep *oxic* subsurface horizon—a horizon generally very high in clay-size particles dominated by hydrous oxides of iron and aluminum. Weathering and intense leaching have removed a large part of the silica from silicate minerals in this horizon, leaving a high proportion of the oxides of iron and aluminum. Some quartz and 1:1-type silicate clay minerals remain, but the hydrous oxides are dominant. The clay content of these soils is very high, but the clays are of the nonsticky type. The depth of weathering in Oxisols is much greater than for most of the other soils—50 or more feet having been observed.

Those soils which in recent years have been termed Latosols and some of those called Ground Water Laterites are included among the Oxisols. They occupy old land surfaces and occur mostly in the tropics. Relatively less is known of the Oxisols than of most of the other soil orders. They occur in large geographic areas, however, and millions of people in the tropics depend upon them for their food and fiber production.

The largest known areas of Oxisols occur in South America and Africa (see Plate 2). Orthox (normal Oxisols) occur in northern Brazil and neighboring countries. An area of Ustox (hot, dry summers) nearly as large occurs in Brazil to the south of the Orthox. Oxisols are found in the southern two thirds of Africa, being located on old land surfaces of this area.

Relatively less is known about the management of Oxisols than of any other soil order. Most of them have not been cleared of their native vegetation or used for modern cultivation. They are mostly either still covered with native vegetation or have been tilled by primitive methods. The few instances where modern farming techniques have been used met with mixed success. Heavy fertilization, especially with phosphorus-rich materials, is required. Deficiencies of micronutrients have also been observed commonly. In some areas torrential rainfall makes practices which leave the soil bare extremely hazardous.

Although extensive research will be needed to better utilize these soils, experience up to now indicates that their potential for food and fiber production is far in excess of that currently being realized. In both Brazil and central Africa selected areas of these soils have been demonstrated to be high in productivity when they are properly managed.

HISTOSOLS (Greek *histos*, tissue). The last of the soil orders includes the organic soils (Bog soils) as well as some of the half-bog soils. These soils have developed in a water-saturated environment. They contain a minimum of 20 percent organic matter if the clay content is low and 30 percent if the clay content is more than 50 percent. In virgin areas, the organic matter retains much of the original plant tissue form. Upon drainage and cultivation of the area, the original plant tissue form tends to disappear.

Less progress has been made on the classification of these soils than of the other orders. However, four suborders have been tentatively established and are being tested in the field (see Table 12:4). These soils are of great practical importance in local areas, being among the most productive, especially for vegetable crops. Their characteristics are given more detailed consideration in Chapter 13.

The soil map of the United States based on the comprehensive soil classification system in Plate 1 merits careful study to identify regions in which the various soil orders and suborders are found. Likewise, the recently prepared generalized world soil map (Plate 2) shows the probable location of the different soil orders throughout the world and requires equal attention.

TABLE 12:4. *Soil Orders and Suborders in the Comprehensive Soil Classification System*

Note that the ending of the suborder names identifies the order in which the soils are found.

Order	Suborder	Order	Suborder
Entisol	Aquent Arent Fluvent Orthent Psamment	Mollisol	Alboll Aquoll Boroll Rendoll Udoll Ustoll Xeroll
Vertisol	Torrert Udert Ustert Xerert	Alfisol	Aqualf Boralf Udalf Ustalf Xeralf
Inceptisol	Andept Aquept Ochrept Plaggept Tropept Umbrept	Ultisol	Aquult Humult Udult Ustult Xerult
Aridisol	Agrid Orthid		
Spodosol	Aquod Ferrod Humod Orthod	Oxisol	Aquox Humox Orthox Torrox Ustox
		Histosol	Fibrists Hemists Saprists Folists

12:8. SOIL SUBORDERS, GREAT GROUPS, AND SUBGROUPS

SUBORDERS. The ten orders just described are subdivided into suborders as shown in Table 12:4. The characteristics used as a basis for subdividing into the suborders are those which give the class the greatest genetic homogeneity. Thus, soils formed under wet conditions are generally identified under separate suborders (for example, Aquents, Aquerts, Aquept), as are the drier soils (for example, Ustalfs, Ustults). This arrangement also provides a convenient device for grouping soils outside the classification system (for example, the *wet* and *dry* soils).

TABLE 12:5. *Formative Elements in Names of Suborders* (*Comprehensive Soil Classification System*)

Formative Elements	Derivation of Formative Element	Connotation of Formative Element
alb	L. *albus*, white	Presence of albic horizon (a bleached eluvial horizon)
and	Modified from Ando	Ando-like
aqu	L. *aqua*, water	Characteristics associated with wetness
ar	L. *arare*, to plow	Mixed horizons
arg	Modified from argillic horizon; L. *argilla*, white clay	Presence of argillic horizon (a horizon with illuvial clay)
bor	Gk. *boreas*, northern	Cool
ferr	L. *ferrum*, iron	Presence of iron
fibr	L. *fibra*, fiber	Least decomposed stage
fluv	L. *fluvius*, river	Flood plains
hem	Gk. *hemi*, half	Intermediate stage of decomposition
hum	L. *humus*, earth	Presence of organic matter
lept	Gk. *leptos*, thin	Thin horizon
ochr	Gk. base of *ochros*, pale	Presence of ochric epipedon (a light surface)
orth	Gk. *orthos*, true	The common ones
plag	Modified from Ger. *plaggen*, sod	Presence of plaggen epipedon
psamm	Gk. *psammos*, sand	Sand textures
rend	Modified from Rendzina	Rendzina-like
sapr	Gk. *sapros*, rotten	Most decomposed stage
torr	L. *torridus*, hot and dry	Usually dry
trop	Modified from Gk. *tropikos*, of the solstice	Continually warm
ud	L. *udus*, humid	Of humid climates
umbr	L. *umbra*, shade	Presence of umbric epipedon (a dark surface)
ust	L. *ustus*, burnt	Of dry climates, usually hot in summer
xer	Gk. *xeros*, dry	Annual dry season

To determine the relationship between suborder names and soil characteristics, reference should be made to Table 12:5. Here the formative elements for suborder names are identified as is their connotation. Thus, the *Boroll* suborder (Greek *boreas*, northern) (Table 12:4) is seen to include *Mollisols* found in cool climates. Likewise, soils in the *Udult* suborder (Latin *udus*, humid) are Ultisols of the more humid climates. Identification of the primary characteristics of each of the other suborders can be made by cross reference to Tables 12:4 and 12:5. Note that some of these suborder names were used in the previous section to identify classes of soil orders.

GREAT GROUPS. The *great groups* are subdivisions of suborders. They are defined "largely on the presence or absence of diagnostic horizons and the arrangements of those horizons." These horizon designations are included in the list of formative elements for the names of great groups shown in Table 12:6. Note that these formative elements refer to epipedons such as mollic and orchic (see Table 12:1), to subsurface horizons such as argillic and natric, and to pans such as duripan and fragipan. Remember that the great group names are made up of these formative elements attached as prefixes to the names of suborders in which the great groups occur. Thus, a *Ustoll* with a *natric* horizon (high in sodium) belongs to the *Natrustoll* great group.

TABLE 12:6. *Formative Elements for Names of Great Groups and Their Connotation*

These formative elements combined with the appropriate suborder names give the great group names.

Formative Element	Connotation	Formative Element	Connotation
acr	Extreme weathering	moll	Mollic epipedon
agr	Agric horizon	nadur	See *natr* and *dur*
alb	Albic horizon	natr	Natric horizon
and	Ando-like	ochr	Ochric epipedon
anthr	Anthropic epipedon	pale	Old development
aqu	Wetness	pell	Low chroma
arg	Argillic horizon	plac	Thin pan
calc	Calcic horizon	plag	Plaggen horizon
camb	Cambic horizon	plinth	Plinthite
chrom	High chroma	quartz	High quartz
cry	Cold	rend	Rendzina-like
dur	Duripan	rhod	Dark-red colors
dystr, dys	Low base saturation	sal	Salic horizon
eutr, eu	High base saturation	sider	Free iron oxides
ferr	Iron	sphagno	Sphagnum moss
frag	Fragipan	torr	Usually dry
fragloss	See *frag* and *gloss*	trop	Continually warm
gibbs	Gibbsite	ud	Humid climates
gloss	Tongued	umbe	Umbric epipedon
hal	Salty	ust	Dry climate, usually hot in summer
hapl	Minimum horizon		
hum	Humus	verm	Wormy, or mixed by animals
hydr	Water	vitr	Glass
hyp	Hypnum moss	xer	Annual dry season
luo, lu	Illuvial	sombr	Dark horizon

As might be expected, the number of great groups is high, more than 200 having been identified. The names of selected great groups from three orders are given in Table 12:7. This list illustrates again the utility of the

Comprehensive Soil Classification System, especially the nomenclature it employs. The names identify the suborder and order in which the great groups are found. Thus, Argiudolls are Mollisols of the Udoll suborder characterized by the presence of an argillic horizon. Cross reference to Table 12:6 identifies the specific characteristics separating the great group classes from each other. Careful study of these two tables will show the utility of this classification system.

TABLE 12:7. *Examples of Names of Great Groups of Selected Suborders of the Mollisol, Alfisol, and Ultisol Orders*

The suborder name is identified as the italicized portion of the great group name.

	Great Group Characterized by:		
	Argillic Horizon	Minimum Horizon	Old Development
Mollisols			
1. Aquoll (wet)	Argi*aquoll*	Hapl*aquoll*	—
2. Udoll (moist)	Argi*udoll*	Hapl*udoll*	Pale*udoll*
3. Ustoll (dry)	Argi*ustoll*	Hapl*ustoll*	Pale*ustoll*
4. Xeroll (very dry)	Argi*xeroll*	Haplo*xeroll*	Pale*xeroll*
Alfisols			
1. Aqualfs (wet)	—		
2. Udalfs (moist)	Arg*udalf*	Hapl*udalf*	Pale*udalf*
3. Ustalfs (dry)	—	Hapl*ustalf*	Pale*ustalf*
4. Xeroll (very dry)	—	Haplo*xeralf*	Pale*xeralf*
Ultisols			
1. Aqults	—	—	—
2. Udults	—	Hapl*udult*	Pale*udult*
3. Ustults	—	Hapl*ustult*	Pale*ustult*
4. Xerults	—	Haplo*xerult*	Pale*xerult*

SUBGROUPS. Subgroups are subdivisions of great groups. They permit characterization of the core concept of a given great group and of gradations from that central concept to other units of the classification system. The subgroup most nearly representing the central concept of a great group is termed *orthic*. Thus the *Orthic Haplaltolls* subgroup typifies the Haplaltoll great group. A Haplaltoll with evidence of poor drainage (mottling) would be classed as an *Aquic Haplaltoll*. One with a natric horizon would fall in the *Natric Haplaltoll* subgroup.

Some intergrades may have properties in common with other orders or with other great groups. Thus, the *Entic Haplaltoll* subgroup intergrades toward the Entisol order. The subgroup concept illustrates very well the flexibility of this classification system.

12:9. SOIL FAMILIES, SERIES, PHASES, ASSOCIATIONS, AND CATENAS

FAMILIES. The family category of classification is based on properties important to the growth of plant roots. The criteria used vary from one subgroup to another but include broad textural classes, mineralogical classes, and temperature classes. Terms such as fine, fine loamy, sandy, and clayey are used to identify the broad textural classes. Terms used to describe the mineralogical classes include montmorillonitic, kaolinitic, siliceous, and mixed. For temperature classes, terms such as frigid, mesic, and thermic are used. Thus, a Typic Argiudall from Iowa, fine in texture, having a mixture of clay minerals, and being located in a temperate region, might be classed in the *fine, mixed, mesic* family. In contrast, a sandy-textured Typic Cryorthod, high in quartz and located in eastern Canada, might be considered in the *sandy, siliceous, frigid* family.

Family categories are in the process of being further developed and tested. The placement of most soils into families will have to await this development and testing.

SERIES. The families are subdivided into lesser groups called *series*. These units are quite distinct. The soils of any one series have similar profile characteristics. In more precise terms, a *series* is a group of soils developed from the same kind of parent material by the same genetic combination of processes, and whose horizons are quite similar in their arrangement and general characteristics. Ideally, the only differences of agricultural importance that should exist between the various soils of any given series are in the textures of the surface layer, and even here the range should not be great.

DIFFERENTIATING AND NAMING OF SERIES. Series are of course established on the basis of profile characteristics. This requires a careful study of the various horizons as to number, order, thickness, texture, structure, color, organic content, and reaction (acid, neutral, or alkaline). Such features as hardpan at a certain distance below the surface, a distinct zone of calcium carbonate accumulation at a certain depth, or striking color characteristics greatly aid in series identification.

In the United States each series is given a name, usually from some city, village, river, or county, such as Fargo, Muscatine, Cecil, Mohave, Ontario, and the like.

SOIL PHASE. A *phase* is a subdivision on the basis of some important deviation such as surface texture, erosion, slope, stoniness, or soluble salt content. It really marks a departure from the normal or key series already established. Thus, a Cecil sandy loam, 3 to 5 percent slope, and a Hagerstown silt loam, stony phase, are examples of soils where distinctions are made in respect to the phase.

SOIL ASSOCIATIONS AND CATENAS. Before proceeding with a consideration of soil survey methods, two more means of classifying associated soils should be mentioned. In the field, soils of different kinds are commonly found together. Such an *association* of soils may consist of a combination from different soil orders (see Fig. 12:9). Thus, a shallow Entisol on steep uplands may be found alongside a well-drained Alfisol or a soil developed from recent alluvium. The only requirement is that the soils be found together

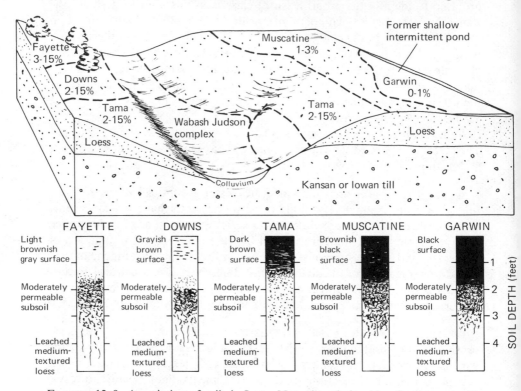

FIGURE 12:9. Association of soils in Iowa. Note the relationship of soil type to (1) parent material, (2) vegetation, (3) topography, and (4) drainage. Two Alfisols (Fayette and Downs) and three Mollisols (Tama, Muscatine, and Garwin) are shown. [*From Riecken and Smith (4)*.]

in the same area. Soil associations are important in a practical way since they help determine combinations of land-use patterns which must be utilized to support a profitable agriculture.

Well-drained, imperfectly drained, and poorly drained soils, all of which have developed from the same parent materials under the same climatic conditions, are often found closely associated under field conditions. This association on the basis of drainage or of differences in relief is known as a *catena* and is very helpful in practical classification of soils in a given region.

The relationship can be seen by referring to Fig. 12:10 where the Bath–Mardin–Volusia–Alden catena is shown. Although all four or five members of a catena are seldom found in a given area, the diagram illustrates the relationship of drainage to topography.

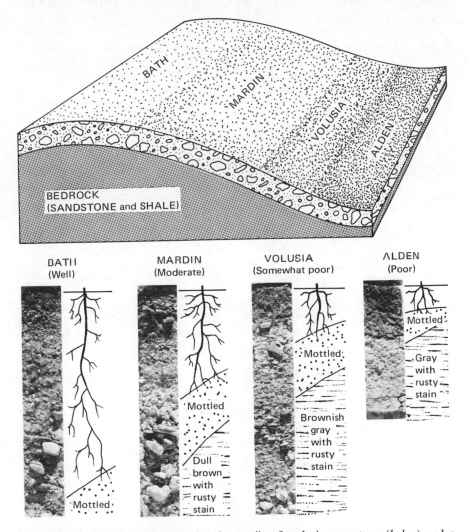

FIGURE 12:10. Monoliths showing four soils of a drainage catena (*below*) and a diagram showing their topographic association in the field (*above*). Note the decrease in the depth of the well-aerated zone (above the mottled layers) from the Bath soil (*left*) to the Alden (*right*). The latter remains poorly aerated throughout the growing season. These soils are all developed from the same parent material and differ only in drainage and topography. The Volusia soil as pictured was cultivated, while the others were located on virgin sites.

The practical significance of the soil catena can be seen in Fig. 12:11, where the tolerance of several crops to differences in soil drainage is shown. Recommendations for crop selection and management are commonly made on the basis of soil drainage class. Such recommendations along with careful study of farm soil survey maps are invaluable to farmers in making the best choices of soil and crop compatibility.

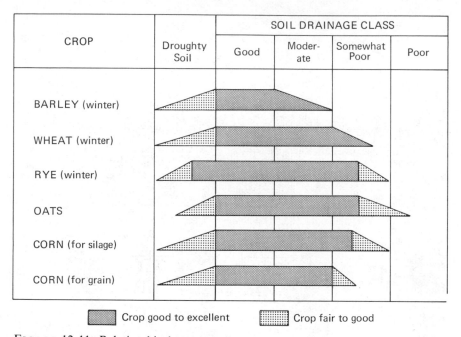

FIGURE 12:11. Relationship between soil drainage class and the growth of certain grain crops. [*From "Cornell Recommends"* (1).]

12:10. SOIL CLASSIFICATION—1949 SYSTEM

The system of soil classification used in the United States prior to the acceptance of the new comprehensive system will now be considered. By means of this classification, soils are grouped in three orders: (a) zonal, (b) intrazonal, and (c) azonal (see Table 12:8).

The characteristics of *zonal* soils are determined primarily by the climate in which they have developed. Differences in rock formation and geological origin are largely masked or have been rendered subordinate by dominating climate influences. Local features such as drainage and topography permit or even encourage the maximum influence of climate and vegetation. As the name *zonal* indicates, these soils are of such wide expanse as to be considered more or less regional in extent.

TABLE 12:8. *Classification of Soils into Orders, Suborders, and Great Soil Groups*[a]

Each great soil group is subdivided into numerous soil series and soil types.

Order	Suborder	Great Soil Groups
Zonal soils	1. Soils of the cold zone	Tundra
	2. Light-colored podzolized soils of timbered regions	Podzol soils Brown Podzolic soils Gray-Brown Podzolic soils Red-Yellow Podzolic soils Gray Podzolic or Gray Wooded soils
	3. Soils of forested warm-temperate and tropical regions	A variety of latosols are recognized; they await detailed classification
	4. Soils of the forest–grassland transition	Degraded Chernozem soils Noncalcic brown or Shantung brown soils
	5. Dark-colored soils of semi-arid, subhumid, and humid grasslands	Prairie soils (semipodzolic) Reddish prairie soils Chernozem soils Chestnut soils Reddish chestnut soils
	6. Light-colored soils of arid regions	Brown soils Reddish brown soils Sierozem soils Red Desert soils
Intrazonal soils	1. Hydromorphic soils of marshes, swamps, flats, and seepage areas	Humic-gley soil (includes wiesenboden) Alpine Meadow soils Bog soils Half-bog soils Low-humic Gley soils Planosols Ground-water Podzols Ground-water Latosols
	2. Halomorphic (saline and alkali) soils of imperfectly drained arid regions, littoral deposits	Solonchak soils (saline soils) Solonetz soils (alkali soils) Soloth soils
	3. Calcimorphic soils	Brown forest soils (Braunerde) Rendzina soils
Azonal soils	(No suborders)	Lithosols Regosols (includes dry sands) Alluvial

[a] Modified from Thorp and Smith (9).

As seen in Table 12:8, the *order* of zonal soils is further broken down into *suborders* on the basis of specific climate and vegetative regions. Each of these suborders in turn is divided into *great soil groups*, such as the Podzol and Prairie. These great soil groups, because they are an expression of more specific conditions, tend to be the central core of this classification system.

The classification of zonal soils outlined in Table 12:8 is presented somewhat more simply in Fig. 12:12. Here the relationship among climate, vegetation, and some of the important zonal soils is illustrated. The diagrams

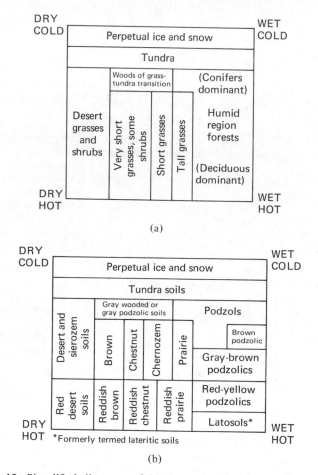

(a)

(b)

FIGURE 12:12. Simplified diagrams of (a) the variation of natural vegetation with climate and (b) the relationship of zonal soils to climate. Note that soil boundaries correspond rather closely to those of the natural vegetation. The right-to-left successions of soils and vegetation shown in these diagrams approximate those that are encountered as one travels from the east coast of North America to the Sierra Nevada and Cascade mountains. Typical desert conditions occur in the intermountain regions west of the Rockies.

are particularly helpful in the study of the soils of the United States since the general climatic differences shown in Fig. 12:14 approximate those found in this country when going from the Rocky Mountain area eastward.

Associated with zonal soils are the two other orders—*intrazonal* and *azonal*. The former includes those soils that, in spite of climate and vegetation, reflect the influence of some local conditions such as poor drainage or alkali salts. Many of their properties are, of course, similar to those of the zonal soils with which they are associated, but the characteristics resulting from local conditions are dominant. Since such soils cross zonal boundaries, they are termed intrazonal. Azonal soils, in marked contrast, are those without horizon differentiation. Profile layers resulting from soil development are not apparent. Most soils developed on Recent alluvial and colluvial deposits belong in this order.

12:11. SOIL SURVEY AND ITS UTILIZATION

The classifications as outlined in the preceding pages are susceptible to enough detail and precision to make them valuable in soil survey. In fact, they were developed for just such a purpose. The function of a soil survey is to classify, locate on a base map, and describe the nature of soils as they occur in the field. The soils in the United States are classified into *series* and *phases* on the basis of their profile characteristics. The field man, since his work is localized, concerns himself mostly with series and phase separations, giving the broader regional distinctions only general consideration.

FIELD MAPPING. As the series and phase identification progresses in the survey of any area, the location of each soil unit is shown on a suitable base map. The maps used in the United States are aerial base maps such as the one shown in Fig. 12:13. They have considerable advantage over the ordinary contour maps in that land cover and field boundaries show very clearly. Thus, the surveyor can quickly locate his position on the map and can indicate readily the soil boundaries (see Fig. 12:14).

Using the procedure outlined above, a field map is thus obtained which gives not only the soil series but also the phase, which furnishes information as to slope and severity of erosion. Thus, the map becomes of even greater practical value.

The field map, after the separations have been carefully checked and correlated, is now ready for reproduction. In older maps this reproduction was made in color. For more recent maps a reproduction is made of the aerial photograph with the appropriate soil symbols shown on it. When published, the map accompanies a bulletin containing a description of the topography, climate, agriculture, and soils of the area under consideration. Each soil series is minutely described as to profile and suggestions as to practical management are usually made. Such a field map is shown in

FIGURE 12:13. Soil survey maps are made using aerial photographs such as that above from Winneshiek County, Iowa. The field soil scientist is able to visualize quickly the topographic features of the section in which he is working. [*Photo courtesy Soil Survey Division, U. S. Soil Conservation Service.*]

FIGURE 12:14. Soil survey maps are prepared by soil scientists examining soils in the field (*left*) using a soil auger and other diagnostic tools. Mapping units are outlined on a topographic map (*right*) first in the field and later more permanently in the map room. [*Photos courtesy Soil Survey Division, U. S. Soil Conservation Service.*]

Fig. 12:15, where the relationship of topography to soil associations can be seen readily.

USE OF SOIL SURVEYS. Soil survey bulletins and maps are useful as a basis for other scientific work. Crop research of all kinds is facilitated if a soil survey has previously been made. Land evaluation and appraisal, statistical studies, and sociological investigations are other interests served.

There is a growing tendency to make use of soil survey maps and bulletins in a more practical way. The extension specialist and the county agricultural agent find the survey maps and bulletins a guide in making suggestions and recommendations. They afford information on the soils and on the area surveyed that is impossible to obtain elsewhere.

In some cases simpler and more practical bulletins follow the rather technical ones. The value of an illustrative means of showing the relationships among soil series within a given region can be seen by referring to Fig. 12:9. A glance at this diagram shows the meaning and need for classification. It also suggests that soil management must be accommodated to the soil series insofar as possible.

FIGURE 12:15. Field examination makes possible the delineation of soil boundaries, which are then indicated on the topographic map shown in Fig. 12:13. The soil legend identifies the soil name (first two letters), the slope (second capital letter) and the degree of erosion (the number). Thus, OsC2 is an Orwood silt loam (Os) that has a 5–9 percent slope (C) and is moderately eroded (2). [*Photo courtesy Soil Survey Division, U. S. Soil Conservation Service.*]

Engineers and hydrologists also use soil survey information in a practical way. Prospective roadbeds can be selected from soil survey maps. Estimates of water runoff and infiltration can be made on the basis of soil characteristics enumerated in soil survey bulletins. Predictions can be made of hydrologic changes in relation to modifications in land-use patterns.

The soil survey in a practical way is perhaps of greatest value in land classification for agricultural and other uses. The classification of greatest importance to agriculturists is that in use by the U. S. Soil Conservation Service because of its widespread application.

12:12. LAND CAPABILITY CLASSIFICATION

At this point it may be well to point out the difference between *soil* and *land*. Soil is the more restrictive term, referring to a collection of natural bodies with depth as well as breadth whose characteristics may be only indirectly related to their current vegetation and use. Land is a broader term which includes among its characteristics not only the soil but other physical attributes, such as water supply, existing plant cover, and location with respect to cities, means of transportation, etc. Thus, we have *forest land*, *bottom land*, and *grasslands*, which may include a variety of soils.

Soil survey maps and reports have become two of the bases for a system of land capability classification. This system requires that every acre of land be used in accordance with its *capability* and *limitations*. Land is classified according to the most suitable sustained use that can be made of it while providing for adequate protection from erosion or other means of deterioration. Thus, an area where the soils are deep, well drained, and have a stable surface structure, and where the slope is only 1 to 2 percent may be cropped intensively almost indefinitely with little danger of erosion or loss in productivity. Such an area has great capabilities and few limitations in the use to which it can be put. In contrast, an area on which shallow or poorly drained soils are found or wherein steep slopes are prevalent has limited capabilities and many limitations as to its use. It can easily be seen how the characteristics of soils are one of the criteria for identifying the best land use.

CAPABILITY CLASSES. Under the system set up by the U. S. Soil Conservation Service, eight land capability classes are recognized (3). These classes are numbered from I to VIII. Soils having greatest capabilities for response to management and least limitations in the ways they can be used are in class I. Those with least capabilities and greatest limitations are found in class VIII (see Figs. 12:16 and 12:17). A brief description of the characteristics and safe use of soils in each class follows.

CLASS I. Soils found in this land class have few limitations that restrict their use. They can be cropped intensively, used for pasture, range, woodlands,

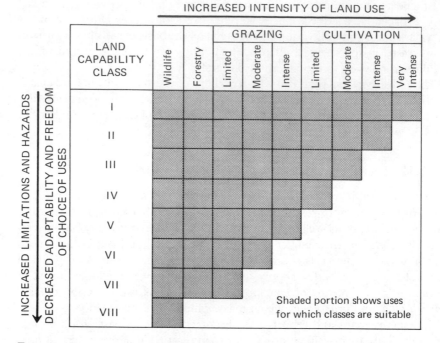

FIGURE 12:16. Intensity with which each land capability class can be used with safety. Note the increasing limitations on the uses to which the land can safely be put as one moves from class I to class VIII. [*Modified from Hockensmith and Steele* (2).]

or even for wildlife preserves. The soils are deep, well drained, and the land is nearly level. They are either naturally fertile or have characteristics which encourage good response of crops to applications of fertilizer.

The water-holding capacity of soils in class I is high. In arid and semiarid areas soils having all the other favorable characteristics mentioned above may be in class I if they are irrigated by a permanent irrigation system.

The soils in class I need only ordinary crop management practices to maintain their productivity. These include the use of fertilizer and lime and the return of manure and crop residues, including green manures. Crop rotations are also followed.

CLASS II. Soils in this class "have some limitations that reduce the choice of plants or require moderate conservation practices." These soils may be used for the same crops as class I. However, they are capable of sustaining less intensive cropping systems, or, with the same cropping systems, they require some conservation practices.

The use of soils in class II may be limited by one or more factors, such as (a) gentle slopes, (b) moderate erosion hazards, (c) inadequate soil

FIGURE 12:17. Several land capability classes in San Mateo County, California. A range is shown from the nearly level land in the foreground (class I), which can be cropped intensively, to that of the badly eroded hillsides (classes VII and VIII). Although topography and erosion hazards are emphasized here, it should be remembered that other factors—drainage, stoniness, droughtiness—also limit soil usage and help determine the land capability class. [*Photo courtesy U. S. Conservation Service.*]

depth, (d) less than ideal soil structure and workability, (e) slight to moderate alkali or saline conditions, and (f) somewhat restricted drainage.

The management practices that may be required for soils in class II include terracing, strip cropping, contour tillage, rotations involving grasses and legumes, and grassed waterways. In addition, those practices which are used on class I land are also generally required for soils in class II.

CLASS III. "Soils in class III have severe limitations that reduce the choice of plants or require special conservation practices or both." The same crops can be grown on class III land as on classes I and II. The amount of clean, cultivated land is restricted, however, as is the choice of the particular crop to be used. Crops which provide soil cover such as grasses and legumes must be more prominent in the rotations used.

Limitations in the use of soils in class III result from factors such as (a) moderately steep slopes, (b) high erosion hazards, (c) very slow water permeability, (d) shallow depth and restricted root zone, (e) low water-holding capacity, (f) low fertility, (g) moderate alkali or salinity, and (h) unstable soil structure.

Soils in class III often require special conservation practices. Those mentioned for class II land must be employed, frequently in combination with restrictions in kinds of crops. Tile or other drainage systems may also be needed.

CLASS IV. Soils in this class can be used for cultivation, but there are very severe limitations on the choice of crops. Also, very careful management may be required. The alternative uses of these soils are more limited than for class III. Close-growing crops must be used extensively, and row crops cannot be grown safely in most cases. The choice of crops may be limited by excess moisture as well as by erosion hazards.

The most limiting factors on these soils may be one or more of the following: (a) steep slopes, (b) severe erosion susceptibility, (c) severe past erosion, (d) shallow soils, (e) low water-holding capacity, (f) poor drainage, and (g) severe alkalinity or salinity. Soil conservation practices must be applied more frequently than on soils in class III. Also, they are usually combined with severe limitations in choice of crop.

CLASS V. Soils in classes V to VIII are generally not suited to cultivation. Those in class V are limited in their safe use by factors other than erosion hazards. Examples of such limitations follow: (a) subject to frequent stream overflow, (b) growing season too short for crop plants, (c) stony or rocky soils, and (d) ponded areas where drainage is not feasible. Oftentimes, pastures can be improved on this class of land.

CLASS VI. Soils in this class have extreme limitations that restrict their use largely to pasture or range, woodland, or wildlife. The limitations are the same as those for class IV land, but they are more rigid.

CLASS VII. Soils in class VII have very severe limitations which restrict their use to grazing, woodland, or wildlife. The physical limitations are the same as VI except they are so strict that pasture improvement is impractical.

CLASS VIII. In this land class are soils that should not be used for any kind of commercial plant production. Their use is restricted to "recreation, wildlife, water supply, or aesthetic purposes." Examples of kinds of soils or land forms included in class VIII are sandy beaches, river wash, and rock outcrop.

SUBCLASSES. In each of the land capability classes are subclasses which have the same kind of dominant limitations for agricultural uses. The four kinds of limitations recognized in these subclasses are risks of erosion (e); wetness, drainage, or overflow (w); root-zone limitations (s); and climatic limitations (c). Thus, a soil may be found in class III (e), indicating that it is in class III because of risks of erosion.

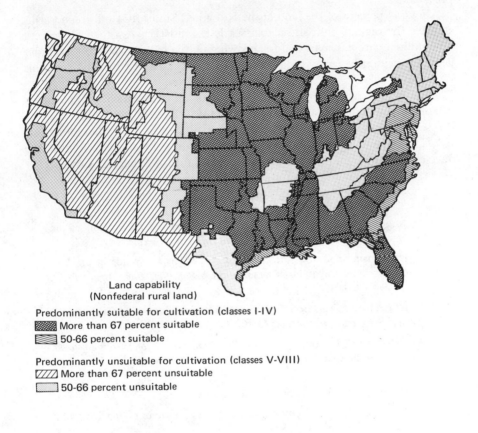

Land capability
(Nonfederal rural land)

Predominantly suitable for cultivation (classes I–IV)
▨ More than 67 percent suitable
▧ 50–66 percent suitable

Predominantly unsuitable for cultivation (classes V–VIII)
▨ More than 67 percent unsuitable
▨ 50–66 percent unsuitable

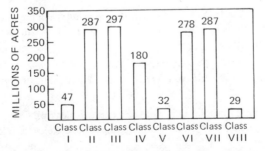

FIGURE 12:18. Geographic distribution (*upper*) and quantitative distribution (*lower*) of land capability classes in the United States. [*From U. S. Department of Agriculture* (10).]

LAND USE CAPABILITY IN THE U. S. In 1967 the U. S. Department of Agriculture made a national inventory of soil and water conservation needs for the United States. This inventory included information on land capability classification. In Figure 12:18 summary information from this inventory is presented.

About 44 percent of the land in the 50 United States (631 million acres) is suitable for regular cultivation (classes I, II, and III). Another 12 percent (180 million acres) is marginal for growing cultivated crops. The remainder is suited primarily for grasslands and forests and is used mostly for this purpose.

This land classification scheme illustrates the use which can be made of soil surveys in a practical way. The many soils delineated on a map by the soil surveyor are viewed in the light of their safest and best longtime use. The eight land capability classes have become the starting point in the development of farm plans so useful to thousands of American farmers.

REFERENCES

(1) Anonymous, "Cornell Recommends," an annual publication of the New York State College of Agriculture, Cornell University, 1962.

(2) Hockensmith, R. D., and Steele, J. G., "Recent Trends in the Use of the Land-Capability Classification," *Soil Sci. Soc. Amer. Proc.*, **14**:383–88, 1949.

(3) Klingebiel, A. A., "Soil Survey Interpretation-Capability Groupings," *Soil Sci. Soc. Amer. Proc.*, **22**:160–63, 1958.

(4) Riecken, F. F., and Smith, G. D., *Principal Upland Soils of Iowa, Their Occurrence and Important Properties*, Agron. 49 rev., Iowa Agr. Exp. Sta., 1949.

(5) Simonson, R. W., "Concept of Soil," *Adv. Agron.*, **20**:1–47, 1968.

(6) Smith, G. D., "Objectives and Basic Assumptions of the New Soil Classification System," *Soil Science*, **96**:6–16, 1963.

(7) Soil Survey Staff, *Soil Classification—A Comprehensive System— 7th Approximation* (Washington, D.C.: U. S. Department of Agriculture, Aug. 1960).

(8) Soil Survey Staff, *Supplement to Soil Classification System—7th Approximation* (Washington, D.C.: U. S. Department of Agriculture, Mar. 1967).

(9) Thorp, J., and Smith, G. D., "Higher Categories of Soil Classification: Order, Suborder, and Great Soil Groups," *Soil Science*, **67**:117–26, 1949.

(10) U. S. Department of Agriculture, *National Inventory of Soil and Water Conservation Needs, 1967* (Washington, D.C.: USA Statistical Bulletin No. 461, 1971).

Chapter 13

ORGANIC SOILS (HISTOSOLS): THEIR NATURE, PROPERTIES, AND UTILIZATION

ON the basis of organic content, two general groups of soils are commonly recognized—*mineral* and *organic*. The mineral soils vary in organic matter from a mere trace to as high as 20 or even 30 percent. Those soils with higher organic matter are arbitrarily termed organic soils and are classed as *Histosols* in the comprehensive soil classification system. Those which are cultivated probably average about 80 percent organic matter.

Organic soils are by no means so extensive as are mineral soils, yet their total acreage is quite large. Their use in favored localities for the intensive production of crops, particularly vegetables, is very important. As the development of organic crop lands accelerates, they will be investigated more closely. In the past, organic soils did not receive the study they deserved. As a result, less exact knowledge is available regarding their physical, chemical, and biological characteristics than is available for mineral soils. In general, however, the same edaphological principles hold in a broad way for both.

13:1. GENESIS OF ORGANIC DEPOSITS

Marshes, bogs, and swamps provide conditions suitable for the accumulation of organic deposits. The highly favorable environment in and adjacent to such areas has encouraged the growth of many plants, such as pondweed, cattails, sedges, reeds and other grasses, mosses, shrubs, and also trees. These plants in numberless generations thrive, die, and sink down to be covered by the water in which they have grown. The water shuts out the air, prohibits rapid oxidation, and thus acts as a partial preservative. The decay that does go on is largely through the agency of fungi, anaerobic bacteria, algae, and certain types of microscopic aquatic animals. They break down the organic tissue, liberate gaseous constituents, and aid in the synthesis of humus.

LAYERING OF PEAT BEDS. As one generation of plants follows another, layer after layer of organic residue is deposited in the swamp or marsh (see Fig. 13:1). The constitution of these successive layers changes as time goes on since a sequence of different plant life is likely to occur. Thus, deep-water plants may be supplanted by reeds and sedges, these by various mosses, and these in turn by shrubs, until finally forest trees, either hardwoods or conifers, with their characteristic undergrowth, may gain a foothold. The succession is by no means regular or definite, as a slight change in climate or water level may alter the sequence entirely.

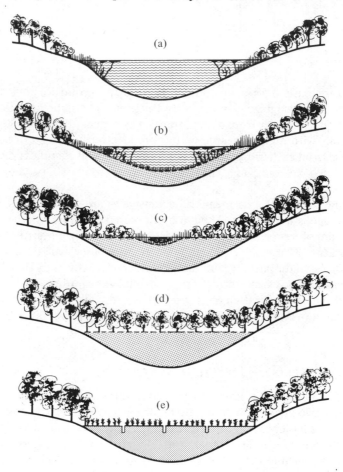

FIGURE 13:1. Four stages in the development of a typical woody peat bog (a–d) and the area after clearing and draining (e). Nutrient runoff from the surrounding uplands encourages aquatic plant growth, especially around the pond edges (a). Organic debris fills in the bottom of the pond (b and c), and eventually trees cover the entire area (d). When the land is cleared and a drainage system installed, the area becomes a most productive muck soil (e).

The profile of an organic deposit is therefore characterized by layers, differing not only as to their degree of decomposition but also as to the nature of the original plant tissue. In fact, these layers later may have become soil horizons. Their final character is determined in part by the nature of the original materials and in part by the type and degree of decomposition. Thus, the profile characteristics of organic soils, as with those dominantly mineral in nature, are in part inherited and in part acquired.

13:2. AREA AND DISTRIBUTION OF PEAT ACCUMULATIONS

As might be expected, peat deposits are found all over the world wherever the conditions are favorable. Some 60 percent of the world's organic deposits are in Russia. Canada has vast deposits covering 300 million acres, mostly undeveloped. In Germany, Holland, Norway, Sweden, Russia, Poland, Ireland, England, and Scotland, economic use has long been made of peat and peat products. Organic deposits in Germany occupy perhaps 5 million acres; in Sweden 12 million; and in Ireland, 3 million acres.

PEAT IN NONGLACIATED UNITED STATES. Organic deposits of the nature described occur in many parts of the United States. In Florida the Everglades, spread out over an extensive plain, contain considerably over 2 million acres of saw-grass (sedge) accumulations. Along the Atlantic coastal plain great marsh deposits are found, especially in North Carolina. In California there are the tule-reed beds of the great central valley, approximately 300,000 acres in extent. Louisiana alone possesses about 3 million acres of organic soils. All of these are southward and outside the glaciated areas of the United States. They are related only indirectly to the glaciation through a change in climate and a rise in ocean level due to the melting of the ice.

PEAT IN GLACIATED UNITED STATES. Northward in the regions covered by glacial debris of various kinds, organic deposits are even more extended. In fact, about 75 percent of the peat deposits of continental United States occur in the glaciated areas. Minnesota, Wisconsin, and Michigan are especially favored in this respect, their combined acreage running well above 12 million acres. Washington, with about 2 million acres, ranks highest of all the western states. Indiana, Massachusetts, New York, and New Jersey fall into the 300,000- to 500,000-acre class. Other states, especially Iowa, Illinois, and Maine, contain smaller but often very important areas.

The total acreage of peat deposits in the United States proper is between 25 and 30 million acres. In Alaska and northern Canada, and in northern Asia and Europe as well, there occur great areas of peat derived from sedges and mosses called *muskeg*. In many cases the subhorizons of such accumulations are perpetually frozen. They are an important feature of the tundra soils that characterize these regions (see p. 341).

The Pleistocene glaciation, by impeding drainage, led to the formation of swamps and bogs. As the climate became milder and gradually attained its present status, conditions were ideal for swamp vegetation to flourish. As a result, certain parts of the glaciated region are liberally dotted with organic accumulations ranging from a few inches to 50 feet in depth. Some areas, a number of square miles in extent, are solidly occupied as in Minnesota, while in other localities the peat lies in isolated patches or in long ribbons, as in southeastern Wisconsin and in central New York.

13:3. PEAT PARENT MATERIALS

Peat, regardless of its stage of decomposition, may conveniently be classified according to its parent materials and under three general heads as follows:

Classification of Peat[1]

1. Sedimentary peat { Mixtures of water lilies, pondweed, hornwort, pollen plankton, etc.

2. Fibrous peat { Sedges of various kinds; mosses—sphagnum, hypnum, and others; reeds and other grasses; cattails, both latifolia and angustifolia } and their mixtures.

3. Woody peat { Deciduous and coniferous trees } and their undergrowth.

SEDIMENTARY PEAT. Sedimentary peat usually accumulates in comparatively deep water and therefore generally is found well down in the profile. Sometimes it is intermixed with the other types of peat nearer the surface. Sedimentary peat seems to be derived from plant materials that humify rather freely. Owing to the nature of the original tissue and perhaps also to the type of decay, a highly colloidal and characteristically compact and *rubbery* substance develops; its moisture capacity is high, perhaps four or five times its dry weight. Water thus imbibed is held tenaciously and therefore this peat dries out very slowly. The colloidal materials of sedimentary peat dry irreversible—that is, when once dry, this peat absorbs water very slowly and persistently remains in a *hard* and *lumpy* condition.

---

[1] In Germany peats are classified in a general way into *high-moor* and *low-moor* in reference to the shape of the deposits. The high-moors are convex, that is, raised in the center; the low-moors are concave. The former is usually quite acid and low in calcium, the latter less acid and quite high in exchangeable calcium. In England the corresponding terms are *moor* and *fen*. The term *heathland* refers to a shallow acid peat.

Sedimentary peat is thus very undesirable as a soil because its unfavorable physical condition renders it unsatisfactory for use in the growing of plants. Even small amounts in the furrow slice lower the desirability of the peat for agricultural purposes. Fortunately, in most cases, it occurs well down in the profile and ordinarily does not appear above the plowline. Therefore, its presence usually is unnoticed or ignored unless it obstructs drainage or otherwise interferes with the agricultural utilization of the peat deposit.

FIBROUS PEAT. A number of fibrous peats occur, often in the same swamp deposit. They are all high in water-holding capacity and may exhibit varying degrees of decomposition. They differ among themselves, especially as to their filamentous or fibrous physical nature. Undecomposed moss and sedge are fine enough to be used in greenhouses and nurseries and as a source of organic matter for gardens and flower beds. Reed and cattails, however, are somewhat coarse, especially the latter.

Fibrous materials as they decay may make satisfactory field soils, although their productivity will vary. Moss peats are almost invariably quite acid and relatively low in ash and nitrogen. The sedges are intermediate in these respects, while cattail peats are not so acid and have a better nutrient balance. Fibrous peats may occur at the surface of the organic accumulation of which they are a part or well down in the profile. They usually lie above the sedimentary deposit when this type of peat is present. Nevertheless, if just the right fluctuation of conditions has occurred, a stratum of sedimentary peat may lie embedded within the fibrous peat horizons and rather near the surface.

WOODY PEAT. Since trees are the vegetation present in many swamp deposits, woody peat is usually at the surface of the organic accumulation. This is not an invariable rule, however, since a rise in water level might kill the trees and so favor reed, sedge, or cattail as to give a layer of fibrous material over the woody accumulation. It is not particularly surprising, therefore, to find subhorizon layers of woody peat.

Woody peat is brown or black in color when wet, according to the degree of humification. It is loose and open when dry or merely moist and decidedly nonfibrous in character. Virgin deposits often are notably granular. It is thus easily distinguished from the other two types of peat unless the samples are unusually well disintegrated and decomposed.

Woody peat develops from the residues not only of deciduous and coniferous trees but also from the shrubs and other plants that occupy the forest floor of the swamp. Maple, elm, tamarack, hemlock, spruce, cedar, pine, and other trees occur as the climax vegetation in swamps of temperate regions. In spite of the great number of plants that contribute to its accumulation, woody peat is rather homogeneous unless it contains admixtures of fibrous materials.

The water capacity of woody peat is somewhat lower than that of sedge peat, which in turn is much less than that of moss peat. For that reason, woody peat is less desirable than the others for use in greenhouses and nurseries, where such materials are used as a means of moisture control and a compost conditioner. Woody residues produce a field soil, however, that is quite superior and much prized for the growing of vegetables and other crops. Unfortunately, such woody peat in the United States is confined mostly to Wisconsin, Michigan, and New York.

13:4. Uses of Peat

NURSERIES, GREENHOUSES, AND LAWNS. Peat is utilized in a number of ways, depending on the nature of the material. In the United States, Canada, and Europe the more or less undecomposed products are very commonly employed as a source of organic matter. This is true of moss and sedge peat. When incorporated with mineral soil in sufficient quantities, peat not only ensures a good physical condition but markedly increases the water capacity of the mixture. Soils for potting and other purposes are greatly benefited physically by its use. It is thus valuable in greenhouses, gardens, flower beds, and nurseries both as a soil amendment and as a mulch around growing plants of all kinds. Peat is likewise useful in the preparation of lawn soils and golf greens, and in numerous other ways where tilth and organic matter are important factors.

PEAT AS FUEL. In Holland, Germany, Belgium, Ireland, and other countries of Europe, peat is dug out in brickette form, dried thoroughly, and used as fuel. Thus, vast amounts of Dutch and German peat have not only been used locally but also were shipped considerable distances for industrial purposes. In lands where wood is scarce and coal expensive, peat deposits are an exceedingly valuable natural resource. In Holland the excavation of peat has been so regulated as to leave the site well drained and in such a condition that field crops can be grown, aided by the portions of the organic matter that, by law, must be left behind. Some of the most productive agricultural areas in Holland lie on the site of these reclaimed and properly exploited bogs. Such soils must be heavily treated, of course, with commercial fertilizer.

FIELD SOILS. In the United States, the most extensive use of peat is as a field soil, especially for vegetable production. Thousands of acres are now under cultivation, often producing two crops a year. Peats for such use should be well decomposed, the decay often having gone so far as to make it almost impossible to identify with certainty the various plants that have contributed to the accumulation. As the following pages indicate, the edaphological interest in peat relates mainly to its use as a natural soil. At the same time, its use as a soil supplement for greenhouse, nursery, and home gardening has definite edaphological implications.

13:5. CLASSIFICATION OF ORGANIC SOILS

Two types of classification are employed, one using traditional field terminology and the other the terminology of the Comprehensive Soil Classification System. In each case, considerable emphasis is placed on the stage of breakdown of original plant materials.

PEAT VS. MUCK. As a practical matter, organic soils are commonly differentiated on the basis of their state of decomposition. Those deposits that are slightly or nondecayed are termed *peat*; those that are markedly decomposed are called *muck*. In peat soils, one can identify the kinds of plants in the original deposit, especially in the upper horizons. By contrast, mucks are generally decomposed to the point where the original plant parts cannot be identified.

Peat soils are coarse or fine-textured depending on the nature of the deposited plant residues. Well-decomposed mucks, on the other hand, are usually quite fine since the original plant material has broken down. When dry, they may be quite powdery and subject to wind erosion.

COMPREHENSIVE CLASSIFICATION SYSTEM. In this system organic soils are identified as the order *Histosols* (see p. 332). This order contains a minimum of 20 percent organic matter if the clay content is low and a minimum of 30 percent if as much as 50 percent clay is present.

The detailed classification scheme for Histosols is being developed. Three major suborders have been established. The *Fibrists* suborder includes soils in which the undecomposed fibrous organic materials are easily identified. Their bulk densities are low, commonly less than 0.1. They have high water-holding capacities and are brown and yellow in color. They usually develop under cold climates and other environmental conditions which discourage decomposition.

The most highly decomposed organic materials are found in soils classed in the *Saprist* suborder. The original plant fibers have mostly disappeared. The bulk densities of these soils are relatively high, commonly 0.2 or more. Water-holding capacities are the lowest for organic soils. Their color is commonly very dark gray to black and they are quite stable in their physical properties.

The *Hemist* suborder includes soils intermediate in their properties between those of the Fibrists and Saprists. They are more decomposed than the Fibrists but less so than the Saprists. They have intermediate values for bulk density (between 0.1 and 0.2) and water-holding capacity. Their colors are also intermediate between those of the other two suborders.

In addition to visual appearance, a chemical test is used to ascertain the degree of decomposition of the original plant material in organic soils. An alkaline solution of sodium pyrophosphate dissolves certain organic acids which are the products of decomposition. The amount of the acids dissolved by the agent is related to tissue decomposition. This method provides some quantification of an otherwise subjective judgment of decomposition.

The great group and family categories for organic soils have been established tentatively. However, their definitive designation will entail further research and field examination.

Drawings of organic soil profiles representative of the Saprist and Fibrists suborders are shown in Fig. 13:2. The Fibrist soil has a high percentage of undecomposed fibrous materials and is very acid. The cultivated Saprist soil has a high pH, low undecomposed fiber percentage, and is darker in color.

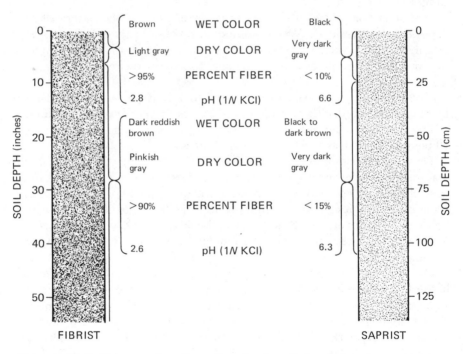

FIGURE 13:2. Two organic soil profiles showing distinguishing characteristics of each. The uncultivated soil on the left is of the Fibrist suborder, with a high percentage of undecomposed fiber and high acidity. In contrast, the example of the Saprist suborder is a cultivated soil with only 10–20 percent undecomposed fibers and reasonably low acidity. Both soils are found in Minnesota. [*Modified from Farnham and Finney* (1).]

13:6. PHYSICAL CHARACTERISTICS OF FIELD PEAT SOILS

COLOR. The color of a typical cultivated organic soil, dark brown or intensely black, is perhaps the first physical characteristic that attracts attention. Although the original materials may be gray, brown, or reddish brown, dark humic compounds appear as decomposition advances. In general, the changes that the organic matter undergoes seem to be somewhat similar to those occurring to the organic residues of mineral soils in spite of the restricted aeration of the peat deposit.

BULK DENSITY. The second outstanding characteristic is the light weight of the representative organic soil when dry. The bulk density compared with mineral surface soils is surprisingly low, 0.20 to 0.30 being common for a well-decomposed organic soil. A cultivated surface mineral soil will usually fall within the limits 1.25 to 1.45. One cubic foot of organic soil will contain from 8 to 20 pounds of dry matter, depending on its original source, the condition of the layer, and the admixture of mineral materials. An acre–furrow slice of the depth common in mineral soils, while variable, may be considered to weigh 400,000 to perhaps 500,000 pounds when dry. Compared with the 2 million or $2\frac{1}{2}$ million pounds ordinarily considered as the dry weight of an acre–furrow slice of a representative mineral soil, such figures seem small indeed.

WATER CAPACITY. A third important property of organic soil is its high water-holding capacity. While a dry mineral soil will adsorb and hold from one fifth to two fifths its weight of water, an organic soil will retain two to four times its dry weight of moisture. Undecayed or only slightly decomposed moss or sedge peat, in contrast with soils developed from these materials, has a much greater water-holding capacity. Not uncommonly, such materials can hold water to the extent of 12, 15, or even 20 times their dry weights. This explains in part their value in greenhouse and nursery operations.

It must not be hastily assumed, however, that organic soils in the field greatly surpass mineral soils in their capacity to supply plants with water. Two conditions militate against the organic soils. In the first place, their amounts of unavailable water are much higher proportionately than that of mineral soils. Also, since the comparative figures quoted are on the basis of dry weight, the peat soils with their low bulk densities are at a considerable disadvantage. When considered on the volume basis, a given layer of peat soil at optimum moisture will supply only slightly more water to plants than a comparable mineral soil.

STRUCTURE. A fourth outstanding characteristic of a typical woody or fibrous organic soil is its almost invariably good physical condition. While decayed organic matter is largely colloidal and possesses high adsorptive powers, its cohesion and plasticity are rather low. An organic soil of good quality is therefore porous, open, and easy to cultivate. These characteristics make it especially desirable for vegetable production. However, during dry periods, light, loose peat, whose structure has been destroyed by cultivation, may drift badly in a high wind and extensive crop damage may result. Peat may also become ignited when dry. Such a fire often is difficult to extinguish and may continue for several years.

In suggesting that the surface layer of an arable peat soil is likely to be granular, it is not to be inferred that the whole profile is in this structural condition also. Far from it. All sorts of physical arrangements are encountered in the subhorizons, such as laminated, vertical, fragmental, fibrous, and rubbery. The variability from horizon to horizon is determined not only by the character of the original materials but also by the nature and degree of decomposition.

13:7. COLLOIDAL NATURE OF ORGANIC SOILS

The colloidal characteristics of organic soils are more pronounced than those exhibited by most mineral soil colloids. Their surface area exceeds even that of montmorillonite clay two to four times, and cation exchange capacities are correspondingly greater. The cation exchange capacity of organic soils is in the neighborhood of 3 meq per gram of dry organic matter.

FORMULA AND CALCIUM CONTENT. The same graphic formula employed for mineral soils

can be used to represent the colloidal complex of peat, and the same metallic cations are adsorbed by the two types of soil in the same order of magnitude: $Ca > Mg > K$ or Na (see pp. 96–97). Some important differences are to be noted, however. The amount of adsorbed calcium is much greater than is the case with mineral soils. Also, a high proportion of the total mineral elements are exchangeable in organic soils. These two facts are shown by the data of Table 13:1. The total calcium content of the low-lime peat is 1.4 percent of calcium oxide and its cation exchange capacity is 184 meq. The corresponding figures for the high-lime soil are 6.0 percent of calcium oxide and 265 meq cation exchange capacity.

The data of Table 13:1 show that most of the calcium present in organic soils and a very large proportion of the magnesium are in an exchangeable condition. Moreover, approximately one third of the total potassium is exchangeable. All this is strikingly in contrast with a mineral soil, where usually less than 35 percent of the calcium, less than 10 percent of the magnesium, and less than 1 percent of the total potassium are exchangeable (see Fig. 2:3).

TABLE 13:1. *Cationic Condition of Two New York Woody Peat Soils—One Low and the Other High in Lime*[a]

Cation	Exchangeable Metallic Cations (meq/100 g)		Percent of Cations Exchangeable	
	Low-Lime Peat	High-Lime Peat	Low-Lime Peat	High-Lime Peat
Calcium	39.8	159.7	80	75
Magnesium	9.7	12.0	100	51
Potassium	0.87	1.16	34	27
Sodium	0.80	0.87	—	—

[a] Calculated from Wilson and Staker (3).

STRENGTH OF ACIDS. The pH of organic soils, as with mineral soils, is controlled by the colloidal complex, its percentage base saturation being a major factor (p. 104). The ratio of the metallic cation and the nature of the micelle also exert an influence, the latter being especially important. In general, the colloidal complex of organic colloids when saturated with hydrogen will develop a lower soil pH than will acid mineral clays similarly charged. In other words, the organic complex is the stronger acid. This means that at the same percentage base saturation, peat is somewhat more acid than mineral soils.

BUFFERING. Since the buffering of a soil is determined in large degree by the magnitude of its cation exchange capacity, organic soils in general show an unusually marked resistance to a change in pH, much greater than do mineral soils. Consequently, considerably more sulfur or lime is necessary to change the pH of an organic soil than to effect a similar modification in a mineral soil at a corresponding pH. However, the buffer curve obtained by plotting pH against percentage base saturation is of the same general order as that for mineral soils (p. 385) except that the pH at 50 percent base saturation is somewhat less.

EXCHANGE. In order that the cation exchange properties may be visualized a little more concretely, further data are offered about the two woody peat surface soils already discussed, one comparatively low in lime, the other rather high. The exchange data, already quoted on page 105 for a representative humid region mineral soil are offered in contrast (see Table 13:2).

TABLE 13:2. *Exchange Data for Two Woody Peat Surface Soils[a] and for a Representative Humid Region Mineral Surface Soil*

Exchange capacity values are in milliequivalents per 100 grams.

Exchange Characteristics	Low-Lime Peat	High-Lime Peat	Humid Region Mineral Soil
Exchangeable Ca	39.8	159.7	6 to 9
Other exchangeable bases, M	21.7	43.4	2 to 3
Exchangeable H	122.3	62.0	4 to 6
Cation exchange capacity	183.8	265.1	12 to 18
Percentage base saturation	33.5	76.6	66.6
pH	4.0	5.1	5.6–5.8

[a] From Wilson and Staker (3).

13:8. CHEMICAL COMPOSITION OF ORGANIC SOILS

Organic soils are quite variable chemically but they have some characteristics in common, as is shown by the data in Table 13:3 from fifteen organic soils. Also, a representative analysis of the surface layer of an arable organic soil is given in Table 13:4 but with the understanding that is merely suggestive. The chemical analysis of the upper layer of a representative humid region mineral soil is cited for comparison and contrast.

TABLE 13:3. *Chemical Analyses of Certain Representative Peat Soils[a]*

Values in percent based on dry matter.

Source	Nature	Organic Matter	N	P_2O_5	K_2O	CaO
Minnesota	Low lime	93.0	2.22	0.18	0.07	0.40
Minnesota	High lime	79.8	2.78	0.24	0.10	3.35
Michigan	Low lime	77.5	2.10	0.26	—	0.16
Michigan	High lime	85.1	2.08	0.25	—	6.80
German	Low lime	97.0	1.20	0.10	0.05	0.35
German	High lime	90.0	2.50	0.25	0.10	4.00
Austria	Low lime	93.3	1.40	0.10	0.06	0.45
Austria	High lime	83.8	2.10	0.18	0.13	2.38
New York	Low lime	94.2	1.26	0.15	0.10	0.60
New York	High lime	81.5	2.56	0.19	0.28	6.51
Minnesota	Low organic	59.7	2.35	0.36	0.17	2.52
Minnesota	High organic	94.0	1.70	0.16	0.04	0.31
Florida	Sawgrass peat	87.1	2.79	0.41	0.04	5.20
Canada	Peat soil	74.3	2.19	0.20	0.16	—
Washington	Woody sedge peat	89.2	3.52	0.43	0.09	1.29

[a] From Lyon et al. (2).

TABLE 13:4. *Analysis for a Representative Peat and a Mineral Surface Soil*

Values in percent based on dry matter.

Constituent	Peat Surface Soil	Mineral Surface Soil
Organic matter	80.00	4.00
Nitrogen (N)	2.50	0.15
Phosphorus (P)	0.09	0.04
Potassium (K)	0.08	1.70
Calcium (Ca)	2.80	0.40
Magnesium (Mg)	0.30	0.30
Sulfur (S)	0.60	0.04

NITROGEN AND ORGANIC MATTER. The high nitrogen and organic content of organic soils are self-evident and need no further emphasis here. But two interrelated features in respect to these constituents justify consideration. First, organic soils have a high carbon–nitrogen ratio, the minimum being in the neighborhood of 20. In contrast, this equilibrium ratio for a cropped mineral soil is usually around 10 to 12.

Second, organic soils show exceedingly vigorous nitrification in spite of their high carbon–nitrogen ratio. In fact, the nitrate accumulation, even in a low-lime peat, is usually greater than that of a representative mineral soil. This can only be explained on the basis of the large amount of nitrogen carried by the peat, the presence of adequate calcium, and the inactivity of part of the carbon. Thus, the *effective* carbon–nitrogen ratio of organic soils may be as narrow as that of mineral soils. As a result, the multiplication of the competitive general purpose heterotrophic organisms is not excessively encouraged. The nitrifiers, therefore, are given ample opportunity to oxidize the ammoniacal nitrogen.

PHOSPHORUS AND POTASSIUM. The phosphorus and potassium of peat are both low, the latter exceedingly so in comparison with a mineral soil. Even the phosphorus is actually less in pounds per acre–furrow slice. On this basis, the representative mineral soil is cited as containing 800 pounds of phosphorus. An equivalent layer of organic soil (say 500,000 pounds) would furnish only 450 pounds of phosphorus, or about one half as much. This explains why, in the growing of crops on an organic soil, phosphorus as well as potash must be applied in large amounts.

CALCIUM AND pH. The high calcium content of organic soils is easily explained. Much of the water entering swamps is from seepage and has had ample opportunity to dissolve lime in its passage through the subsoil and substratum of the surrounding uplands. Since decaying organic matter is highly adsorptive and calcium ions plentiful, the resultant organic horizons cannot avoid the presence of large amounts of exchangeable calcium ions. Nor is leaching, as with mineral soils, such an important factor in robbing the surface layers of lime. High lime, most of which is exchangeable, is an outstanding characteristic of many organic soils, especially those of woody origin.

In spite of this high lime content, the majority of organic soils are distinctly acid, often very markedly so. It is not at all uncommon to find organic soils with pH values of 5.5 or less even though the calcium oxide content is as high as 4 percent. So great are the cation adsorption capacities of organic soils that they may be at a low percentage base saturation and yet be carrying exceptionally large amounts of exchangeable calcium. At the same time, the percentage base saturation is such as to assure a decidedly acid condition.

MAGNESIUM AND SULFUR. The percentage of magnesium in organic soil is usually no greater than that of a mineral soil. The actual amount, however, is much less because of the low dry weight of peat by volume (see p. 361). The situation is thus much the same as that of phosphorus, although it is alleviated somewhat by the high proportion of the magnesium that is held in an exchangeable condition. Organic soils, long intensively cropped, may develop a magnesium deficiency unless fertilizers carrying this constituent have been used. Since most organic soils are seldom limed, there is little chance of adding magnesium in this particular way.

The abundance of sulfur in organic soils is not at all surprising. Plant tissue always contains considerable sulfur. Consequently, organic deposits such as peat should be comparatively high in this constituent. When sulfur oxidation is vigorous, as is usually the case with arable organic soils, sulfates may accumulate. Sulfur is abundant enough in most organic soils to reduce the possibility of this element being a limiting factor in plant growth.

ANOMALOUS FEATURES. Organic soils, in comparison with mineral soils, have been shown to exhibit three somewhat anomalous features. They are worthy of restatement. First, the representative organic soil possesses a high carbon–nitrogen ratio and yet supports a very vigorous nitrification. Second, organic soils are usually comparatively high in calcium and still may be definitely acid, often highly so. And third, in the presence of a high hydrogen ion concentration, nitrate accumulation takes place far beyond that common in mineral soils with the same low pH. This last feature indicates that in many organic soils the hydrogen ion concentration does not in itself impede this very important biochemical transformation. Apparently, the high calcium content and the low content of iron, aluminum, and manganese in organic soils account for this anomaly.

13:9. BOG LIME—ITS IMPORTANCE

In many cases, organic soils are underlain at varying depths by a soft, impure calcium carbonate called *bog lime* or *marl*. Marl, as correctly used by the geologist, refers to a calcareous clay of variable composition. Bog lime, when it contains numerous shells, is often termed *shell marl*. Such a deposit may come from the shells of certain of the Mollusca which have inhabited the basin, or from aquatic plants such as mosses, algae, and species of *Chara*. These organisms have the power of precipitating the calcium as insoluble calcium carbonate. It seems that these plants and animals occupied the basin before or in some cases during the formation of the peat, the marl resulting from the accumulation of their residues on the bottom of the basin or farther up in the profile.

In cases in which marl is present, it may not only supply calcium to the circulating waters, but if high in the profile, may actually become mixed with the surface peat. As a result, such organic soils are likely to be low in acidity or even alkaline. In general, an alkaline peat is not considered as highly desirable for intensive culture, especially the growing of vegetables, as are moderately acid peats.

13:10. FACTORS THAT DETERMINE THE VALUE OF PEAT AND MUCK SOILS

DRAINAGE AND WATER TABLE. The value of peat agriculturally will depend on a number of factors. Of first consideration is the possibility of drainage, that is, a more or less permanent lowering and control of the water table sufficient to allow an adequate aeration of the root zone during the growing season. Moreover, it may be advantageous to raise or lower the water table of organic soils at various times during the season. For instance, celery at setting is benefited by plenty of moisture. As the crop develops, the water table should be gradually lowered to accommodate the root development. The probable cost of drainage and of seasonal control of the moisture may be such as to make the reclamation of a peat bed economically unwise. Since peat is very often covered with a forest growth, the cost of clearing also must be reckoned with. In some cases this cost may be excessively high.

DEPTH AND QUALITY. Peat settles considerably during the first few years after drainage and cultivation have become operative, and may continue to shrink appreciably in after-years. If a depth adequate for cropping does not remain, the expense of reclamation is more or less thrown away. Three or four feet of organic material are desirable, especially if calcareous clay or marl underlies the deposit.

The quality of the peat is of special importance not only as to the degree of decay but also as to the nature of the original plant materials. The presence of the rubbery sedimentary type in the furrow slice is especially deplorable (see p. 356). Woody peat, on the other hand, is generally considered more desirable than that coming from cattails, reeds, and other plants, and it usually is highly prized wherever it occurs. Much of the acreage of this type of peat soil in the United States lies in Wisconsin, Michigan, and New York.

13:11. PREPARATION OF PEAT FOR CROPPING

Since organic soils are often forested or covered with shrubs and other plants, the first step is to drain and clear the land. Drainage may then be facilitated further and the water table lowered and brought under control. The utilization of the area as pasture for a few years is sometimes practiced. The roots and stumps are thus given time to decay, their removal being relatively easy when the land is finally fitted for cultivation.

If the peat area is burned over to remove the brush and other debris, it should be done early in the season while the land is wet and not likely to catch fire. Such a fire is often difficult to extinguish and may ruin the deposit by destroying the more fertile surface layer. If, however, the surface soil is fibrous with a more desirable layer below, it may be advantageous to burn this part, the resultant ash serving to mineralize the newly exposed layers.

After breaking the peat soil, preferably with heavy plows drawn by a tractor powerful enough to mash down second-growth brush or sapling trees, it is advisable to grow such crops as corn, oats, or rye for a year or two, as they do well on raw and uneven peat lands. Once the peat is adequately weathered, freed of roots and stumps, and all hummocks eliminated, it is ready for vegetable production. More thorough drainage is now required and is usually obtained by a system of ditches. Sometimes tile drains or even mole drains are used.

13:12. MANAGEMENT OF PEAT SOILS

All sorts of vegetable crops may be grown on organic soils—celery, lettuce, spinach, onions, potatoes, beets, carrots, asparagus, and cabbage perhaps being the most important, along with such specialized crops as peppermint. In some cases, peat is used for field crops, a more or less definite rotation being followed. Sugar beets, corn, oats, rye, buckwheat, flax, clover, timothy, and other field crops give good yields when suitable fertilization is provided. Moreover, certain nursery stocks do well on peat. In many cases, especially in Europe, organic soils are used extensively for pasture and meadows. In fact, almost any crop will grow on organic soil if properly managed. For a view of peat under cropping see Fig. 13:3.

STRUCTURAL MANAGEMENT. Plowing is ordinarily unnecessary every year as the peat is porous and open unless it contains considerable silt and clay.

FIGURE 13:3. Windbreaks, such as these in Michigan, help protect valuable muck land from blowing. The unprotected field in the lower left has been wetted by sprinkler to prevent its blowing. [*Photo courtesy U. S. Soil Conservation Service.*]

In fact, a cultivated organic soil generally needs packing rather than loosening. The longer a peat has been cropped the more important compaction is likely to be. Cultivation tends to destroy the original granular structure, leaving the soil in a powdery condition when dry. It is then susceptible to wind erosion, a very serious problem in some sections (see Fig. 13:3).

For this reason, a roller or packer is an important implement in the management of such land. The compacting of organic soils allows the roots to come into closer contact with the soil and facilitates the rise of water from below. It also tends to reduce the blowing of the soil during dry weather, although a windbreak of some kind is much more effective. The cultivation of peat, while easier than for mineral soils, is carried on in much the same way and should be shallow, especially after the root development of the crop has begun.

USE OF LIME. Lime, which so often must be used on mineral soils, ordinarily is less necessary on organic soils since they are usually adequately supplied with calcium. On acid mucks containing appreciable quantities of inorganic matter, however, the situation is quite different. The highly acid conditions

result in the dissolution of iron, aluminum, and manganese to the extent
that they become present in toxic quantities. Under these conditions, large
amounts of lime may be necessary to obtain normal plant growth.

COMMERCIAL FERTILIZERS. Of much greater importance than lime are
commercial fertilizers. In fact, complete reliance is placed on these materials
in the production of most crops, especially vegetables. As organic soils are
very low in phosphorus and potassium, these elements must be added.
Since vegetables usually are rapid-growing plants, succulence often being an
essential quality, large amounts of readily available nitrogen are necessary.
The nitrogen of newly cleared peat is often available rapidly enough to
supply this need, especially for oats, rye, corn, wheat, and similar crops.
Such land requires at the beginning only a small amount of nitrogen with the
phosphoric acid and potash. This is especially true of woody peat and is taken
advantage of in the fertilization of such soils. After organic soils have been
cropped for a few years, decay and nitrification are frequently too slow to
meet the crop demand for nitrogen. Under such conditions this element is
needed in larger amounts and a fertilizer containing more nitrogen as well
as phosphoric acid and potash is usually recommended. The amount of any
given fertilizer applied will depend upon the kind of crop to be grown, the
chemical and physical nature of the peat, its drainage, its previous fertilization, and the length of time it has been under cultivation.

TRACE ELEMENTS. Peat soils not only are in need of potassium, phosphorus,
and nitrogen, but often some of the trace elements as well. On New York
woody peats, copper sulfate has given good results in the control of certain
diseases of lettuce and has aided in the coloration of onions. In Florida and
elsewhere, not only copper sulfate but salts of manganese and zinc are used
to better the physiological condition of both peat and muck soils. Boron
deficiencies are also becoming evident. Michigan peat in general need both
boron and copper. Common salt in addition seems to give good results on
Michigan peats, especially for beets, celery, and cabbage. Perhaps both the
sodium and the chlorine play important rôles in crop nutrition on these soils.

13:13. ORGANIC VERSUS MINERAL SOILS

In assigning organic soils to a separate chapter, one is encouraged to
think of them as distinctly and even radically different from most mineral
soils. In many respects this certainly is true. Yet, fundamentally, the same
types of change occur in the two groups; nutrients become available in
much the same way and their management is based upon the same principles
of fertility and water management.

REFERENCES

(1) Farnham, R. S., and Finney, H. R., "Classification of Organic Soils," *Adv. in Argon.,* **17**:115–62, 1966.

(2) Lyon, T. L., Buckman, H. O., and Brady, N. C., *The Nature and Properties of Soils* (New York: Macmillan, 1952), p. 381.

(3) Wilson, B. D., and Stoker, E. V., *Ionic Exchange of Peat Soils,* Mem. 172, Cornell Univ. Agr. Exp. Sta., 1935.

Chapter 14

SOIL REACTION: ACIDITY
AND ALKALINITY

O NE of the outstanding physiological characteristics of the soil solution is its reaction. Since microorganisms and higher plants respond so markedly to their chemical environment, the importance of soil reaction and the factors associated with it have long been recognized. Three conditions are possible: acidity, neutrality, and alkalinity.

Soil acidity is common in all regions where precipitation is high enough to leach appreciable amounts of exchangeable bases from the surface layers of soils. So widespread is its occurrence and so pronounced is its influence on plants that it has become one of the most discussed properties of soils.

Alkalinity occurs when there is a comparatively high degree of base saturation. The presence of salts, especially calcium, magnesium, and sodium carbonates, also gives a preponderance of hydroxyl ions over hydrogen ions in the soil solution.[1] Under such conditions the soil is alkaline, sometimes very strongly so, especially if sodium carbonate is present, a pH of 9 or 10 being common. Alkaline soils are, of course, characteristic of most arid and semiarid regions. Their discussion will follow that of acid soils.

14:1. SOURCE OF HYDROGEN IONS

As previously indicated (p. 103), two adsorbed cations are largely responsible for soil acidity—hydrogen and aluminum. The mechanisms by which these two ions exert their influence differ, however. This difference is related to the source and nature of the charge to which each of these ions is attracted.

SOURCE OF NEGATIVE CHARGES. Two types of negative charges have been identified on soil colloidals: (a) permanent and (b) pH dependent. The first is associated primarily with the silicate clays. It is due to the electrostatic

---

[1] When salts of strong bases and weak acids, such as Na_2CO_3, K_2CO_3, and $MgCO_3$, go into solution, they undergo hydrolysis and develop alkalinity. For Na_2CO_3 the reaction is as follows:

$$2Na^+ + CO_3^{--} + 2HOH \rightleftharpoons 2Na^+ + 2OH^- + H_2CO_3$$

Since the dissociation of the NaOH is greater than that of the weak H_2CO_3, a domination of OH^- ions results.

forces resulting from isomorphic substitutions within the clay crystals (see p. 85). The charge sites are located mostly on the internal surfaces. Cations are exchangeable at all pH levels at these permanent charge sites. Permanent charges are highest on the 2:1-type clay where ionic substitution is greatest.

The second type of charge is not permanent but is related directly to soil pH. The charge is low in very acid soils and increases as the pH rises. This charge is thought to have several sources. First, there are SiOH and AlOH groups at the broken edges and external surfaces of silicate clays (see p. 84). Also, there are carboxyl (COOH) and phenol (phenyl—OH) groups on the humus colloids (see p. 95). These groups each contain covalent-bonded hydrogen which is not dissociated at low pH values. As the pH increases, however, the hydrogen dissociates, leaving a negative charge on the colloid. In nature the hydrogen is replaced by metallic cations, which are, in turn, exchangeable.

In acid soils, complex aluminum and iron hydroxy ions are tightly adsorbed within the crystal units of certain 2:1-type clays, particularly vermiculites. These ions block some of the negative charge sites of the colloid, thereby reducing its cation exchange capacity. As the pH is raised, the complex ions are removed, forming insoluble $Al(OH)_3$ and $Fe(OH)_3$ and thereby releasing the exchange sites. In this way the pH-dependent charge is increased.

The relative proportion of permanent and pH-dependent charges will depend on the kind of colloid present. The 2:1-type clays are generally high in the permanent-type charge, whereas humus is dominated by the pH-dependent types (see Fig. 4:11). Kaolinite is intermediate between the two.

STRONGLY ACID SOILS. Under very acid soil conditions much aluminum becomes soluble and is present in the form of aluminum or aluminum hydroxy cations. These become adsorbed even in preference to hydrogen by the permanent electrostatic charges of clay minerals—charges that result from ionic substitutions within the crystal lattice (see p. 86).

The adsorbed aluminum is in equilibrium with aluminum ions in the soil solution. The latter contribute to soil acidity through their tendency to hydrolize. A simplified reaction illustrates how adsorbed aluminum can increase acidity in the soil solution:

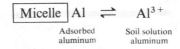

$$\boxed{\text{Micelle} \ \Big| \ \text{Al}} \ \rightleftarrows \ \text{Al}^{3+}$$

Adsorbed aluminum   Soil solution aluminum

The aluminum ions in the soil solution are then hydrolyzed in a manner such as the following:

$$Al^{3+} + H_2O \rightarrow Al(OH)^{++} + H^+$$

The hydrogen ions thus released give a very low pH value in the soil solution and are perhaps the major source of hydrogen in most very acid soils.

Adsorbed hydrogen is a second source of hydrogen ions in very acid soils. However, under these conditions, much of the hydrogen, along with some iron and aluminum, is held by covalent bonds in the organic matter and on clay crystal edges. In this condition the hydrogen is so tightly adsorbed that it contributes little to the soil solution. On only the strong acid groups of humus and some of the permanent charge exchange sites of the clays is the hydrogen held in an exchangeable form. This hydrogen is in equilibrium with the soil solution. A simple equation to show the release of adsorbed hydrogen to the soil solution follows:

$$\boxed{\text{Micelle}\;\big|\;\text{H}} \quad \rightleftharpoons \quad \text{H}^+$$

Adsorbed   Soil solution
hydrogen   hydrogen

Thus, it can be seen that the effect of both adsorbed hydrogen and aluminum is to increase the hydrogen ion concentration in the soil solution.

MODERATELY ACID SOILS. Aluminum and hydrogen compounds also account for soil solution hydrogen ions in these soils, but again by different mechanisms. These soils have somewhat higher percentage base saturations and pH values. The aluminum can no longer exist as Al^{3+} ions but has been converted to aluminum hydroxy ions by reactions such as these:

$$Al^{3+} + OH^- \rightarrow Al(OH)^{++}$$

$$Al(OH)^{++} + OH^- \rightarrow Al(OH)_2{}^+$$
Aluminum
hydroxy ions

The actual aluminum hydroxy ions are likely much more complex than those shown. Formulas such as $[Al_6(OH)_{12}]^{6+}$ and $[Al_{10}(OH)_{22}]^{8+}$ with the possibility of ring configurations have been postulated (2).

Some of the aluminum hydroxy ions are adsorbed and act as exchangeable cations. As such, they are in equilibrium with the soil solution just as was the Al^{3+} ion in very acid soils. In the soil solution, they are able to produce hydrogen ions by the following hydrolysis reactions, using again as examples the most simple of the aluminum hydroxy ions:

$$Al(OH)^{++} + H_2O \rightarrow Al(OH)_2{}^+ + H^+$$

$$Al(OH)_2{}^+ + H_2O \rightarrow Al(OH)_3 + H^+$$

In some 2:1-type clays, particularly vermiculite, the aluminum hydroxy ions (as well as iron hydroxy ions) play another role. They move in between the crystal units and become very tightly adsorbed. In this form they prevent intercrystal expansion and block some of the exchange sites. Their removal, which can be accomplished by raising the soil pH, results in the release of these exchange sites. In this way they are partly responsible for the "pH-dependent" charge of soil colloids.

In moderately acid soils adsorbed hydrogen also makes a contribution to the soil solution hydrogen. The readily exchangeable hydrogen held by the permanent charges contributes in the same manner shown for very acid soils. In addition, with the rise in pH, some hydrogen ions which have been held tenaciously through covalent bonding by the organic matter and clay are now subject to release. These are associated with the pH-dependent sites previously mentioned. Their contribution to the soil solution might be illustrated as follows:

$$\boxed{\text{Micelle}} \begin{array}{l} \text{H} \\ \text{H} \end{array} \quad + \quad Ca^{++} \quad \rightarrow \quad \boxed{\text{Micelle}}\, Ca^{++} \quad + \quad 2H^{+}$$

Bound hydrogen   Exchangeable   Soil solution
(not dissociated)   calcium   hydrogen

Again, the colloidal control of soil solution pH has been demonstrated, as has the dominant rôle of the calcium and aluminum ions.

NEUTRAL TO ALKALINE SOILS. Soils that are neutral to alkaline in reaction are no longer dominated by either hydrogen or aluminum ions. The permanent charge exchange sites are now occupied primarily by exchangeable bases, both the hydrogen and aluminum hydroxy ions having been largely replaced. The aluminum hydroxy ions have been converted to gibbsite by reactions such as the following:

$$Al(OH)_2{}^{+} \;+\; OH^{-} \;\rightarrow\; Al(OH)_3$$

Insoluble
gibbsite

More of the pH-dependent charges have become available for cation exchange, and the hydrogen released therefrom moves into the soil solution. Its place on the exchange complex is taken by calcium, magnesium, and other bases. The reaction is the same as that shown for the moderately acid soils.

Figure 14:1 presents the distribution of ions in a hypothetical soil as affected by pH. Study it carefully, keeping in mind that for any particular soil the distribution of ions might be quite different.

The effect of pH on the distribution of bases and of hydrogen and aluminum in a muck and in a soil dominated by 2:1 clays is shown in Fig. 14:2. Note that permanent charges dominate the exchange complex of the mineral soil, whereas the pH-dependent charges account for most of the adsorption in the muck soil. Kaolinite and related clays have a distribution intermediate between that of the two soils shown.

In Fig. 14:2 two forms of hydrogen are shown. That tightly held by the pH-dependent sites (covalent bonding) is termed *bound* hydrogen. The

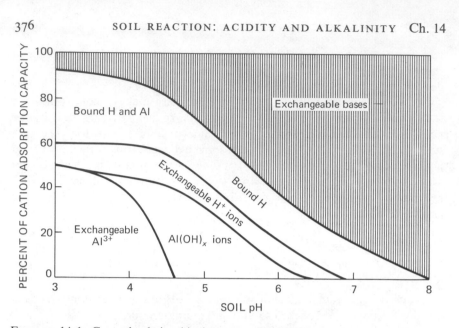

FIGURE 14:1. General relationship between soil pH and the cations held by soil colloids. Under very acid conditions exchangeable aluminum ions and bound H and Al dominate. At higher pH values the exchangeable bases predominate, while at intermediate values aluminum-hydroxy ions such as $Al(OH)^{++}$ and $Al(OH)_2^+$ are prominent. This diagram is for average conditions. Any particular soil would likely give a modified distribution.

hydrogen ions associated with permanent electrostatic charges is exchangeable.

It is obvious that the factors responsible for soil acidity are far from simple. At the same time, there are two dominant groups of elements in control. Aluminum and hydrogen generate acidity, and most of the other cations combat it. This simple statement is worth remembering.

SOURCE OF HYDROXYL IONS. If adsorbed hydrogen and aluminum are replaced from acid soils by cations such as calcium, magnesium, and potassium, the hydrogen ion concentration in the soil solution will decrease. The concentration of hydroxyl ions will simultaneously increase since there is an inverse relationship between the hydrogen and hydroxyl ions. Thus, the "base-forming" cations become sources of hydroxyl ions merely by replacing the adsorbed hydrogen.

The metallic cations such as calcium, magnesium, and potassium also have a more direct effect on the hydroxyl ion concentration of the soil solution. A definite alkaline reaction results from the hydrolysis of colloids saturated with these cations. An example of such a reaction is as follows:

$$\text{Ca} \boxed{\text{Micelle}} + 2H_2O \rightleftharpoons {}^{H}_{H}\boxed{\text{Micelle}} + Ca^{++} + 2OH^-$$

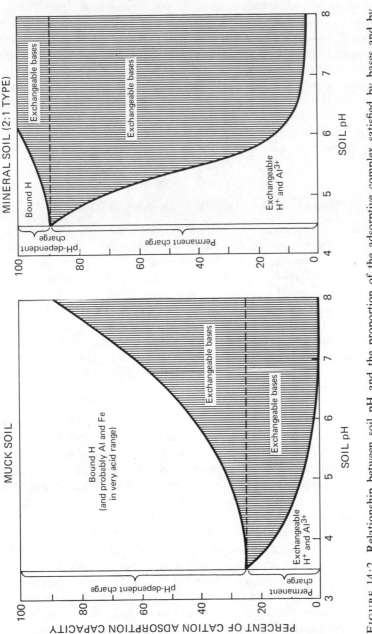

FIGURE 14:2. Relationship between soil pH and the proportion of the adsorptive complex satisfied by bases and by hydrogen and aluminum. The mineral soil (*right*) is of the White Store series in North Carolina. Its properties tends to be dominated by 2:1-type clays. The muck soil (*left*) is also from North Carolina. Note the dominance of the permanent charge in the mineral soil and the large pH-dependent charge in the muck. Soils dominated by 1:1-type colloids have distributions intermediate between these two extremes. [*Redrawn from A. Mehlich* (6).]

In a calcium-saturated soil, the tendency for the metallic cations to encourage hydroxyl ion formation is obvious. In a soil containing hydrogen and aluminum as well as calcium ions the same tendency is there, but its effect is not so obvious because it is countered by the effect of the adsorbed hydrogen and aluminum ions. Under natural conditions the reactions to furnish hydrogen and hydroxyl ions to the soil solution occur simultaneously; that is, hydrogen and aluminum ions and the basic cations are held at one time by the same micelle (see Fig. 14:1). The pH of the soil solution therefore will depend upon the relative amounts of adsorbed hydrogen and aluminum compared to adsorbed metallic cations. Where the effect of the hydrogen and aluminum is dominant, acidity results. Excess bases yield alkalinity, whereas at just the right balance the pH of the soil solution will be 7 (see Fig. 2:5).

14:2. COLLOIDAL CONTROL OF SOIL REACTION

PERCENTAGE BASE SATURATION. The relative proportions of the adsorbed hydrogen and aluminum and the exchangeable bases of a colloidal complex are shown by the *percentage base saturation* (see p. 103). Obviously, a low percentage base saturation means acidity, whereas a percentage base saturation approaching 100 will result in neutrality or alkalinity. In general, humid region soils dominated by *silicate clays* and *humus* are acid if their percentage base saturation is much below 80. When such soils have a percentage base saturation of 80 or above, they usually are neutral or alkaline. The exact pH value in any case is determined by at least two other factors in addition to the percentage base saturation: (a) the nature of the micelle, and (b) the kind of adsorbed bases.

NATURE OF THE MICELLE. At the same percentage base saturation, different types of colloids will have different pH values. This is because the various colloidal materials differ in their ability to furnish hydrogen ions to the soil solution. For example, the organic complex contains enough strong acid exchange sites to give very low pH values when the degree of base saturation is low. Even as bases are added, the degree of hydrogen ionization at the pH-dependent sites is sufficiently rapid to give lower pH values than are found commonly among mineral soils of comparable base saturation.

In contrast, the dissociation of the adsorbed hydrogen from the iron and aluminum hydrous oxides is relatively low. Consequently, soils dominated by this type of colloid have relatively high pH values for a given percentage base saturation. The dissociation of absorbed hydrogen from silicate clays is intermediate between that from humus and from the hydrous oxides.

The relative abilities to supply soil solution hydrogen can be seen by comparing the pH values found when the various colloids are about 50 percent saturated with bases. The organic colloids would have pH values of

4.5 to 5.0, the silicate clays 5.2 to 5.8, and the hydrous oxides 6.0 to 7.0. These figures verify the importance of type of colloid in determining the pH of a soil.

The diverse acid silicate clays—the kaolinite, montmorillonite, and hydrous mica types—evidently supply hydrogen ions in somewhat different degrees, the kaolinite least and the montmorillonite greatest (see p. 106). Likewise, the organic colloids exhibit considerable variety among themselves. In spite of this, however, the organic group apparently has a lower pH value than any of the clays when at corresponding percentage base saturations.

KIND OF ADSORBED BASES. Another factor which influences the pH of a soil is the comparative amounts of the *particular* bases present in the colloidal complex. Sodium-saturated soils have much higher pH values than those dominated by calcium and magnesium. Thus, at a percentage base saturation of 90, the presence of calcium, magnesium, potassium, and sodium ions in the ratio 10:3:1:1 would certainly result in a lower pH than if the ratio were 4:1:1:9. In the one instance, calcium is dominant—in the other, a sodium–calcium complex is dominated by sodium.

Since the reaction of the soil solution is influenced by three distinct and uncoordinated factors—percentage base saturation, nature of the micelle, and the ratio of the exchangeable bases—a close correlation would not be expected between percentage base saturation and pH when comparing soils at random. Yet with soils of similar origin, texture, and organic content, a rough correlation does exist.

14:3. MAJOR CHANGES IN SOIL pH

Two major groups of factors cause large changes in soil pH: (a) those resulting in increased adsorbed hydrogen and aluminum, and (b) those which increase the content of adsorbed bases.

ACID-FORMING FACTORS. In the process of organic matter decomposition, both organic and inorganic acids are formed. The simplest and perhaps the most widely found is carbonic acid (H_2CO_3), which results from the reaction of carbon dioxide and water. The solvent action of H_2CO_3 on the mineral constituents of the soil is exemplified by its dissolution of limestone or calcium carbonate (p. 283). The long-time effects of this acid have been responsible for the removal of large quantities of bases by solution and leaching. Because carbonic acid is relatively weak, however, it cannot account for the low pH values found in many soils.

Inorganic acids such as H_2SO_4 and HNO_3 are potent suppliers of hydrogen ions in the soil. In fact, these acids, along with the stronger organic acids, are responsible for the development of moderately and strongly acid conditions. Sulfuric and nitric acids are formed, not only by the organic decay processes but also from the microbial action on certain fertilizer materials such as sulfur and ammonium sulfate. In the latter case, both nitric and sulfuric acids are formed (see p. 520).

A good example of a process by which strong *organic* acids are formed is that of "podzolization," whereby acid leaching occurs. The organic debris is attacked largely by fungi which have among their important metabolic end products relatively complex but strong organic acids. As these are leached into the mineral portion of the soil, they not only supply hydrogen for adsorption but also replace bases and encourage their solution from the soil minerals.

Leaching also encourages acidity. Consequently, bases which have been replaced from the colloidal complex or which have been dissolved by percolating acids are removed in the drainage waters. This process encourages the development of acidity in an indirect way by removing those metallic cations which might compete with hydrogen and aluminum on the exchange complex. The effect of leaching on acidity of soils developed under grassland and forests is shown in Fig. 14:3.

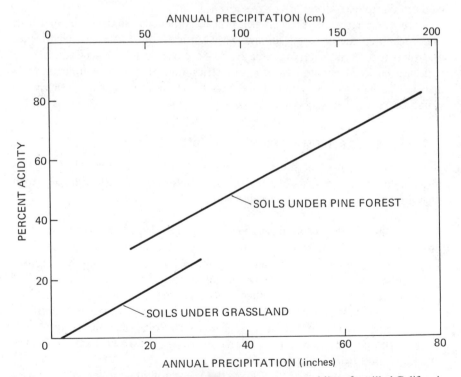

FIGURE 14:3. Effect of annual precipitation on the percent acidity of untilled California soils under grassland and pine forests. Note that the degree of acidity goes up as the precipitation increases. Also note that the forest produced a higher degree of soil acidity than did the grassland. [*From Jenny et al. (3).*]

BASE-FORMING FACTORS. Any process which will encourage the maintenance or buildup of the exchangeable bases such as calcium, magnesium, potassium, and sodium will contribute toward a reduction in acidity and an increase in alkalinity. Of great significance are the weathering processes, which release these exchangeable cations from minerals and make them available for absorption. The addition of base-containing materials such as limestone is a common procedure which is used to furnish metallic cations in order to augment nature's supply. Irrigation waters also frequently contain various kinds of salts, the cations being adsorbed by soil colloids, thus increasing soil alkalinity, sometimes excessively.

Conditions which permit the exchangeable bases to remain in the soil will encourage high pH values. This accounts for the relatively high pH of soils of the semiarid and arid regions. Leaching waters do not remove most of the metallic cations as they are weathered from soil minerals. Consequently, the percentage base saturation of these soils remains high. In general, this situation is favorable for crop production. Only when the pH is too high or when sodium is the dominant cation is plant growth unfavorably affected.

14:4. MINOR FLUCTUATIONS IN SOIL pH

Not only do soil solutions suffer major changes in hydrogen ion concentration but they also exhibit minor fluctuations. For instance, the drying of soils, especially above field temperatures, will often cause a noticeable increase in acidity. This is probably due to a change in the organization of the colloidal matter and should be kept in mind in preparing soil samples for pH determination.

The pH of mineral soils declines during the summer, especially if under cultivation, as a result of the acids produced by microorganisms. The activities of the roots of higher plants, particularly with regard to acidic exudates, may also be a factor. In winter and spring an increase in pH often is noted because biotic activities during this time are considerably slower.

14:5. HYDROGEN ION HETEROGENEITY OF THE SOIL SOLUTION

In considering the hydrogen concentration of the soil solution, it must not be inferred that this is an ordinary homogeneous solution. Differences in pH are noted from one portion of soil to that only a few inches away. Such variation results from local microbial action and the uneven distribution of organic residues in the soil.

The variability of the soil solution is important in many respects. For example, it affords microorganisms and plant roots a great variety of solution environments. Organisms unfavorably influenced by a given hydrogen ion concentration may find. at an infinitesimal distance away. another that is more satisfactory. This may account in part for the many different floral species present in normal soils.

Even at a given location in the soil—in fact around a given micelle—there are marked differences in the distribution of hydrogen and aluminum ions (see Fig. 14:4). These cations are concentrated near the surfaces of the colloid and become less numerous as the distance from the micelle is increased. They are least concentrated in the soil solution. Furthermore, equilibrium conditions exist between the adsorbed and soil solution ions, permitting the ready movement from one form to another. This fact is of great practical importance since it provides the basis for the suffering capacity of soils. The mechanisms by which adsorbed aluminum and hydrogen ions supply H^+ ions to the soil solution has been discussed (see p. 373).

14:6. ACTIVE VERSUS EXCHANGE ACIDITY

Figure 14:4 illustrates the presence in acid soils of two kinds of acidity. The hydrogen ion concentration of the soil solution is designated *active acidity*. Those hydrogen and aluminum ions held on the soil colloids are referred to as the *reserve* or *exchange acidity* of the soil. This situation is shown graphically by Fig. 14:4 and by the following equation:

$$\text{Adsorbed H (and Al) ions} \rightleftharpoons \text{Soil solution } H^+ \text{ (and Al) ions}$$

(Reserve acidity)   (Active acidity)

Since the adsorbed hydrogen and aluminum move outward and become active when the acidity of the soil solution is reduced, the term "reserve" is particularly significant. Distinct as the two groups apparently are, they grade into each other as progress outward from the colloidal interface is made.

AMOUNTS OF ACTIVE AND RESERVE ACIDITY. The relative magnitude of the two types of acidity—active versus reserve—is both interesting and of vital practical importance. In referring to the acidity of a soil as high or very high, the impression may be given that the active acidity under certain conditions may be exceedingly great. Actually, the reverse is true. For example, only about $\frac{1}{50}$ *pound* of calcium carbonate would be required to neutralize the active acidity in an acre–furrow slice of an average mineral soil having a pH of 6. This assumes that the calcium carbonate could be brought in contact with the soil solution and that only the hydrogen ions actually in this solution would react with the calcium carbonate. A soil moisture content of 20 percent was assumed in making this calculation. If the same soil should possess a pH of 5, $\frac{1}{5}$ pound of calcium carbonate would be adequate; if the pH should be lowered to 4, 2 pounds would be ample. Thus, the *active* acidity is evidently ridiculously small even at its maximum.

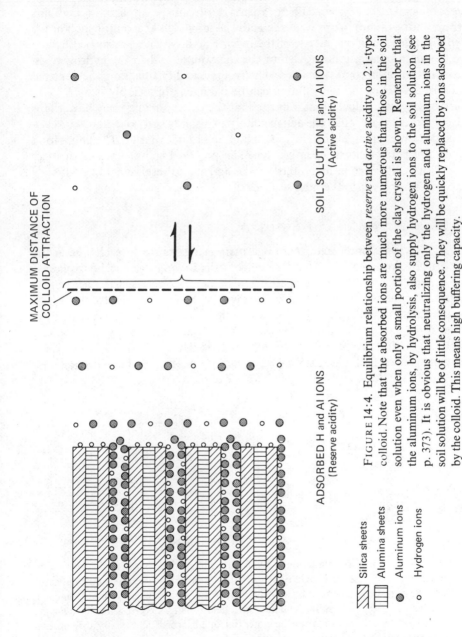

FIGURE 14:4. Equilibrium relationship between *reserve* and *active* acidity on 2:1-type colloid. Note that the absorbed ions are much more numerous than those in the soil solution even when only a small portion of the clay crystal is shown. Remember that the aluminum ions, by hydrolysis, also supply hydrogen ions to the soil solution (see p. 373). It is obvious that neutralizing only the hydrogen and aluminum ions in the soil solution will be of little consequence. They will be quickly replaced by ions adsorbed by the colloid. This means high buffering capacity.

Since limestone at the rate of 1, 2, or even 4 tons to the acre is often recommended, this neutralizing agent is obviously applied in amounts thousands of times in excess of the active acidity. The reason for such heavy applications is the magnitude of the reserve acidity. These adsorbed hydrogen and aluminum ions move into the soil solution when the hydrogen ion concentration becomes depleted. The reserve acidity thus must be depleted before the pH of the soil solution can be changed appreciably.

Conservative calculations indicate that the reserve acidity may be perhaps 1,000 times greater than the active acidity in a sandy soil, and 50,000 or even 100,000 times greater for a clayey soil high in organic matter. The figure for a peat soil is likely to be even greater. The practical significance of this tremendous difference in magnitude of the active and reserve acidities of soils becomes apparent in the next section.

14:7. BUFFERING OF SOILS

As previously indicated, there is a distinct resistance to a change in the pH of the soil solution. This resistance, called *buffering*, can be explained very simply considering the equilibrium that exists between the active and reserve acidities (see Fig. 14:5). Removal of hydrogen ions from the soil solution results in their being largely replenished from the reserve acidity. The resistance to change in hydrogen ion concentration (pH) of the soil solution is thus established (see equation p. 382).

If just enough liming material was added to neutralize the hydrogen ions

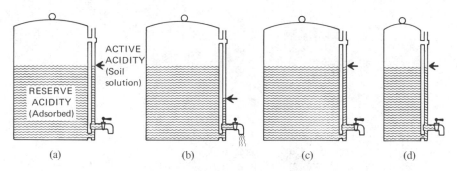

FIGURE 14:5. The buffering action of a soil can be likened to that of a coffee dispenser depicted above. (a) The active acidity, which is represented by the coffee in the indicator tube on the outside of the urn, is small in quantity. (b) When hydrogen ions are removed, this active acidity falls rapidly. (c) The active acidity is quickly restored to near the original level by movement from the potential or absorbed acidity. By this process there is considerable resistance to the change of active acidity. A second soil with the same active acidity (pH) level but a much smaller reserve acidity (d) would have a lower buffering capacity.

in the soil solution, the above reaction would immediately be shifted to the right, resulting in more hydrogen ions moving out into the soil solution. Consequently, the resulting pH rise would be negligibly small and remain so until enough lime had been added to deplete appreciably the reserve acidity.

This resistance to pH change is equally important in preventing a rapid lowering of the pH of soils. When hydrogen ions are added to a soil or they result from certain biochemical changes, a temporary increase in the hydrogen ions in the soil solution occurs. In this case the equilibrium reaction above would immediately shift to the left and more hydrogen ions would become adsorbed on the micelle. Again, the resultant pH change, this time a lowering, in the soil solution would be very small.

These two examples indicate clearly the principles involved in buffering. In addition, they show that the basis of buffer capacity lies in the adsorbed cations of the complex. Hydrogen and aluminum ions, together with the adsorbed metallic cations, not only indirectly control the pH of the soil solution but also determine the quantity of lime or acidic constituents necessary to bring about a given pH change.

14:8. BUFFER CAPACITY OF SOILS AND RELATED PHASES

The higher the exchange capacity of a soil, the greater will be its buffer capacity, other factors being equal. This is because more reserve acidity must be neutralized to effect a given rise or lowering of the percentage base saturation. In practice this is fully recognized in that the heavier the texture of a soil and the higher its organic content, the larger must be the application of lime to force a given change in pH.

BUFFER CURVES. Another important phase of buffering logically presents itself at this point. Is the buffer capacity of soils the same throughout the percentage base saturation range? This is best answered by reference to an average theoretical titration curve presented by Peech (7) for a large number of Florida soils (see Fig. 14:6).

Three things are clearly obvious from the curve. First, there is a correlation between the percentage base saturation of these soils and their pH, as previously suggested (p. 103). Second, the generalized curve indicates that the degree of buffering varies, being lowest at the extreme base saturation values. Between these extremes where the curve is flatter, the buffering reaches a maximum. Theoretically, the greatest buffering occurs at about 50 percent base saturation, a situation extremely important both technically and practically. Third, the buffering, as indicated by the curve, is uniform over the pH range 4.5 to 6.5. This is vitally significant, indicating that under field conditions approximately the same amount of lime will be required to change the soil pH from 5.0 to 5.5 as from 5.5 to 6.0.

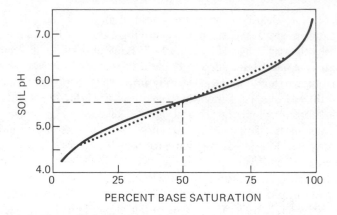

FIGURE 14:6. Theoretical titration curve for a large number of Florida soils. The dotted line indicates the zone of greatest buffering. The maximum buffering should occur at approximately 50 percent base saturation. [*From Peech* (7).]

Before leaving this phase, note that the curve of Fig. 14:6 is a composite and represents the situation in respect to any particular soil only in a general way. That the titration curves for individual soils will deviate widely is to be expected since the colloidal complex of different soils varies with the kinds and amounts of clay and humus present.

VARIATION IN TITRATION CURVES. A significant indication of the difference between the clay groups is shown in Mehlich's results (5). He found that the pH values of montmorillonite and the hydrous micas were between 4.5 and 5.0 when these clays were at 50 percent base saturation. The pH values of kaolinite and halloysite under comparable conditions were found within the range 6.0 to 6.5.

The influence of the different clays upon the titration curves of individual soils is nicely shown by the data in Fig. 14:7. Samples from the B horizon were used to avoid the complicating effects of organic matter. In light of the figures quoted from Mehlich, it is easy to see from these curves that the Cecil subsoil with a pH of about 6.4 at 50 percent base saturation must be highly kaolinitic. The corresponding pH for the Iredell is about 4.4, indicating that clays of the montmorillonite or similar types are dominant in this soil. Obviously, the clays of the Iredell develop stronger acids than those of the Cecil and, under comparable conditions, exhibit a higher buffer capacity.

14:9. IMPORTANCE OF BUFFERING

STABILIZATION OF SOIL pH. A marked change in pH clearly indicates a radical modification in soil environment, especially in respect to the availability of plant nutrients. And if this environment should fluctuate too widely,

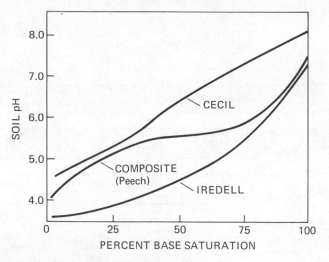

FIGURE 14:7. Titration curves for the B horizon of two North Carolina soils. The Cecil is dominated by kaolinitic minerals (1:1 types) and the Iredell by montmorillonitic clays (2:1 types). Note the difference in the trends of the two curves and how they in turn differ from the composite ideal as set up by Peech (Fig. 14:6). This is to be expected. [*From Mehlich (5).*]

higher plants and microorganisms undoubtedly would suffer seriously before they could make adequate adjustments. Not only would they be affected directly by the change in hydrogen ion concentration, but the indirect influences on nutrient elements might prove to be exceedingly unsatisfactory (see Figs. 14:8 and 14:9). The stabilization of soil pH through buffering seems to be an effective guard against these difficulties.

QUANTITIES OF AMENDMENTS REQUIRED. Obviously, the greater the buffering capacity of a soil, the larger the amounts of lime or sulfur necessary to effect a given change in pH. Therefore, in deciding the amount of lime to apply to a soil of known pH, texture and organic content are among the important soil factors to be considered. These properties give a rough idea of the adsorptive capacity of a soil and hence of its buffering (see p. 103). Chemical tests, such as a determination of cation exchange capacity, are also helpful in a practical way.

14:10. SOIL-REACTION CORRELATIONS

EXCHANGEABLE CALCIUM AND MAGNESIUM. It has already been shown that as the exchangeable calcium and magnesium are lost by leaching, the acidity of the soil gradually increases. Consequently, in humid region soils there is a fairly definite correction between the pH and the amounts of these two

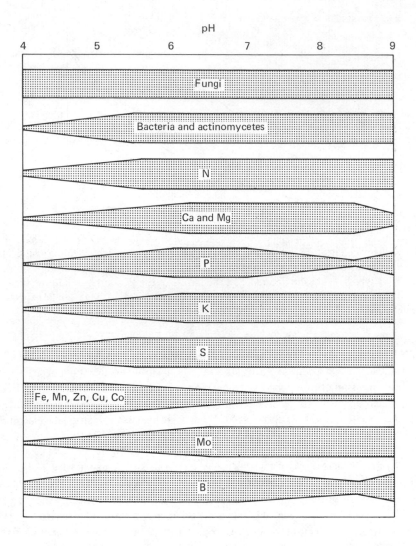

FIGURE 14:8. Relationships existing in mineral soils between pH on the one hand and the activity of microorganisms and the availability of plant nutrients on the other. The wide portions of the bands indicate the zones of greatest microbial activity and the most ready availability of nutrients. Considering the correlations as a whole, a pH range of approximately 6 to 7 seems to promote the most ready availability of plant nutrients. In short, if soil pH is suitably adjusted for phosphorus, other plant nutrients, if present in adequate amounts, will be satisfactorily available in most cases.

constituents present in exchangeable form (see Fig. 14:8). In arid regions, the same general relationships hold except under conditions where a substantial amount of sodium is adsorbed (see pp. 396–99).

ALUMINUM AND THE TRACE ELEMENTS. The relationships between soil reaction and the activity of aluminum, iron, and manganese are shown in Fig. 14:8. When the pH of a mineral soil is low, appreciable amounts of these three constituents are soluble, so much that they may become extremely toxic to certain plants. However, as the pH is increased, precipitation takes place and the amounts of these ions in solution become less and less until at neutrality or somewhat above certain plants may suffer from a lack of available manganese and iron. This is especially true if a decidedly acid sandy soil is suddenly brought to a neutral or alkaline condition by an excessive application of lime.

While deficiencies of manganese and iron are not widespread, they do occur in certain areas, particularly on overlimed sandy soils or alkaline arid region soils. If the soil reaction is held within a soil pH range of 6.0 to 7.0, the toxicity of the aluminum, iron, and manganese may satisfactorily be suppressed. At the same time their unavailability will be avoided unless these elements are decidedly lacking in the soil. Copper and zinc are affected in the same way by a rise in pH, the critical point being near pH 7, above which their availability definitely declines (see Fig. 14:8).

With boron, the situation is somewhat different and more complicated. Although neither the soil untreated nor the lime alone appreciably precipitate boron, the two in combination fix it markedly. Also, the excess of calcium, in spite of its solubility, may hinder in some way the movement of boron into the plant. Too much calcium in the plant cells might even interfere with boron metabolism, even if plenty of the latter were present. It has also been suggested that lime may create serious competition for boron by stimulation of soil microorganism activity.

Molybdenum availability is significantly dependent on pH. In strongly acid soils it is quite unavailable. As the pH is raised to 6 and above, its availability increases. The correlation between molybdenum availability and pH is so strong that some researchers believe the main reason for liming is to increase the molybdenum supply.

AVAILABLE PHOSPHORUS. The kind of phosphate ion present varies with the pH of the soil solution. When the soil is alkaline, the HPO_4 ion apparently is the most common form. As the pH is lowered and the soil becomes slightly to moderately acid, both the HPO_4 ion and the H_2PO_4 ion prevail. At higher acidities H_2PO_4 ions tend to dominate. Because of the formation of insoluble compounds, a soil reaction which yields a mixture of HPO_4 and H_2PO_4 ions is usually preferred.

The activity of the soil phosphorus is indirectly related to pH in another way. It has already been explained that as soil acidity increases, there is a rise in the activity of the iron, aluminum, and manganese. Under such conditions soluble phosphates are markedly fixed as very complex and insoluble compounds of these elements. This fixation is most serious when the soil pH is below 5.0. The situation is shown graphically in Fig. 17:3.

If the pH of a mineral soil is raised much above 7, the phosphate nutrition of higher plants is disturbed in other ways. In the first place, at these high pH values, complex insoluble calcium phosphates are formed. Thus, the solubility of both the native and applied phosphorus may be very seriously impaired. Furthermore, at pH values above 7, the excess calcium may hinder phosphorus absorption and utilization of plants.

The correlation of phosphorus availability and soil reaction would be sadly incomplete without an examination of the situation in the intermediate pH range of mineral soils from 6 to 7. Between these two pH limits phosphorus fixation is at a *minimum*; conversely, phosphorus availability is at a maximum as far as most plants are concerned (see Fig. 14:8). In the regulation of the phosphorus nutrition of crops, it is therefore important that soil pH be kept within, or very near, the conservative limits of 6 to 7. Even then, higher plants often do not absorb one half or even one third of the available phosphorus currently supplied by the application of superphosphate or other fertilizers carrying phosphorus. (For further consideration of phosphorus fixation and availability in soils, see pp. 462–66.)

SOIL ORGANISMS AND pH.   It is well known that soil organisms are influenced by fluctuations in the reaction of the soil solution. This may be due in extreme cases to the hydrogen ion itself, but in most soils it must be attributed to the factors correlated with soil pH.

In general, it is recognized that bacteria and actinomycetes function better in mineral soils at intermediate and higher pH values, the activity being sharply curtailed when the pH drops below 5.5. Fungi, however, are particularly versatile, flourishing satisfactorily at all soil reactions. In normal soils, therefore, fungi predominate at the lower pH values, but at intermediate and higher ranges they meet strong competition from the bacteria and actinomycetes and must yield the field to some degree (see Fig. 14:8).

Nitrification and nitrogen fixation take place vigorously in mineral soils only at pH values well above 5.5. However, mineralization, although curtailed, will still proceed with considerable intensity at lower pH values because most fungi are able to effect these enzymic transfers at high acidities. This is fortunate since higher plants growing on very acid soils are provided with at least ammoniacal nitrogen.

All in all, a soil in the intermediate pH range presents the most satisfactory biological regime. Nutrient conditions are favorable without being extreme and phosphorus availability is at a maximum.

One very significant exception to the generalized correlation of bacteria with soil reaction should be mentioned. The organisms that oxidize sulfur to sulfuric acid (p. 451) seem to be markedly versatile. Apparently they not only function vigorously in soils at medium to higher pH values but under decidedly acid conditions as well. This is extremely important since it is therefore possible to apply sulfur to soils and develop highly acid conditions through the activity of these bacteria. If these organisms were at all sensitive to low pH and its correlated factors, their activity would soon be retarded and finally brought to a halt by their own acidic products. Under such conditions sulfur would be relatively ineffective as a soil acidifier.

14:11. RELATION OF HIGHER PLANTS TO SOIL REACTION

Because of the many physiological factors involved, it is often difficult to correlate the optimum growth of plants on mineral soils with their pH. On the other hand, the general relationships of higher plants to strong acidity or alkalinity can be established and in a practical way are just as significant. With this relationship as a basis, some of the important plants are rated in Fig. 14:9.

CROP RATINGS. With such crops as alfalfa and sweet clover, calcium seems to be a very important factor. These plants are calcium loving and are adjusted to the physiological conditions of a high-lime soil. Except under especially favorable conditions, humid region mineral soils must be limed in order to grow crops of this group satisfactorily.

Native rhododendrons and azaleas are at the other end of the scale and apparently require a considerable amount of iron. This constituent is abundantly available only at low pH values and consequently at low percentage base saturations. Undoubtedly, highly acid soils also present other conditions physiologically favorable for this type of plant. Also, soluble aluminum apparently is not detrimental to these plants as it is to certain plants higher up the scale. If the pH and the percentage base saturation are not low enough, plants of the low-lime type will show chlorosis and other symptoms indicative of an unsatisfactory nutritive condition.

Because arable soils in a humid region are usually somewhat acid, it is fortunate that most cultivated crop plants not only grow well on moderately to slightly acid soils but seem to prefer the physiological conditions therein (see Fig. 14:9). Since pasture grasses, many legumes, small grains, intertilled field crops, and a large number of vegetables are included in this broadly tolerant group, soil acidity is not such a detriment as it was once considered. In terms of pH, a range from 5.8 to 6.0 to slightly above 7.0 is most suitable for this group. (The significance of this as it relates to liming is presented on pp. 417–19.)

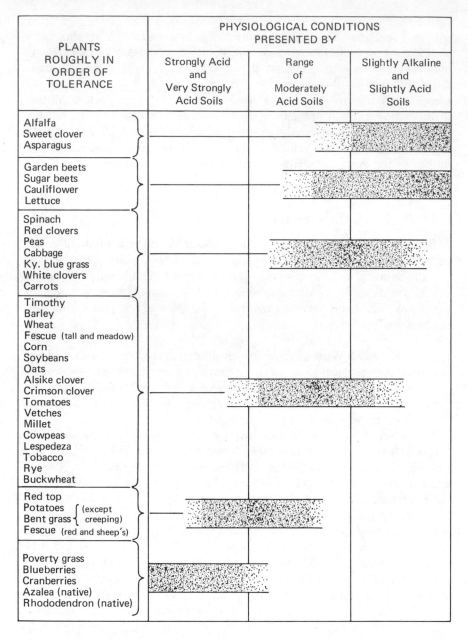

FIGURE 14:9. Relation of higher plants to the physiological conditions presented by mineral soils of different reactions. Note that the correlations are very broad and are based on pH ranges. The fertility level will have much to do with the actual relationship in any specific case. Such a chart is of great value in deciding whether or not to add lime and the rate of application, if any.

Forest trees seem to grow well over a wide range of soil pH values. They are particularly tolerant of acid soils, however. Many species, particularly the conifers, tend to intensify soil acidity. Forests exist as natural vegetation in regions of acid soils. This may not be a direct response to soil pH but rather to the climatic environment, which incidentally encourages the development of acid soils. Also, there are some isolated areas of high soil pH in humid regions where the natural vegetation is grass rather than trees. The "black-belt" soil areas of Alabama and Mississippi are examples. These areas are surrounded by acid soils on which forests are the natural vegetation. This indicates that the trees are better competitors on the more acid soil areas.

14:12. DETERMINATION OF SOIL pH

The importance of pH measurements as a tool in liming and similar problems is obvious. In fact, pH is a diagnostic figure of unique value and as a result its determination has become one of the routine tests made on soils. Moreover, its determination is easy and rapid.

ELECTROMETRIC METHOD. The most accurate method of determining soil pH is by a pH meter. In this electrometric method the hydrogen concentration of the soil solution is balanced against a standard hydrogen electrode or a similarly functioning electrode. Although the instrument gives very consistent results, its optimum use requires a skilled operator since the mechanism is rather complicated.

DYE METHODS. A second method, very simple and easy but somewhat less accurate than the electrometric, consists in the use of certain indicators (see Fig. 14:10). Many dyes change color with an increase or decrease of pH, making it possible, within the range of the indicator, to estimate the approximate hydrogen ion concentration of a solution. By using a number of dyes, either separately or mixed, a range of pH from 3 to 8 is easily covered. In making such a pH determination on soil, the sample is saturated with the dye, and after standing in contact a few minutes a drop of the liquid is run out and its color observed in thin layer. By the use of a suitable color chart the approximate pH may be ascertained. When properly manipulated, the indicator method is accurate within about 0.2 pH unit.

LIMITATIONS OF pH VALUES. Because of the precision with which pH readings can be duplicated electrometrically, interpretations may be made which the pH measurement cannot justify. There are several reasons for this. First, there is considerable variation in the pH from one spot in a given field to another. Even at a given location there are seasonal variations in pH. Localized effects of fertilizers may give sizable pH variations within the

FIGURE 14:10. The indicator method for determining pH is widely used in the field. It is simple and is accurate enough for most purposes. [*Photo courtesy N. Y. State College of Agriculture and Life Sciences, Cornell University.*]

space of a few inches. Last, the pH of a given soil sample will vary depending on the amount of water used in wetting the soil prior to the measurement. Obviously, standardization against field performance must be obtained.

Because of these limitations, it might seem peculiar that so much reliance is placed on soil pH. In the first place, it is easily and quickly determined. More important, however, is its susceptibility to certain broad correlations that are of great practical significance (Figs. 14:8 and 14:9). Thus, a great deal may be inferred regarding the physiological condition of a soil from its pH value, much more than from any other single analytical datum. Furthermore, the variations in pH value from one local soil area to another in a given field indicate that great accuracy in estimating soil acidity may not be so necessary from a practical point of view.

14:13. SOIL ACIDITY PROBLEMS

Other than the maintenance of fertility in general, two distinct procedures are often necessary on acid soils, especially those at intermediate pH values. One is the intensification of the acidity to encourage such plants as azaleas and rhododendrons. The other is the application of lime, usually in amounts to raise the pH at least to 6.0, 6.5, or 7.0. This modifies the physiological conditions, thereby favoring alfalfa, sweet clover, red clover, and other lime-loving crops.

Since liming is such an important agricultural feature, its consideration will be reserved for a later and fuller discussion (Chapter 15). However, the methods of intensifying the acidity of the soil are briefly discussed next.

14:14. METHODS OF INTENSIFYING SOIL ACIDITY

A reduction of the pH of soils is often desirable, for several reasons. For example, this is done to favor such plants as rhododendrons and azaleas as previously suggested and also to discourage certain diseases, especially the actinomycetes that produce potato scab. In arid regions treatments are sometimes made to reduce the high pH of alkali soils sufficiently to allow common field plants to grow and to eliminate deficiencies of iron, manganese, and zinc in other soils (see p. 490).

ACID ORGANIC MATTER. When dealing with ornamental plants, acid organic matter may be mixed with the soil already at hand to lower the pH of the latter. Leafmold, pine needles, tanbark, sawdust, and moss peat, if highly acid, are quite satisfactory in preparing such a compost. Farm manure, however, may be alkaline and consequently should be used with caution for this purpose.

USE OF CHEMICALS. When the above methods are not feasible, chemicals may be used. For rhododendrons, azaleas, and other plants that require considerable iron, such as blueberries and cranberries, ferrous sulfate[2] is sometimes recommended. This salt by hydrolysis develops sulfuric acid, which drastically lowers the pH and liberates some of the iron already present in the soil. At the same time, soluble and available iron is being added. Such a chemical thus serves a double purpose in effecting a change in the physiological condition of a soil.

Another material that is even better in some respects is flowers of sulfur. This usually undergoes vigorous microbial oxidation in the soil (see p. 451) and under favorable conditions is four or five times more effective pound for pound in developing acidity than is ferrous sulfate. Moreover, it is comparatively inexpensive and easy to obtain and is a material often used on the farm for other purposes.

No definite recommendation can be made as to the amounts of ferrous sulfate or sulfur that should be applied, since the buffering of soils and their

---

[2] Aluminum sulfate will acidify soil just as satisfactorily as the ferrous sulfate, but it carries no iron. Moreover, the aluminum thereby introduced may be toxic to plants that might later occupy the soil.

original pH are so variable. However, for rhododendrons and azaleas 1 to 2 pounds of sulfur per 100 square feet for each $\frac{1}{2}$ pH unit which a medium-textured soil is to be lowered is not too much. The dosage must be varied according to the texture of the soil and its organic content. In any case the sulfur should be well mixed with the surface soil.

CONTROL OF POTATO SCAB. Sulfur is also effective in the control of potato scab since the actinomycetes responsible are discouraged by acidity. Ordinarily when the pH is lowered to 5.3, their virulence is much reduced. In using sulfur to increase soil acidity, the management of the land, especially the rotation, should be such that succeeding crops are not unfavorably affected.

The amount of sulfur to apply to control potato scab will vary depending on circumstances. The buffering capacity of the soil and the original soil pH will be the determining factors. The results of a given treatment both on the pH and the crop should be checked and succeeding applications changed to conform with the influence exerted by the previous dosage.

14:15. REACTION OF SOILS OF ARID REGIONS

Arid region zonal soils occur in areas where the rainfall is seldom more than 20 inches per year. Lack of extensive leaching leaves the base status of these soils high. A fully and normally developed profile usually carries at some point in its profile (usually in the C horizon) a calcium carbonate accumulation greater than that of its parent material. The lower the rainfall, the nearer the surface this layer will be (see Fig. 12:9).

As a result, these soils may have alkaline subsoils and alkaline or neutral surface layers. When enough leaching has occurred to free the solum of calcium carbonate, a mild acidity may develop in the surface horizons. Such a situation exists in certain mollisols. The genetic classification of soils of arid regions with a pertinent description of each group has already been presented (p. 326).

14:16. REACTION OF SALINE AND SODIC SOILS

When the drainage of arid region soils is impeded and the surface evaporation becomes excessive, soluble salts accumulate in the surface horizon. Such intrazonal soils are designated "halomorphic" (see p. 341) and have been classified (8) under three headings: saline, saline–sodic, and sodic.

SALINE SOILS.[3] These soils contain a concentration of neutral soluble salts sufficient to seriously interfere with the growth of most plants. The electrical conductivity of a saturated extract (ECe) is greater than 4 mmhos/cm. Less

Salinization is the term used in reference to the natural processes that result in the accumulation of neutral soluble salts in soils.

than 15 percent of the cation exchange capacity of these soils is occupied by
sodium ions and the pH usually is below 8.5. This is because the soluble
salts present are mostly neutral, and because of their domination, only a
small amount of exchangeable sodium is present.

Such soils are sometimes called *white alkali* soils because a surface in-
crustation, if present, is light in color (see Fig. 14:11). The excess soluble
salts, which are mostly chlorides and sulfates of sodium, calcium, and magnesium, can readily be leached out of these soils with no appreciable rise in
pH. This is a very important practical consideration in the management of
these soils. Care must be taken to be certain that the leaching water is low
in sodium.

SALINE–SODIC SOILS. The saline–sodic soils contain appreciable quantities
of neutral soluble salts and enough adsorbed sodium ions to seriously
affect most plants. Although more than 15 percent of the total exchange
capacity of these soils is occupied by sodium, their pH is likely to be below
8.5. This is because of the repressive influence of the neutral soluble salts,
as in the saline soils previously described. The electrical conductivity of a
saturated extract is more than 4 mmhos/cm.

But unlike the saline soils, leaching will markedly raise the pH of saline–
sodic soils unless calcium or magnesium salt concentrations are high in the
soil or in the irrigation water. This is because the exchangeable sodium,
once the neutral soluble salts are removed, readily hydrolyzes and thereby
sharply increases the hydroxyl ion concentration of the soil solution. In
practice, this is detrimental since the sodium ions disperse the mineral
colloids, which then develop a tight, impervious soil structure. At the same
time, sodium toxicity to plants is increased.

SODIC SOILS. These soils do not contain any great amount of neutral soluble
salts, the detrimental effects on plants being largely due to the toxicity of the
sodium as well as of the hydroxyl ions. The high pH is largely due to the
hydrolysis of sodium carbonate, which occurs as follows:

$$2Na^+ + CO_3^{--} + 2H_2O \rightleftharpoons 2Na^+ + 2OH^- + H_2CO_3$$

The resulting hydroxyl ions give pH values of 10 and above. Also, the sodium
complex undergoes hydrolysis as follows:

$$Na\boxed{Micelle} + HOH \rightleftharpoons H\boxed{Micelle} + Na^+ + OH^-$$

The exchangeable sodium, which occupies decidedly more than 15 percent
of the total exchange capacity of these soils, is free to hydrolyze because the
concentration of neutral soluble salts is rather low. The electrical conductivity
of a saturated extract is less than 4 mmhos/cm. Consequently, the pH is
above 8.5, often rising as high as 10.0. Owing to the deflocculating influence
of the sodium, such soils usually are in an unsatisfactory physical condition.

FIGURE 14:11. (*Upper*) White "alkali" spot in a field of alfalfa under irrigation. Because of upward capillarity and evaporation, salts have been brought to the surface where they have accumulated in amounts toxic to plants. [*Photo courtesy U. S. Soil Conservation Service.*] (*Lower*) White salt crust on a saline soil from Colorado. The white salts are in contrast with the darker colored soil (*left foreground*) underneath. (Scale in inches and centimeters is shown at the bottom of photo.)

As already stated, the leaching of a saline–sodic soil will readily change it to a characteristic sodic soil.

Because of the extreme alkalinity resulting from the Na_2CO_3 present, the surface of sodic soils usually is discolored by the dispersed humus carried upward by the capillary water—hence the name *black alkali* is frequently used. These soils are often located in small areas called *slick spots* surrounded by soils that are relatively productive.

14:17. GROWTH OF PLANTS ON HALOMORPHIC SOILS

Saline and saline–sodic soils with their relatively low pH (usually less than 8.5) detrimentally influence plants largely because of their high soluble-salt concentration (see Fig. 14:11). It is common knowledge that when a water solution containing a relatively large amount of dissolved salts is brought into contact with a plant cell, it will cause a shrinkage of the protoplasmic lining. This action, called *plasmolysis*, increases with the concentration of the salt solution. The phenomenon is due to the osmotic movement of the water, which passes from the cell toward the more concentrated soil solution. The cell then collapses. The nature of the salt, the species, and even the individuality of the plant, as well as other factors, determine the concentration at which the individual succumbs. The adverse physical condition of the soils, especially the saline–sodic, may also be a factor.

Sodic soils, dominated by active sodium, exert a detrimental effect on plants in three ways: (a) caustic influence of the high alkalinity induced by the sodium carbonate and bicarbonate, (b) toxicity of the bicarbonate and other anions, and (c) the adverse effects of the active sodium ions on plant metabolism and nutrition.

14:18. TOLERANCE OF HIGHER PLANTS TO HALOMORPHIC SOILS

The capacity of higher plants to grow satisfactorily on salty soils depends on a number of interrelated factors. The physiological constitution of the plant, its stage of growth, and its rooting habits certainly are among them. It is interesting to note that old alfalfa is more tolerant than young alfalfa and that deep-rooted legumes show a greater resistance than those with shallow rootage.

Concerning the soil the nature of the various salts, their proportionate amounts, their total concentration, and their distribution in the solum must be considered. The structure of the soil and its drainage and aeration are also important.

As a result, it is difficult to forecast accurately the tolerance of crops. Only carefully controlled trials will answer this question and even then with no great degree of certainty. Perhaps the best comparative data are those presented by Richards (8) (see Table 14:1), who stipulates that tolerance shall mean the ability to produce fairly satisfactory crop yields.

TABLE 14:1. *Relative Tolerance of Certain Plants to Salty Soils*[a]

The placings are more or less tentative because of the number of complicating factors involved.

High Tolerance	Medium Tolerance	Low Tolerance
Alkali sacation	Alfalfa	Alsike clover
Bermuda grass	Barley	Apples
Canadian wild rye	Birdsfoot trefoil	Cabbage
Cotton	Carrots	Celery
Date palm	Figs	Ladino clover
Garden pea	Grapes	Lemon
Kale	Lettuce	Orange
Milo	Oats	Peach
Rape	Olives	Pear
Rescue grass	Onions	Peas
Rhodes grass	Rye	Plum
Salt grass	Sweet clover	Potatoes
Sugar beets	Sudan grass	Red clover
Western wheat grass	Tomatoes	White clover

[a] From Richards (8).

14:19. MANAGEMENT OF SALINE AND SODIC SOILS

There are three general ways in which saline and alkali lands may be handled in order to avoid injurious effects to plants (9, 4). The first is *eradication*; the second is a *conversion* of some of the salts to less injurious forms; the third may be designated *control*. In the first two methods, an attempt is made actually to eliminate by various means some of the salts or to render them less toxic. In the third, soil management procedures are utilized which keep the salts so well distributed throughout the soil solum that there is no toxic concentration within the root zone.

ERADICATION. The most common methods used to free the soil of excess salts are (a) underdrainage, and (b) leaching or flushing. A combination of the two, flooding after tile drains have been installed, is the most thorough and satisfactory. When this method is used in irrigated regions, heavy and repeated applications of water can be made. The salts that become soluble are leached from the solum and drained off through the tile. The irrigation water used must be relatively free of silt and salts, especially those containing sodium.

The leaching method works especially well with pervious saline soils, whose soluble salts are largely neutral and high in calcium and magnesium.

Of course, little exchangeable sodium should be present. Leaching saline–sodic soils (and even sodic soils if the water will percolate) with waters very high in salt but low in sodium may be effective. Conversely, treating sodic and saline–sodic soils with water low in salt may intensify their alkalinity because of the removal of the neutral soluble salts. This allows an increase in the percent sodium saturation thereby increasing the concentration of hydroxyl ions in the soil solution. This may be avoided, as explained next, by converting the toxic sodium carbonate and bicarbonate to sodium sulfate by first treating the soil with heavy applications of gypsum or sulfur. Leaching will then render the soil more satisfactory for crops.

CONVERSION. The use of gypsum on sodic soils is often recommended for the purpose of changing part of the caustic alkali carbonates into sulfates (1). Several tons of gypsum per acre are usually necessary. The soil must be kept moist to hasten the reaction, and the gypsum should be cultivated into the surface, not plowed under. The treatment may be supplemented later by a thorough leaching of the soil with irrigation water to free it of some of its sodium sulfate. The gypsum reacts with both the Na_2CO_3 and the adsorbed sodium as follows:

$$Na_2CO_3 + CaSO_4 \rightleftharpoons CaCO_3 + \underset{\text{Leachable}}{Na_2SO_4} \downarrow$$

$$\overset{Na}{\underset{Na}{}}\boxed{\text{Micelle}} + CaSO_4 \rightleftharpoons Ca\boxed{\text{Micelle}} + \underset{\text{Leachable}}{Na_2SO_4} \downarrow$$

It is also recognized that sulfur can be used to advantage on salty lands, especially where sodium carbonate abounds. The sulfur upon oxidation yields sulfuric acid, which not only changes the sodium carbonate to the less harmful sulfate but also tends to reduce the intense alkalinity. The reactions of the sulfuric acid with the compounds containing sodium may be shown as follows:

$$Na_2CO_3 + H_2SO_4 \rightleftharpoons CO_2 + H_2O + \underset{\text{Leachable}}{Na_2SO_4} \downarrow$$

$$\overset{Na}{\underset{Na}{}}\boxed{\text{Micelle}} + H_2SO_4 \rightleftharpoons \overset{H}{\underset{H}{}}\boxed{\text{Micelle}} + \underset{\text{Leachable}}{Na_2SO_4} \downarrow$$

Not only is the sodium carbonate changed to sodium sulfate, a mild neutral salt, but the carbonate radical is entirely eliminated. When gypsum is used, however, the carbonate remains as a calcium salt.

CONTROL. The retardation of evaporation is an important feature of the control of salty soils. This will not only save moisture but will also retard the translocation upward of soluble salts into the root zone. As previously indicated, there are no inexpensive methods of reducing evaporation from large acreages. Consequently, other control practices must be explored.

Where irrigation is practiced, an excess of water should be avoided unless it is needed to free the soil of soluble salts. Frequent light irrigations are often necessary, however, to keep the salts sufficiently dilute to allow normal plant growth.

The timing of irrigation is extremely important on salty soils, particularly during the spring planting season. Since young seedlings are especially sensitive to salts, irrigation often precedes or follows planting to move the salts downward. After the plants are well established, their salt tolerance is somewhat greater.

The use of salt-resistant crops is another important feature of the successful management of saline and alkali lands. Sugar beets, cotton, sorghum, barley, rye, sweet clover, and alfalfa are particularly advisable (see Table 14:1). Moreover, a temporary alleviation of alkali will allow less-resistant crops to be established. Farm manure is very useful in such an attempt. A crop, such as alfalfa, once it is growing vigorously, may maintain itself in spite of the salt concentrations that may develop later. The root action of tolerant plants is exceptionally helpful in improving sodic soils which have a poor physical condition.

14:20. CONCLUSION

Although the discussion closes with pertinent suggestions regarding the management of saline and sodic soils, the major theme of this chapter is soil reaction or soil pH. And it is obvious that as many features of practical concern stem from soil reaction as from any other single soil characteristic. Material in the following chapters emphasizes this point even more decisively.

REFERENCES

(1) Bower, C. A., *Chemical Amendments for Improving Sodium Soils,* Agr. Inf. Bull. 195 (Washington, D.C.: U. S. Department of Agriculture, 1959).

(2) Hsu, P. H., and Bates, T. F., "Fixation of Hydroxyl-Aluminum Polymers by Vermiculite," *Soil Sci. Soc. Amer. Proc.,* **28**:763–69, 1964.

(3) Jenny, H., et al., "Interplay of Soil Organic Matter and Soil Fertility with State Factors and Soil Properties," in *Organic Matter and Soil Fertility,* Pontificiae Acadamiae Scientiarum Scripta Varia 32 (New York: Wiley, 1968).

(4) Kelley, W. P., *Alkali Soils, Their Formation Properties and Reclamation* (New York: Van Nostrand Reinhold, 1951).

(5) Mehlich, A., "Base Unsaturation and pH in Relation to Soil Type," *Soil Sci. Soc. Amer. Proc.,* **6**:150–56, 1941.

(6) Mehlich, A., "Influence of Sorbed Hydroxyl and Sulfate on Neutralization of Soil Acidity," *Soil Sci. Soc. Amer. Proc.,* **28**:492–96, 1964.

(7) Peech, M., "Availability of Ions in Light Sandy Soils as Affected by Soil Reaction," *Soil Sci.,* **51**:473–86, 1941.

(8) Richards, L. A., Ed., *Diagnosis and Improvement of Saline and Alkali Soils* (Riverside, Calif.: U. S. Regional Salinity Lab., 1947).

(9) Thorne, D. W., and Peterson, H. B., *Irrigated Soils* (New York: McGraw-Hill, Blakiston Division, 1954).

Chapter 15

LIME AND ITS SOIL–PLANT RELATIONSHIPS

SOIL acidity and the nutritional conditions that accompany it result when there is a deficiency of adsorbed metallic cations relative to hydrogen. To increase acidity the hydrogen must be replaced by metallic cations. This is commonly done by adding oxides, hydroxides, or carbonates of calcium and magnesium.

These compounds are referred to as *agricultural limes*. They are relatively inexpensive, comparatively mild to handle, and leave no objectionable residues in the soil. They are vital to successful agriculture in most humid regions.

15:1. LIMING MATERIALS

OXIDE OF LIME. Commercial oxide of lime is normally referred to as *burned lime*, *quicklime*, or often simply as the *oxide*. It is usually in a finely ground state and is marketed in paper bags. Oxide of lime is more caustic than limestones and may be somewhat disagreeable to handle.

Burned lime is produced by heating limestone in large commercial kilns. Carbon dioxide is driven off and the impure calcium and magnesium oxides are left behind. The essential reactions that occur when limestones containing calcite and dolomite are heated are as follows:

$$CaCO_3 + heat \rightarrow CaO + CO_2\uparrow$$
$$\text{Calcite}$$

$$CaMg(CO_3)_2 + heat \rightarrow CaO + MgO + 2CO_2\uparrow$$
$$\text{Dolomite}$$

The purity of burned lime sold for agricultural purposes varies from 85 to 98 percent, 95 percent being an average figure. The impurities of burned lime consist of the original impurities of the limestone, such as chert, clay, and iron compounds. In addition to the calcium oxide and magnesium oxides are small amounts of the hydroxides since the oxides readily take up water from the air and slake to some extent even when bagged. Also, contact with the carbon dioxide of the atmosphere will tend to produce carbonates.

HYDROXIDE OF LIME. This form of lime is commonly, though improperly, referred to as the *hydrate*. And since it is produced by adding water to burned lime, the hydroxides that result are often called *slaked lime*. The reaction is as follows:

$$CaO + MgO + 2H_2O \rightarrow Ca(OH)_2 + Mg(OH)_2$$

Hydroxide of lime appears on the market as a white powder and is more caustic than burned lime. Like the oxide, it requires bagging. Representative samples generally show a purity of perhaps 95 or 96 percent.

To maintain the concentration of this form of lime at a high point, the slaking often is not carried to completion. As a result, considerable amounts of the oxides are likely to remain. Moreover, hydroxide of lime carbonates rather readily. The carbonation of calcium and magnesium hydroxides occurs as follows:

$$Ca(OH)_2 + CO_2 \rightarrow CaCO_3 + H_2O$$

$$Mg(OH)_2 + CO_2 \rightarrow MgCO_3 + H_2O$$

Carbonation is likely to occur if the bag is left open and the air is moist. Besides the impurities, six important lime compounds are usually present: the oxides, the hydroxides, and the carbonates of calcium and magnesium. The hydroxides, of course, greatly predominate.

CARBONATE OF LIME. There are a number of sources of carbonate of lime. Of these, pulverized or ground limestone is the most common. There are also bog lime or marl, oyster shells, and precipitated carbonates. Also lime carbonates are by-products of certain industries. All of these forms of lime are variable in their content of calcium and magnesium.

The two important minerals carried by limestones are *calcite,* which is mostly calcium carbonate $(CaCO_3)_2$, and *dolomite*, which is primarily calcium magnesium carbonate $[CaMg(CO_3)_2]$. These occur in varying proportions. When little or no dolomite is present, the limestone is referred to as calcic. As the magnesium increases, this grades into a *dolomitic limestone* and, finally, if very little calcium carbonate is present and the stone is almost entirely composed of calcium–magnesium carbonate and impurities, the term *dolomite* is used. Most of the crushed limestone on the market is calcic and dolomitic, although ground dolomite is available in certain localities.

Ground limestone is effective in increasing crop yields (see Fig. 15:1) and is used to a greater extent than all other forms of lime combined. It varies in purity from approximately 75 to 99 percent. The average purity of the representative crushed limestone is about 94 percent.

FIGURE 15:1. Alfalfa was seeded in this field trial. In the foreground the soil was unlimed (pH 5.2). In the background 4 tons of limestone per acre were applied before seeding.

15:2. CHEMICAL GUARANTEE OF LIMING MATERIALS

Since the various forms of lime are sold on the basis of their chemical composition, the commercial guarantees in this respect become a matter of some importance. The oxide and hydroxide forms may bear composition guarantees stated in one or more of the following ways—the *conventional oxide content,* the *calcium oxide equivalent,* the *neutralizing power,* and *percentages of calcium and magnesium.* To facilitate the explanation and comparison of the various methods of expression, composition figures for commercial burned and hydroxide of lime are shown in Table 15:1.

CONVENTIONAL OXIDE AND CALCIUM OXIDE EQUIVALENT. Since the *oxide* form of expression is so commonly used, this type of guarantee is designated here as the *conventional* method. The *calcium oxide equivalent,* as the term implies, is a statement of the strength of the lime in one figure, calcium oxide. The magnesium oxide is expressed in terms of calcium oxide equivalent and this figure is added to the percentage of calcium oxide present. This may be conveniently done by means of conversion factors. Thus, for the commercial oxide of Table 15:1, 18 percent magnesium oxide is equivalent to 25 percent calcium oxide ($18 \times 1.389 = 25$) and $77 + 25 = 102$, the calcium oxide

TABLE 15:1. *Composition of a Representative Commercial Oxide and Hydroxide of Lime Expressed in Different Ways*

Forms of Lime	Conventional Oxide Content (%)	Calcium Oxide Equivalent	Neutraliz- ing Power	Element (%)
Commercial oxide	CaO = 77	102.0	182.1	Ca = 55.0
	MgO = 18			Mg = 10.8
Commercial hydroxide	CaO = 60	76.7	136.9	Ca = 42.8
	MgO = 12			Mg = 7.2

equivalent. This means that every 100 pounds of the impure burned lime is equivalent in neutralizing capacity to 102 pounds of *pure* calcium oxide.

To express one lime in chemically equivalent amounts of another, simply multiply by the appropriate ratio of molecular weights. Thus, the calcium oxide equivalent of magnesium oxide is obtained by multiplying by the molecular ratio of CaO/MgO, or $56/40.3 = 1.389$. Other appropriate factors are:

$CaCO_3$ to CaO	0.560	$MgCO_3$ to Mg	0.288
CaO to $CaCO_3$	1.786	$CaCO_3$ to Ca	0.400
$MgCO_3$ to CaO	0.664	MgO to Mg	0.602
$MgCO_3$ to $CaCO_3$	1.186	CaO to Ca	0.714

NEUTRALIZING POWER AND ELEMENTAL EXPRESSION. The neutralizing power, as the term is arbitrarily used in respect to lime, is nothing more than a statement of its strength in terms of calcium carbonate—its *calcium carbonate equivalent*. By multiplying by 1.786 in the case above, the calcium oxide equivalent of 102 becomes 182.1, the calcium carbonate equivalent. This means that every 100 pounds of the impure burned lime is equivalent in neutralizing capacity to 182.1 pounds of pure calcium carbonate.

The *elemental* method of expression, while not so common as the others, is required by law in some states. It may be readily calculated from the conventional oxide guarantee. Or, if given alone, the other forms of statement may be derived from it.

Commercial limes practically always carry the conventional oxide guarantee and sometimes the elemental. Thus, the amount of magnesium as well as calcium present is indicated. This is an important consideration. In addition, one or even both of the other forms of guarantee may be given.

LIMESTONE GUARANTEES. The guarantees on ground limestone differ in two respects from those of the oxide and hydroxide forms. Usually the separate percentages of calcium and magnesium carbonates are given. These are almost always accompanied by the percentage of *total carbonates* and sometimes by the neutralizing power. In addition, one or more of the other three modes of guarantee may be used. By way of illustration the six methods of expression are presented in Table 15:2 for a representative ground limestone.

TABLE 15:2. *Composition of a Representative Commercial Ground Limestone Expressed in Different Ways*

Separate Carbonates (%)	Total Carbonates (%)	Neutral- izing Power	Conventional Oxide (%)	Calcium Oxide Equivalent	Element (%)
$CaCO_3 = 80$			$CaO = 44.80$		$Ca = 32.00$
	94	96.6		54.10	
$MgCO_3 = 14$			$MgO = 6.70$		$Mg = 4.03$

TOTAL CARBONATES VERSUS NEUTRALIZING POWER. The total carbonate method of guarantee has the advantages of being simple and of requiring

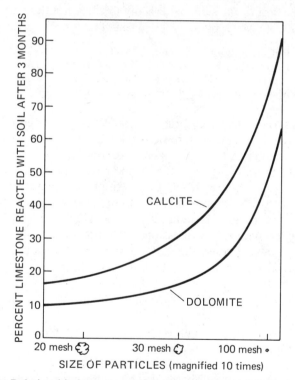

FIGURE 15:2. Relationship between particle size of calcite and dolomite and the rate of reaction of these minerals with the soil. Note that calcite particles of a given size react more rapidly than do corresponding dolomite particles. The coarse fractions of both minerals neutralize the soil acidity very slowly. [*Data calculated from Schollenberger and Salter* (3).]

no chemical explanation. Furthermore, the total carbonates give perhaps a truer indication of the immediate comparative field value of limestones.

Those favoring the neutralizing-power method of expression stress its accuracy and the fact that it can be determined very easily and quickly. It gives a measure of the long-term neutralizing ability of limestones and provides recognition of the higher neutralizing power of the dolomite-containing stones. Since the rate of reaction of dolomite is significantly slower than calcite, however, there is some question as to whether this recognition is appropriate (see Fig. 15:2).

15:3. FINENESS GUARANTEE OF LIMESTONE

The application to the soil of two different liming materials in chemically equivalent quantities does not necessarily mean that equivalent results will be attained. If the two materials are both limestones, the particles contained therein are likely to be quite variable in size as well as hardness. This is important since the finer the division of any material, the more rapid is likely to be its solution and rate of reaction. The oxide and hydroxide of lime usually appear on the market as almost impalpable powders; consequently their fineness is always satisfactory. Therefore, to rate a limestone as to its probable effects in comparison with other limes, a *fineness guarantee* is desirable. This is usually one of the requirements of laws controlling the sale of agricultural limestone. A mechanical analysis is made by the use of screens of different mesh, a 10-mesh sieve, for example, having ten openings to the linear inch.[1] The proportion of the limestone that will pass through the various screens used constitutes the guarantee.

INTERPRETATION OF GUARANTEE. Figure 15:3 shows that the finer grades of limestone are much more effective than the coarser grades in benefiting plant growth. Other data indicate that while the coarser lime is less rapid in its action, it remains in the soil longer and its influence should be effective for a greater period of years. Other investigators have published results that substantiate these conclusions. The proportionate responses are somewhat different, however. This is to be expected since limestones of different hardness applied to various soil types are sure to respond rather diversely.

FINE LIMESTONE. Everything considered, a pulverized limestone, *all of which will pass a 10-mesh screen, and at least 50 percent of which will pass a*

---

[1] The diameter of the individual openings will, of course, be much less than 0.1 inch, the exact size depending on the wire used in making the sieve. Usually the diameter of the openings in inches is a little more than one-half of the quotient obtained by dividing 1 by the mesh rating. For instance, the openings of a 10-mesh screen are approximately 0.07 in. in diameter; those of a 50-mesh screen, 0.0122 in.; and those of a 100-mesh screen, approximately 0.0058 in. Unfortunately there is no standardization of the sieves used for grading agricultural limestone.

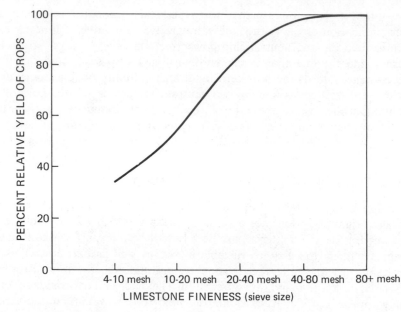

LIMESTONE FINENESS (sieve size)

FIGURE 15:3. Relationship between the particle size fraction of limestone and yield of crops in nine field experiments. [*Data from several sources; summarized by Barber* (1).]

100-mesh sieve, should give excellent results and yet be cheap enough to encourage its use. Such a lime is sufficiently pulverized to rate as a *fine* limestone. Some limestones are finer than this, 50 or 60 percent passing a 200-mesh screen, but the cost of grinding the stone to this very fine condition seldom can be justified. The fine limestone described above has enough of the finer particles to give quick results and yet a sufficient amount of the coarser fragments to make it last over the period of the rotation.

A limestone which does not approximate the fineness designated above should be discounted to the extent to which it falls short. It may be necessary, for example, to consider 3,000 pounds of one limestone as equal to 1 ton of another, even though their chemical analyses are the same. Considerable judgment in the interpretation of fineness guarantee is necessary in order that such an adjustment be made correctly. Much limestone that falls below the standard set above is now being used because, as a by-product, it is very cheap. When the amounts put on the land are properly adjusted, as good results may be expected from the coarser as from the finer limestones and the effects should last considerably longer.

15:4. CHANGES OF LIME ADDED TO THE SOIL

In considering the changes which lime undergoes after it is added to the soil we must always keep in mind that (a) the calcium and magnesium compounds applied undergo solution under the influence of a variable partial pressure of carbon dioxide, and (b) an acid colloidal complex will adsorb considerable amounts of calcium and magnesium ions.

REACTION WITH CARBON DIOXIDE. When lime—whether the oxide, hydroxide, or the carbonate—is applied to an acid soil, the movement as solution occurs is toward the bicarbonate form. This is because the carbon dioxide partial pressure in the soil, usually several hundred times greater than that in the atmospheric air, generally is high enough to cause a reaction with the hydroxide and even the carbonate. The reactions, written only for the purely calcium limes, follow:

$$CaO + H_2O \rightarrow Ca(OH)_2$$
$$Ca(OH)_2 + 2H_2CO_3 \rightarrow Ca(HCO_3)_2 + 2H_2O$$
$$CaCO_3 + H_2CO_3 \rightarrow Ca(HCO_3)_2$$

REACTION WITH SOIL COLLOIDS. All liming materials will react with acid soils, the calcium and magnesium replacing hydrogen on the colloidal complex. The adsorption in respect to calcium may be indicated as follows:

$$\begin{matrix} H \\ H \end{matrix} \boxed{\text{Micelle}} + Ca(OH)_2 \rightleftharpoons Ca \boxed{\text{Micelle}} + 2H_2O$$

$$\begin{matrix} H \\ H \end{matrix} \boxed{\text{Micelle}} + \underset{\text{In solution}}{Ca(HCO_3)_2} \rightleftharpoons Ca \boxed{\text{Micelle}} + 2H_2O + 2CO_2\uparrow$$

$$\begin{matrix} H \\ H \end{matrix} \boxed{\text{Micelle}} + \underset{\text{Solid phase}}{CaCO_3} \rightleftharpoons Ca \boxed{\text{Micelle}} + H_2O + CO_2\uparrow$$

As the above reactions of limestone proceed, carbon dioxide is freely evolved. In addition, the adsorption of the calcium and magnesium raises the percentage base saturation of the colloidal complex and the pH of the soil solution is pushed up correspondingly.

DISTRIBUTION OF CALCIUM AND MAGNESIUM IN LIMED SOIL. Calcium and magnesium, if supplied by a limestone of average fineness, will exist in the soil at least for a time, in three forms: (a) as solid calcium and calcium–magnesium carbonates, (b) as exchangeable bases adsorbed by the colloidal matter, and (c) as dissociated cations in the soil solution mostly in association with bicarbonate ions. When the calcium and calcium–magnesium carbonates have all dissolved, the system becomes somewhat simpler, involving only the exchangeable cations and those in soil solution, both of which are subject to loss by leaching.

DEPLETION OF CALCIUM AND MAGNESIUM. As the soluble calcium and magnesium compounds are removed, the percentage base saturation and pH are gradually reduced and eventually another application of lime is necessary. This then is the type of cycle through which much of the calcium and magnesium added to arable soils swings in humid regions.

15:5. Loss of Lime from Arable Soils

Calcium and magnesium are lost from soils in three ways: (a) by erosion, (b) by crop removal, and (c) by leaching. Since these three types of lime loss have already been discussed and evaluated in Chapter 9, it is a simple matter to draw such data together in tabular form (Table 15:3). Although the values shown do not represent losses from soils in general, they are quite significant when used comparatively.

TABLE 15:3. *Lime Losses from Soil*

Values in pounds per acre per year.

Manner of Removal	Calcium		Magnesium	
	CaO	CaCO$_3$	MgO	MgCO$_3$
By erosion, Missouri experiments, 4 percent slope	120	214	48	102
By the average crop of a standard rotation	56	100	50	105
By leaching from a representative silt loam	140	250	33	70
Total		564		277

The data for erosion removals are for a silt loam in Missouri cropped to a rotation of corn, wheat, and clover (see Table 9:6). The acre annual removal by crops was calculated for a standard and representative rotation, assuming reasonable yields (see Table 9:2). The leaching losses are for a postulated representative silt loam under a rainfall of perhaps 36 to 40 inches and cropped to a standard farm rotation. All data are in pounds per acre per year expressed in the conventional *oxide* form and also in the more practical terms of *calcium* and *magnesium carbonates*.

The greater loss of calcium than magnesium no doubt is because the soil colloidal matter almost always carries a much larger amount of the calcium in an exchangeable condition. And since the average liming material supplies several times more calcium than magnesium, this loss ratio will in general be maintained and even accentuated in arable lands as liming proceeds.

This does not mean, however, that the magnesium in lime is of minor importance. Far from it. In fact, judging from the figures of Table 15:3, there should always be at least one third as much magnesium as calcium in the lime applied in order to proportionately meet the outgo of the two constituents. Other things being equal, it is generally wise to select a magnesium-containing limestone.

The figures quoted in Table 15:3 indicate that 500 to 600 pounds of limestone per acre per year may be required to meet the loss from cropped soils in humid regions. This amounts to about 1 to 1.5 tons of carbonate of lime during the period of the average rotation, depending upon the kind of soil and other factors. Such a conclusion verifies the importance of lime in any scheme of fertility management in areas of medium to heavy rainfall.

15:6. EFFECTS OF LIME ON THE SOIL

It has already been emphasized that the changes of lime in the soil are many and complicated. Therefore, the following presentation must necessarily be more or less general in nature. The better known effects of liming are (a) physical, (b) chemical, and (c) biological.

PHYSICAL EFFECTS. A satisfactory granular structure is somewhat encouraged in an acid soil by the addition of any form of lime, although the influence is largely indirect. For example, the effects of lime upon biotic forces are significant, especially those concerned with the decomposition of the soil organic matter and the synthesis of humus. The genesis of the humus as well as its persistence greatly encourages granulation (see p. 58). The stimulating effect of lime on deep-rooted plants, especially legumes, is significant.

CHEMICAL EFFECTS. If a soil at pH 5.0 is limed to a more suitable pH value of about 6.5, a number of significant chemical changes occur. Many of these were described in Chapter 14 and will be outlined here to reemphasize their importance.

1. The concentration of hydrogen ions will decrease.
2. The concentration of hydroxyl ions will increase.
3. The solubility of iron, aluminum, and manganese will decline.
4. The availability of phosphates and molybdates will increase.
5. The exchangeable calcium and magnesium will increase.
6. The percentage base saturation will increase.
7. The availability of potassium may be increased or decreased depending on conditions.

Of the specific chemical effects of lime mentioned, the reduction in acidity is one commonly recognized. However, the indirect effects on nutrient availability and on the toxicity of certain elements are probably more important. Liming of acid soils enhances the availability and plant uptake of elements such as molybdenum, phosphorus, calcium, and magnesium. At the same time, it drastically reduces the concentration of iron, aluminum, and manganese, which under very acid conditions are likely to be present in toxic quantities.

BIOLOGICAL EFFECTS. Lime stimulates the general purpose, heterotrophic soil organisms. This stimulation not only favors the formation of humus but also encourages the elimination of certain organic intermediate products that might be toxic to higher plants.

Most of the favorable soil organisms, as well as some of the unfavorable ones such as those that produce potato scab, are encouraged by liming. The formation of nitrates and sulfates in soil is markedly speeded up by increasing the pH. The bacteria that fix nitrogen from the air, both nonsymbiotically and in the nodules of legumes, are especially stimulated by the application of lime. The successful growth of most soil microorganisms so definitely depends upon lime that satisfactory biological activities cannot be expected if calcium and magnesium levels are low.

15:7. CROP RESPONSE TO LIMING

Figure 14:9 specifies in a general way the lime levels at which various plants seem to grow most satisfactorily, thereby indicating whether or not a particular crop is likely to be benefited by liming. Of the lime-loving plants, alfalfa, sweet clover, red clover, asparagus, cauliflower, and lettuce are representative. However, a surprisingly large proportion of crop plants is quite tolerant to the conditions presented by moderately acid soils. If such plants respond to lime, it is because of the stimulating influence of lime upon the legume which preceded them in the rotation.

REASONS FOR RESPONSE. When plants are benefited by lime, a number of possible reasons may be suggested: (a) the direct nutritive or regulatory action of the calcium and magnesium, (b) the removal or neutralization of toxic compounds of either an organic or inorganic nature; (c) the retardation of plant diseases, (d) an increased availability chemically of plant nutrients, and (e) an encouragement of microorganic activities favorable in a nutritive way. Since several of these factors undoubtedly function concurrently, crop response to liming is a complicated phenomenon, and only the broadest conclusions may be drawn.

The growth of a number of plants is definitely retarded by liming, prominent among which are cranberries, blueberries, watermelons, laurel, and certain species of azaleas and rhododendrons. It is, therefore, advantageous not only to know the condition of the soil but also to understand the influence of lime on the crop to be grown. Lime is too often used as a cure-all, little attention being paid to the widely differing responses exhibited by various crops.

15:8. OVERLIMING

This leads to the question of *overliming*, the addition of lime until the pH of the soil is above that required for optimum plant growth on the soil in question. Under such conditions many crops that ordinarily respond to lime are detrimentally affected, especially during the first season following the application. With heavy soils and when only moderate amounts of lime are used, the danger is negligible. But on sandy soils low in organic matter and therefore lightly buffered, it is easy to injure certain crops, even with a relatively moderate application of lime.

The detrimental influences of excess lime have been mentioned (see p. 389). Consequently, they are merely outlined here for convenience:

1. Deficiencies of available iron, manganese, copper, and/or zinc may be induced.
2. Phosphate availability may decrease because of the formation of complex and insoluble calcium phosphates.
3. The absorption of phosphorus by plants and especially its metabolic use may be restricted.
4. The uptake and utilization of boron may be hindered.
5. The drastic change in pH may in itself be detrimental.

With so many possibilities and with such complex biocolloidal interrelations to handle, it is easy to see why overliming damage in many cases has not been satisfactorily explained.

The use of lime in a practical way raises three questions. If the first is answered in the affirmative, the other two present themselves in logical sequence. The three questions are: (a) Shall lime be applied? (b) Which form shall be used? (c) What shall be the rate of application?

15:9. SHALL LIME BE APPLIED?

The old concept concerning lime was that of a *cure-all*—that it was almost certain to be beneficial, no matter what the problem. Such an opinion should now be discarded since it may lead to a waste of money and in some cases to overliming. The use of lime must be based on measured soil acidity and on crop requirements.

In determining the desirability of applying lime, the chemical condition of the soil itself should be examined. For this a pH determination commonly is made, either by means of a pH meter or by the less accurate indicator-dye method (p. 393). Representative subsoil as well as surface samples should be examined. The pH is correlated fairly closely with percentage base saturation and is an indicator of the probable activity of the calcium, magnesium, and other elements in the soil. Besides, the test is very easy and the results are quickly determined, two reasons why the test is one of the most popular now available for soil diagnosis.

Before a recommendation can be made, however, the general lime needs of the crop or crops to be grown should be considered. The final decision depends upon the proper coordination of these two types of information. A grouping of crops such as that shown in Fig. 14:8 will greatly aid in deciding whether or not to lime.

15:10. FORM OF LIME TO APPLY

On the basis of the ideas already presented, it is evident that five major factors should be considered in deciding on a specific brand of lime to apply:

1. Chemical guarantees of the limes under consideration.
2. Cost per ton applied to the land.
3. Rate of reaction with soil.
4. Fineness of the limestone.
5. Miscellaneous considerations (handling, storage, bag or bulk, and so on).

GUARANTEE AND COST RELATIONS. By a purely arithmetical calculation based on factors 1 and 2, the cost of equivalent amounts of lime as applied to the land can be determined. These factors will show which lime will furnish the greatest amount of total neutralizing power for every dollar expended.

For instance, the *neutralizing power* ($CaCO_3$ equivalent) of two limes, a hydroxide and a ground limestone, are guaranteed at 135 and 95, respectively. The cost of applying a ton of each to the land (all charges, including trucking and spreading) for purposes of calculation will be considered to be $15.00 for the hydrate and $8.00 for the carbonate. Obviously, it will require only $\frac{95}{135}$, or 0.7 ton of the hydroxide to equal 1 ton of the carbonate. The cost of equivalent amounts of neutralizing power based on 1 ton of limestone will therefore be $10.50 for the hydroxide of lime and $8.00 for the limestone. Unless a rapid rate of reaction were desired, the advantage in this case would definitely be with the limestone. Economic considerations account for the fact that about 95 percent of the agricultural liming material applied in the United States is ground limestone.

RATE OF REACTION WITH SOIL. Burned and hydrated limes react with the soil much more rapidly than do the carbonate forms. For this reason, these caustic materials may be preferred where immediate reaction with the soil is required. In time, however, this initial advantage is nullified because of the inevitable carbonation of the caustic forms of lime.

In comparing carbonate-containing materials, it should be remembered that highly dolomitic limestones generally react more slowly with the soil than do those which are highly calcic. This difference is due to the comparatively slow rate of reaction of dolomite which is supplied along with calcite in dolomitic limestones (see Fig. 15:2). When rapid rate of reaction is not a factor, however, dolomitic limestone is often preferred because significant quantities of magnesium are supplied by this material. Over a period of two or three rotations, soils treated with highly calcic limestone have been known to develop a magnesium deficiency even though the pH was maintained near 7.

The fineness of the limestones under consideration is important, especially if the material is high in dolomite. If it is not sufficiently pulverized to rate as a *fine* lime (see the rule on pp. 409–10), allowance must be made for the lack of rapid acting material by increasing the rate of application.

MISCELLANEOUS FACTORS. Several miscellaneous factors may at times be important. The handling of the caustic limes, even when bagged, is somewhat more disagreeable than working with limestone. The necessity for storage also comes in since sometimes it is desirable to carry lime from one season to another. Limestone has the advantage here since it does not change in storage as the others do.

The question of purchasing the lime, especially the limestone, in bags or in bulk has become increasingly important in recent years. Spreading limestone in bulk by trucks has greatly reduced handling costs (see Fig. 15:4). This method tends, however, to limit the choice of when the material may be applied, since wet or plowed fields cannot be serviced by heavy machinery. The decision as to what method to use should not be finally approved until the nature of the soil and the probable response of the crops have again been reviewed.

15:11. AMOUNTS OF LIME TO APPLY

The amount of lime to apply is affected by a number of factors, including the following:

1. Soil:
   Surface { pH.
   Texture and structure.
   Amount of organic matter.
   Subsoil: pH, texture, and structure.
2. Crops to be grown.
3. Kind and fineness of lime used.
4. Economic returns in relation to cost of lime.

SOIL CHARACTERISTICS. The pH test is invaluable in making decisions because it gives some idea of the percentage base saturation of the soil and the need for lime. The texture and organic matter also are important since they are indicative of the adsorptive capacity of the soil and the strength of

FIGURE 15:4. Bulk application of limestone by specially equipped trucks is becoming more and more common. Because of the weight of such machinery, much limestone is applied to sod land and plowed under. In many cases this same method is used to spread commercial fertilizers. [*Photo courtesy Harold Sweet, Agway Inc., Syracuse, N.Y.*]

buffering (see Fig. 15:5). Naturally the higher the buffer capacity of a soil, the greater must be the amount of lime applied to attain a satisfactory change in pH.

The subsoil also should be tested for pH and examined as to texture and structure. A subsoil pH markedly above or below that of the furrow slice may justify a reduction or an increase as the case may be in the acre rate of lime application. Advice as to the cropping of any soil should not be given without knowledge of subsoil conditions.

OTHER CONSIDERATIONS. The other factors listed above have been discussed; besides, their importance is self-evident. As to the kinds of lime, the three forms in respect to their effects on the soil are roughly in the ratio 1 ton of representative finely ground limestone to 0.7 ton of commercial hydroxide to a little over 0.5 ton of representative oxide. The fineness of limestone is as important as chemical composition. The experience factor emphasizes taking advantage of lime knowledge wherever found.

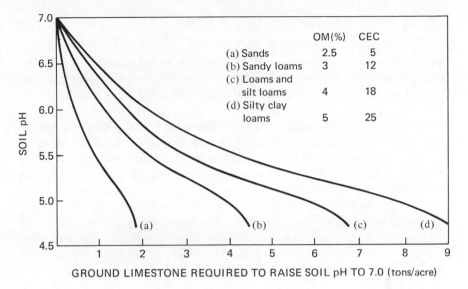

FIGURE 15:5. Relationship between soil texture and the amount of limestone required to raise the pH of New York soils to 7.0. Representative organic matter (OM) and cation exchange capacity (CEC) levels are shown. [*From Peech* (2).]

SUGGESTED AMOUNTS. In ordinary practice it is seldom economical to apply more than 3 to 4 tons of finely ground limestone to the acre of mineral soil at any one time unless it is very acid and the promise for increased crop yield exceptionally good. The data given in Table 15:4 serve as guides in practical liming operations for a four- or five-year rotation with average mineral soils. It is assumed that a crop with a medium lime requirement is the principal legume of the rotation. The recommendations are in terms of finely ground limestone. If the lime is coarser than the minimum quoted (see pp. 409–10), or if an oxide or hydroxide is used, due allowance should be made.

TABLE 15:4. *Suggested Total Amounts of Finely Ground Limestone That Should Be Applied per Acre–Furrow Slice of Mineral Soil for Alfalfa in Rotation[a]*

| Need for Lime | Limestone (lb/acre) | |
	Sandy Loam	Silt Loam
Moderate	2,000–3,000	3,000–6,000
High	3,000–5,000	6,000–8,000

[a] In calculating these rates, the assumption is made that the amounts suggested will be used as *initial* applications. After the pH of the soil has been raised to the desired level, smaller *maintenance* rates may be satisfactory.

15:12. METHODS OF APPLYING LIME

To obtain quick action, lime is best applied to plowed land and worked into the soil as the seedbed is prepared. It should be mixed thoroughly with the surface half of the furrow slice. Top-dressing with lime and leaving it on the surface is seldom recommended except on permanent meadows and pastures.

However, it is often much more convenient and in the long run just as effective to apply the lime on the surface and plow it under. The spreading is usually done in the fall on sod land that is to be turned later that autumn or the next spring. This practice provides a longer interval during which lime may be applied and usually results, in the case of a sod, in a minimum of soil packing if heavy machinery is used. It also makes possible the bulk spreading of lime, using large trucks. This is becoming increasingly popular (see Fig. 15:4).

The time of year at which lime is applied is immaterial, the system of farming, the type of rotation, and related considerations being the deciding factors. Winter application may even be practiced.

EQUIPMENT USED. Limestone is commonly spread in bulk form using large trucks (see Fig. 15:4). A smaller lime distributor may also be used, especially if the amount to be applied is not unusually large. Small amounts of lime may be distributed by means of the fertilizer attachment on a grain drill. The evenness of distribution is as important as the amount of lime used and should not be neglected.

The addition of small amounts of limestone, 300 to 500 pounds per acre, often gives remarkable results when drilled in with the crop being seeded. Even though the lime is not mixed thoroughly with the soil and there is little change in the pH of the furrow slice as a whole, the influence on the crop may be very favorable. Apparently the lime in this case is functioning more as a fertilizer and as a means of rectifying conditions within the crop and at its root–soil interfaces than as an amendment for the whole furrow slice.

PLACE IN ROTATION. Lime should be applied with or ahead of the crop that gives the most satisfactory response. Thus, in a rotation of corn, oats, fall wheat, and two years of alfalfa and timothy, the lime is often applied when the wheat is seeded in the fall. It can then be spread on the plowed ground and worked in as the seedbed is prepared. Its effect is thus especially favorable on the new legume seeding made in the wheat.

However, application to plowed land may result in some compaction of the soil if heavy machinery is used and certainly more power is necessary than for a soil in sod. For that reason, spreading on sod land is often favored. In practice, the place of lime in the rotation is often determined by expediency since the vital consideration is, after all, the application of lime regularly and in conjunction with a suitable rotation of some kind.

15:13. LIME AND SOIL FERTILITY MANAGEMENT

The maintenance of satisfactory soil fertility levels in humid regions is dependent upon the judicious use of lime (see Fig. 15:6). The pH and associated nutritional condition of the soil is determined by the lime level. Furthermore, the heavy use of acid-forming nitrogen-containing fertilizers increases soil acidity, making even more important the maintenance of a favorable base status in soils.

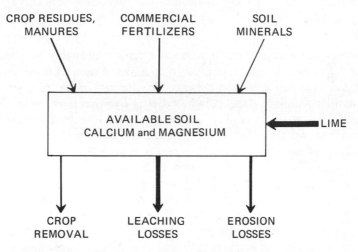

FIGURE 15:6. Important ways by which *available* calcium and magnesium are supplied to and removed from soils. The major losses are through leaching and erosion. These are largely replaced by lime and fertilizer applications. Additions in the latter form are much higher than is generally realized, owing to the large quantities of calcium contained in superphosphates.

REFERENCES

(1) Barber, S. A., "Liming Materials and Practices," Pearson, R. W., and Adams, F., Eds., in *Soil Acidity and Liming* (*Agronomy,* **12**:125–60) (Madison, Wisc.: American Society of Agronomy, 1967).

(2) Peech, M., "Lime Requirements vs. Soil pH Curves for Soils of New York State," mimeographed (Ithaca, N. Y.: Agronomy, Cornell University, 1961).

(3) Schollenberger, C. J., and Salter, R. M., "A Chart for Evaluating Agricultural Limestone," *Jour. Amer. Soc. Agron.* **35**:955–66, 1943.

Chapter 16

NITROGEN AND SULFUR ECONOMY OF SOILS

O F the various plant nutrients, nitrogen probably has been subjected to the greatest amount of study and still receives much attention. And there are very good reasons. The amount in the soil is small, while the quantity withdrawn annually by crops is comparatively large. At times, the soil nitrogen is too readily soluble and is lost in drainage; at other times, it suffers volatilization; at still other times it is definitely unavailable to higher plants. Moreover, its effects on plants usually are very marked and rapid. Thus, overapplication, which may be harmful, sometimes occurs.

Some plants, such as the legumes, have associated with their roots soil organisms which "fix" atmospheric nitrogen into forms which they can use. Others, such as the grasses, are largely dependent upon outside sources— either through nonsymbiotic fixation or through the addition of fertilizers or other combined forms of nitrogen. All in all, nitrogen is a potent nutrient element that should not only be conserved but carefully regulated.

16:1. INFLUENCE OF NITROGEN ON PLANT DEVELOPMENT

FAVORABLE EFFECTS. Of the macronutrients usually applied in commercial fertilizers, nitrogen seems to have the quickest and most pronounced effect. It tends primarily to encourage aboveground vegetative growth and to impart to the leaves deep green color. With cereals, it increases the plumpness of the grain and their percentage of protein. With all plants, nitrogen is a regulator that governs to a considerable degree the utilization of potassium, phosphorus, and other constituents. Moreover, its application tends to produce succulence, a quality particularly desirable in such crops as lettuce and radishes.

Plants receiving insufficient nitrogen are stunted in growth and possess restricted root systems. The leaves turn yellow or yellowish green and tend to drop off. The addition of available nitrogen will cause a remarkable change, indicative of the unusual activity of this element within the plant.

OVERSUPPLY. Because of the immediate effect of nitrogen on plants, higher applications than are necessary are sometimes made. This is unfortunate since nitrogen is expensive and is easily lost from the soil. Very dark green soft sappy leaves are an indication of an oversupply of nitrogen.

This oversupply may delay crop maturation by encouraging excessive vegetative growth. This results in the weakening of the stems and subsequent lodging of grains. An oversupply of nitrogen may adversely affect fruit and grain quality, as in apples, peaches, and barley. Also, resistance to some diseases is reduced by excess nitrogen.

It must not be inferred, however, that all plants are detrimentally affected by large amounts of nitrogen. Many crops, such as the grasses and vegetables, should have plenty of this element for their best and most normal development. Detrimental effects are not to be expected with such crops unless exceedingly large quantities of nitrogen are applied. Nitrogenous fertilizers may be used freely in such cases, the cost of the materials in respect to the value of the crop increases being the major consideration.

16:2. FORMS OF SOIL NITROGEN

There are three major forms of nitrogen in mineral soils: (a) organic nitrogen associated with the soil humus, (b) ammonium nitrogen fixed by certain clay minerals, and (c) soluble inorganic ammonium and nitrate compounds.

Most of the nitrogen in soils is associated with the organic matter. In this form it is protected from rapid microbial release, only 2–3 percent a year being mineralized under normal conditions. About half the organic nitrogen is known to be in the form of amino compounds. The form of the remainder is uncertain.

Some of the clay minerals have the ability to fix ammonium nitrogen between their crystal units. The amount fixed varies depending on the nature and amount of clay present. Up to 8 percent of the total nitrogen in surface soils and 40 percent of that in subsoils has been found to be in the "clay-fixed" form. In most cases, however, both these figures would be considerably lower. Even so, the nitrogen so fixed is only slowly available to plants and microorganisms.

The amount of nitrogen in the form of soluble ammonium and nitrate compounds is seldom more than 1–2 percent of the total present, except where large applications of inorganic nitrogen fertilizers have been made. This is fortunate since inorganic nitrogen is subject to loss from soils by leaching and volatilization. Only enough is needed to supply the daily requirements of the growing crops.

16:3. THE NITROGEN CYCLE[1]

In all soils there is considerable intake and release loss of nitrogen in the course of a year accompanied by many complex transformations. Some of

[1] This section is based on a review by Bartholomew and Clark (5).

these changes may be controlled more or less by man while others are beyond his command. This interlocking succession of largely biochemical reactions constitutes what is known as the *nitrogen cycle* (see Fig. 16:1). It has attracted scientific study for years, and its practical significance is beyond question.

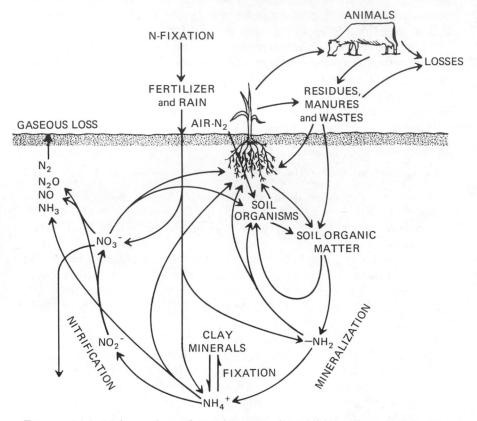

FIGURE 16:1. Main portions of the nitrogen cycle. Additions of chemical fertilizer make up an increasing important source of this element.

The nitrogen income of arable soils is derived from such materials as commercial fertilizers, crop residues, green and farm manures, and ammonium and nitrate salts brought down by precipitation. In addition, there is the fixation of atmospheric nitrogen accomplished by certain microorganisms. The depletion is due to crop removal, drainage, erosion, and to loss in a gaseous condition.

Much of the nitrogen added to the soil undergoes many transformations before it is removed. The nitrogen in organic combination is subjected to especially complex changes. Proteins are converted into various decomposition products, and finally some of the nitrogen appears in the nitrate form.

Even then it is allowed no rest since it is either appropriated by microorganisms and higher plants, removed in drainage, or lost by volatilization. And so the cyclic transfer goes on and on. The mobility of nitrogen is remarkable, rivaling carbon in its ease of movement.

MAJOR DIVISIONS OF THE NITROGEN CYCLE. At any one time, the great bulk of the nitrogen in a soil is in organic combinations protected from loss but largely unavailable to higher plants. For this reason much scientific effort has been devoted to the study of organic nitrogen, how it is stabilized, and how it may be released to forms usable by plants. The process of tying up nitrogen in organic forms is called *immobilization* and its slow release is called *mineralization* (see Fig. 16:2).

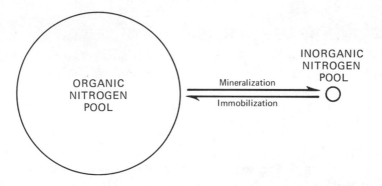

FIGURE 16:2. At any one time most of the nitrogen in soils is *immobilized* in organic combination and as such is not available for plant growth, leaching, or gaseous loss. Through the process of *mineralization* microorganisms change a small amount of nitrogen each year to inorganic forms that higher plants can use.

IMMOBILIZATION. During the process of microbial decomposition of plant and animal residues, especially those low in nitrogen, much of the inorganic nitrogen is converted to organic forms (see p. 152). Initially the nitrogen is probably tied up in microbial tissue. If the residues are not sufficiently high in inorganic nitrogen, soil nitrate and ammonium ions will be assimilated. As the rate of microbial activity subsides, some of this immobilized nitrogen will be mineralized and ammonium and nitrate ions will again appear in the soil solution. However, much of the immobilized nitrogen remains in the organic form.

The mechanism by which simple nitrogen compounds are changed to organic combinations that resist breakdown is still obscure (see p. 149). Somehow the immobilized nitrogen in the microbial tissue becomes an integral part of the soil organic matter. In this form it is only slowly mineralized to compounds usable by higher plants.

MINERALIZATION. Tagged nitrogen experiments have demonstrated that only 2–3 percent immobilized nitrogen may be expected to be mineralized annually. Even so, this release of nitrogen to inorganic forms has long supplied a significant portion of our crop needs.

Although the general term "mineralization" covers a whole series of reactions, the net effect can be rather simply visualized. Heterogenous soil organisms, both plant and animal, attack the organic nitrogen compounds. As a result of enzymatic digestion, the more complex proteins and allied compounds are simplified and hydrolyzed. The end product is ammonia. The enzymic process may be indicated as follows, using an amino compound as an example of the nitrogen source:

$$\underset{\substack{\text{Amino} \\ \text{combination}}}{R\!\!-\!\!NH_2} + HOH \xrightarrow[\text{hydrolysis}]{\text{enzymic}} R\!\!-\!\!OH + NH_3 + \text{energy}$$

$$2NH_3 + H_2CO_3 \rightarrow (NH_4)_2CO_3 \rightleftharpoons 2NH_4 + CO_3{}^-$$

Mineralization seems to proceed to the best advantage in well-drained aerated soils with plenty of basic cations present. It will take place to some extent under almost any conditions, however, because of the great number of different organisms capable of accomplishing such a change. This is one of the advantages of a general purpose flora and fauna.

UTILIZATION OF AMMONIUM COMPOUNDS. The fate of the ammoniacal nitrogen (as shown in Fig. 16:1) is fourfold. First, considerable amounts are appropriated by organisms capable of using this type of compound. Mycorrhizal fungi (see p. 125) undoubtedly are able to absorb ammoniacal nitrogen and pass it on in some form to their host.

Second, higher plants are able to use this form of nitrogen, often very readily. Young plants of almost all kinds are especially capable in this respect, although they seem to grow better if some nitrate nitrogen is also available. Azaleas, laurel, and other plants requiring a low-lime soil are additional examples. Still other plants, such as lowland rice, even prefer ammoniacal nitrogen to the nitrate.

Third, ammonium ions are subject to fixation by some of the clay minerals and organic matter. In this fixed form, the nitrogen is not subject to rapid oxidation, although in time it may become available. This will receive attention in the next section.

Finally, when plant and animal syntheses temporarily are satisfied, the remaining ammonium nitrogen may go in a third direction. It is readily oxidized by certain special purpose forms of bacteria which use it not only as a source of nitrogen but also as a source of energy. Thus, a much discussed and perhaps comparatively overemphasized phase of biochemistry—*nitrification*—is reached. It is so named because its end product is nitrate nitrogen, and it will receive attention following a consideration of ammonia fixation.

16:4. AMMONIA FIXATION

Both the organic and inorganic soil fractions have the ability to "fix" ammonia in forms relatively unavailable to higher plants or even micro-organisms. Since different mechanisms and compounds are involved in these two types of fixation, they will be considered separately.

FIXATION BY CLAY MINERALS. Several clay minerals with a 2:1-type structure have the capacity to "fix" ammonium and potassium ions. Vermiculite has the greatest capacity, followed by illite and montmorillonite.

It will be remembered that these minerals have internal negative charges which attract cations to internal surfaces between crystal units (see p. 82). Most cations which satisfy these charges can move freely into and out of the crystal lattice. In other words, they are exchangeable. Ammonium and potassium ions, however, are apparently just the right size to fit into the "cavities" between crystal units, thereby becoming fixed more or less as a rigid part of the crystal (see Fig. 4:6). They prevent the normal expansion of the crystal lattice and in turn are held in a nonexchangeable form, from which they are only slowly released to higher plants and microorganisms. The relationship of the various forms of ammonium might be represented as follows:

$$ NH_4^+ \quad \rightleftharpoons \quad NH_4^+ \quad \rightleftharpoons \quad NH_4^+ $$

$$ \text{(Soil solution)} \qquad \text{(Exchangeable)} \qquad \text{(Fixed)} $$

Ammonium fixation by clay minerals is generally greater in subsoils than in topsoil because of the higher clay content of subsoils. In some cases this fixation may be considered to be an advantage since it is a means of conserving soil nitrogen. In others, the rate of release of the fixed ammonium is too slow to be of much practical value.

FIXATION BY ORGANIC MATTER. Anhydrous ammonia or other fertilizers which contain free ammonia or which form it when added to the soil can react with soil organic matter to form compounds which resist decomposition. In this sense the ammonia can be said to be "fixed" by the organic matter. The exact mechanism by which the fixation occur is not known, although reactions with aromatic compounds and quinones are suspected (8). The reaction takes place most readily in the presence of oxygen and at high pH values.

The practical significance of organic fixation depends upon the circum-stances. In organic soils with a high fixing capacity it could be serious and would dictate the use of fertilizers other than those which supply free ammonia. In normal practice on mineral soils, however, organic fixation should not be too disadvantageous. In the first place, the fertilizer is usually banded, thereby contacting a relatively small portion of the entire soil mass and minimizing opportunities for fixation. Furthermore, the fixed ammonia is subject to subsequent slow release by mineralization.

16:5. Nitrification

Nitrification is a process of enzymic oxidation brought about by certain special purpose bacteria (1). It seems to take place in two coordinated steps, two distinct groups of bacteria being involved. As shown below, the first step is the production of nitrous acid by one group of bacteria, apparently followed immediately by its oxidation to the nitrate form by another (see Fig. 16:3 for this organism). The enzymic changes are represented very simply as follows:

$$2NH_4^+ + 3O_2 \xrightarrow[\text{oxidation}]{\text{enzymic}} 2NO_2^- + 2H_2O + 4H^+ + \text{energy}$$

$$2NO_2^- + O_2 \xrightarrow[\text{oxidation}]{\text{enzymic}} 2NO_3^- + \text{energy}$$

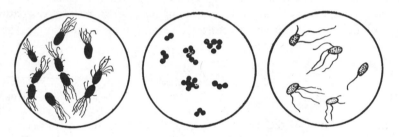

FIGURE 16:3. Some soil organisms especially important in the nitrogen cycle. (*Left to right*) *Azotobacter*, nitrate bacteria, and nodule organisms of alfalfa.

Under most conditions favoring the two reactions, the second transformation is thought to follow the first so closely as to prevent any great accumulation of the nitrite. This is fortunate since this ion in any concentration is toxic to higher plants. In very alkaline soils or where large fertilizer ammonia applications have been made, there is some evidence that the second reaction may be delayed until after the ammonium ion concentration is reduced to a relatively low level. This may result in nitrite accumulation of sufficient magnitude to affect adversely plant growth or to encourage gaseous losses of nitrogen (see p. 431).

ORGANISMS CONCERNED. The organisms in this case are all special purpose and autotrophic (see p. 129), in contrast to the general purpose and heterotrophic organisms concerned with immobilization and mineralization. They obtain most of their energy by the oxidation of inorganic compounds. Like higher plants, they acquire carbon from carbon dioxide while respiring quantities of this gas at the same time.

Collectively, the nitrifying organisms are called nitrobacteria. Of these, the *Nitrosomonas* are involved in the conversion of ammonia into nitrites. The organisms having to do with the oxidation of nitrites to nitrates are designated *Nitrobacter*. In common usage these two groups of bacteria are simply referred to as nitrite and nitrate organisms.

However, other soil organisms are probably able to produce nitrate by-products. A large number of heterotrophic bacteria, fungi, and actino-mycetes have demonstrated abilities to oxidize nitrogen compounds. It is as yet uncertain that these organisms contribute significantly to nitrate formation in soils. It is also possible that photochemical oxidation may take place in tropical soils. How extensive such forms of nitrification are cannot be stated. Apparently, two-stage oxidation—first to nitrite, then to nitrate—is the most important in a practical way.

RATE OF NITRIFICATION. Under ideal temperature, soil, and moisture conditions, nitrification occurs at a very rapid rate. Daily rates of from 6 to 22 pounds of nitrogen per 2 million pounds of soil have been found when 100 pounds of nitrogen in the ammonium form was added (7). Much higher rates occurred when a larger application of ammonium compounds was made. Obviously, under ideal conditions the nitrifying organisms can supply nitrates at a rate which more than meets the needs of crop plants.

16:6. SOIL CONDITIONS AFFECTING NITRIFICATION

The nitrifying bacteria are extremely sensitive to their environment, much more so than most heterotrophic organisms. Consequently, soil conditions that influence the vigor of nitrification deserve practical consideration. They are (a) aeration, (b) temperature, (c) moisture, (d) active lime, (e) fertilizer salts, and (f) the nitrogen–carbon ratio.

AERATION. Since nitrification is a process of oxidation, any procedure that increases the aeration of the soil should, up to a point, encourage it. Plowing and cultivation, especially if granulation is not impaired, are recognized means of promoting nitrification.

TEMPERATURE. The temperature most favorable for the process of nitrifica-tion is from 80 to 90°F. At a temperature of 125°F, nitrification practically ceases. At freezing or below, nitrification will not take place, but at about 30 to 40°F it begins and slowly increases in intensity until the optimum temperature is reached. The lateness with which nitrification attains its full vigor in the spring is well known, application of fertilizer nitrogen being used to offset the delay.

MOISTURE. The rate with which nitrification proceeds in a soil is governed to a marked extent by the water content, the process being retarded by both very low and very high moisture conditions. In practice, it is safe to assume that the optimum moisture as recognized for higher plants is also optimum for nitrification. One reservation must be made, however. Nitrification will progress appreciably at soil moisture contents at or even below the wilting coefficient.

EXCHANGEABLE BASES. Nitrification requires an abundance of exchangeable bases. This accounts in part for the weak nitrification in acid mineral soils and the seeming sensitiveness of the organisms to a low pH. However, within reasonable limits acidity itself seems to have little influence on nitrification when adequate bases are present. This is especially true of peat soils. At pH values even below 5 these soils may show remarkable accumulations of nitrates (see p. 366).

FERTILIZERS. Small amounts of many kinds of salts, even those of the trace elements, stimulate nitrification. A reasonable balance of nitrogen, phosphorus, and potassium has been found to be helpful. Apparently, the stimulation of the organisms is much the same as is that of higher plants.

Applications of large quantities of ammonium nitrogen to strongly alkaline soils have been found to depress the second step in the nitrification reaction. Apparently the ammonia is toxic to the *Nitrobacter* under these conditions but does not affect adversely the *Nitrosomonas*. Consequently, nitrite accumulation may occur in toxic quantities when ammonium-containing compounds are added to soils very high in pH. Similarly, on such soils adverse effects may result from a compound such as urea, which supplies ammonium ions in the soil by hydrolysis.

CARBON–NITROGEN RATIO. The significance of the carbon–nitrogen ratio has been rather fully considered (pp. 151–54), so the explanation here may be brief. When microbes decompose plant and animal residues with high C/N ratios, they incorporate into their bodies all the inorganic nitrogen. It is thereby immobilized. Nitrification is thus more or less at a standstill because of a lack of ammoniacal nitrogen, this also having been swept up by the decay organisms. A serious competition with higher plants for nitrogen is thereby initiated.

After the carbonaceous matter has partially decomposed so that emergizing material is no longer abundant, some of the immobilized nitrogen will be mineralized and ammonium compounds will again appear in the soil. Conditions are now favorable for nitrification and nitrates may again accumulate. Thus, the carbon–nitrogen ratio, through its selective influence on soil microorganisms, exerts a powerful control on nitrification and the presence of nitrate nitrogen in the soil (see Fig. 16:4).

16:7. FATE OF NITRATE NITROGEN

The nitrate nitrogen of the soil, whether added in fertilizers or formed by nitrification, may go in four directions (see Fig. 16:1). It may (a) be used by microorganisms and (b) by higher plants, (c) be lost in drainage, and (d) escape from the nitrogen cycle in a gaseous condition (3).

USE BY SOIL ORGANISMS AND PLANTS. Both plants and soil microorganisms readily assimilate nitrate nitrogen. However, if microbes have a ready food supply (for example, carbonaceous organic residues) they utilize the nitrates more rapidly than do higher plants. Thus, the higher plants must be satisfied with what is left by the microorganisms or must await the subsequent release of the nitrogen when microbial activity slows down (see p. 152). This is one of the reasons crops often are able during a growing season to recover only about one half the fertilizer nitrogen added. Fortunately, much of the immobilized nitrogen is released during the following growing seasons.

LEACHING AND GASEOUS LOSS. The amount of nitrate nitrogen lost in drainage water depends upon the climate and cultural conditions. In arid and semiarid regions, such losses are minimal since water loss by leaching is low or nonexistent. In humid areas and where irrigation is practiced, losses of nitrate nitrogen by leaching are significant. Heavy nitrogen fertilization, especially for vegetables and other cash crops grown on coarse-textured soils, accentuates loss by this means.

Under certain conditions, especially those of poor drainage and aeration, nitrate compounds in the soil may be *reduced*—their nitrogen at least in part escaping in gaseous form. The various phases of reduction will be considered in the next section.

16:8. GASEOUS LOSSES OF SOIL NITROGEN

The conditions under which the nitrates may be changed to gaseous forms are not well understood, but most authorities agree that this loss is greatly encouraged by poor drainage and lack of aeration. The maintenance of the soil in a bare condition and the presence of excessive amounts of mineral nitrogen compounds may also be important factors. The meager data available in respect to such volatilization indicate that this form of loss is of considerable magnitude even from well-managed cropped soils.

REDUCTION BY ORGANISMS. The biochemical reduction of nitrate nitrogen to gaseous compounds is called *denitrification* and is thought to be the most widespread type of volatilization. The microorganisms involved are common facultative anaerobic forms. They prefer elemental oxygen but under inadequate aeration can use the combined oxygen in nitrates and some of their reduced products. The exact mechanisms by which the reductions take place are not known. However, the general trend of the reactions may be represented as follows:

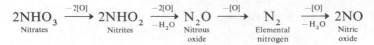

$$2NHO_3 \xrightarrow{-2[O]} 2NHO_2 \xrightarrow[-H_2O]{-2[O]} N_2O \xrightarrow{-[O]} N_2 \xrightarrow[-H_2O]{-[O]} 2NO$$

Nitrates   Nitrites   Nitrous oxide   Elemental nitrogen   Nitric oxide

Under field conditions, nitrous oxide is the gas lost in largest quantities, although elemental nitrogen is also lost under some conditions (see Fig. 16:4). Nitric oxide loss is generally not great and apparently occurs most readily under acid conditions.

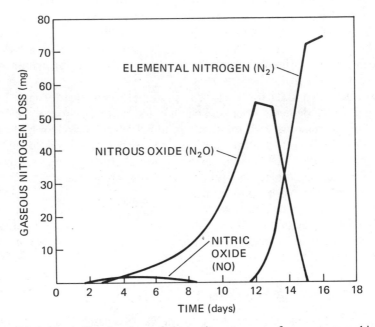

FIGURE 16:4. Denitrification loss of three nitrogen gases from an anaerobic acid Norfolk sandy loam at 12.5 percent moisture. A closed system was used. Apparently N_2 was formed from the reduction of N_2O. Under field conditions much of the N_2O would likely have gone off as a gas and would not have remained to produce elemental N_2. [*From Cady and Bartholomew* (9).]

CHEMICAL REDUCTION. There are other ways by which nitrogen may be lost in the gaseous form. For instance, nitrites in a slightly acid solution will evolve gaseous nitrogen when brought in contact with certain ammonium salts, with simple amines such as urea, and even with nonnitrogenous sulfur compounds and carbohydrates. The following reaction is suggestive of what may happen to urea:

$$2HNO_2 + CO(NH_2) \rightarrow CO_2 + 3H_2O + 2N_2\uparrow$$
$$\underset{\text{Nitrite}}{} \quad \underset{\text{Urea}}{}$$

This type of gaseous loss is strictly chemical and does not require either the presence of microorganisms or adverse soil conditions. Its practical importance, of course, will depend upon the amount of nitrogen that thereby escapes from the soil.

Large, often rapid losses of gaseous nitrogen have been observed upon making heavy applications of urea or ammonium fertilizers. In cases where the fertilizers were applied on the surface, moved to the surface through capillarity, or were inadequately incorporated into the soil, losses as ammonia have occurred, especially in alkaline soils. Even where proper soil coverage and contact was obtained losses as elemental nitrogen and nitrous oxide have been found. Apparently a high concentration of ammonia is toxic to the second step of the nitrification process, resulting in an unusual buildup of nitrites. Under acid conditions these nitrites are converted to gaseous elemental nitrogen or nitrous oxide. The exact mechanism by which these losses take place is not known, but they occur under conditions of good as well as poor drainage and are not always dependent upon microbiological activity.

QUANTITY OF NITROGEN LOST IN GASEOUS FORM. Since gaseous nitrogen is unavailable to higher plants, any loss in this form is serious. As might be expected, the exact magnitude of the losses will depend upon the cultural and soil conditions. In well-drained humid region soils that are not too heavily fertilized, the gaseous losses are probably less than that from leaching. Where the drainage is restricted, and where large applications of ammonia and urea fertilizer are made, especially if they are not well incorporated, substantial losses might be expected—20 to 40 percent of the nitrogen added not being too uncommon.

Lysimeter experiments (2) show a loss of 10 to 20 percent of the nitrogen added, while field experiments (6) in a dryland area showed even greater losses. If one were to take an average of all soil and cultural conditions, perhaps 10 to 15 percent of the annual nitrogen additions might be a reasonable estimate of gaseous losses.

In general, these losses are largely unavoidable. Some control can be exerted by keeping a crop on the land, by providing for adequate drainage and tilth, and by avoiding an excess of fertilizer nitrogen.

The part of the nitrogen cycle yet to be considered is concerned with the acquisition by the soil of nitrogen from various sources. Four ways of addition are recognized in arable soils: (a) nitrogen fixation by legume bacteria; (b) free fixation or azofication; (c) additions in rainwater and snow; and (d) application of nitrogen in fertilizers, farm manure, and green manures. They will be considered in order.

16:9. FIXATION OF ATMOSPHERIC NITROGEN BY LEGUME BACTERIA

It has been recognized for centuries that certain crops, such as the clovers, alfalfa, peas, and beans, improve the soil in some way, making it possible to grow larger yields of cereals after these plants have occupied the land. Within the last century the benefit has been traced to the fixation of nitrogen through the agency of bacteria contained in nodules on the roots of certain host plants. Most of the specific plants so affected belong to the family of legumes, although a significant number of nonlegume species are nodulated (see p. 438).

NODULE FORMATION. The legume organisms live in the root nodules, take free nitrogen from the soil air, and synthesize it into complex forms. The nodules evidently are the result of an irritation of the root surface much as a gall is caused to develop on a leaf or on a branch of a tree by an insect. The entrance of the organisms normally is effected through the hairs. The infection tube ultimately extends the entire length of the root hair and into the cortex cells of the rootlets, where the growth of the nodule starts and where the fixation of nitrogen occurs (see Figs. 16:3 and 16:5).

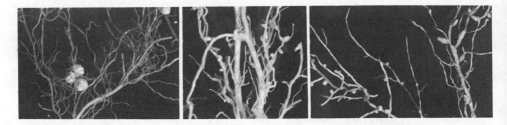

FIGURE 16:5. Nodules on the roots of legumes. (*Left to right*) Red clover, alfalfa, and beans. Although the nitrogen-fixing bacteria are segregated in the nodules, they apparently freely use the juices of the host plant as a source of energy. The nitrogen fixed may go in three directions—(1) appropriated by the host; (2) used by nonlegumes growing in close association; (3) left in the soil when the nodules slough off and decompose.

How the plant absorbs this nitrogen after it has been secured by the bacteria is not well understood, nor is it known in exactly what form the nitrogen is at first fixed, although amino and amide forms very soon appear. It seems likely that some of the nitrogen compounds produced within the bacterial cells are diffused through the cell wall and absorbed by the host plant.

ORGANISMS INVOLVED. Since the plants are able to use some of the nitrogen fixed by the microbes, the relationship thus established is often spoken of as *symbiotic*. The organism is a *Rhizobium* and there are a number of strains, depending on the host plants. They may, therefore, be classified according to the host as *Rhizobium meliloti* for those of alfalfa and sweet clover, *Rhizobium trifolii* for the clovers, *Rhizobium japonicum* for soybeans, and so on. As a group, they are often referred to as the *nodule* or *legume organisms* instead of by the more technical terms.

The organisms from one species of legumes are not necessarily well adapted to the ready production of effective nodules on other leguminous species. Certain cross inoculations are successful in practice, however. For example, the organisms seem to be readily interchangeable within the clovers, the vetches, and the beans. Those from sweet clover and bur clover will inoculate alfalfa, and the bacteria may be transferred from vetch to field pea or from cowpea to velvet bean (13). The part played by the plant is doubtless to furnish the carbohydrates that supply energy to the nitrogen-assimilating bacteria.

A partial listing of the groups within which cross inoculation may easily be made follows:

Group 1. (Alfalfa group) Alfalfa, bur clovers, white, yellow and hubam sweet clovers, fenugreek, yellow trefoil, and others.

Group 2. (Clover group) Mammoth and red clover, alsike clover, crimson clover, hop clover, white clover, zigzag clover, ladino clover, and others.

Group 3. (Cowpea group) Lespedezas, acacia, kudzu, cowpea, peanut, partridge pea, jack bean, velvet bean, lima bean, and others.

Group 4. (Pea and vetch group) Garden pea, sweet pea, horse bean, lentil, Canada field pea, hairy vetch, common vetch, purple vetch, and others.

Group 5. (Soybean group) All varieties of soybeans.

Group 6. (Bean group) Garden bean (numerous varieties), pinto bean, and scarlet runner.

Group 7. (Lupine group) Lupine, serradella, blue lupine, yellow lupine, and white lupine.

EFFECT OF INOCULATION. The influence of inoculation upon the crop yield and percentage of nitrogen in the legume is very striking. Data bearing upon the latter phase are shown in Table 16:1. Analyses are given for both the tops and roots of sweet clover and alfalfa grown on soil inoculated and not inoculated. The soil originally was not known to carry the legume organisms.

NUTRITIONAL REQUIREMENTS. It is difficult to separate requirements of symbiotic nitrogen fixers from those of the host plants. However, a few generalizations can be made.

1. The organisms are sensitive to an excess of H^+ ions, although there is considerable variation among groups of organisms in their tolerance to acidity.
2. Apart from any effect on soil acidity, calcium affects favorably nodule formation and possibly fixation.

TABLE 16:1. *Effect of Inoculation with Legume Organisms on the Nitrogen Content of Sweet Clover and Alfalfa*[a]

Calculated as percentages based on dry matter

	Sweet Clover		Alfalfa	
Soil Treatment	Tops	Roots	Tops	Roots
Inoculated	2.29	2.01	2.56	2.14
Not inoculated	1.37	0.88	1.51	0.71

[a] From Arny and Thatcher (4).

3. The organisms apparently require phosphorus, potassium, and sulfur for normal functioning.
4. The micronutrients molybdenum, boron, cobalt, and perhaps iron seem to play specific roles in the fixation process.
5. Nitrogen from combined sources (for example, nitrates and ammonia) tends to reduce both nodule formation and nitrogen fixation.

16:10. AMOUNT OF NITROGEN FIXED BY LEGUME BACTERIA

The amount of nitrogen fixed by the legume bacteria depends on many factors. The conditions of the soil, especially aeration, drainage, moisture, pH, and the amount of active calcium, are of prime importance. The sensitivity of legumes to soil acidity is largely attributed to the failure of nodule bacteria to function under these conditions. Even when the above factors are favorable, a large amount of readily available nitrogen in the soil will discourage nodulation and thereby reduce fixation (see Fig. 16:6).

Some legume field crops, such as alfalfa and sweet clover, facilitate a fixation of large amounts of nitrogen. The clovers are less effective and some legumes can be credited with but a scanty acquisition. Lyon and Bizzell (17) in a ten-year experiment at Ithaca, New York, report the following magnitude of fixation in pounds per acre per year:

Alfalfa	251	Soybeans	105
Sweet clover	168	Hairy vetch	68
Red clover	151	Field beans	58
Alsike clover	141	Field peas	48

A fixation ranging from 188 to 260 pounds per acre per year was recorded by Collison et al. (11) at Geneva, New York, for alfalfa. Perhaps for an average crop of alfalfa, 200 to 250 pounds of nitrogen per acre would be a conservative figure for the first and possibly for the second year. The corresponding estimate for red clover might well be 100 to 150 pounds per acre.

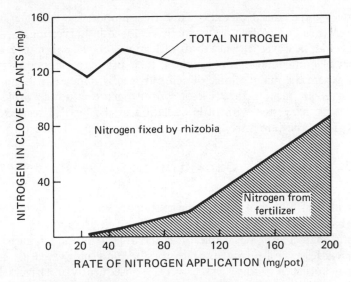

FIGURE 16:6. Influence of added inorganic nitrogen on the total nitrogen in clover plants, the proportion supplied by the fertilizer and that fixed by the rhizobium organisms associated with the clover roots. Increasing the rate of nitrogen application decreased the amount of nitrogen fixed by the organisms in this greenhouse experiment. [*From T. W. Walker et al. (21).*]

16:11. FATE OF THE NITROGEN FIXED BY LEGUME BACTERIA

The nitrogen fixed by the nodule organisms may go in three directions. First, it may be used by the host plant, the latter benefiting greatly by the symbiosis. Second, the nitrogen may pass into the soil itself, either by excretion or more probably by the sloughing off of the roots, and especially of their nodules. The crop in association with the legume may thereby benefit. This influence is a matter of common observation. The vigorous development of a bluegrass lawn in association with clover is a well-known example. Third, when a legume sod is turned under, some of the nitrogen becomes available to the succeeding crop. Because of the narrow carbon–nitrogen ratio of such residues, some of their nitrogen may swing through the nitrogen cycle with remarkable ease and may quickly appear in the ammoniacal and nitrate forms.

16:12. DO LEGUMES ALWAYS INCREASE SOIL NITROGEN?

Some assume that legumes always leave the soil on which they have been growing definitely richer in nitrogen. Actually, the result can be just the opposite, especially when the soil already is high in nitrogen or the tops are harvested and removed from the land. Beans and peas in particular are likely to leave the land depleted because they support low nitrogen fixation. Moreover, their roots often are partially removed in harvesting.

In general, however, the net draft of legumes on the soil nitrogen is less than that of nonlegumes; consequently, their use is to be encouraged in a rotation where the maintenance of soil nitrogen is important. Legumes are usually so economical in their use of soil nitrogen that a high-protein crop may often be harvested with little depletion of the land in respect to this element. Thus, they are nitrogen *savers*. This is an important fertility axiom.

16:13. FIXATION BY ORGANISMS IN SYMBIOSIS WITH NONLEGUMES

A significant number of nonlegume species are known to develop nodules, and the bacteria existing therein are proved nitrogen fixers (20). The species are mostly angiosperms (see Table 16:2) and are often found in nature under conditions of low soil nitrogen. Since these are not cultivated crop plants, they have not been studied extensively. Consequently, relatively little is known species by species about their annual fixation rate, mode of organism infection, and so on. From the meager data available, however, rates of fixation comparable to those of legume bacteria have been observed.

Recently the presence of nodules on the roots of desert plants such as sage, cactus, and rabbitbrush has been noted in the western part of the

TABLE 16:2. *Number and Distribution of Nodulated Nonlegume Species*[a]

In comparison there are nearly 9,000 legume species.

Family	Genus	Species Nodulated	Geographical Distribution
Betulaceae	*Alnus*	15	Cool regions of the Northern Hemisphere
Elaeagnaceae	*Elaeagnus*	9	Asia, Europe, North America
	Hippophae	1	Asia and Europe, from Himalayas to Arctic Circle
	Shepherdia	2	Confined to North America
Myricaceae	*Myrica*	7	Temperate regions of both hemispheres
Coriariaceae	*Coriaria*	3	Widely separated regions, chiefly Japan, New Zealand, Central and South America, and the Mediterranean region
Rhamnaceae	*Ceanothus*	7	Confined to North America
Casuarinaceae	*Casuarina*	12	Tropics and subtropics, extending from East Africa to the Indian Archipelago, Pacific Islands, and Australia

[a] From Stevenson (20).

United States. Nitrogen fixation has been demonstrated on at least three species (14), and the relatively high nitrogen content of the foliage of others suggests the possibility that nitrogen fixation may be occurring. These findings likely account, at least in part, for the fairly high animal carrying capacity of some desert areas.

There is some evidence that nitrogen fixation occurs in the rhizosphere of certain plants, including rice grown under flooded conditions. The exuded organic materials near the root surfaces are thought to supply energy for the nitrogen-fixing organisms much as is the case in legume nodules. The practical significance of this mode of nitrogen fixation is yet to be determined.

16:14. NONSYMBIOTIC FIXATION OF ATMOSPHERIC NITROGEN

There exist in soils and water certain free-living microorganisms that are able to fix elemental nitrogen from the soil air into their body tissue. Since these organisms are not directly associated with higher plants as are the legume bacteria, the transformation is often referred to as *nonsymbiotic* or *free fixation*.

SPECIFIC ORGANISMS INVOLVED. Several different groups of bacteria, blue-green algae, and fungi are able to acquire atmospheric nitrogen nonsymbiotically (15). In upland mineral soils the major fixation apparently is brought about by two groups of heterotrophic bacteria. One of these is the aerobic *Azotobacter* and related bacteria such as the *Beijerinckia* (common in tropical soils) (see Fig. 16:3). The other is an anaerobic, or perhaps a faculative, bacterium called *Clostridium butyricum*. Because of pockets of low oxygen supply in most soils even when they are in the best of tilth, these two groups of bacteria probably work side by side in the fixation of the nitrogen of the soil air.

The *Azotobacter* and *Clostridium butyricum* do not acquire all of their nitrogen from atmospheric sources. Ammoniacal and nitrate nitrogen are readily used by these organisms. In fact, in a soil high in such available nitrogen, it is doubtful whether a great deal of free fixation takes place.

Where soils are used for lowland rice culture, nitrogen fixation by *blue-green algae* may occur. These organisms carry on photosynthesis and thus require no outside source of organic matter. Their ability to fix nitrogen seems to be enhanced by the presence of growing rice plants.

Blue-green algae are also known to grow in association with lichens. In desert areas and in the initial stages of rock disintegration, these algae fix nitrogen, which ultimately can become available to higher plants.

FACTORS INFLUENCING NONSYMBIOTIC NITROGEN FIXATION. The heterotrophic fixers are encouraged by low available soil nitrogen and organic matter which supplies ready energy. Sod land presents almost ideal conditions.

Azotobacter are notably sensitive to soil pH, being most active at a neutral reaction. In mineral soils the free fixation of nitrogen often begins to lag noticeably at pH 5.6 and at pH values below 5.0 becomes more or less negligible.

Clostridium are more tolerant of acid conditions, although they too perform best at a near-neutral reaction. *Beijerinckia*, common in tropical soils, seem to be tolerant of a wide range of pH levels.

The blue-green algae require light and a high moisture level or even water logging. They do best at a neutral to slightly alkaline reaction and fix nitrogen only in the absence of nitrate or ammonia nitrogen.

16:15. Amount of Nitrogen Fixed by Nonsymbiosis

It is difficult to determine accurately the exact amount of nitrogen fixed by nonsymbiosis in soils because of the other processes involving nitrogen that are taking place simultaneously. However, estimates indicate that considerable quantities of nitrogen are fixed annually in this manner. For example, experiments in several areas indicate that 20 to 100 pounds of nitrogen per acre per year may be fixed by nonsymbiotic organisms (see Table 16:3). In these experiments carbonaceous plant residues were generally returned to the soil. Not unexpectedly, fixation would have been at a maximum under these conditions.

Research with isotopic forms of nitrogen has shown wide variations in the amount of nitrogen fixed by free-living organisms (12). The range in values found is from a few pounds to perhaps 60 pounds per acre per year. In most cases, evidence indicates that no more than about 25 pounds of nitrogen would be added in this manner and that the amount may be much less.

TABLE 16:3. *Nitrogen Gains Attributed to Nonsymbiotic Fixation (Field Experiments)*[a]

Location	Period (years)	Description	Nitrogen Gain (lb/acre/yr)
Utah	11	Irrigated soil and manure	44
Missouri	8	Bluegrass sod	102
California	10	Lysimeter experiment	48
California	60	Pinus Ponderosa stand	56
United Kingdom	20	Monoculture tree stands	52
Australia	3	Solonized soil	22
Nigeria	3	Latosolic soil	80
Michigan	7	Straw mulch	50

[a] From Moore (19).

16:16. ADDITION OF NITROGEN TO SOIL IN PRECIPITATION

Nitrogen occurring in rain and snow generally is in the nitrate and ammoniacal forms and consequently is readily available to plants. The amounts thus brought down are variable, usually fluctuating markedly with season and location. It is well known that the additions are greater in the tropics than in humid temperate regions and larger in the latter than under semiarid climates. In Table 16:4 will be found some of the more important data regarding the amounts of nitrogen thus added to the soil in various parts of the world. Also, the figures are for the most part from temperate regions.

Apparently the ammoniacal nitrogen added to the soil in precipitation is, at least in temperate regions, always larger in amount than that in the nitrate form. The nitrate nitrogen is about the same for most locations, but the ammonium form shows wide variations. The figures in Table 16:4 suggest that in a humid temperate climate, an average of about $4\frac{1}{2}$ pounds of ammoniacal and $1\frac{1}{2}$ pounds of nitrate nitrogen fall on every acre of land yearly in rain and snow. Allowing for some loss in runoff, perhaps 5 pounds of nitrogen an acre actually enters the soil each year. This annual acquisition of nitrogen in a readily available form to each acre of land affords some aid in the maintenance of soil fertility.

TABLE 16:4. *Amounts of Nitrogen Brought Down in Precipitation*[a]

Location	Years of Record	Rainfall (in.)	Pounds/Acre/Year	
			Ammoniacal Nitrogen	Nitrate Nitrogen
Harpenden, England	28	28.8	2.64	1.33
Garford, England	3	26.9	6.43	1.93
Flahult, Sweden	1	32.5	3.32	1.30
Gröningen, Holland	—	27.6	4.54	1.46
Bloemfontein and Durban, South Africa	2	—	4.02	1.39
Ottawa, Canada	10	23.4	4.42	2.16
Ithaca, N.Y.	11	29.5	3.65	0.69

[a] From Lyon et al. (18).

16:17. REACTIONS OF NITROGEN FERTILIZERS

Nitrogen applied in fertilizers undergoes the same kinds of reactions as does nitrogen released by biochemical processes from plant residues. Most of the fertilizer nitrogen will be present in one or more of three forms: (a) nitrate, (b) ammonia, and (c) urea. The fate of each of these forms has already been discussed briefly. Thus, urea nitrogen is subject to ammonification, nitrification, and utilization by microbes and higher plants. Ammonium fertilizers can be oxidized to nitrates, fixed by the soil solids, or they can be utilized without change by higher plants or microorganisms. And nitrate salts can be lost by volatilization or leaching or they can be absorbed by plants or microorganisms.

HIGH CONCENTRATIONS. One important fact should be remembered when dealing with the reaction of fertilizer nitrogen. The added fertilizer salts will almost invariably be localized and in higher concentrations than expected in unfertilized soils. For this reason, the usual reactions are sometimes subject to modification.

When anhydrous ammonia, ammonium-containing salts, or even urea are added to highly alkaline soils, there is a possibility of some nitrogen loss in the form of free ammonia. Also, under these conditions, the nitrification process is inhibited, only the first step proceeding normally. Nitrites may thus accumulate in this situation until essentially all the ammonium form has been oxidized. Only then will the second step, that of nitrate formation, take place at normal speed.

The addition of large amounts of nitrate-containing fertilizers may affect the processes of free fixation and gaseous nitrogen loss. In general, fixation by free living organisms is depressed by adequate mineral nitrogen. Gaseous losses, on the other hand, are often encouraged by abundant nitrates. Heavy nitrate fertilization would thus tend to increase losses of nitrogen from the soil.

In most soil situations the effects of higher localized concentrations of fertilizer materials on nitrogen transformations are not serious. Except for a few such situations, it can be assumed that fertilizer nitrogen will be changed in soils in a manner very similar to that of nitrogen released by biological transformations.

SOIL ACIDITY. Ammonium-containing fertilizers and those which form ammonia upon reacting in the soil have a tendency to increase soil acidity (see p. 520 for a more thorough discussion of this). The process of nitrification (see the equations on p. 428) releases hydrogen ions which become adsorbed on the soil colloids. For best crop growth, continued and substantial use of acid-forming fertilizers in humid regions must be accompanied by applications of lime.

The nitrate component of fertilizers does not increase soil acidity. In fact, nitrate fertilizers containing cations in the molecule (for example, sodium nitrate) have a slight alkalizing effect.

16:18. PRACTICAL MANAGEMENT OF SOIL NITROGEN

The problem of nitrogen control is twofold: (a) the maintenance of an adequate supply in the soil, and (b) the regulation of the turnover to assure a ready availability to meet crop demands.

NITROGEN BALANCE SHEET. Major gains and losses of available soil nitrogen are diagrammed in Fig. 16:7. While the relative additions and losses by various mechanisms will vary greatly from soil to soil, the principles illustrated in the diagram are valid.

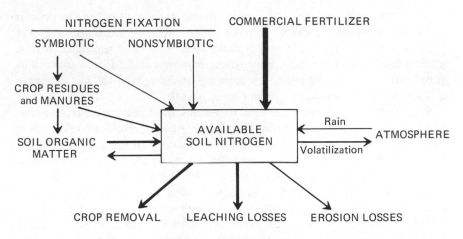

FIGURE 16:7. Major gains and losses of available soil nitrogen. The widths of the arrows indicate roughly the magnitude of the losses and the addition often encountered. It should be emphasized that the diagram represents average conditions only and that much variability is to be expected in the actual and relative quantities of nitrogen involved.

The major loss of nitrogen from most soils is that removed in crop plants. A good crop of wheat or cotton may remove only 125 pounds per acre, nearly half of which may be returned to the soil in the stalks or straw. A bumper silage corn crop, in contrast, may contain over 200 pounds per acre and a good yield of alfalfa or of well-fertilized grass hay more than 300 pounds. It is obvious that modern yield levels require nutrient inputs far in excess of those of a generation ago.

Erosion, leaching, and volatilization losses are determined to no small degree by water management practices. Their magnitudes are so dependent upon specific situations that generalizations are difficult. However, soil and crop management practices which give optimum crop yields will likely hold these sources of nitrogen loss to a satisfactory minimum.

MEETING THE DEFICIT. In practice, nitrogen deficits are met in four ways—crop residues, farm manure, legumes, and commercial fertilizers. On dairy farms and beef ranches, much of the deficit will be met by the first three methods, fertilizers being used as a supplementary source. On most other types of farms, however, fertilizers will play a major role. Where vegetables and other cash crops with high nutrient requirements are grown, the nitrogen deficit will be met almost entirely with commercial fertilizers. Even with the general field crops, modern yield levels can be maintained only through the extensive use of fertilizers.

TURNOVER REGULATIONS. By all odds, the more difficult of the two general problems of nitrogen control is the regulation of this element after it enters the soil. Availability at the proper time and in suitable amounts, with a minimum of loss, is the ideal. Even where commercial fertilizers are used to supply much of the nitrogen, maintaining an adequate but not excessive quantity of available nitrogen is not an easy task.

Soils under any given climate tend to assume what may be called a *normal* or *equilibrium content* of nitrogen. Thus, under ordinary methods of cropping and manuring, any attempt to raise the nitrogen content to a point materially higher than this normal will be attended by an unnecessary waste due to drainage and other losses. At the same time, the nitrogen should be kept suitably active by the use of legumes and other organic materials with a narrow C/N ratio and by the applications of lime and commercial fertilizers.

This is essentially the recommendation already made for soil organic matter (p. 163), and it is known to be both economical and effective. In short, the practical problem is to supply adequate nitrogen to the soil, to keep it mobile, and to protect it from excessive losses caused by leaching, volatilization, and erosion.

16:19. IMPORTANCE OF SULFUR

Sulfur has long been recognized as essential for plant and animal growth. Although much is yet to be learned about the functions of this element, it is already known to be indispensable for many reactions in every living cell. Sulfur is a constituent of the amino acids methionine and cystine, deficiencies of which result in serious human malnutrition. The vitamins biotin and thiamine contain sulfur, and the structure of proteins is determined to a considerable extent by sulfur groups. The properties of certain protein enzymes are thought to be attributable to the type of sulfur linkages present. As with the other essential elements, sulfur plays a unique role in plant and animal metabolism.

Plants that are sulfur deficient are characteristically small and spindly. The younger leaves are light green to yellowish and in the case of legumes, nodulation of the roots is reduced. The maturity of fruits and seeds is delayed in the absence of adequate sulfur.

DEFICIENCIES OF SULFUR. It is only in recent years that deficiencies of sulfur have become common. Since it was first manufactured in 1840, sulfur-bearing superphosphate has helped supply the needs for this element. Likewise, ammonium sulfate, long a significant constituent of fertilizers, has been an important sulfur source. Atmospheric sulfur dioxide, a by-product of the combustion of sulfur-rich coals and residual fuel oils, has supplied large quantities of this element to both plants and soils. Thus, by seemingly incidental means the sulfur needs of crops in the past have been largely satisfied, especially in areas near industrial centers.

In recent years, the trend to the use of high-analysis fertilizers has forced manufacturers to use alternatives to superphosphate and ammonium sulfate (see Table 16:5). As a result, many sulfur-free fertilizers are on the market and the average sulfur content of fertilizers has decreased. Even sulfur-containing pesticides so commonly used a few years ago have been replaced largely by organic materials free of sulfur.

TABLE 16:5. *Changes in the Production and Use of Two Important Sulfur-Containing Fertilizers*[a]

Item	1953–1954	1962–1963
Percent of the world's phosphorus fertilizer production from superphosphate	66	47
Percent of U. S. nitrogen fertilizer consumption from ammonium sulfate	17	8

[a] From Coleman (10).

The replacement of wood and coal for domestic heating by natural gas, electricity, and low-sulfur fuel oil has affected the amount and distribution of sulfur dioxide in the atmosphere. Intensified efforts in the United States to reduce air pollution in and around cities and industrial areas will likely further reduce the quantity of sulfur in the atmosphere. The recognition that clean air is a primary goal will necessitate finding alternative means of supplying sulfur for plant growth.

Coupled with these reductions in the supply of sulfur to soils and plants is the greater removal of this element in harvested crops. Yields have increased markedly during the past 10–15 years, and much of the sulfur removed in crops have not been returned. The quantity of sulfur thus removed is about the same as that of phosphorus. It is not surprising, there-fore, that increased attention is being given to sulfur.

AREAS OF DEFICIENCY. In the United States, deficiencies of sulfur are most common in the Southeast, the Northwest, California, and the Great Plains. In the Northeast and in other areas with heavy industry and large cities, sulfur deficiencies do not seem to be widespread (see Fig. 16:8).

Crops vary in their sulfur requirements. Legume crops such as alfalfa, the clovers, and soybeans have high sulfur requirements, as do cotton, sorghum, sugar beets, cabbage, turnips, and onions. Grasses and cereals generally have lower sulfur requirements, although wheat in the Northwest is often quite responsive to sulfur applications.

16:20. NATURAL SOURCES OF SULFUR

There are three major natural sources from which plants can be supplied with available sulfur: (a) soil minerals, (b) sulfur gases in the atmosphere, and (c) organically bound sulfur. These will be considered in order.

SOIL MINERALS. There are several soil minerals in which sulfur is combined and from which it may be released for growing plants. For example, sulfides of iron, nickel, and copper are found in many soils, especially those with restricted drainage. They are quite abundant in soils of tidal marsh areas. Upon oxidation, the sulfides are changed to sulfates, which are quickly available for plant use.

In regions of low rainfall, sulfate minerals are common in soils. Accumulations of gypsum in the lower horizons of Mollisols and Aridisols (see p. 327) are examples. Accumulations of soluble salts, including sulfates in the surface layers, are characteristic of saline soils (see pp. 396–97) of arid and semiarid regions. When the soils are dry, the salt accumulations are visible.

The accumulation of sulfates in subsoils is not limited to areas of low rainfall. For example, in the Southeast, higher sulfate contents are often found in the subsoils as compared to topsoils. Apparently, the sulfates are absorbed by the clays which are usually present in larger quantities in the subsoil.

ATMOSPHERIC SULFUR. The combustion of fuels, especially coal, releases sulfur dioxide and other sulfur compounds into the atmosphere. In fact, the content of sulfur in the air is generally directly related to the distance from industrial centers (see Fig. 16:9). Except near seashores, where salt spray adds sulfur in significant quantities, and near marshes, which are sources of hydrogen sulfide, the combustion of coal is the most significant source of atmospheric sulfur.

Atmospheric sulfur becomes part of the soil–plant system in three ways. Some of it is absorbed directly from the atmosphere by growing plants. A considerable quantity is absorbed directly by the soil from the atmosphere, and a similar amount is added with precipitation.

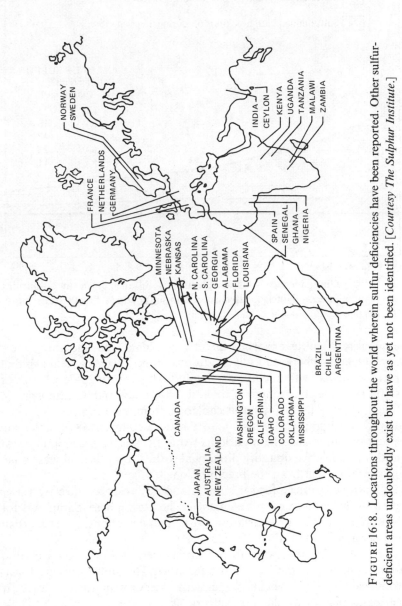

FIGURE 16:8. Locations throughout the world wherein sulfur deficiencies have been reported. Other sulfur-deficient areas undoubtedly exist but have as yet not been identified. [*Courtesy The Sulphur Institute.*]

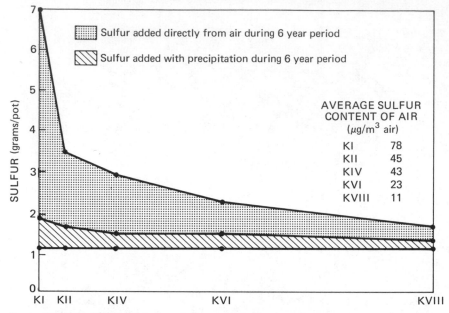

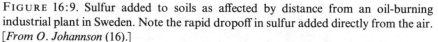

FIGURE 16:9. Sulfur added to soils as affected by distance from an oil-burning industrial plant in Sweden. Note the rapid dropoff in sulfur added directly from the air. [*From O. Johannson* (16).]

The quantity of sulfur absorbed directly by plants from the atmosphere will vary with atmospheric and soil conditions. Experiments have shown that even plants supplied with adequate soil sulfate can absorb 25–35 percent of their sulfur from the atmosphere. If the soil sulfur is low and the atmospheric sulfur high, most of the plant sulfur can come from the atmosphere.

Data presented graphically in Fig. 16:9 show a comparison of direct soil absorption of sulfur and the addition of this element in precipitation. At least in this instance the quantity directly absorbed from the soil was somewhat greater than that from the precipitation.

The quantity of sulfur added in precipitation or absorbed directly varies according to the content of this element in the atmosphere. Samplings for sulfur in precipitation show variations in annual accretions from less than 1 pound per acre to more than 100 (22).

It is not surprising that soils absorb atmospheric sulfur directly or that this element moves in through rain and snow. The sulfur dioxide forms sulfurous acid (H_2SO_3) when in contact with water or with water vapor. Any sulfur trioxide present would form sulfuric acid by the same process. These strong acids are readily absorbed by soils.

ORGANIC BOUND SULFUR. In most humid region surface soils, a major portion of the sulfur is in the organic form. Just as is the case with nitrogen, however, all too little is known of the specific organic sulfur compounds present. The sulfur added in plant residues is mostly in the form of proteins, which are normally subject to rather ready microbial attack. In some manner, the sulfur (along with the nitrogen) is stabilized during humus formation. In this stable form or forms, it is protected from rapid release and thus cannot be lost or taken up by higher plants. The similarity between the general behavior of organic nitrogen and sulfur fractions is unique.

Although the number of specific sulfur compounds associated with soil organic matter is not known, the presence of some forms is strongly suspected. For example, amino acids and other compounds with direct C—S linkages are thought to be present, as are organic sulfates.

FORMS OF SULFUR. This discussion has identified three major forms of sulfur in soils and fertilizers: *sulfides, sulfates,* and *organic forms.* To these must be added the fourth important form, *elemental sulfur,* the starting point of most of the manmade chemical sulfur compounds. The following sections will show relationships among these forms.

16:21. THE SULFUR CYCLE

The major transformations that sulfur undergoes in soils are shown in Fig. 16:10. The inner circle shows the relationships among the four major forms of this element in soils and in fertilizers. The outer portions show the most important sources of sulfur and how this element is lost from the system.

Some similarity between the sulfur and nitrogen cycles is evident (see pp. 423–24). In each case, the atmosphere is an important source of the element in question. Each is held largely in the organic fraction of the soil and each is dependent to a considerable extent upon microbial action for its various transformations.

Figure 16:10 should be referred to frequently as a more detailed examination is made of sulfur in plants and soils, beginning with the behavior of this element in soils.

16:22. BEHAVIOR OF SULFUR COMPOUNDS IN SOILS

MINERALIZATION AND IMMOBILIZATION. It is not surprising that sulfur behaves much like nitrogen as it is absorbed by plants and microorganisms and moves through the sulfur cycle. The organic forms of sulfur must be mineralized by soil organisms if the sulfur is to be used by plants. The rate at which this occurs is dependent upon the same environmental factors as affect nitrogen mineralization, including moisture, aeration, temperature, and pH. When conditions are proper for general microbial activity, sulfur

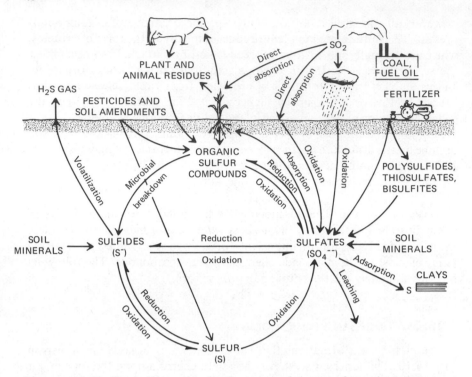

FIGURE 16:10. The sulfur cycle, showing some of the transformations that occur as this element changes form in soils, plants, and animals. It is well to keep in mind that, except for certain soils in arid areas, the great bulk of the sulfur is in the form of organic compounds.

mineralization occurs. The mineralization reaction might be expressed as follows:

$$\text{Organic sulfur} \rightarrow \text{Decay products} \rightarrow \text{Sulfates}$$

Proteins and other organic combinations H_2S and other sulfides are simple examples

Immobilization of inorganic forms of sulfur occurs when low-sulfur, energy-rich organic materials are added to soils not plentifully supplied with inorganic sulfur. The mechanism is thought to be the same as for nitrogen. The energy-rich material stimulates microbial growth and the inorganic sulfur is synthesized into microbial tissue. Only when the microbial activity subsides does the inorganic sulfate form again appear in the soil solution.

These facts suggest that, like nitrogen, sulfur in soil organic matter may be associated with organic carbon in a reasonably constant ratio. The ratio among carbon, nitrogen, and sulfur for a number of soils on three different continents is given in Table 16:6. There appears to be some definite relationship among the contents of these elements, a relationship which is in accord with their behavior in soils.

TABLE 16:6. *Mean Ratios of Carbon to Nitrogen to Sulfur in a Variety of Soils Throughout the World*[a]

Location	Description and Number of Soils	Ratio C/N/S
North Scotland	Agricultural, noncalcareous (40)	147:10:1.4
Minnesota	Chernozems (6)	114:10:1.55
Minnesota	Podzolic soils (24)	132:10:1.22
Oregon	Agricultural, varied (16)	145:10:1.01
Eastern Australia	Acid soils (128)	152:10:1.21
Eastern Australia	Alkaline soils (27)	140:10:1.52

[a] From Whitehead (22).

SULFUR OXIDATION AND REDUCTION. During the microbial decomposition of organic sulfur compounds, sulfides are formed along with other incompletely oxidized substances such as elemental sulfur, thiosulfates, and polythionates. These reduced substances are subject to oxidation, just as are the ammonium compounds formed when nitrogenous materials are decomposed. The oxidation reactions may be illustrated as follows, using hydrogen sulfide and elemental sulfur as examples of the sulfur substances which are oxidized:

$$H_2S + 2O_2 \rightarrow H_2SO_4$$
$$2S + 3O_2 + 2H_2O \rightarrow 2H_2SO_4$$

The oxidation of some sulfur compounds, such as sulfites (SO_3^{--}) and sulfides (S^{--}), can occur by strict chemical reactions. However, most of the sulfur oxidation occurring in soils is thought to be *biochemical* in nature. It is accomplished by a number of autotrophic bacteria of the genus *Thiobacillus*, five species of which have been characterized. Since the environmental requirement and tolerances of these five species vary considerably, the process of sulfur oxidation occurs over a wide range of soil conditions. For example, it occurs at pH values ranging from less than 2 to higher than 9. This is in contrast to the comparable nitrogen oxidation process, nitrification, where a rather narrow pH range near neutral is required.

Like nitrates, sulfates tend to be unstable in anaerobic environments. They are reduced to sulfides by a number of bacteria of two genera, *Desulfovibro* (five species) and *Desulfotomaculum* (three species). The organisms use the combined oxygen in sulfate to oxidize organic materials. A representative reaction follows:

$$2(R—CH_2OH) + SO_4^{--} \rightarrow 2(R—COOH) + 2H_2O + S^{--}$$

Organic alcohol   Sulfate   Organic acid   Sulfide

In soils, the sulfide ion would likely react immediately with iron, which in anaerobic conditions would be present in the ferrous form. This reaction might be expressed as follows:

$$Fe^{++} + S^{--} \rightarrow FeS$$

Iron
sulfide

Sulfur reduction takes place with compounds other than sulfates. For example, sulfites (SO_3^{--}), thiosulfates ($S_2O_3^{--}$), and elemental sulfur (S) are rather easily reduced to the sulfide form by bacteria and other organisms.

The oxidation and reduction of inorganic sulfur compounds is of great importance to growing plants. In the first place, these reactions determine to a considerable extent the quantity of sulfate present in soils at any one time. Since this is the form taken up by plants, the nutrient significance of sulfur oxidation and reduction is obvious. Second, the state of sulfur oxidation determines to a marked degree the acidity of a soil, a fact which will be considered further in the next section.

SULFUR OXIDATION AND ACIDITY. Sulfur oxidation is an acidifying process. The reactions on page 451 illustrate this point. For every sulfur atom oxidized, two hydrogen ions result. This effect is utilized by adding sulfur to reduce the extreme alkalinity of certain alkali soils of arid regions and to reduce the pH of certain soils for the control of diseases such as potato scab (see p. 396).

The acidifying effect of sulfur oxidation can bring about extremely acid soil conditions. For example, this has been known to occur when land is drained after it has been submerged for some time under brackish water or sea water. During the submerged period, sulfates in the water are reduced to sulfides, in which form they are stabilized generally as iron sulfides. Since there is a continuous supply of sulfates under these conditions, high levels of sulfides are built up. If there are periods of partial drying, elemental sulfur can form by partial oxidation of the sulfides. The sulfide and elemental contents are hundreds of times higher than would be found in comparable upland soils.

When these areas are drained, the sulfides and/or elemental sulfur are quickly oxidized, forming sulfuric acid. The soil pH may drop to levels as low as 1 or 2—levels unknown in normal upland soils. Obviously plant growth cannot occur under these conditions. Furthermore, the quantity of limestone needed to neutralize the acidity is so high as to make this remedy completely uneconomical.

Sizable areas of these kinds of soils, called *cat-clays*, are found in southeast Asia. They also occur in the tideland areas along the coasts of several other areas, including the southeastern part and the West Coast of the United States. So long as these soils are kept submerged, the soil reaction does not drop prohibitively. Consequently, production of paddy rice is sometimes possible under these conditions.

SULFATE RETENTION. Most soils will retain sulfate, although the quantity held is generally small and its strength of retention is low compared to that of phosphate. The retentive capacity is generally higher in the subsoil than in the surface layers. This is because of the high sulfate-retentive capacity of certain compounds which tend to accumulate in the lower horizons. These include hydrous oxides of iron and aluminum and silicate clays, especially those high in kaolinite. Soils of the southeast are commonly high in these sulfate-retaining substances. Consequently, sulfate retention in soils of this region is likewise high.

The mechanism of absorbing sulfate is thought to involve hydroxyl groups in the hydrous oxide and silicate clays. Hydroxyl groups held by aluminum ions in the oxides or silicates are replaced by sulfate or acid sulfate ions. A generalized equation will illustrate how this may occur:

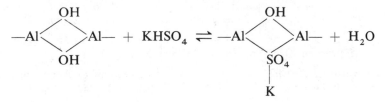

The addition of hydroxyl ions (increasing pH) would tend to drive the above reaction to the left, releasing the absorbed sulfate. For this reason, sulfate retention is generally lowered by the liming of acid soils.

16:23. SULFUR AND SOIL FERTILITY MAINTENANCE

The problem of maintaining adequate quantities of sulfur for minerals is decidedly less than that of phosphorus. This element is added to soils through adsorption from the atmosphere and as an incidental component of many fertilizers. All in all, chances for widespread sulfur deficiencies are generally less than for the three "fertilizer" elements.

Even though areas of sulfur deficiency are not widespread, situations where crops respond to sulfur applications are becoming more and more common. Less "incidental" fertilizer sulfur is being added in some areas, and additions of this element from the atmosphere will continue to decrease as air-pollution-abatement efforts are accelerated. Furthermore, steadily increasing crop yields are removing proportionately larger quantities of this and other elements. Although crop residues and farmyard manures can help replenish the sulfur removed, greater and greater dependence must be placed on fertilizer additions. Regular sulfur applications are necessary now for good crop yields in large areas of the southeast and northwest and in parts of the Great Plains far removed from industrial plants. There will certainly be an increased necessity for and use of sulfur in the future.

REFERENCES

(1) Alexander, M., "Nitrification," in *Soil Nitrogen* (*Agronomy*, **10**) (Madison, Wisc.: American Society of Agronomy, 1965).

(2) Allison, F. E., "The Enigma of Soil Nitrogen Balance Sheets," *Adv. in Agron., 7*:213–50, 1955.

(3) Allison, F. E., "The Rate of Nitrogen Applied to Soils," *Adv. in Agron., 18*:219–58, 1966.

(4) Arny, A. C., and Thatcher, R. W., "The Effect of Different Methods of Inoculation on the Yield and Protein Content of Alfalfa and Sweet Clover," *Jour. Amer. Soc. Agron., 9*:127–37, 1917.

(5) Bartholomew, W. V., and Clark, F. E., Eds., *Soil Nitrogen* (*Agronomy*, **10**) (Madison, Wisc.: American Society of Agronomy, 1965).

(6) Bracken, A. F., and Greaves, J. E., "Losses of Nitrogen and Organic Matter from Dry-Land Soils," *Soil Sci., 51*:1–15, 1941.

(7) Broadbent, F. E., and Tyler, K. B., "Nitrification of Ammoniacal Fertilizers in Some California Soils," *Hilgardia, 27*:247–67, 1957.

(8) Broadbent, F. E., et al., "Factor Influencing the Reaction Between Ammonia and Soil Organic Matter," *Trans. 7th Int. Cong. Soil Sci., 2*:509–16, 1961.

(9) Cady, F. B., and Bartholomew, W. V., "Sequential Products of Anaerobic Denitrification in Norfolk Soil Material," *Soil Sci. Soc. Amer. Proc., 24*:477–82, 1960.

(10) Coleman, R., "The Importance of Sulfur as a Plant Nutrient in World Crop Production," *Soil Sci., 101*:230–39, 1966.

(11) Collison, R. C., et al., *Lysimeter Investigations. III Mineral and Water Relations and Final Nitrogen Balance in Legume and Nonlegume Crop Rotations for a Period of 16 Years,* Bull. 212, N.Y. State Agr. Exp. Sta., 1933.

(12) Delwiche, C. C., and Wyler, J., "Nonsymbiotic Nitrogen Fixation in Soils," *Plant and Soil, 7*:113–29, 1956.

(13) Erdman, L. W., *Legume Inoculation: What It Is—What It Does,* Farmers Bull. 2003 (Washington, D.C.: U. S. Department of Agriculture, 1959).

(14) Farnsworth, R. B., "New Species of Nodulated Non-legumes on Range and Forest Soils," Summary in *Agronomy Abstracts* (Madison, Wisc.: American Society of Agronomy, 1972).

(15) Jensen, H. L., "Nonsymbiotic Nitrogen Fixation," in *Soil Nitrogen* (*Agronomy*, **10**) (Madison, Wisc.: American Society of Agronomy, 1965).

(16) Johannson, O., "On Sulfur Problems in Swedish Agriculture," *Kgl. Lanabr. Ann.,* **25**:57–169, 1960.

(17) Lyon, T. L., and Bizzell, J. A., "A Comparison of Several Legumes with Respect to Nitrogen Secretion," *Jour. Amer. Soc. Agron.,* **26**: 651–56, 1934.

(18) Lyon, T. L., Buckman, H. O., and Brady, N. C., *The Nature and Properties of Soils* (New York: Macmillan, Inc., 1952), p. 467.

(19) Moore, A. W., "Non-symbiotic Nitrogen Fixation in Soil and Soil–Plant Systems," *Soils and Fertilizers,* **29**:113–28, 1966.

(20) Stevenson, F. J., "Origin and Distribution of Nitrogen in Soil," in *Soil Nitrogen* (*Agronomy*, **10**) (Madison, Wisc.: American Society of Agronomy, 1965).

(21) Walker, T. W., et al., "Fate of Labeled Nitrate and Ammonium Nitrogen when Applied to Grass and Clover Grown Separately and Together," *Soil Sci.,* **81**:339–52, 1956.

(22) Whitehead, D. C., "Soil and Plant-Nutrition Aspects of the Sulfur Cycle," *Soils and Fertilizers,* **27**:1–8, 1964.

Chapter 17

SUPPLY AND AVAILABILITY OF PHOSPHORUS AND POTASSIUM

CONSIDERABLE nitrogen can be added to soils through biochemical fixation brought about by microorganisms. If the proper legume is chosen, for example, the organisms will often fix this element from the air in quantities sufficient to temporarily increase the nitrogen already present. With other nutrient elements, such as phosphorus and potassium, however, there is no such microbial aid. Consequently, other sources must be depended upon to meet the demands of plants.

There are at least four main sources of phosphorus and potassium from which these demands can be met: (a) commercial fertilizer; (b) animal manures; (c) plant residues, including green manures; and (d) native compounds of these elements, both organic and inorganic, already present in the soil. Since the first three sources are to be considered in later chapters, attention now will be focused on the ways and means of utilizing the body of the soil as source of these mineral elements.

17:1. IMPORTANCE OF PHOSPHORUS

With the possible exception of nitrogen, no other element has been as critical in the growth of plants in the field as has phosphorus. A lack of this element is doubly serious since it may prevent other nutrients from being acquired by plants. For example, prior to the extensive usage of commercial fertilizers, most soil nitrogen was indirectly dependent upon the supply of phosphorus. This was due to the vital influence of phosphorus on legume growth. Today the demand for phosphorus by nitrogen-yielding legumes is universally recognized.

The need of plants for phosphorus has been especially considered in the formulation of commercial fertilizers. This element, in the form of superphosphate, was the first to be supplied as a manufactured product. Until fairly recently the amount of "phosphoric acid" in mixed fertilizers almost invariably exceeded that of nitrogen or potash. Even today the total tonnage of phosphorus, expressed as P_2O_5, is exceeded only by that of nitrogen.

456

17:2. INFLUENCE OF PHOSPHORUS ON PLANTS

It is difficult to state in detail the functions of phosphorus in the economy of even the simplest plants. Only the more important functions will be considered here. Phosphorus makes its contribution through its favorable effect on the following:

1. Cell division and fat and albumin formation.
2. Flowering and fruiting, including seed formation.
3. Crop maturation, thus counteracting the effects of excess nitrogen applications.
4. Root development, particularly of the lateral and fibrous rootlets.
5. Strength of straw in cereal crops, thus helping to prevent lodging.
6. Crop quality, especially of forages and of vegetables.
7. Resistance to certain diseases.

17:3. THE PHOSPHORUS PROBLEM

Although the amount of total phosphorus in an average mineral soil compares favorably with that of nitrogen, it is much lower than potassium, calcium, or magnesium (see Table 2:3). Of even greater importance, however, is the fact that most of the phosphorus present in soils is currently unavailable to plants. Also, when soluble sources of this element are supplied to soils in the form of fertilizers, their phosphorus is often "fixed" or rendered insoluble or unavailable to higher plants, even under the most ideal field conditions (see p. 465).

Fertilizer practices in many areas exemplify the problem of phosphorus availability. As already emphasized, the tonnage of phosphorus-supplying materials used as fertilizers definitely exceeds all except the nitrogen carriers. The removal of phosphorus from soils by crops, however, is low compared to that of nitrogen and potassium, often being only one third or one fourth that of the latter elements. The necessity for high fertilizer dosage when relatively small quantities of phosphorus are being removed from soils indicates that much of the added phosphates becomes unavailable to growing plants.

The influence of this situation on fertilizer practice is clearly shown when considering the additions of fertilizer phosphorus in comparison with crop removal.

In the United States, phosphorus added in fertilizers exceeds that removed by crops by more than 24 percent (see Table 17:1). In some areas, notably the eastern seaboard states, additions of phosphorus more than triple the removal of this element by crops. Since phosphorus is lost only sparingly by leaching, the inefficiency of utilization of phosphate fertilizers is obvious.

Briefly, then, the overall phosphorus problem is threefold: (a) a small total amount present in soils, (b) the unavailability of such native phosphorus, and (c) a marked "fixation" of added soluble phosphates. Since crop removal of phosphorus is relatively low and world phosphate supplies are huge, problem (a), that of supplying sufficient total phosphorus, is not serious. Increasing the availability of native soil phosphorus and the retardation of fixation or reversion of added phosphates are, therefore, the problems of greatest importance. These two phases will be discussed following a brief review of the phosphorus compounds present in soils.

TABLE 17:1. *Nutrients Removed by Crops in the United States Compared to That Added in Fertilizers* (1965)[a]

	N	P	K
Removed in crops (thousands of tons)	8,838	1,207	4,152
Added in fertilizers (thousands of tons)	4,580	1,499	2,313
Addition as percent of removal	52	124	56

[a] Nutrient removal figures calculated from White (25). Fertilizer additions from Tennessee Valley Authority (23).

17:4. Phosphorus Compounds in Soils[1]

Both inorganic and organic forms of phosphorus occur in soils and both are important to plants as sources of this element. There is a serious lack of information, however, on the relative amounts of these two forms in different soils. Data available from Oregon, Iowa, and Arizona (Table 17:2) give some idea of their relative proportions. Despite the variation which occurs, it is evident that a consideration of soil phosphorus would not be complete unless some attention were given to both forms (see Fig. 17:1).

INORGANIC COMPOUNDS. Most inorganic phosphorus compounds in soils fall into one of two groups: (a) those containing *calcium*, and (b) those containing *iron* and *aluminum*. The calcium compounds of most importance are listed in Table 17:3. Fluorapatite, the most insoluble and unavailable of the group, usually is an original mineral. It is found in even the more weathered soils, especially in their lower horizons. This fact is an indication of the extreme insolubility and consequent unavailability of the phosphorus contained therein. The simpler compounds of calcium, such as mono and dicalcium phosphate, are readily available for plant growth. Except on recently fertilized soils, however, these compounds are present in extremely small quantities only since they easily revert to the more insoluble forms.

Much less is known of the exact constitution of the iron and aluminum phosphates contained in soils. The compounds involved are probably hydroxy phosphates such as dufrenite, wavellite, strengite, and variscite (10). These compounds are most stable in acid soils and are extremely insoluble.

[1] For a review of soil phosphorus, see Larsen (17).

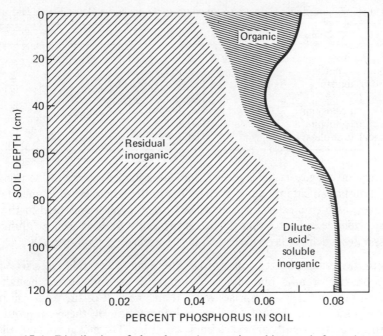

FIGURE 17:1. Distribution of phosphorus in organic and inorganic forms in an Iowa soil. The dilute-acid-soluble inorganic phosphorus is more readily available than the residual inorganic forms. In heavily fertilized soils the upper horizons would likely be much higher in inorganic phosphorus. [*From Black* (6).]

TABLE 17:2. *Total Phosphorus Content of Soils from Three States and the Percentage of Total Phosphorus in the Organic Form*

Soils	Number of Samples	Total P (ppm)	Organic Fraction (%)
Western Oregon soils			
Hill soils	4	357	65.9
Old valley-filling soils	4	1,479	29.4
Recent valley soils	3	848	25.6
Iowa soils			
Prairie soils	2	613	41.6
Gray-brown podzolic soils	2	574	37.3
Planosols	2	495	52.7
Arizona soils			
Surface soils	19	703	36.0
Subsoils	5	125	34.0

[a] Figures for Oregon from Bertramson and Stephenson (5), for Iowa from Pearson and Simonson (19), and for Arizona from Fuller and McGeorge (13).

TABLE 17:3. *Inorganic Calcium Compounds of Phosphorus Often Found in Soils*

Compound	Formula		
Fluor apatite	$3 Ca_3(PO_4)_2 \cdot CaF_2$		
Carbonate apatite	$3 Ca_3(PO_4)_2 \cdot CaCO_3$		
Hydroxy apatite	$3 Ca_3(PO_4)_2 \cdot Ca(OH)_2$	Solubility	increases
Oxy apatite	$3 Ca_3(PO_4)_2 \cdot CaO$		
Tricalcium phosphate	$Ca_3(PO_4)_2$		
Dicalcium phosphate	$CaHPO_4$		
Monocalcium phosphate	$Ca(H_2PO_4)_2$		

Many investigators have shown that phosphates react with certain iron or aluminum silicate minerals such as kaolinite. There is some uncertainty, however, as to the exact form in which this phosphorus is held in the soil. Most evidence indicates that it, too, is probably fixed as iron or aluminum phosphates, such as those described in the preceding paragraph.

ORGANIC PHOSPHORUS COMPOUNDS. There has been relatively less work done on the organic phosphorus compounds in soils, even though this fraction in some cases comprises more than one half of the total soil phosphorus. One of the reasons for lack of information on these compounds is because they apparently are exceedingly complex. The meager data available, however, indicate that the three main groups of organic phosphorus compounds[2] found in plants are also present in soils (7). These are (a) phytin and phytin derivatives, (b) nucleic acids, and (c) phospholipids. There are likely other organic phosphorus compounds present in soils; some investigators doubt that those listed account for all the organic phosphorus.

17:5. FACTORS THAT CONTROL THE AVAILABILITY OF INORGANIC SOIL PHOSPHORUS

The availability of inorganic phosphorus is largely determined by the following factors: (a) soil pH; (b) soluble iron, aluminum, and manganese; (c) presence of iron-, aluminum-, and manganese-containing minerals; (d) available calcium and calcium minerals; (e) amount and decomposition of organic matter; and (f) activities of microorganisms. The first four factors are interrelated because their effects are largely dependent upon soil pH.

17:6. pH AND PHOSPHATE IONS

As indicated in Chapter 14 (pp. 389–90), the availability of phosphorus to plants is determined to no small degree by the ionic form of this element

---

[2] Phytin is a calcium–magnesium salt of inositol phosphoric acid and is rather widely distributed in plants, especially in the seeds. Nucleic acids are found in both plants and animals and are even more complex than is phytin. They are apparently polymeric combinations of phosphoric acid, carbohydrates, and bases such as pyrimidine and purine.

(see Fig. 17:2). The ionic form in turn is determined by the pH of the solution in which the ion is found. Thus, in highly acid solutions only the H_2PO_4 ions are present. If the pH is increased, first the HPO_4 ions and finally PO_4 ions dominate. This situation is shown by the following equations:

$$H_2PO_4^- \underset{}{\overset{OH^-}{\rightleftharpoons}} H_2O + HPO_4^{--} \underset{}{\overset{OH^-}{\rightleftharpoons}} H_2O + PO_4^{3-}$$

(Very acid
solutions)

(Very alkaline
solutions)

At intermediate pH levels two of the phosphate ions may be present simultaneously. Thus, in solutions at pH 6.0, both H_2PO_4 and HPO_4 ions are found.

In general, the H_2PO_4 ion is considered somewhat more available to plants than is the HPO_4 ion. In soils, however, this relationship is complicated by the presence or absence of other compounds or ions. For example, the presence of soluble iron and aluminum under very acid conditions, or calcium at high pH values, will markedly affect the availability of the phosphorus. Clearly, therefore, the effect of soil pH on phosphorus availability is determined in no small degree by the various cations present. The effect of these ions in acid soils will be discussed first.

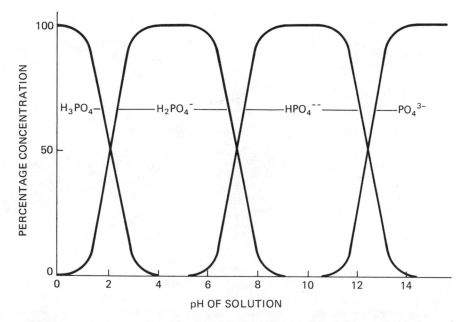

FIGURE 17:2. Relationship between solution pH and the relative concentrations of three soluble forms of phosphate. In the pH range common for soils, the $H_2PO_4^-$ ions predominate.

17:7. Inorganic Phosphorus Availability in Acid Soils

Assume that there is either a nutrient solution or an organic soil very low in inorganic matter (see Fig. 17:3). Assume also that these media are acid in reaction but that they are low in iron, aluminum, and manganese. The H_2PO_4 ions, which would dominate under these conditions, would be readily available for plant growth. Normal phosphate absorption by plants would be expected so long as the pH was not too low.

PRECIPITATION BY IRON, ALUMINUM, AND MANGANESE IONS. If the same degree of acidity should exist in a normal mineral soil, however, quite different results would be expected. Some soluble iron, aluminum, and manganese are usually found in strongly acid mineral soils. Reaction with the H_2PO_4 ions would immediately occur, rendering the phosphorus insoluble and also unavailable for plant growth.

The chemical reactions occurring between the soluble iron and aluminum and the H_2PO_4 ions probably result in the formation of hydroxy phosphates (11). This may be represented as follows, using the aluminum cation as an example:

$$Al^{3+} + H_2PO_4^- + 2H_2O \rightleftharpoons 2H^+ + Al(OH)_2H_2PO_4$$

(Soluble)   (Insoluble)

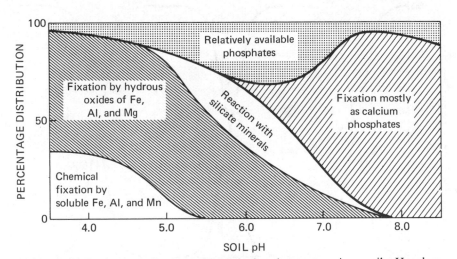

FIGURE 17:3. Inorganic fixation of added phosphates at various soil pH values. Average conditions are postulated and it is not to be inferred that any particular soil would have exactly the same distribution. The actual proportion remaining in an available form will depend upon contact with the soil, time for reaction, and other factors. It should be kept in mind that some of the added phosphorus may be changed to an organic form in which it would be temporarily unavailable.

In most strongly acid soils the concentration of the iron and aluminum ions greatly exceeds that of the H_2PO_4 ions. Consequently, the above reaction moves to the right, forming the insoluble phosphate. This leaves only minute quantities of the H_2PO_4 ion immediately available for plants under these conditions.

An interesting series of reactions occur when fertilizers containing $Ca(H_2PO_4)_2$ are added to soils, even those relatively high in pH (18) (see Fig. 17:4). The $Ca(H_2PO_4)_2$ in the fertilizer granules attracts water from the soil and the following reaction occurs:

$$Ca(H_2PO_4)_2 \cdot H_2O + H_2O \rightarrow CaHPO_4 \cdot 2H_2O + H_3PO_4$$

As more water is attracted, a H_3PO_4-laden solution with a pH of about 1.4 moves outward from the granule. This solution is sufficiently acid to dissolve and displace large quantities of iron, aluminum, and manganese. These ions react with the phosphate to form complex compounds, which later probably revert to the hydroxy phosphates of iron, aluminum, and manganese in acid soils and of calcium in neutral to alkaline soils. In any case, the immediate products of the addition to soils of a water-soluble compound $[Ca(H_2PO_4)_2 \cdot H_2O]$ are a group of insoluble iron, aluminum, manganese, and calcium compounds. Even so, the phosphorus in these compounds is released quite readily for plant growth. It is only after these freshly precipitated compounds are allowed to "age" or to revert to more insoluble forms that availability to plants is greatly reduced.

FIXATION BY HYDROUS OXIDES. It should be emphasized that the H_2PO_4 ion reacts not only with the soluble iron, aluminum, and manganese but also with insoluble hydrous oxides of these elements, such as limonite and goethite. The actual quantity of phosphorus fixed by these minerals in acid soils quite likely exceeds that due to chemical precipitation by the soluble iron, aluminum, and manganese cations (see Fig. 17:3).

The compounds formed as a result of fixation by iron and aluminum oxides are likely to be hydroxy phosphates, just as in the case of chemical precipitation described above (15). Their formation can be illustrated by means of the following equation if the hydrous oxide of aluminum is represented as aluminum hydroxide:

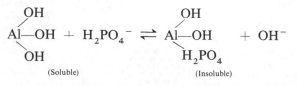

By means of this and similar reactions the formation of several basic phosphate minerals containing either iron or aluminum or both is thought to occur. Since several such compounds are possible, fixation of phosphorus by this mechanism probably takes place over a relatively wide pH range.

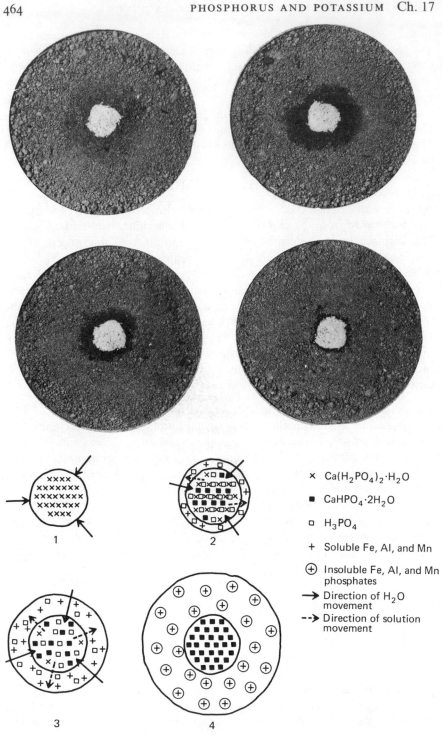

× Ca(H₂PO₄)₂·H₂O

■ CaHPO₄·2H₂O

□ H₃PO₄

+ Soluble Fe, Al, and Mn

⊕ Insoluble Fe, Al, and Mn phosphates

→ Direction of H₂O movement

--→ Direction of solution movement

Also the large quantities of hydrous iron and aluminum oxides present in most soils make possible the fixation of tremendous total amounts of phosphorus by this means.

Thus, as both of the equations above show, the acid condition which would make possible the presence of the readily available H_2PO_4 ion in mineral soils at the same time results in conditions conducive to the vigorous *fixation* or precipitation of the phosphorus by iron, aluminum, and manganese compounds (see Fig. 17:4).

FIXATION BY SILICATE CLAYS. A third means of fixation of phosphorus under moderately acid conditions involves silicate minerals such as kaolinite, montmorillonite, and illite. Although there is some doubt about the actual mechanisms involved, the overall effect is essentially the same as when phosphorus is fixed by simpler iron and aluminum compounds. Some scientists visualize the fixation of phosphates by silicate minerals as a surface reaction between exposed —OH groups on the mineral crystal and the H_2PO_4 ions. Other investigators have evidence that aluminum and iron ions are removed from the edges of the silicate crystals forming hydroxy phosphates of the same general formula as those already discussed. This type of reaction might be expressed as follows:

$$[Al] + H_2PO_4^- + 2H_2O \rightleftharpoons 2H^+ + Al(OH)_2H_2PO_4$$

(In silicate crystal)   (Insoluble)

Thus, even though phosphates react with different ions and compounds in acid soils, apparently the same insoluble iron and aluminum compounds are formed in each case. Major differences from soil to soil are probably due to differences in rate of phosphate precipitation and in the surface area of the phosphates once the reaction has occurred. This will be discussed again later.

ANION EXCHANGE. Part of the phosphate which has reacted with iron and aluminum compounds and with silicate clays is subject to replacement by other anions, such as the hydroxyl ion. Such replacement is called *anion exchange*. It may be illustrated by a reverse of the reaction on page 463.

FIGURE 17:4 [OPPOSITE]. Reaction of $Ca(H_2PO_4)_2 \cdot H_2O$ granules with moist soil. (1) The granule has just been added to the soil and is beginning to absorb water from it. (2) In the moistened granule H_3PO_4 and $CaHPO_4 \cdot 2H_2O$ are being formed and more soil water is being absorbed. (3) The H_3PO_4-laden solution moves into the soil, dissolving and displacing Fe, Al, and Mn and leaving insoluble $CaHPO_4 \cdot 2H_2O$ in the granule. (4) The Fe, Al, and Mn ions have reacted with the phosphate to form insoluble compounds, which, along with the residue of $CaHPO_4 \cdot 2H_2O$, are the primary reaction products. [*Photos courtesy G. I. Terman and National Plant Food Institute, Washington, D.C.*]

Thus,

$$Al(OH)_2H_2PO_4 + OH^- \rightleftharpoons Al(OH)_3 + H_2PO_4^-$$

One anion (OH) has been exchanged for another (H_2PO_4). This reaction shows how anion exchange can take place and illustrates the importance of liming in helping to maintain a higher level of available phosphates.

17:8. Inorganic Phosphorus Availability at High pH Values

In alkaline soils, phosphate precipitation is caused mostly by calcium compounds (see Fig. 17:3). Such soils are plentifully supplied with exchangeable calcium and in most cases with calcium carbonate. Available phosphates will react with both the calcium ion and its carbonate. As an illustration, assume that concentrated superphosphate is added to a calcareous soil. The reactions would be as follows:

$$Ca(H_2PO_4)_2 + 2Ca^{++} \rightleftharpoons Ca_3(PO_4)_2 + 4H^+$$
(Soluble)   (Adsorbed)   (Insoluble)

$$Ca(H_2PO_4)_2 + 2CaCO_3 \rightleftharpoons Ca_3(PO_4)_2 + 2CO_2\uparrow + 2H_2O$$
(Soluble)   (Insoluble)

Although the $Ca_3(PO_4)_2$ thus formed is quite insoluble, it may be converted in the soil to even more insoluble compounds. Hydroxy, oxy, carbonate, and even fluor apatite compounds may be formed if conditions are favorable and if sufficient time is allowed (see Table 17:3).

This type of reversion may occur in soils of the eastern United States which have been heavily limed. It is much more serious, however, in Western soils, owing to the widespread presence of excess $CaCO_3$. The problem of utilizing phosphates in alkaline soils of the arid West is thus fully as serious as it is on highly acid soils in the East.

17:9. pH for Maximum Inorganic Phosphorus Availability

With insolubility of phosphorus occurring at both extremes of the soil pH range (see Fig. 17:3), the question arises as to the range in soil reaction in which minimum fixation occurs. The basic iron and aluminum phosphates have a minimum solubility around pH 3 to 4. At higher pH values some of the phosphorus is released and the fixing capacity somewhat reduced. Even at pH 6.5, however, much of the phosphorus is still probably chemically combined with iron and aluminum. As the pH approaches 6, precipitation as calcium compounds begins; at pH 6.5 the formation of insoluble calcium salts is a factor in rendering the phosphorus unavailable. Above pH 7.0, even more insoluble compounds, such as apatites, are formed.

These facts seem to indicate that maximum phosphate availability to plants is obtained when the soil pH is maintained in the range from 6.0 to 7.0 (see Fig. 17:3). Even in this range, however, the fact should be emphasized that phosphate availability may still be very low and that added soluble phosphates are readily fixed by soils. The low recovery (perhaps 10 to 30 percent) by plants of added phosphates in a given season is partially due to this fixation.

17:10. AVAILABILITY AND SURFACE AREA OF PHOSPHATES

When soluble phosphates are added to soils two kinds of compounds form immediately: (a) fresh precipitates of calcium, or iron and aluminum phosphates; and (b) similar compounds formed on the surfaces of either calcium carbonate or iron and aluminum oxide particles. In each case, the total surface area of the phosphate is high, and consequently the availability of the phosphorus contained therein is reasonably rapid. Thus, even though the water-soluble phosphorus in superphosphate may be precipitated in the soil in a matter of a few days, the freshly precipitated compounds will release much of their phosphorus to growing plants.

EFFECTS OF AGING. With time, changes take place in the reaction products of soluble phosphates and soils. These changes generally result in a reduction in surface area of the phosphates and a similar reduction in their availability. An increase in the crystal size of precipitated phosphates occurs in time. This decreases their surface area. Also, there is a penetration of the phosphorus held by calcium carbonate and iron or aluminum oxide particles into the particle itself (see Fig. 17:5). This leaves less of the phosphorus near the surface where it can be made available to growing plants. By these

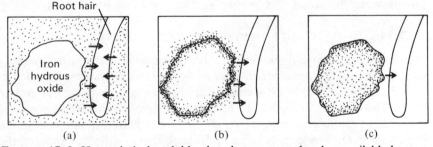

FIGURE 17:5. How relatively soluble phosphates are rendered unavailable by compounds such as hydrous oxide. (a) The situation just after application of a soluble phosphate. The root hair and the hydrous iron oxide particle are surrounded by soluble phosphates. Within a very short time (b) most of the soluble phosphate has reacted with the surface of the iron oxide crystal. The phosphorus is still fairly readily available to the plant roots since most of it is located at the surface of the particle where exudates from the plant can encourage exchange. In time (c) the phosphorus penetrates the crystal and only a small portion is found near the surface. Under these conditions its availability is low.

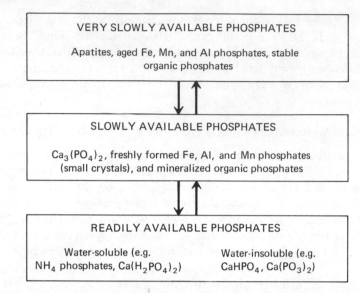

FIGURE 17:6. Classification of phosphate compounds in three major groups. Fertilizer phosphates are generally in the "readily available phosphate" group but are quickly converted to the slowly available forms. These can be utilized by plants at first but upon aging are rendered less available and are then classed as very slowly available. At any one time perhaps 80–90 percent of the soil phosphorus is in "very slowly available" forms. Most of the remainder is in the slowly available form since perhaps less than 1 percent would be expected to be readily available.

processes of aging, phosphate availability is reduced. Thus, the supply of available phosphorus to plants is determined not only by the kinds of compounds which form but also by their surface areas (see Fig. 17:6).

17:11. PHOSPHORUS-FIXING POWER OF SOILS

In light of the above discussion it is interesting to note the actual quantity of phosphorus which soils are capable of fixing. Data from three New Jersey soils presented in Table 17:4 emphasize the tremendous power of certain soils in this respect. For example, to satisfy the phosphorus-fixing power of the unlimed Collington soil, nearly 47 tons of superphosphate containing 20 percent P_2O_5 would be required. Although liming definitely reduced the fixing capacity, the quantity of phosphorus fixed even on the limed soils is enormous. Thus, over 25 tons of superphosphate would be required to completely satisfy the phosphorus-fixing power of the limed Collington soil.

One Coastal Plain soil was reported (4) to have a phosphate-fixing capacity of 125 tons of 20 percent superphosphate per acre–furrow slice. Although such values are somewhat higher than usual because of the nature and amounts of the iron and aluminum compounds in the soils, they do not overemphasize the problem of phosphate fixation.

TABLE 17:4. *Phosphorus-Adsorbing Power of Three New Jersey Soils Limed and Unlimed*[a]

Expressed as pounds of 20 percent superphosphate per acre–furrow slice.

Soil	Treatment	pH	Phosphorus Fixing Power (lb 20% super/A.F.S.)
Sassafras	No lime	3.6	28,400
Sassafras	Lime	6.5	13,916
Collington	No lime	3.2	93,720
Collington	Lime	6.5	50,268
Dutchess	No lime	3.8	68,728
Dutchess	Lime	6.5	44,020

[a] From Toth and Bear (24).

17:12. INFLUENCE OF SOIL ORGANISMS AND ORGANIC MATTER ON THE AVAILABILITY OF INORGANIC PHOSPHORUS

In addition to pH and related factors, organic matter and microorganisms strikingly affect inorganic phosphorus availability. Just as was the case with nitrogen, the rapid decomposition of organic matter and consequent high microbial population results in the *temporary* tying up of inorganic phosphates in microbial tissue.

Products of organic decay such as organic acids and humus are thought to be effective in forming complexes with iron and aluminum compounds. This engagement of iron and aluminum reduces inorganic phosphate fixation to a remarkable degree. The exact importance of this effect has not as yet been completely ascertained. The ability of humus and lignin to reduce phosphate fixation, however, is shown in Fig. 17:7. Both materials were effective in releasing phosphorus after it had been fixed as basic iron phosphate. Thus, organic decomposition products undoubtedly play an important role in organic phosphorus availability.

17:13. AVAILABILITY OF ORGANIC PHOSPHORUS

Only meager information has been obtained on the factors affecting the availability to higher plants or organic phosphorus compounds. It has been established that both *phytin* and *nucleic acids* can be utilized as sources of phosphorus (20). Apparently the phytin is absorbed directly by the plants while the nucleic acids probably are broken down by enzymes at the root surfaces and the phosphorus is adsorbed in either the organic or inorganic form. In spite of the readiness with which these compounds may be assimilated, however, plants commonly suffer from a phosphorus deficiency even in the presence of considerable quantities of organic forms of this element. Just as with inorganic phosphates, the problem is one of availability.

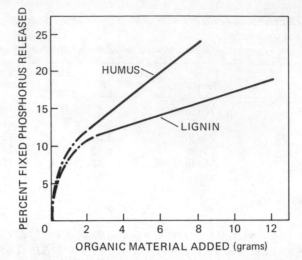

FIGURE 17:7. Effect of added organic materials on the release of phosphorus previously fixed by iron compounds. Both humus and lignin are effective, the humus to the greater degree. These results suggest that mineral fixation of phosphorus may be lower in soils comparatively high in organic matter. [*After Swenson et al.* (22).]

Phytin behaves in the soil much as do the inorganic phosphates (8), forming iron, aluminum, and calcium phytates. In acid soils the phytin is rendered insoluble and thus unavailable because of reaction with iron and aluminum. Under alkaline conditions calcium phytate is precipitated and the phosphorus carried is rendered unavailable.

The fixation of nucleic acids involves an entirely different mechanism, but the end result—low phosphorus availability—is the same. Evidently, nucleic acids are strongly adsorbed by clays, especially montmorillonite.

This adsorption is particularly pronounced under acid conditions and results in a marked decrease in the rate of decomposition of the nucleic acids. Consequently, the available phosphorus supply from this source is low, especially in acid soils which contain appreciable amounts of montmorillonite.

The judicious application of lime to acid soils is thus fully as important in organic phosphorus nutrition as it is in rendering inorganic compounds available. Whether we are dealing with inorganic soil phosphates, added fertilizers, or organic materials, the importance of lime as a controlling factor in phosphate availability is clearly evident.

17:14. PRACTICAL CONTROL OF PHOSPHORUS AVAILABILITY

From a practical standpoint, the phosphorus-utilization picture is not too encouraging. The inefficient utilization of applied phosphates by plants has long been known. The experimental use of radioactive phosphorus materials has emphasized this point even more thoroughly. By adding fertilizers containing traceable phosphorus, it has been possible to determine the proportion of the applied phosphates absorbed during the year of application.

The results of experiments on corn, soybeans, and potatoes are shown in Table 17:5. Even though on some soils marked responses were obtained from the addition of phosphate fertilizers, the efficiency of phosphorus utilization was very low. Apparently maximum yields were obtained only by supplying much more phosphorus than the plants absorbed in a given season.

TABLE 17:5. *Recovery of Applied Fertilizer Phosphates During the First Crop Year*

Soil	Crop	Fertilizer Phosphorus Recovered the First Year (%)
Bladen	Corn	11.8
Bladen	Potatoes	7.6
Bladen	Soybeans	18.2
Webster	Corn	6.5
Clarion	Corn	3.4

[a] Data for bladen soil from Krantz, et al. (16) and for other soils from Stanford and Nelson (21).

LIMING AND PLACEMENT OF FERTILIZERS. The small amount of control that can be exerted over phosphate availability seems to be associated with *liming, fertilizer placement,* and *organic matter maintenance.* By holding the pH of soils between 6.0 and 7.0, the phosphate fixation can be kept at a minimum (see Figs. 14:7 and 17:2). In order to prevent rapid reaction of phosphate fertilizers with the soil, these materials are commonly placed in localized bands. In addition, phosphatic fertilizers are quite often pelleted or aggregated to retard still more their contact with the soil. The effective utilization of phosphorus in combination with animal manures is evidence of the importance of organic matter in increasing the availability of this element.

In spite of these precautions, a major portion of the added phosphates still reverts to less available forms (see Fig. 17:8). It should be remembered, however, that the reverted phosphorus is not lost from the soil and through the years undoubtedly is slowly available to growing plants. This becomes an important factor, especially in soils which have been heavily phosphated for years.

In summary, maintaining sufficient available phosphorus in a soil largely narrows down to a twofold program: (a) the addition of phosphorus-containing fertilizers, and (b) the regulation in some degree of the fixation in the soil of both the added and the native phosphates.

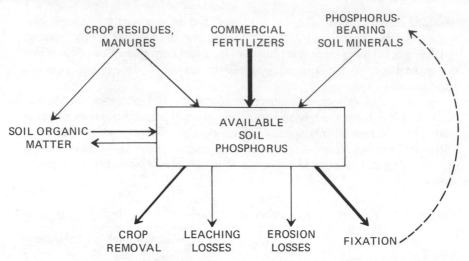

FIGURE 17:8. How the available phosphorus level in a soil is depleted and replenished. Note that the two main features are the addition of phosphate fertilizers and the fixation of this element in insoluble forms. It should be remembered that the amount of available phosphorus in the soil at any one time is relatively small, especially when compared to that of calcium, magnesium, and potassium.

17:15. POTASSIUM—THE THIRD "FERTILIZER" ELEMENT

The history of fertilizer usage in the United States shows that nitrogen and phosphorus received most of the attention when commercial fertilizers first appeared on the market. Although the role played by potassium in plant nutrition has long been known, the importance of potash fertilization has received full recognition only in comparatively recent years.

The reasons that a widespread deficiency of this element did not develop earlier are at least twofold. First, the supply of available potassium originally was so high in most soils that it took many years of cropping for a serious depletion to appear. Second, even though the potassium in certain soils may have been insufficient for optimum crop yields, production was much more drastically limited by a lack of nitrogen and phosphorus. With an increased usage of fertilizers carrying these latter elements, crop yields have been correspondingly increased. As a consequence, the drain on soil potassium has been greatly increased. This, coupled with considerable loss by leaching, has enhanced the demand for potassium in commercial fertilizers.

17:16. EFFECTS OF POTASSIUM ON PLANT GROWTH

The presence of adequate available potassium in the soil has much to do with the general tone and vigor of the plants grown. Moreover, by increasing crop resistance to certain diseases and by encouraging strong root systems, potassium tends to prevent the undesirable "lodging" of plants and to counteract the damaging effects of excessive nitrogen. In delaying maturity, potassium works against undue ripening influences of phosphorus. In a general way, it exerts a balancing effect on both nitrogen and phosphorus, and consequently is especially important in a mixed fertilizer.

Potassium is essential for photosynthesis and for starch formation and the translocation of sugars. It is necessary in the development of chlorophyll, although it does not, like magnesium, enter prominently into its molecular structure. This element is important to cereals in grain formation, as it aids in the development of plump, heavy kernels. Abundant available potassium also is absolutely necessary for tuber development. Therefore, the percentage of this element usually is comparatively high in mixed fertilizers recommended for potatoes. All root crops respond to liberal applications of potassium. As with phosphorus, it may be present in large quantities in the soil and yet exert no harmful effect on the crop.

The leaves of crops suffering from a potassium deficiency (1) appear dry and scorched at the edges, and the surfaces are irregularly chlorotic. In such plants as red and alsike clover, alfalfa, and sweet clover, these symptoms are preceded by the appearance of small dots arranged more or less regularly around the edges of the leaves. As a result of such deterioration, photosynthesis is much impaired and the synthesis of starch is practically brought to a standstill.

In considering potassium plant nutrition it should be noted that sodium has been found to partially take the place of potassium in the nutrition of certain plants. When there is a deficiency of potash, the native sodium of the soil, or that added in such fertilizers as nitrate of soda, may be very useful.

17:17. THE POTASSIUM PROBLEM

AVAILABILITY OF POTASSIUM. In contrast to the situation regarding phosphorus, most mineral soils, except those of a sandy nature, are comparatively high in *total* potassium. In fact, the total quantity of this element is generally greater than that of any other major nutrient element. Amounts as great as 40,000 to 60,000 pounds of potassium oxide per acre–furrow slice are not at all uncommon (see p. 24). Yet the quantity of potassium held in an easily exchangeable condition at any one time often is very small. Most of this element is held rigidly as part of the primary minerals or is fixed in forms that are at best only moderately available to plants. Also competition by microorganisms for this element contributes at least temporarily to its unavailability to higher plants. Thus, the situation in respect to potassium utilization parallels that of phosphorus and nitrogen in at least one way. A very large proportion of all three of these elements in the soil is insoluble and relatively unavailable to growing plants.

LEACHING LOSSES. Unlike the situation with respect to nitrogen and particularly phosphorus, however, much potassium is lost by leaching. An examination of the drainage water from mineral soils on which rather liberal fertilizer applications have been made will usually show considerable quantities of potash. In extreme cases, the magnitude of this loss may approach that of potash removal by the crop. For example, heavily fertilized sandy soils on which crops such as vegetables or tobacco are grown may suffer serious losses by leaching. Even on a representative humid region soil receiving only moderate rates of fertilizer, the annual loss of potash by leaching is usually about 20 pounds per acre (see Table 9:5).

CROP REMOVAL. The third phase of the potassium problem concerns the quantity of this element in plants—in other words, its removal by growing crops. Under ordinary field conditions and with an adequate nutrient supply, potassium removal by crops is high, often being three to four times that of phosphorus and equaling that of nitrogen. The removal of 100 to 125 pounds potassium oxide per acre by a 20-ton silage corn crop is not at all unusual. Moreover, this situation is made even more critical by the fact that plants tend to take up soluble potassium far in excess of their needs if sufficiently large quantities are present. This tendency is termed *luxury consumption*, because the excess potassium absorbed apparently does not increase crop yields to any extent.

EXAMPLES OF LUXURY CONSUMPTION. The principles involved in luxury consumption are shown by the graph of Fig. 17:9. For many crops there is a more or less direct relationship between the available potassium in the soil and the removal of this element by plants. The available potassium would, of course, include both that added and that already present. A certain amount of this element is needed for optimum yields and this is termed *required potassium*. All potassium above this critical level is considered a *luxury*, the removal of which is decidedly wasteful.

Under field conditions, luxury consumption becomes particularly serious. For example, to save labor one may be tempted to supply potassium (a) only once in a three- or four-year rotation, or (b) during the first year of a three- or four-year perennial hay crop. Much of the potassium thus added is likely to be absorbed wastefully by the first crop of the rotation sequence, or in the case of the hay, even in the first cutting. Consequently, little of the added potassium would remain for subsequent crops.

In summary, then, the problem of potassium economy in its most general terms is at least threefold: (a) a very large proportion of this element at a given time is relatively unavailable to higher plants; (b) because of the solubility of its available forms, it is subject to wasteful leaching losses; and (c) the removal of potassium by crop plants is high, especially when luxury quantities of this element are supplied. With these ideas as a background,

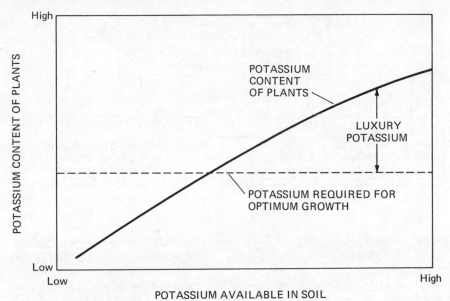

FIGURE 17:9. General relationship between the potassium content of plants and the available soil potassium. If excess quantities of potash fertilizers are applied to a soil, the plants will absorb potassium in excess of that required for optimum yields. This luxury consumption may be wasteful, especially if the crops are completely removed from the soil.

the various forms of potassium in soils and their availabilities will now be considered.

17:18. FORMS AND AVAILABILITY OF POTASSIUM IN SOILS

For convenience, the various forms of potassium in soils can be classified on the basis of availability in three general groups: (a) *unavailable*, (b) *readily available*, and (c) *slowly available*. Although most of the soil potassium is in the first of these three forms, from an immediate practical standpoint the latter two are undoubtedly of greater significance.

The relationship among these three general categories is shown diagrammatically in Fig. 17:10. Equilibrium tendencies presented therein are of vital practical importance, especially when dealing with the slowly and readily available forms. A slow change from one form to another can and does occur. This makes possible a fixation and conservation of added soluble potassium and a subsequent slow release of this element when the readily available supply is reduced.

RELATIVELY UNAVAILABLE FORMS. By far the greatest part (perhaps 90 to 98 percent) of all soil potassium in a mineral soil is in relatively unavailable

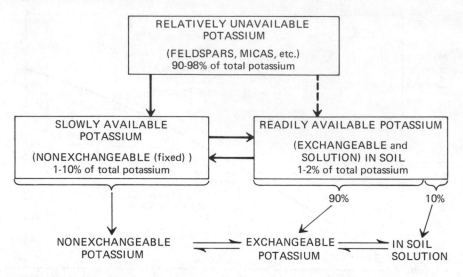

FIGURE 17:10. Relative proportions of the total soil potassium in unavailable, slowly available, and readily available forms. Only 1 to 2 percent is rated as readily available. Of this, approximately 90 percent is exchangeable and only 10 percent appears in the soil solution at any time. [*Figures modified from Attoe and Truog* (2).]

forms (see Fig. 17:10). The compounds containing most of this form of potassium are the feldspars and the micas. These minerals are quite resistant to weathering and probably supply relatively insignificant quantities of potassium during a given growing season. However, their accumulative contribution from year to year to the overall available potassium in the soil undoubtedly is of considerable importance. Potassium is gradually released to more available forms through the action of solvents such as carbonated water. Also, the presence of acid clay is of some significance in the breakdown of these primary minerals with the subsequent release of potassium and other bases (see p. 31).

READILY AVAILABLE FORMS. The readily available potassium constitutes only about 1 or 2 percent of the total amount of this element in an average mineral soil. It exists in soils in two forms: (a) potassium in the soil solution, and (b) exchangeable potassium adsorbed on the soil colloidal surfaces. Although most of this available potassium is in the exchangeable form (approximately 90 percent), soil solution potassium is somewhat more readily absorbed by higher plants and is, of course, subject to considerable leaching loss.

As represented in Fig. 17:10, these two forms of readily available potassium are in dynamic equilibrium. Such a situation is extremely important from a practical standpoint. In the first place, absorption of soil solution potassium by plants results in a temporary disruption of the equilibrium. Then to restore the balance, some of the exchangeable potassium immediately moves into the soil solution until the equilibrium is again established. On the other hand, when water-soluble fertilizers are added, the reverse of the above adjustment occurs.

It is not to be inferred from this, however, that it is necessary for an element to be in the soil solution before absorption by plants can take place. Direct absorption from the colloidal surfaces is also thought to occur, which undoubtedly plays a vital role in the nutrition of many plants.

SLOWLY AVAILABLE FORMS. In the presence of vermiculite, illite, and other 2:1-type minerals, the potassium of such fertilizers as muriate of potash not only becomes adsorbed but also may become definitely "fixed" by the soil colloids (see Fig. 17:11). The potassium as well as ammonium ions (see p. 479) fit in between crystal units of these normally expanding clays

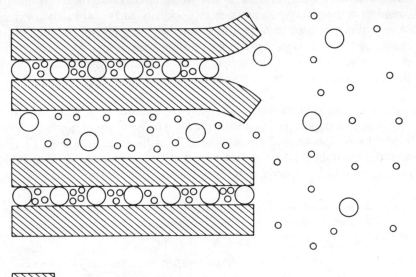

◨ 2:1-type silicate clay layer

◯ Potassium and ammonium ions

○ Other smaller cations (H$^+$, Na$^+$, Ca^{++}, etc.)

FIGURE 17:11. Clay minerals of the 2:1 type such as illite have the ability to fix ammonium and potassium ions. Smaller ions, such as H$^+$, Na$^-$, and Ca^{++}, can move in and out of the internal adsorption surface and are thus exchangeable. Potassium and ammonium ions are of such size as to fit snugly between crystals, thereby holding them together. At the same time, these larger ions are rendered at least temporarily *non-exchangeable* or *fixed*.

and become an integral part of the crystal. Potassium in this form cannot be replaced by ordinary exchange methods and consequently is referred to as *nonexchangeable potassium*. As such, this element is not readily available to higher plants. This form is in equilibrium, however, with the available forms and consequently acts as an extremely important reservoir of slowly available potassium. The entire equilibrium may be represented as follows (see Fig. 17:10):

$$\text{Nonexchangeable K} \rightleftharpoons \text{Exchangeable K} \rightleftharpoons \text{Soil solution K}$$

The importance of the adjustment shown to practical agriculture should not be overlooked. It is of special value in the conservation of added potassium. For example, most potash-containing fertilizers are added at planting time and are usually applied in localized bands. Immediately upon dissolving, a large proportion of the soluble (soil solution) potassium becomes attached to the colloids. As a result, the above equilibria shift to the left and some exchangeable ions are converted to the nonexchangeable form. Although this renders more potassium at least temporarily unavailable, the overall effect apparently is beneficial. In this form the potassium is not subject to leaching, and thus a significant conservation is attained. Also, since the fixed potassium is slowly reconverted to the available forms later, it is by no means completely lost to growing plants.

RELEASE OF FIXED POTASSIUM. The above reaction is perhaps of even greater practical importance in another way. The quantity of nonexchangeable or "fixed" potassium in some soils is quite large (see Fig. 17:10). The fixed potassium in such soils is continually released to the exchangeable form in amounts large enough to be of great practical importance. For

TABLE 17:6. *Potassium Removal by Very Intensive Cropping and the Amount of This Element Coming from the Nonexchangeable Form*

Soil	Total Potassium Used by Crops (lb/acre)	Percentage of the Potassium Coming from Nonexchangeable Form
Wisconsin soils[a]		
Carrington silt loam	119	75
Spencer silt loam	59	80
Plainfield sand	88	25
Mississippi soils[b]		
Robinsonville fine silty loam	108	33
Houston clay	57	47
Ruston sandy loam	42	24

[a] Average of six consecutive cuttings of Ladino clover, from Evans and Attoe (12).
[b] Average of eight consecutive crops of millet, from Gholston and Hoover (14).

example, Bray and DeTurk (9) found a release of as much as 288 pounds of potassium per acre from an Illinois soil in a six-month period. Similar results have been observed. The data in Table 17:6 give an idea of the magnitude of the release of nonexchangeable potassium from certain soils. In several cases cited, the potassium removed by crops was supplied largely from nonexchangeable forms.

17:19. FACTORS AFFECTING POTASSIUM FIXATION IN SOILS

Although the exact mechanism or mechanisms that govern potassium fixation and release are not clearly understood, it is definitely known that several soil conditions markedly influence the amounts fixed. Among the factors are (a) the nature of the soil colloids, (b) wetting and drying, (c) freezing and thawing, and (d) the presence of excess lime.

EFFECTS OF COLLOIDS, MOISTURE, AND TEMPERATURE. The ability of the various soil colloids to fix potassium varies widely. For example, 1:1-type clays such as kaolinite and soils in which these clay minerals are dominant fix little potassium. On the other hand, clays of the 2:1-type, such as vermiculite and illite, fix potassium very readily and in large amounts.

The mechanisms for potassium fixation is probably the same as that for fixation of the ammonium ion (p. 427). These two ions are of such size as to fit snugly into the cavities which exist between the respective silica sheets of adjoining crystal units in 2:1-type clays (see Fig. 17:11). Because of this, these ions become trapped as a part of the rigid crystal structure, thereby preventing normal crystal lattice expansion and reducing the cation exchange capacity of the clay.

The potassium and ammonium ions are attracted between the crystal units by the same negative charges responsible for the internal adsorption of these and other cations. The tendency for fixation is greatest in minerals where the major source of negative charge is in the silica (tetrahedral) sheet. For that reason vermiculite and illite have fixing capacities which far exceed that of montmorillonite (see Table 4:4 for formulas for these minerals).

Alternate freezing and thawing has been shown to result in the release of fixed potassium under certain conditions. Although the practical importance of this is recognized, the mechanism by means of which it occurs is not as yet understood.

INFLUENCE OF LIME. Applications of lime sometimes result in an increase in potassium fixation of soils. Under normal liming conditions this may be more beneficial than detrimental because of the conservation of the potassium so affected. Thus, potassium in well-limed soils is not as likely to be leached out as drastically as is that in acid soils. This is illustrated by data in Table 17:7.

TABLE 17:7. *Effect of Percentage Base Saturation on the Leaching Losses of Exchangeable Potassium from a Creedmoor Coarse Sandy Loam*[a]

Soil pH	Base Saturation (%)	Exchangeable Potassium Loss by Leaching (% of total)
4.83	28	70
5.30	40	49
5.63	50	26
7.03	72	16

[a] From Baver (3).

The sandy soil studied was limed to increase the base saturation from 28 to 72 percent. When 6.2 inches of water was percolated through the original soil (28 percent base saturation), about 70 percent of the exchangeable potassium was removed. The limed sample (72 percent base saturation) when similarly treated sustained a potassium loss of only 16 percent. The effect of lime in preventing potash loss undoubtedly is of considerable practical importance in many cases.

There are conditions, however, under which the effects of lime on the availability of potash are so drastic as to become quite undesirable. For example, in certain soils potassium deficiency apparently results from the presence of excess calcium carbonate. The unavailability of potassium in such soils is thought to be due in some way to lime-induced fixation.

17:20. PRACTICAL IMPLICATIONS IN RESPECT TO POTASSIUM

FREQUENCY OF APPLICATION. One very important suggestion that is evident from the facts thus far considered is that frequent light applications of potassium are usually superior to heavier and less frequent ones. Such a conclusion is reasonable when considering the luxury consumption by crops, the ease with which this element is lost by leaching, and the fact that excess potassium is subject to fixation. Although the latter phenomenon has definite conserving features, these are in most cases entirely outweighed by the disadvantages of leaching and luxury consumption.

POTASSIUM-SUPPLYING POWER OF SOILS. A second very important suggestion is that full advantage should be taken of the potash-supplying power of soils. The idea that each pound of potassium removed by plants or through leaching must be returned in fertilizers may not always be correct. In some soils the large quantities of moderately available forms already present can be utilized. Where slowly available forms are not found in significant quantities, however, supplementary additions are necessary. Moreover, the importance of lime in reducing leaching losses of potassium should not be overlooked as a means of effectively utilizing the power of soils to furnish this element.

POTASSIUM LOSSES AND GAINS. The problem of maintaining soil potassium is outlined diagrammatically in Fig. 17:12. Crop removal of potash generally exceeds that of the other essential elements, with the possible exception of nitrogen. Annual losses from plant removal as great as 100 pounds of potassium per acre are not uncommon, particularly if the crop is a legume and is cut several times for hay. As might be expected, therefore, the return crop residues and manures are very important in maintaining soil potash. As an example, 10 tons of average animal manure supply about 100 pounds of potassium oxide, fully equal to the amount of nitrogen thus supplied.

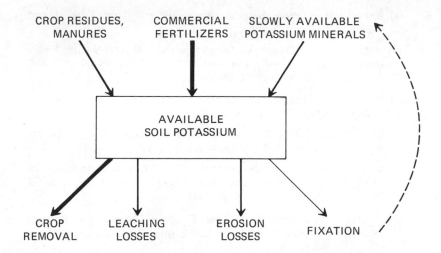

FIGURE 17:12. Gains and losses in *available* soil potassium under average field conditions. The approximate magnitude of the changes is represented by the width of the arrows. For any specific case the actual amounts of potassium added or lost undoubtedly may vary considerably from the above representation. As was the case with nitrogen and phosphorus, it is well to note the increasing importance of commercial fertilizers in meeting crop demands. Also, it should be pointed out that the fixation and release by potash-bearing minerals will be much greater if soils contain appreciable quantities of minerals such as illite, a clay that is especially high in total and exchangeable potash.

The annual losses of available potassium by leaching and erosion greatly exceeds those of nitrogen and phosphorus. They are generally not as great, however, as the corresponding losses of available calcium and magnesium. The available potassium depletion by erosion assumes great importance considering that total potassium removal in this manner generally exceeds that of any other major nutrient element. The loss of potential sources of available potassium (soil minerals) cannot but eventually be serious.

INCREASING USE OF POTASH FERTILIZERS. In the past, potash in fertilizers was added to supplement that returned in crop residues and that obtained from the slowly available forms of potassium carried by soil minerals. Fertilizer potash is now depended upon to supply much of the potassium needed for crop production. This is essentially true in cash-crop areas and in regions where sandy soils are prominent. Even in some heavier soils, the release of potassium from mineral form is much too slow to support maximum plant yields. Consequently, increased usage of commercial potash must be expected if yields are to be increased or even maintained.

REFERENCES

(1) Anonymous, *Hunger Signs in Crops*, 3rd ed. (New York: David McKay), 1964.

(2) Attoe, O. J., and Truog, E., "Exchangeable and Acid Soluble Potassium as Regards Availability and Reciprocal Relationships," *Soil Sci. Soc. Amer. Proc.*, **10**:81–86, 1945.

(3) Baver, L. D., "Practical Applications of Potassium Interrelationships in Soils and Plants," *Soil Science*, **55**:121–26, 1943.

(4) Bear, F. E., and Toth, S. J., "Phosphate Fixation in Soil and Its Practical Control," *Ind. and Eng. Chem.*, **34**:49–52, 1942.

(5) Bertramson, B. R., and Stephenson, R. E., "Comparative Efficiency of Organic Phosphorus and of Superphosphate in the Nutrition of Plants," *Soil Science*, **53**:215–26, 1942.

(6) Black, C. A., *Behavior of Soil and Fertilizer Phosphorus in Relation to Water Pollution*, Jour. Paper J6373, Iowa Agr. and Home Econ. Exp. Sta., 1971.

(7) Bower, C. A., "Separation and Identification of Phytin and Its Derivatives from Soils." *Soil Sci.*, **59**:277–85, 1945.

(8) Bower, C. A., *Studies on the Forms and Availability of Soil Organic Phosphorus, Iowa*, Res. Bull. 362, Iowa Agr. Exp. Sta., Aug. 1949.

(9) Bray, R. H., and DeTurk, E. E., "Release of Potassium from Non-replaceable Forms in Illinois Soils," *Soil Sci. Soc. Amer. Proc.*, **3**:101–6, 1939.

(10) Chang, S. C., and Jackson, M. L., "Fractionation of Soil Phosphorus," *Soil Sci.*, **84**:133–44, 1957.

(11) Cole, C. V., and Jackson, M. L., "Solubility Equilibrium Constant of Dehydroxy Aluminum Dehydrogen Phosphate Relating to a Mechanism of Phosphate Fixation in Soils," *Soil Sci. Soc. Amer. Proc.*, **15**:84–89, 1950.

(12) Evans, C. E., and Attoe, O. J., "Potassium Supplying Power of Virgin and Cropped Soils," *Soil Science*, **66**:323–34, 1948.

(13) Fuller, W. H., and McGeorge, W. T., "Phosphates in Calcareous Arizona Soils, II. Organic Phosphorus Content," *Soil Science*, **71**:45–50, 1951.

(14) Gholston, L. E., and Hoover, C. D., "The Release of Exchangeable and Nonexchangeable Potassium from Several Mississippi and Alabama Soils Upon Continuous Cropping," *Soil Sci. Soc. Amer. Proc.*, **13**:116–21, 1948.

(15) Kittrick, J. A., and Jackson, M. L., "Electron-Microscope Observations of the Reaction of Phosphate with Minerals, Leading to a Unified Theory of Phosphate Fixation in Soils," *Soil Sci.*, **7**:81–90, 1956.

(16) Krantz, B. A., et al., "A Comparison of Phosphorus Utilization by Crops," *Soil Sci.*, **68**:171–78, 1949.

(17) Larsen, S., "Soil Phosphorus," *Adv. in Agron.*, **19**:151–210, 1967.

(18) Lindsay, W. L., and Stephenson, H. F., "Nature of the Reactions of Monocalcium Phosphate Monohydrate in Soils: I. The Solution That Reacts with the Soil; and II Dissolution and Precipitation Reactions Involving Fe, Al, Mu, and Ca," *Soil Sci. Soc. Amer. Proc.*, **23**:12–22, 1959.

(19) Pearson, R. W., and Simonson, R. W., "Organic Phosphorus in Seven Iowa Profiles: Distribution and Amounts as Compared to Organic Carbon and Nitrogen," *Soil Sci. Soc. Amer. Proc.*, **4**:162–67, 1939.

(20) Rogers, H. T., et al., "Absorption of Organic Phosphorus by Corn and Tomato Plants and the Mineralization Action of Exo-enzyme Systems of Growing Roots," *Soil Sci. Soc. Amer. Proc.*, **5**:285–91, 1941.

(21) Stanford, G., and Nelson, L. B., "Utilization of Phosphorus as Affected by Placement: I. Corn in Iowa," *Soil Sci.*, **68**:129–36, 1949.

(22) Swenson, R. M., et al., "Fixation of Phosphate by Iron and Aluminum and Replacement by Organic and Inorganic Ions," *Soil Sci.*, **67**:3–22, 1949.

(23) Tennessee Valley Authority, *Fertilizer Summary Data by States and Geographic Regions* (Muscle Shoals, Ala.: TVA, 1966).

(24) Toth, S. J., and Bear, F. E., "Phosphorus Adsorbing Capacities of Some New Jersey Soils," *Soil Sci.*, **64**:199–211, 1947.

(25) White, O. E., "1965 Harvest, U.S.A.," *Plant Food Rev.*, **11**, No. 4: 16, 1965.

Chapter 18

MICRONUTRIENT ELEMENTS

O F the seventeen elements known to be essential for plant and micro-organism growth, eight are required in such small quantities that they are called *micronutrients* or trace elements. These are iron, manganese, zinc, copper, boron, molybdenum, cobalt, and chlorine. Other elements, such as silicon, vanadium, and sodium, appear to be helpful for the growth of certain species. Still others, such as iodine and fluorine, have been shown to be essential for animal growth but are apparently not required by plants. As better techniques of experimentation are developed and as purer salts are made available, it is likely that these and other elements may be added to the list of essential nutrients.

Micronutrients have become of widespread concern during the past fifteen years (see Table 18:1). The following conditions have likely contributed to this concern: (a) crop removal of the trace elements has in some cases lowered their concentration in the soil below that required for normal growth, (b) improved crop varieties and macronutrient fertilizer practices have greatly increased the level of crop production and thereby the micronutrient removal, (c) the trend toward high analysis fertilizers has reduced the use of impure salts which formerly contained some micronutrients, and (d) increased knowledge of plant nutrition has helped in the diagnosis of trace element deficiencies which previously may have gone unnoticed. Demands for efficiency of crop production will undoubtedly continue to encourage attention to these elements. Their field deficiencies have been noted in many states (see Table 18:1).

18:1. DEFICIENCY VERSUS TOXICITY

One common characteristic of all the micronutrients is that they are required in very small amounts (Fig. 18:1). Also, they are all harmful when the available forms are present in the soil in large quantities. Thus the range of concentration of these elements in which plants will grow satisfactorily is not too great. Molybdenum, for example, may be beneficial if added at rates as little as $\frac{1}{2}$ to 1 *ounce* per acre, and applications of 3 or 4 pounds of available molybdenum per acre may be toxic to most plants.

484

TABLE 18:1. *Number of States Reporting Field Deficiencies of Six Micronutrients in Important Crops of the United States*

Crop Group	Number of States Reporting Deficiency					
	B	Cu	Zn	Mn	Fe	Mo
Forage legumes	36	1	1	2	2	15
Edible legumes	0	2	9	9	5	4
Soybeans	0	0	5	12	5	9
Corn and sorghum	3	5	26	8	8	0
Forage grasses	0	4	0	0	9	0
Small grains	0	4	3	11	4	0
Crucifers and various leafy vegetables	26	7	2	9	1	9
Solanaceous crops	6	3	1	6	1	1
Root and bulb crops	18	6	6	9	1	2
Cucurbits	1	0	0	4	1	2
Tree fruits	22	3	17	8	16	1
Small fruits	3	0	1	2	8	0
Nut crops	2	1	12	1	4	0
Ornamentals	0	0	2	0	18	1
Cotton	6	0	1	1	0	0
Tobacco	2	0	0	0	1	0
Other	0	3	4	1	5	0

[a] From a report of the Soil Test Commission of the Soil Science Society of America (2) and Berger (3).

Even those quantities present under natural soil conditions are in some cases excessive for normal crop growth. Although somewhat larger amounts of the other micronutrients are required and can be tolerated by plants, control of the quantities added, especially in maintaining nutrient balance, is absolutely essential. The concepts of deficiency, adequacy, and toxicity are illustrated in Fig. 18:2 for several micronutrients.

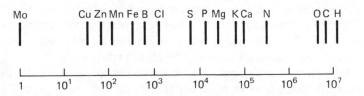

FIGURE 18:1. Relative number of atoms of the essential elements in alfalfa at bloom stage, expressed logarithmically. Note that there are more than 10 million hydrogen atoms for each molybdenum atom. Even so, normal plant growth would not occur without molybdenum. Cobalt is generally present in even smaller quantities in plants than is molybdenum. [*Modified from Viets* (8).]

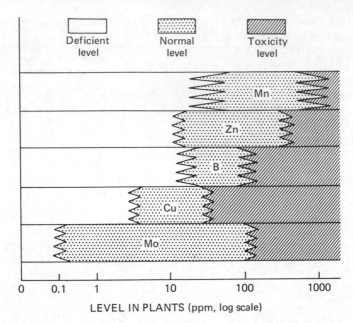

FIGURE 18:2. Deficient, normal, and toxic levels in plants for five micronutrients. [*Levels taken from Allaway* (1).]

18:2. ROLE OF THE MICRONUTRIENTS

The specific role of the various micronutrients in plant and microbial growth processes is not well understood. However, these elements are known to be associated with certain essential processes, some of which are listed in Table 18:2.

The meager information on the role of micronutrients suggests that several of the trace elements are effective through certain *enzyme systems*. For example, copper, iron, and molybdenum are capable of acting as "electron carriers" in enzyme systems which bring about oxidation–reduction reactions in plants. Apparently such reactions, essential to plant development and reproduction, will not take place in the absence of these micronutrients. Zinc and manganese also function in enzyme systems which are necessary for important reactions in plant metabolism.

Molybdenum and manganese have been found to be essential for certain nitrogen transformations in microorganisms as well as in plants. Molybdenum is thought to be essential for the process of nitrogen fixation, both symbiotic and nonsymbiotic. Also, it must be present in plants if nitrates are to be metabolized into amino acids and proteins. In each case molybdenum apparently is an essential part of the respective enzyme system which facilitates the nitrogen change.

TABLE 18:2. *Functions in Higher Plants of Several Micronutrients*

Micronutrient	Functions in Higher Plant Processes
Zinc	Formulation of growth hormones, promotion of protein synthesis, seed and grain maturation and production
Iron	Chlorophyll synthesis, oxidation–reduction in respiration, constituent of certain enzymes and proteins
Copper	Catalyst for respiration, enzyme constituent, chlorophyll synthesis, carbohydrate and protein metabolism
Boron	Protein synthesis, nitrogen and carbohydrate metabolism, root system development, fruit and seed formation and water relations
Manganese	Nitrogen and inorganic acid metabolism, carbon dioxide assimilation (photosynthesis) carbohydrate breakdown, formation of carotene, riboflavin, and ascorbic acid
Molybdenum	Symbiotic nitrogen fixation and protein synthesis

Zinc is thought to be concerned in the formation of some growth hormones and in the reproduction process of certain plants. Copper is involved in respiration and in the utilization of iron. A boron deficiency decreases the rate of water absorption and of translocation of sugars in plants—and iron is essential for chlorophyll formation and for the synthesis of proteins contained in the chloroplasts. It is obvious that the place of the trace elements in plant metabolism is a complicated one. As research uncovers new facts, the role of these micronutrients will be more fully understood and undoubtedly more adequately respected.

Chlorine and cobalt are the elements whose essentiality has been determined most recently. The role of chlorine is still somewhat obscure; however, both root and top growth seem to suffer if it is absent.

Cobalt is essential for the symbiotic fixation of nitrogen (see Fig. 18:3). This element is a component of vitamin B_{12}, which is thought to be necessary for the formation of a type of hemoglobin in nitrogen-fixing nodule tissue. Legumes and other plants are thought to have a cobalt requirement independent of nitrogen fixation, although the amount required is small compared to that for the nitrogen-fixation process.

18:3. SOURCE OF MICRONUTRIENTS

Parent materials tend to influence in a practical way the micronutrient contents of soils, perhaps even more so than that of the macronutrients. Deficiencies of trace elements can frequently be related to low contents of the micronutrients in the parent rocks or transported parent material. Similarly, toxic quantities are commonly related to abnormally large amounts in the soil-forming rocks and minerals.

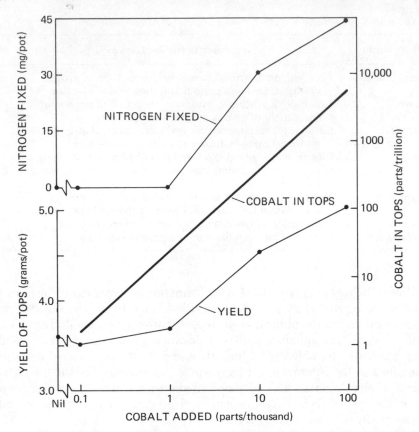

FIGURE 18:3. Effect of cobalt concentration in a nutrient solution (parts per thousand) on nitrogen fixation and on the yield and cobalt content of the tops of nodulated alfalfa plants. [*Redrawn from Wilson and Reisenauer* (9).]

INORGANIC FORMS. The scant data available indicate that sources of the eight micronutrients vary markedly from area to area. Also, because of the extremely small quantities of some of these elements present in soils and in rocks, little is known about the specific compounds in which they are found. Except for iron and manganese, which are present in many soils in large total quantities, widespread accurate analyses for micronutrients in soils and rocks have not been made (see Table 18:3). For these reasons only generalizations can be drawn concerning micronutrient sources.

All of the micronutrients have been found in varying quantities in igneous rocks. Two of them, iron and manganese, have prominent structural positions in certain of the original silicate minerals. Others such as cobalt and zinc may also occupy structural positions as minor replacements for the major constituents of silicate minerals including clays.

TABLE 18:3. *Major Natural Sources of the Eight Micronutrients and their Suggested Contents in a Representative Humid Region Surface Soil*

Element	Major Forms in Nature	Suggested Analysis of a Surface Soil[a] (ppm)
Iron	Oxides, sulfides, and silicates	25,000
Manganese	Oxides, silicates, and carbonates	2,500
Zinc	Sulfides, oxides, and silicates	100
Copper	Sulfides, hydroxy carbonates	50
Boron	Borosilicates, borates	50
Molybdenum	Sulfides, molybdates	2
Chlorine	Chlorides	50
Cobalt	Silicates	8

[a] Analysis of a given soil would likely give values severalfold higher or lower than those listed. Consequently, these more or less average figures should be used with caution.

As mineral decomposition and soil formation occur, the mineral forms of the micronutrients are changed just as macronutrients. Oxides and in some cases sulfides of elements such as iron, manganese, and zinc are formed (see Table 18:3). Secondary silicates, including the clay minerals, may contain considerable quantities of iron and manganese and smaller quantities of zinc and cobalt. The micronutrient cations released as weathering occurs are subject to colloidal adsorption just as are the calcium or hydrogen ions. Anions such as the borate and molybdate may undergo adsorption or reaction in soils similar to that of the phosphates. Chlorine, by far the most soluble of the group, is added to soils in considerable quantities each year through rainwater. Its incidental addition in fertilizers and in other ways helps prevent the deficiency of chlorine under field conditions.

ORGANIC FORMS. Organic matter is an important secondary source of some of the trace elements. They seem to be held as complex combinations by the organic colloids. Copper is especially tightly held. In uncultivated profiles there is a somewhat greater concentration of micronutrients in the surface soil, much of it presumably in the organic fraction. Correlations between soil organic matter and copper, molybdenum, and zinc have been noted. Although the elements thus held are not always readily available to plants, their release through decomposition is undoubtedly an important fertility factor.

18:4. GENERAL CONDITIONS CONDUCIVE TO MICRONUTRIENT DEFICIENCY

Micronutrients are most apt to limit crop growth under the following conditions: (a) highly leached acid sandy soils, (b) muck soils, (c) soils very high in pH; and (d) soils which have been very intensively cropped and heavily fertilized with macronutrients only.

Strongly leached acid sandy soils are low in micronutrients for the same reasons they are deficient in most of the macronutrients. Their parent materials were originally deficient in the elements and acid leaching has removed much of the small quantity of micronutrients originally present. In the case of molybdenum, acid soil conditions also have a marked depressing effect on availability.

The micronutrient contents of organic soils are dependent upon the extent of the washing or leaching of these elements into the bog area as the deposits formed. In most cases, this rate of movement was too slow to give deposits as high in micronutrients as are the surrounding mineral soils. Intensive cropping of muck soils and their ability to bind certain elements, notably copper, also accentuates trace-elements deficiencies. Much of the harvested crops, especially vegetables, are removed from the land. Eventually the micro- as well as macronutrients must be supplied in the form of fertilizers if good crop yields are to be maintained. Intensive cropping of heavily fertilized mineral soils can also hasten the onset of micronutrient shortage, especially if the soils are coarse in texture.

The soil pH has a decided influence on the availability of all the micronutrients except chlorine. Under very acid conditions, molybdenum is rendered unavailable; at high pH values all the cations are unfavorably affected. Overliming or a naturally high pH is associated with deficiencies of iron, manganese, zinc, copper, and even boron. Such conditions occur in nature in many of the calcareous soils of the west.

18:5. FACTORS INFLUENCING THE AVAILABILITY OF THE MICRONUTRIENT CATIONS

Each of the five micronutrient cations (iron, manganese, zinc, copper, and cobalt) is influenced in a characteristic way by the soil environment. However, certain soil factors have the same general effects on the availability of all of them.

SOIL pH. The micronutrient cations are most soluble and available under acid conditions. In very acid soils there is a relative abundance of the ions of iron, manganese, zinc, and copper. In fact, under these conditions the concentrations of one or more of these elements is often sufficiently high to be toxic to common plants. As indicated in Chapter 15 (p. 413), one of the primary reasons for liming acid soils is to reduce the concentration of these ions.

As the pH is increased, the ionic forms of the micronutrient cations are changed to the hydroxides or oxides. An example follows, using the ferrous ion as typical of the group:

$$\underset{\text{(Soluble)}}{Fe^{++}} \quad + \ 2OH^- \ \rightarrow \ \underset{\text{(Insoluble)}}{Fe(OH)_2}$$

All of the hydroxides of the trace element cations are insoluble, some more so than others. The exact pH at which precipitation occurs varies from element to element and even between oxidation states of a given element. For example, the higher valent states of iron and manganese form hydroxides much more insoluble than their lower valent counterparts. In any case, the principle is the same—at low pH values the solubility of micronutrient cations is at a maximum, and as the pH is raised, their solubility and availability to plants decrease. The desirability of maintaining an intermediate soil pH is obvious.

Zinc fixation at high pH values is favored by a second phenomenon only indirectly related to soil pH. Particles of limestone, particularly if they contain dolomite as well as calcite, strongly adsorb the zinc ions. It is likely that zinc may react with the magnesium-containing limestone, replacing the magnesium ion in the mineral framework.

OXIDATION STATE AND pH. Three of the trace element cations are found in soils in more than one valent state. These are iron, manganese, and copper. The lower valent states are encouraged by conditions of low oxygen supply and relatively higher moisture level. They are responsible for the subdued subsoil colors, grays and blues, in contrast to the bright reds, browns, and yellows of well-drained mineral soils.

The changes from one valent state to another are in most cases brought about by microorganisms and organic matter. In some cases the organisms may obtain their energy directly from the inorganic reaction. For example, the oxidation of manganese from the two-valent manganous form (Mn^{++}) to MnO_2 can be carried on by certain bacteria and fungi (see Fig. 18:4). In other cases, organic compounds formed by the microbes may be responsible for the oxidation or reduction. In general, high pH values favor oxidation, whereas acid conditions are more conducive to reduction.

The oxidized states of iron, manganese, and copper are generally much less soluble at pH values common in soils than are the reduced states. The hydroxides (or hydrous oxides) of these high-valent forms precipitate at lower pH values and are extremely insoluble. For example, the hydroxide of trivalent ferric iron precipitates at pH values near 3.0, whereas ferrous hydroxide does not precipitate until a pH of 6.0 or higher is reached.

The interaction of soil acidity and aeration in determining micronutrient availability is of great practical importance. The micronutrient cations and molybdenum are somewhat more available under conditions of restricted drainage. Flooded soils generally show higher availabilities than well-aerated

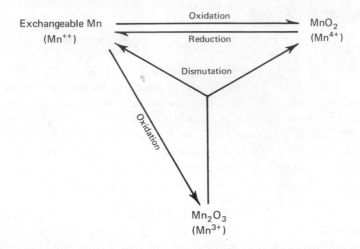

FIGURE 18:4. Relationship among the various forms of manganese in soils. Exchangeable Mn^{++} is available to plants but is subject to oxidation by microorganisms to the trivalent form (Mn_2O_3) or to the more insoluble four-valent form (MnO_2). The Mn_2O_3 can be changed (dismutation) into Mn^{++} and MnO_2. High pH and good oxidation conditions encourage formation of MnO_2 and low manganese availability. [*Modified from Dion and Mann (5).*]

soils. Very acid soils that are poorly drained often supply toxic quantities of iron and manganese. Such toxicity is less apt to occur under well-drained conditions. At the high end of the soil pH range, good drainage and aeration often have the opposite effect. Well-oxidized calcareous soils are sometimes deficient in available iron, zinc, or manganese even though there are adequate total quantities of these trace elements present. The hydroxides of the high-valent forms of these elements are too insoluble to supply the ions needed for plant growth.

OTHER INORGANIC REACTIONS. Micronutrient cations interact with silicate clays in two ways. First, they may be involved in cation exchange reactions much the same as calcium or hydrogen. Also, they may be more tightly bound or fixed to certain silicate clays, especially of the 2:1 type. Zinc, manganese, cobalt, and iron ions are found as integral elements in these clays. Depending on the conditions, they may be released from the clays or fixed by them. The fixation may be serious in the case of cobalt and sometimes zinc since the total amounts of these two elements in soil is small (see Table 18:3).

The application of large quantities of phosphate fertilizers can adversely affect the supply of some of the micronutrients. The uptake of both iron and zinc may be reduced in the presence of excess phosphates. From a practical standpoint, phosphate fertilizers should be used in only those quantities that are required for good plant growth.

Lime-induced chlorosis (iron deficiency) in fruit trees has been found to be encouraged by the presence of the bicarbonate ion. The chlorosis apparently results from iron deficiency on soils with high pH. In some way the bicarbonate ion interferes with iron metabolism.

ORGANIC COMBINATIONS. Each of the four micronutrient cations may be held in organic combination. Microorganisms also assimilate them since they are apparently required for many microbial transformations. The organic compounds in which these trace elements are combined undoubtedly vary considerably, but they include proteins, amino acids, and constituents of humus, including the humic acids (see p. 95) and acids such as citric and tartaric. Among the most important are the *organic complexes*, combinations of the metallic cation and certain organic groups. These complexes may protect the micronutrients from certain harmful reactions, such as the precipitation of iron by phosphates and vice versa (see p. 469). On the other hand, complex formation may reduce micronutrient availability below that necessary for normal plant needs.

On soils high in organic matter, complex formation by copper is thought to be responsible for the deficiency of this element. For example, copper appears to be bound very tightly by the organic matter in certain muck soils. Regular applications of copper-containing salts may be necessary for the normal production of vegetables on these highly organic soils. Zinc deficiencies have also been attributed to reaction with organic matter. Advantage has been taken of the complex-forming tendencies of manganese, copper, zinc, and especially iron in the development of synthetic compounds called *chelates*.

18:6. CHELATES

Chelate is a term derived from a Greek word meaning "claw." As this term indicates, the chelates have a marked tendency to hold tightly certain cations attracted to them. Furthermore, they tend to protect these cations from harmful inorganic reactions. At the same time the chelated nutrients can be utilized by growing plants.

A chelate is an organic compound which combines with and protects certain metallic cations, including iron, manganese, zinc, and copper. The cation–chelate combinations are complex ring structures and the metals so bound essentially lose their usual ionic characteristics (see Fig. 18:5). They are less likely to take part in reactions with other soil constituents, and this may be advantageous. For example, chelated metals remain in solution at much higher pH values than do the inorganic ionic forms. The protected cations are not so subject to precipitation as insoluble hydroxides.

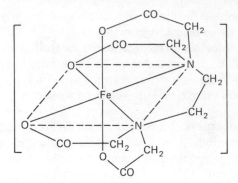

FIGURE 18:5. Structural formula for a common iron chelate, ferric ethylenediamine-tetraacetate (Fe EDTA). The iron is protected and yet can be utilized by plants.

In recent years primary attention has been given to synthetic chelates. However, the protection of heavy metal ions by naturally occurring organic complexes has been known for some time. The movement of iron, zinc, cobalt, and other heavy metals in soils is thought to be associated with the formation of organic chelates. For example, the downward movement of organic matter and of ion and aluminum as Spodosols form is thought to involve chelates. These compounds protect the metal cations from reaction with materials which would otherwise render them insoluble.

Specific examples can be cited showing the protective action of chelating compounds. As indicated in Section 18:5, inorganic iron salts added to a calcareous soil are precipitated and rendered unavailable for plant growth. If ferric sulfate were added, the following reaction would occur, assuming the hydroxyl ions to come from the soil solution:

$$\underset{\text{(Soluble)}}{Fe^{3+}} + 3OH^- \rightleftarrows \underset{\text{(Insoluble)}}{Fe(OH)_3}$$

As the equilibrium arrows indicate, most of the ferric ion would be changed to the insoluble form.

If, instead of the sulfate salt of iron an iron chelate were used, the ferric ion would be protected from this precipitation. As shown in Fig. 18:5, the iron is part of a complex combination which protects the ferric ions. Consequently, there would be little opportunity for the reaction above to take place except where the soil pH is very high. One would expect the chelate to be a better continuing but slow source of iron than an inorganic salt, where there is danger of reaction between the metallic cation (Fe^{3+}) and soil constituents.

Although chelated metals are protected against soil reactions, these forms of the micronutrients are apparently assimilated readily by growing plants. Thus, so long as the nutrients remain in these combinations, they are considered as being in the available form.

The mechanism by which micronutrients from chelates are absorbed by plants is still obscure. The chelating agents in most cases are absorbed by growing plants, but the rate of their absorption is lower than that of the metals they carry. Thus, it would appear that the primary function of the chelate is to keep the metals available in the soil. At the same time, there is evidence that some of the benefit from the chelating agents is through increased translocation of the metals once they are absorbed by the plants.

STABILITY OF CHELATES. The use of chelates to supply iron is quite successful so long as the soil pH is not too high. Application can be made to the soil or as a spray to the foliage. With some of the other nutrients, however, less consistent advantage has been shown for the chelates over other nutrient compounds. This may be due to the fact that for most of the chelates, iron is attracted more strongly by the organic compound than are the other micronutrients. Thus, if a zinc-containing chelate is added to a soil which has significant quantities of available iron, the following reaction may occur:

$$\text{Zn chelate} + \text{Fe}^{++} \rightleftharpoons \text{Fe chelate} + \text{Zn}^{++}$$

The iron chelate is more stable than its zinc counterpart, which explains the tendency of the reaction to go to the right. The released zinc ion is subject to reaction with the soil the same as is zinc from an inorganic salt such as $ZnSO_4$. It is obvious that the added metal-chelate combination must be stable within the soil if it is to have any lasting advantage.

It should not be inferred that only iron chelates are effective. The chelates of other micronutrients, including zinc, manganese, and copper, have been used successfully to supply these nutrients. Apparently, replacement by iron in the soil is sufficiently slow to permit absorption by plants of the added trace element. Also, since spray and banded applications are often used to supply zinc and manganese, the possibility of reaction of these elements with iron in the soil can be reduced or eliminated.

The use of synthetic chelates in the United States is substantial in spite of the fact that they are quite expensive. Much of the use is to meet micronutrient deficiencies of citrus and other fruit trees. Although chelates may not replace the more conventional methods of supplying most micronutrients, they do offer some possibilities in special cases. Agricultural and chemical research will likely continue to increase the opportunities for their use.

The following specific synthetic chelating agents have been found to be effective for the indicated ions:

Fe, Zn, Mn, Cu	EDTA	Ethylenediaminetetraacetic acid
Fe	EDDHA	Ethylenediaminedi-o-hydroxyphenylacetic acid
Fe	DTPA	Diethylenetriaminepentaacetic acid
Zn	NTA	Nitrilotriacetic acid
Fe, Zn	HEDTA	Hydroxyethylethylenediaminetetraacetic acid

18:7. FACTORS INFLUENCING THE AVAILABILITY OF THE MICRONUTRIENT ANIONS

Unlike the cations needed in trace quantities by plants, the anions seem to have relatively little in common. Chlorine, molybdenum, and boron are quite different chemically, so little similarity would be expected in their reaction in soils.

CHLORINE. Chlorine has only recently been found to be essential for plant growth, in spite of the fact that it is used in larger quantities by most crop plants than any of the micronutrients except iron. Two reasons account for man's failure to recognize the essentiality of this nutrient earlier: (a) the wide occurrence of chlorine as an impurity in salts used for research work, and (b) the annual additions to soil of significant quantities of chlorine through precipitation.

Most of the chlorine in soils is in the form of simple, soluble chloride salts such as potassium chloride. The chloride ions are not tightly adsorbed by negatively charged clays and as a result are subject to movement with the water both upward and downward in the profile. In humid regions, one would expect little chlorine to remain in the soil since it would be leached out. In semiarid and arid regions, a somewhat higher concentration might be expected, the amount reaching the point of salt toxicity in some of the poorly drained saline soils. In most well-drained areas, however, one would not expect a high chlorine content in the surface of arid-region soils.

Except under conditions where toxic quantities of chlorine are found in soils, there are apparently no common situations under field conditions which reduce the availability and utilization of this element. Accretions of chlorine from the atmosphere are believed to be in sufficient quantities to meet crop needs. Salt spray alongside ocean beaches evaporates, leaving sodium chloride dust which moves into the atmosphere to be returned later dissolved in snow and rain. The amount added to the soil in this way varies tremendously, depending on the distance from the salty body of water and other factors. It is likely that a figure of about 10 pounds per acre per year is the minimum that can be expected in most situations. An average figure of perhaps 20 pounds would be more representative of the amount added. In any case, this form of accretion plus that commonly added as an incidental component of commercial fertilizers should largely prevent a field deficiency of chlorine.

BORON. The availability and utilization of boron is determined to a considerable extent by pH. Boron is most soluble under acid conditions. It apparently occurs in acid soils in part as boric acid (H_3BO_3) which is readily available to plants. In quite acid sandy soils, soluble boron fertilizers may be leached downward with comparative ease. Evidently, the element is not fixed under these conditions. In heavier soils, especially if they are not too acid, this rapid leaching does not occur.

At higher pH values, boron is less easily utilized by plants. This may be due to lime-induced fixation of this element by clay and other minerals since the calcium and sodium borates are reasonably soluble. In any case, overliming can and often does result in a deficiency of boron.

Boron is held in organic combinations from which it may be released for crop use. The content of this nutrient in the topsoil is generally higher than that in the subsoil. This may in part account for the noticeably greater boron deficiency in periods of dry weather. Apparently, during drought periods plant roots are forced to exploit only the lower soil horizons, where the boron content is quite low. When the rains come, plant roots again can absorb boron from the topsoil, when its concentration is highest.

MOLYBDENUM. Soil conditions affect the availability of molybdenum much the same as they do that of phosphorus. For example, molybdenum is quite unavailable in strongly acid soils (see Fig. 18:6). Under these conditions, this element evidently reacts with soil minerals, such as the silicates and iron and aluminum compounds. The fixed molybdate ion (MoO_4^{3-}) can be replaced by phosphates through anion exchange, which indicates that the same soil compounds may be involved in molybdenum and phosphate fixation.

The liming of acid soils will usually increase the availability of molybdenum. The effect is so striking that some researchers, especially those in

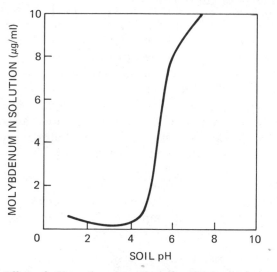

FIGURE 18:6. Effect of pH on the adsorption of molybdenum by an Australian soil. A given quantity of molybdenum was shaken with samples of soil at different pH values and the amount of Mo in solution was determined. [*From Jones* (6).]

Australia and New Zealand, argue that the primary reason for liming very acid soils is to supply molybdenum. Furthermore, in some instances an ounce or so of molybdenum added to acid soils has given about the same increase in the yield of legumes as has the application of several tons of lime. Although these general results have not been consistently verified by experiments in the United States, research has pointed to specific acid soil areas where economical responses can be obtained from molybdenum applications.

The utilization of phosphate by plants seems to favor that of molybdenum and vice versa. For this reason, molybdate salts are often applied along with superphosphate to molybdenum-deficient soils. This practice apparently encourages the uptake of both elements and is a convenient way to add the extremely small quantities of molybdenum required.

A second common anion, the sulfate, seems to have an effect on molybdenum absorption opposite to that of phosphorus. Sulfate has been found to reduce molybdenum uptake, although the specific mechanisms for this antagonism is not yet known.

18:8. NEED FOR NUTRIENT BALANCE

Nutrient balance among the trace elements is essential and is perhaps even more difficult to maintain than is that for the macronutrients. Some of the plant enzyme systems which are dependent upon micronutrients require more than one element. For example, both manganese and molybdenum are needed for the assimilation of nitrates by plants. The mutually beneficial effects of phosphates and molybdenum have already been discussed. Apparently, for some plants zinc and phosphorus are needed for optimum utilization of manganese. The utilization of boron and calcium is dependent upon the proper balance between these two nutrients. A similar relationship exists between potassium and copper and between potassium and iron in the production of good-quality potatoes. Copper utilization is favored by adequate manganese, which in some plants is assimilated only if zinc is present in sufficient amounts. Of course, the effects of these and other nutrients will depend upon the specific plant being grown, but the complexity of the situation can be seen from the examples cited.

ANTAGONISM. Some enzymatic and other biochemical reactions requiring a given micronutrient may be "poisoned" by the presence of a second trace element in toxic quantities. Examples of known antagonistic effects of elements on the absorption of micronutrients follow:

1. Excess copper or sulfate may adversely affect the utilization of molybdenum.
2. Iron deficiency is encouraged by an excess of zinc, manganese, and copper.

3. Excess phosphate may encourage a deficiency of zinc, iron, and copper.
4. Heavy nitrogen fertilization intensifies copper deficiency.
5. Excess sodium or potassium may adversely affect manganese uptake.
6. Excess lime reduces boron uptake.
7. Iron, copper, and zinc may reduce the absorption of manganese.

Some of these antagonistic effects may be utilized effectively in reducing toxicities of certain of the micronutrients. For example, copper toxicity of citrus groves caused by residual copper from insecticide sprays may be reduced by adding iron and phosphate fertilizers. Sulfur additions to calcareous soils containing toxic quantities of soluble molybdenum may reduce the availability and hence the toxicity of molybdenum.

These examples of nutrient interactions, both beneficial and detrimental, emphasize the highly complicated nature of the biological transformations in which micronutrients are involved. Luckily, the total acreage is small where unfavorable nutrient balances require special micronutrient treatment. This acreage is increasing, however, as man makes more intensive use of the soil and the crops grown thereon.

18:9. SOIL MANAGEMENT AND MICRONUTRIENT NEEDS

The characteristics of each micronutrient are just as specific as are those for the macronutrients. Because of the interactions just presented, they may be even more so. Thus, any generalizations with respect to management practices involving all the trace elements should be used with caution. The advice of the expert should be sought in any specific case. He will be acquainted with the particular problem and will be prepared to suggest a solution. And he can point out the necessary precautions to prevent toxicities.

SOIL CHARACTERISTICS. A review of factors determining micronutrient availability gives a clue as to how management practices might help keep these elements available at optimum levels. In ascertaining the cause of plant abnormalities one should keep in mind the conditions wherein micronutrient deficiencies or toxicities are likely to occur. Sandy soils, mucks, and soils having very high or very low pH values would be suspect. Areas of intensive cropping and heavy macronutrient fertilization may be deficient in the micronutrients.

CHANGES IN SOIL ACIDITY. In deciding remedies for a known deficiency or toxicity, soil pH changes should be considered. In very acid soils, one might expect toxicities of iron and manganese and deficiencies of phosphorus and molybdenum. These can be corrected by liming and by appropriate fertilizer additions. Calcereous alkaline soils may have deficiencies of iron, manganese, zinc, and copper, and, in a few cases, a toxicity of molybdenum.

Acid-forming fertilizers or sulfur may be utilized in lowering the soil pH. Large additions of organic matter may not be desirable under these conditions since they encourage the oxidation of iron and manganese to the higher valent and more insoluble forms.

No specific statement can be made concerning the pH value most suitable for all the elements. However, medium-textured soils generally supply adequate quantities of micronutrients when the soil pH is held between 6 and 7. In sandy soils, a somewhat more acid reaction may be justified since the total quantity of micronutrients is low and even at pH 6.0 some deficiencies may occur.

SOIL MOISTURE. Drainage and moisture control can influence micronutrient solubility in soils. Improving the drainage of acid soils will encourage the formation of the oxidized forms of iron and manganese. These are less soluble and, under acid conditions, less toxic than are the reduced forms.

Moisture control at high pH values can have the opposite effect. High moisture levels maintained by irrigation may result in the chemical reduction of high-valent compounds, the oxides of which are extremely insoluble. Flooding a soil will favor the reduced forms, which are more available to growing plants.

FERTILIZER APPLICATIONS. Perhaps the most common management practice is to add chemical nutrients to correct the deficiency or even toxicity, a procedure which has been used successfully for many years. Onion producers on the muck lands of some of the northern states have added

TABLE 18:4. *Tonnage of Five Micronutrient Fertilizers Sold by Region of the United States (1970–1971)[a]*

Region	Quantity of Element Sold (tons)				
	Fe	Mn	Zn	Cu	Mo
New England	3	20	103	13	—
Mid-Atlantic	234	456	942	111	2
South Atlantic	1,024	5,969	4,859	575	11
East North Central	249	5,420	1,553	32	32
West North Central	94	857	4,172	43	7
East South Central	8	69	993	9	5
West South Central	80	160	1,453	5	9
Mountain	368	190	1,281	6	1
Pacific	1,210	362	1,798	76	2
Alaska, Hawaii, Puerto Rico	0.2	2	1	0.1	—
Total United States	3,270	13,505	17,155	868	69

[a] From U. S. Department of Agriculture (7).

copper sulfate as a regular component of their fertilizer for many years. Citrus growers in California and especially in Florida have supplied zinc as a spray for as many as 25 years and have used other micronutrients in their fertilizers for an even longer period. In wide areas throughout the country, boron is added as a regular component of fertilizer for alfalfa. It is also used for certain crops such as beets and celery which have high requirements for this element.

Economic responses to micronutrients are becoming more widespread as intensity of cropping increases. For example, responses of fruits, vegetables, and field crops to zinc and iron applications have been noted in the Rocky Mountain area, the west coast and the northwest, and the Great Plains states. Even on acid soils of the south and the east coast, deficiencies of the elements have been demonstrated. Molybdenum, which has been used for some time for alfalfa and for cauliflower and other vegetables, has received

TABLE 18:5. *Areas of Micronutrient Deficiency, Range in Recommended Rates of Application in the Deficient Areas, and Some Crops Having a High Requirement for Micronutrients*

Micronutrient	Area Deficient in United States[a] (millions of hectares)	Common Range in Recommended Application Rates (kg of element/hectare)	Crops Having a High Requirement
Iron	1.54	0.5–10	Blueberries, cranberries, rhododendron, peaches, grapes, nut trees
Manganese	5.26	5–30	Dates, beans, soybeans, onions, potatoes, citrus
Zinc	2.63	0.5–20	Citrus and fruit trees, soybeans, corn, beans
Copper	0.24	1–20	Citrus and fruit trees, onions, small grains
Boron	4.85	0.5–5	Alfalfa, clovers, sugar beets, cauliflower, celery, apples, other fruits
Molybdenum	0.61	0.05–1	Alfalfa, sweet clover, cauliflower, broccoli, celery

[a] Calculated from Burgess (4).

attention in recent years in soybean-growing areas, especially those of the south. Seed treatments of about $\frac{1}{4}$ ounce per acre have given good responses. These examples, along with those from muck areas and sandy soils of the southeast (particularly Florida), where micronutrients have been used for years, illustrate the need for these elements if optimum yields are to be maintained. Data on micronutrient use in different regions of the United States are shown in Table 18:4.

Marked differences in crop needs for micronutrients make fertilization a problem where rotations are being followed. On general crop farms, vegetables are sometimes grown in rotation with small grains and forages. If the boron fertilization is adequate for a vegetable crop such as red beets or even for alfalfa, the small-grain crop grown in the rotation is apt to show toxicity damage. Such a situation occurs commonly in western New York. These facts emphasize the need for specificity in determining crop nutrient requirements and for care in meeting these needs (see Table 18:5).

Macronutrient deficiencies are more universally apparent than are those of the trace elements. For many years, applications of fertilizers containing nitrogen, phosphorus, and potash have been a means of supplying small quantities of the micronutrients. Superphosphate, for example, may contain up to about 20 parts per million of boron. Chilean nitrate of soda supplies small quantities of several of the trace elements. These and other fertilizers will be considered in the next chapter.

REFERENCES

(1) Allaway, W. H., "Agronomic Controls over the Environmental Cycling of Trace Elements," *Adv. in Agron.*, **20**:235–74, 1968.

(2) Anonymous, "A Survey of Micronutrient Deficiencies in the U.S.A. and Means of Correcting Them," A report of the Soil Test Commission of the Soil Science Society of America (Madison, Wisc., 1965).

(3) Berger, K. C., "Micronutrient Deficiencies in the United States," *Jour. of Agr. and Food Chemistry*, **10**:178–81, 1962.

(4) Burgess, W. D., *The Market for Secondary and Micronutrients* (New York: Allied Chemical Corp., 1966).

(5) Dion, H. G., and Mann, P. J. G., *Jour. Agr. Sci.*, **36**:239–45, 1946.

(6) Jones, L. H. P., "The Solubility of Molybdenum in Simplified Systems and Aqueous Soil Suspension," *Jour. Soil Sci.*, **8**:313–27, 1957.

(7) U. S. Department of Agriculture, *Commercial Fertilizers, Consumption in the United States* (Washington, D.C.: Statistical Reporting Service, USDA, 1972).

(8) Viets, F. G., Jr., "The Plants' Needs For and Use of Nitrogen," in *Soil Nitrogen* (*Agronomy*, **10**) (Madison, Wisc.: American Society of Agronomy, 1965).

(9) Wilson, D. O., and Reisenauer, H. M., "Cobalt Requirements of Symbiotically Grown Alfalfa," *Plant and Soil*, **19**:364–73, 1963.

Chapter 19

FERTILIZERS AND
FERTILIZER MANAGEMENT

ALTHOUGH the use of animal excrement on cultivated soils was common as far back as agricultural records can be traced, mineral salts have been systematically and extensively employed for the encouragement of crop growth hardly more than 100 years. They are now an economic necessity on many soils. Any inorganic salt, such as ammonium nitrate, or an organic substance, such as sewage sludge, purchased and applied to the soil to promote crop development is considered to be a commercial fertilizer.

19:1. THE FERTILIZER ELEMENTS

There are at least fourteen essential nutrient elements that plants obtain from the soil. Two of these, calcium and magnesium, are applied as lime in regions where they are deficient. Although not usually rated as a fertilizer, lime does exert a profound nutritive effect. Sulfur is present in several commercial fertilizers and its influence is considered important, especially in certain localities. This leaves three elements other than the micronutrients —*nitrogen, phosphorus,* and *potassium.* And since they are so commonly applied in commercial fertilizers, they are often referred to as the *fertilizer elements.*

19:2. THREE GROUPS OF FERTILIZER MATERIALS[1]

Fertilizer materials are classified into three major groups on the basis of the material supplied: those supplying (a) nitrogen, (b) phosphorus, or (c) potassium. The classification is not so simple as this grouping would imply, however, since several fertilizer materials carry two of the elements of nutrition. As examples of this overlapping, potassium nitrate and ammonium phosphate may be cited. The situation will be fully apparent when the fertilizer tables presented later are examined and the compounding of mixed goods considered.

[1] For an excellent review of this subject, see Nelson (3).

Each of the three groups will be discussed in a general way, listing the various fertilizers of each and mentioning the outstanding characteristics of the more important ones. Detailed descriptions of the various fertilizers and their properties are available in published form if additional information is desired.

19:3. NITROGEN CARRIERS—TWO GROUPS

Nitrogen fertilizers may be divided for convenience into two groups: (a) *organic*, and (b) *inorganic*. Organic carriers include materials such as cottonseed meal, guano, and fish tankage. Because of their high cost per unit of nitrogen supplied, these nitrogen carriers are not used extensively, supplying less than 2 percent of the total nitrogen added on commercial fertilizers. However, some of them are used as specialty fertilizers for lawns, flower gardens, and potted plants. Nitrogen is released slowly from these organic materials by microbiological action. This helps provide a continuing supply of the element during the warm summer months.

19:4. INORGANIC NITROGEN CARRIERS

GENERAL CONSIDERATION. Many inorganic carriers are used to supply nitrogen in mixed fertilizers. The most important of these, together with their compositions and sources, are listed in Table 19:1. Fortunately, there is a wide range in the nitrogen contents of the materials—from 3 percent in ammoniated superphosphate to 82 percent in anhydrous ammonia. Also, several chemical forms are represented, including ammonium and nitrate compounds as well as materials such as urea and cyanamid. Both of the latter upon hydrolysis in the soil yield NH_4 ions which can be taken up by plants or can be oxidized to nitrates.

The materials listed in Table 19:1 have one thing in common—they can all be produced synthetically starting with atmospheric nitrogen. This fact has far-reaching significance, especially in respect to future nitrogen utilization. In the first place, it means that the quantity of nitrogen available to produce these compounds is limited only by the quantity present in the atmosphere. Second, synthetic methods of supplying nitrogen have played a significant role in reducing the cost of this element, historically so much more expensive than either potassium or phosphorus. In addition, synthetic processes have yielded a wide variety of materials in large enough quantities to make their usage practical. Such a variety was not possible when natural deposits alone were depended upon.

As a result of these and other features, synthetic nitrogen carriers are assuming more and more importance. Considerably more than three fourths of the fertilizer nitrogen used in the United States today is carried by synthetics. Their methods of manufacture are worthy of brief consideration.

TABLE 19:1. *Inorganic Nitrogen Carriers*

Fertilizer	Chemical Form	Source	Approximate Percent Nitrogen
Sodium nitrate	$NaNO_3$	Chile saltpeter and synthetic	16
Ammonium sulfate	$(NH_4)_2SO_4$	By-product from coke and gas, and also synthetic	21
Ammonium nitrate	NH_4NO_3	Synthetic	33
Cal-nitro and A.N.L.	NH_4NO_3 and dolomite	Synthetic	20
Urea	$CO(NH_2)_2$	Synthetic	42–45
Calcium cyanamid	$CaCN_2$	Synthetic	22
Anhydrous ammonia	Liquid NH_3	Synthetic	82
Ammonia liquor	Dilute NH_4OH	Synthetic	20–25
Nitrogen solutions	NH_4NO_3 in NH_4OH or urea in NH_4OH	Synthetic	27–53
Ammonium phosphate	$NH_4H_2PO_4$ and other ammonium salts	Synthetic	11 (48 % P_2O_5)
Diammonium phosphate	$(NH_4)_2HPO_4$	Synthetic	21 (53 % P_2O_5)

AMMONIA AND ITS SOLUTIONS. Perhaps the most important of the synthetic processes is that in which ammonia gas is formed from the elements hydrogen and nitrogen. This may be represented as follows:

$$N_2 + 3H_2 \rightarrow 2NH_3$$

This reaction is extremely important since it yields a compound which is at the present time the least expensive per unit of nitrogen of any listed in Table 19:1. Furthermore, the consumption of this material in the United States far exceeds that of any other nitrogen carrier (see Table 19:2). Of equal importance is the fact that this is the first step in the formation of many other synthetic compounds.

The ammonia gas formed in the reaction above is utilized in at least three ways. First, it may be liquefied under pressure, yielding anhydrous ammonia. This material is employed in the production of ammoniated superphosphate and mixed fertilizers in addition to being used as a separate material for direct application (see Fig. 19:1). Second, ammonia gas may

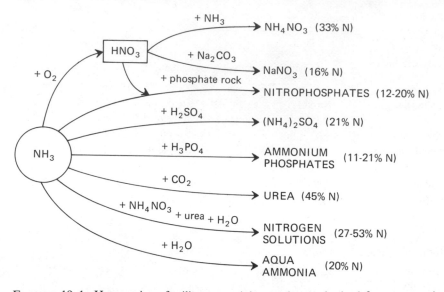

FIGURE 19:1. How various fertilizer materials may be synthesized from ammonia. This gas is obtained as a by-product of coke manufacture and, in even larger quantities, by direct synthesis from elemental nitrogen and hydrogen. In recent years, ammonia and the other synthetic materials shown have supplied most of our fertilizer nitrogen.

be dissolved in water, yielding NH_4OH. This is often used alone (ammonia liquor) but is used more frequently as a solvent for other nitrogen carriers such as NH_4NO_3 and urea, to give the "nitrogen solutions." In recent years, these solutions have become increasingly popular and now furnish a high proportion of the nitrogen in commercial fertilizers.

A third very important use of ammonia gas is in the manufacture of other inorganic nitrogen fertilizer materials. The reactions by means of which some fertilizer compounds are produced synthetically from ammonia are shown diagrammatically in Fig. 19:1. A careful study of this diagram will be very helpful in establishing the relationship between ammonia and the materials derived synthetically from it. These substances will be considered briefly.

SULFATE OF AMMONIA. Sulfate of ammonia is produced synthetically as shown in Fig. 19:1 and is also a by-product of coke manufacture. This material has long been one of the most important nitrogen carriers, especially in the manufacture of mixed fertilizers. Its nitrogen is cheaper than that carried by nitrate of soda but is more expensive than that of the liquid forms. Its use is most satisfactory on soils well supplied with lime since it has a residual acidifying effect (see p. 520).

SODIUM AND AMMONIUM NITRATES. Nitric acid, the manufacture of which is possible from oxidized ammonia, is used in making both ammonium and sodium nitrates. Ammonium nitrate, which has the advantages of supplying both ammonium and nitrate ions, has been employed in increasing quantities in the past few years. When used as a separate material, this salt is best supplied in the form of pellets which are suitably coated or otherwise treated to reduce deliquescence. It must always be remembered that this fertilizer is a hazard in case of fire because of its explosive properties. The percentage of nitrogen in ammonium nitrate carriers ranges from 20 percent for Calnitro and A.N.L. to perhaps 33 percent for the higher grades of ammonium nitrate (Table 19:2).

TABLE 19:2. *Consumption of Selected Fertilizer Materials in the United States in 1967 and Quantities of Plant Nutrients Supplied*[a]

Expressed as N, P_2O_5, and K_2O.

Materials	Thousands of Tons of Nutrients
Nitrogen Carriers (N)	
Anhydrous ammonia[b]	9,747
Nitrogen solutions	1,340
Ammonium nitrate	1,686
Urea	1,016
Ammonium sulfate	545
Phosphorus Carriers (P_2O_5)	
Concentrated superphosphate	1,472
Ordinary superphosphate	1,183
Ammonium phosphates	1,655
Potassium Carriers (K_2O)	
Muriate of potash	3,716
Sulfate of potash	184

[a] Nitrogen and phosphorus data courtesy E. A. Harre, Tennessee Valley Authority. Potash data from American Potash Institute and include Canadian as well as American deliveries in the United States.

[b] About 1.7 million tons of anhydrous ammonia are applied directly. The remainder is used to manufacture other fertilizers.

In addition to the synthetic source mentioned above, sodium nitrate is obtained as a natural product, saltpeter, from salt beds in Chile. Before the coming of synthetics, Chilean nitrate of soda represented essentially the sole source of this compound.

Sodium nitrate has long been an important inorganic source of commercial nitrogen. It supplies nitrogen in a form that immediately stimulates many crops even if the soil is cold. Hence, it is extremely valuable early in the spring and as a side dressing later in the season. Its high cost per unit of nitrogen, however, releases it to a minor position of usage.

UREA. Another promising synthetic is urea, a fertilizer containing almost three times as much nitrogen as nitrate of soda. It readily undergoes hydrolysis in the soil, producing ammonium carbonate as follows:

$$CO(NH_2)_2 + 2H_2O \rightarrow (NH_4)_2CO_3$$

Thus, the immediate effect of this fertilizer is toward alkalinity, although its residual influence tends to lower soil pH. The ammonium carbonate produced is ideal for rapid nitrification, especially if exchangeable bases are present in adequate amounts. Urea thus ultimately presents both ammonium and nitrate ions for plant absorption. Its one serious objection, high deliquescence, has been largely overcome by coating its particles with dry powders.

AMMONIUM PHOSPHATES. Of the synthetics carrying phosphorus in addition to nitrogen, the ammonium phosphates are perhaps most important. Both mono- and diammonium phosphates are available. These compounds are made from phosphoric acid and ammonia (see Fig. 19:1). Since their phosphorus as well as their nitrogen is water soluble, these compounds are in demand where a high degree of water solubility is required.

Ammonium polyphosphates are beginning to find a place especially in liquid fertilizers. These materials are very high in phosphorus (58–61 percent P_2O_5) and yet contain 12–15 percent nitrogen.

OTHER SYNTHETIC NITROGEN CARRIERS. Calcium cyanamid, another synthetic product, has declined in importance in the United States in recent years because it is relatively high in per unit cost of nitrogen. By acidulating rock phosphate with nitric rather than sulfuric or phosphoric acids, fertilizers containing nitraphosphates are formed. They are apparently as effective as other materials in supplying nitrogen but are used primarily in the manufacture of complete fertilizers.

SLOW-RELEASE NITROGEN CARRIERS. For some purposes nitrates and to a lesser degree the other common nitrogen materials have the disadvantage of too ready availability. For example, the home owner wants a material which can be applied to the lawn in the spring with the expectation that it will keep his grass green throughout the summer. Materials have been developed which at least partially meet such slow-release requirements.

Urea-formaldehyde complexes were among the first slow-release synthetic compounds produced. They contain about 38 percent nitrogen which is very slowly available. Other slow-release materials commercially available are floranid (crotonylideneduirea), which contains 28 percent nitrogen and magnesium ammonium phosphate (8 percent N and 17.5 percent P_2O_5). The rate of release of nitrogen with these materials is dependent primarily on particle size, making possible a wide range of rates of release. The major difficulty with these materials as well as experimental products such as oxamid $(CONH_2)_2$ produced by the Tennessee Valley Authority is their relatively high cost. This has up to now limited their use to specialty crops, lawns, and turfs.

Another approach to the problem of too rapid nitrogen release is that of coating conventional fertilizer with materials which slow down their rate of solution and microbial attack. Waxes, paraffin and acrylic resins, and elemental sulfur are among the materials which have been used with some success. They slow down the rate of moisture penetration of the granule and the outward movement of the soluble nitrogen (see Fig. 19:2).

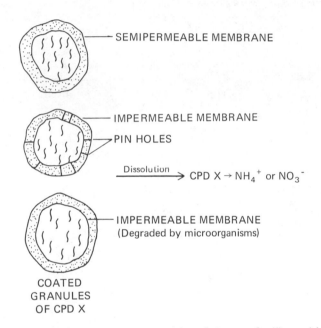

FIGURE 19:2. Three methods of coating granules of nitrogen fertilizer with plastics, paraffins, or elemental sulfur to reduce the rate of release of the nitrogen. (a) Coating with a semipermeable membrane prevents movement of the nitrogen compound into the soil while permitting water movement inward; (b) an impermeable membrane with pin holes permits slow outward movement; (c) a membrane subject to microbial breakdown eventually releases the nitrogen compound.

Materials which delay and prevent nitrification have also been developed. Their purpose is to keep the nitrogen in the ammonium form so as to slow down its rate of assimilation by plants and its leaching from the soil. Two products that have been used are 2-chloro-6 (trichloromethyl)-pyridine ("N-Serve") and potassium azide (KN_3). There are other such compounds aimed at slowing down nitrification.

19:5. PHOSPHATIC FERTILIZER MATERIALS

SUPERPHOSPHATE. The principal phosphorus fertilizer material at the present time is superphosphate (see Table 19:3). The ordinary grades containing 16 to 21 percent available P_2O_5 (7–9 percent P) are made by treating raw rock phosphate with suitable amounts of sulfuric acid (see Fig. 19:3). A large proportion of the phosphorus is thus changed to the primary phosphate form ($Ca(H_2PO_4)_2$), although some is left in the secondary condition ($CaHPO_4$).

TABLE 19:3. *Phosphorus Carriers*

Fertilizer	Chemical Form	Approximate Percentage of Available P_2O_5	Phosphorus (%)
Superphosphates	$Ca(H_2PO_4)_2$ and $CaHPO_4$	16–50	7–22
Ammoniated superphosphate	$\begin{cases} NH_4H_2PO_4 \\ CaHPO_4 \\ Ca_3(PO_4)_2 \\ (NH_4)_2SO_4 \end{cases}$	16–18 (3–4% N)	7–8
Ammonium phosphate	$NH_4H_2PO_4$ mostly	48 (11% N)	21
Ammonium polyphosphates	$(NH_4)_4P_2O_7$ and others	58–60 (12–15% N)	
Diammonium phosphate	$(NH_4)_2HPO_4$	46–53 (21% N)	20–23
Basic slag	$(CaO)_5 \cdot P_2O_5 \cdot SiO_2$	15–25	7–11
Steamed bone meal	$(Ca_3PO_4)_2$	23–30	10–13
Rock phosphate	Fluor and chlor apatites	25–30	11–13
Calcium metaphosphate	$Ca(PO_3)_2$	62–63	27–28
Phosphoric acid	H_3PO_4	54	24
Superphosphoric acid	H_3PO_4 and $H_4P_2O_7$	76	33

By representing the complex raw rock by the simple formula $Ca_3(PO_4)_2$, the following conventional reactions may be used to show the changes that occur during the manufacture of ordinary 16 to 21 percent superphosphate.

$$Ca_3(PO_4)_2 + 2H_2SO_4 \rightarrow Ca(H_2PO_4)_2 + 2CaSO_4 + \text{impurities}$$
(Insoluble)   (Water soluble)

The acid is never added in amounts capable of completing this reaction.

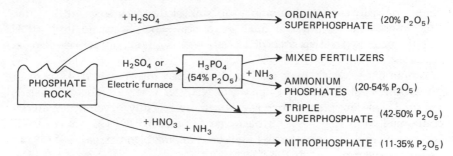

FIGURE 19:3. How several important phosphate fertilizers are manufactured. [*Concept from Travis Hignett, Tennessee Valley Authority*.]

Consequently, some secondary phosphate—$CaHPO_4$, referred to as *citrate soluble phosphoric acid*, is produced.

$$Ca_3(PO_4)_2 + H_2SO_4 \rightarrow 2CaHPO_4 + CaSO_4 + \text{impurities}$$
(Insoluble)

Much of the superphosphate now used is this ordinary grade, which consists of about 31 percent phosphates, 50 percent gypsum, and 19 percent impurities of various kinds. The total *phosphorus* to the hundredweight is rather low, ranging from 7 to 9 pounds, and the sulfur and calcium amount to about 12 and 18 pounds.

The high-analysis treble superphosphate contains 40 to 47 percent available P_2O_5 (17–21 percent P). It differs from the orindary type principally in that it contains more phosphorus and no gypsum.

The *treble* superphosphate is synthesized by treating a high-grade phosphate rock with phosphoric acid:

$$Ca_3(PO_4)_2 + 4H_3PO_4 \rightarrow 3Ca(H_2PO_4)_2 + \text{impurities}$$

This material is becoming more and more widely used since higher analysis fertilizers have become accepted. Ordinary superphosphate, 16 to 21 percent available P_2O_5 (7–9 percent P), is still commonly used, although considerably less so in recent years than the higher analysis fertilizers.

Since superphosphate gives a very acid reaction when tested with indicators, it is generally supposed that it must increase the acidity of soils to which it is added. As a matter of fact, it has practically no effect within the ordinary pH range. But at a low-pH superphosphate tends to reduce acidity, whereas at a pH of 7.5 to 8.5 the modification is in the other direction. The slight influence of this fertilizer on soil reaction probably is due to the vigorous fixation of phosphorus that occurs as soon as it contacts the soil.

AMMONIATED PHOSPHATES. Ammoniated superphosphate contains from 3 to 4 percent nitrogen and 16 to 18 percent phosphoric acid. It affords a chance of easily changing ammonia to a suitable fertilizer form, at the same time improving the physical qualities of the superphosphate itself. It is usually made by treating superphosphate with ammonia liquor or nitrogen solutions.

Ammo-phos, which may analyze 11 percent nitrogen and 48 percent phosphoric acid, is also an economical fertilizer when a higher analysis is required. Diammonium phosphate is a more recently developed material containing up to 21 percent nitrogen and 53 percent P_2O_5 (23 percent P). It is very valuable as one of the materials used in bulk blending (see p. 518) and as a constituent of high-analysis fertilizers.

Ammonium polyphosphates made by ammoniating superphosphoric acid show great potential in manufacturing of liquid fertilizers. They help prevent the precipitation of iron and other impurities and keep most of the micronutrients in solution.

BASIC SLAG AND BONE MEAL. Basic slag, commonly used in Europe, is on the market only to a limited extent in the United States. Because of the alkalinity and the rather ready availability of its phosphoric acid, it is a very desirable phosphatic fertilizer. It seems to be especially effective on acid soils, apparently because of its high content of calcium hydroxide.

Bone meal is an expensive form of phosphoric acid. Moreover, it is rather slowly available in the soil. Bone meal can be applied in large amounts and yet produce no detrimental influence on crop growth.

ROCK PHOSPHATE. Raw rock phosphate, because of its insolubility, must be finely ground if it is to react at all readily when applied to the soil. Its availability is markedly increased by the presence of decaying organic matter. Rock phosphate is the least soluble of the phosphatic fertilizers mentioned, the order being: ammonium phosphates and superphosphate, basic slag, bone meal, and raw rock. Although its chemical formula is given conventionally as $Ca_3(PO_4)_2$, rock phosphate is much more complicated than this formula would suggest. It apparently approaches fluor apatite $[3Ca_3(PO_4)_2 \cdot CaF_2]$ in its molecular makeup (see p. 460). This, no doubt, accounts in part for its slow availability.

Finely ground rock phosphate is most effective when added to soils high in organic matter. Because of its low solubility, however, its use will continue to be as a source for the manufacture of other and more soluble forms.

HIGH-ANALYSIS PHOSPHATES. Mention should be made of two very high analysis phosphate fertilizers not as yet in wide use—calcium metaphosphate $Ca(PO_3)_2$—62 to 63 percent available P_2O_5 (27–28 percent P) and superphosphoric acid containing 76 percent P_2O_5 (33 percent P) (see Table 19:3). Both are more or less in the experimental stage but hold great promise since they seem to be as effective, when used in equivalent amounts, as superphosphate. Their concentrations make them extremely attractive when transportation is a factor.

$Ca(PO_3)_2$, commonly called *meta-phos*, may be made by treating either phosphate rock or limestone with phosphorus pentoxide (Table 19:3). The highly concentrated P_2O_5 can be produced at the mine, shipped to a point where limestone is cheaply available, and the meta-phos manufactured in the territory where it is to be used. Transportation costs are thus cut to a minimum.

Superphosphoric acid, a new synthetic product, is the highest phosphorus-containing material used in fertilizer manufacture today (Table 19:3). Its P_2O_5 content is 76 percent (33 percent P). It is made of a mixture of ortho-phosphoric, pyrophosphoric, and other polyphosphoric acids. This liquid can be used in the manufacture of other liquid fertilizers, or to make a high-analysis superphosphate containing 54 percent P_2O_5 (24 percent P). In the formulation of liquid fertilizers, the polyphosphates help keep iron, aluminum, and micronutrients in solution.

CLASSIFICATION OF PHOSPHATE FERTILIZERS. For purpose of evaluation and sale, the various phosphorus compounds present in phosphatic fertilizers are classified in an arbitrary yet rather satisfactory way as follows:

1. *Water-soluble* $\begin{cases} Ca(H_2PO_4)_2 \\ NH_4H_2PO_4 \\ K \text{ phosphate} \end{cases}$

2. *Citrate-soluble (In 15 percent neutral ammonium citrate)*—$CaHPO_4$ — Available.

3. *Insoluble*
   Phosphate of bone and raw phosphate rock — Unavailable.

It is well to note that the classification of the phosphates is in some degree artificial as well as arbitrary. For instance, *available* phosphates, because of the reversion that occurs, are not strictly available once they contact the soil. The term "available" really refers to those phosphates that readily stimulate plant growth. Those phosphates that are less effective are rated as currently *unavailable*, yet in the soil they may eventually supply a certain amount of phosphorus to crops (see Fig. 19:4).

19:6. FERTILIZER MATERIALS CARRYING POTASSIUM

Potassium is obtained primarily by mining underground salt beds, large deposits being found in France, Germany, Canada, and the United States. Brine from salt lakes is also a source of some importance.

Kainit and manure salts (see Table 19:4) are the most common of the crude potash sources. The high-grade chloride and sulfate of potash originally imported from Germany and France are their refined equivalents. The

WATER-INSOLUBLE PHOSPHATE

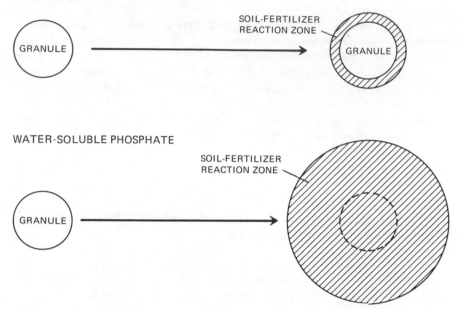

WATER-SOLUBLE PHOSPHATE

FIGURE 19:4. Reaction of water-soluble and water-insoluble phosphates with soil. The water-insoluble granules react with only a small volume of soil in their immediate vicinity. Soluble phosphates move into the soil from water-soluble granules, reacting with iron, aluminum, and manganese in acid soils, with calcium in alkaline soils. [*From Englestad* (2).]

potassium chloride now used so extensively in the United States comes mostly from underground deposits found both in Canada and in the United States.

TABLE 19:4. *Potash Fertilizer Materials*

Fertilizer	Chemical Form	Percentage Expressed as:	
		K_2O	K
Potassium chloride[a]	KCl	48–60	40–50
Potassium sulfate	K_2SO_4	48–50	40–42
Sulfate of potash-magnesia[b]	Double salt of K and Mg	25–30	19–25
Manure salts	KCl mostly	20–30	17–25
Kainit	KCl mostly	12–16	10–13
Potassium nitrate	KNO_3	44 (and 13% N)	37

[a] All of these fertilizers contain other potash salts than those listed.
[b] Contains 25 percent of $MgSO_4$ and some chlorine.

All potash salts used as fertilizers are water soluble and are therefore rated as readily available. Unlike nitrogen salts, most potassium fertilizers, even if employed in large amounts, have little or no effect on the soil pH. Some discrimination is made, however, against potassium chloride (muriate) in respect to potatoes and especially tobacco since large dosages are considered to lower the quality. Hence, when large amounts of potash are to be applied for tobacco, the major part usually is preferred in the sulfate form.

Potassium-magnesium sulfate, although rather low in potash, has attained considerable usage in parts of this country where magnesium is likely to be deficient. In some respects it is a more desirable source of magnesium than is dolomitic limestone or dolomite.

19:7. SULFUR IN FERTILIZERS

Fifteen years ago there was little reason to be concerned with sulfur deficiency in most soils of the United States. Additions of sulfur in rain and snow varied from a few pounds per acre annually to as much as 100 pounds near industrial centers. Just as important was the rather automatic additions of sulfur in the mixed fertilizers then in use. These almost invariably contained two sulfur-containing ingredients, ordinary superphosphate (which contains gypsum or calcium sulfate) and ammonium sulfate.

In recent years two developments have made it necessary to reconsider the needs for fertilizer sulfur. Public concern over atmospheric pollution has resulted in marked reductions in sulfur additions to agricultural lands near cities and industrial sites. Also, the trend in recent years toward high-analysis fertilizers has made it necessary to replace in many mixed fertilizers the superphate and ammonium sulfate with low-sulfur materials such as treble superphosphate, the ammonium phosphates, and anhydrous ammonia. In areas far removed from atmospheric sources of sulfur, such as eastern Washington and Oregon, parts of the southeast, and of the midwest, sulfur deficiencies have been noted. These are being met by adjusting the fertilizer constituents to include sulfur-containing materials. Thus, the automatic supply of sulfur through fertilizers can no longer be taken for granted. This element must be consciously added where deficiencies are apt to occur.

19:8. MICRONUTRIENTS

Micronutrient additions to fertilizers must be much more carefully controlled than is the case for the macronutrients. The difference between the amount of a given micronutrient present when deficiency occurs and when there is a toxicity is extremely small. Consequently, micronutrients should be added only when their need is certain and when the amount required is known.

When a trace-element deficiency is to be corrected, especially if the case is urgent, a salt of the lacking nutrient often is added separately to the soil (see Table 19:5). Copper, manganese, iron, and zinc generally are supplied as the sulfate and boron is applied as borax. Molybdenum is added as sodium molybdate. Iron and in some cases zinc may be supplied as a chelate (see p. 493). Iron, manganese, and zinc are sometimes sprayed in small quantities on the leaves rather than being applied directly to the soil. "Fritted" silicate compounds of boron, manganese, iron, and zinc may also be used to supply these nutrients.

TABLE 19:5. *Salts of Micronutrients Commonly Used in Fertilizers*[a]

Compound	Formula	Nutrient Content
Copper sulfate	$CuSO_4$	25–35% Cu
Basic copper sulfate	$CuSO_4 \cdot 3Cu(OH)_2$	13–53% Cu
Copper carbonate (basic)	$CuCO_3 \cdot Cu(OH)_2$	57% Cu
Zinc sulfate	$ZnSO_4$	23–35% Zn
Zinc sulfate (basic)	$ZnSO_4 \cdot 4Zn(OH)_2$	55% Zn
Manganese sulfate	$MnSO_4$	23% Mn
Manganese sulfate (basic)	$2MnSO_4 \cdot MnO$	40–49% Mn
Sodium borate	$Na_2B_4O_7$	34–44% B_2O_3
Ferrous sulfate	$FeSO_4$	20% Fe
Ferric sulfate	$Fe_2(SO_4)_3$	17% Fe
Sodium molybdate	Na_2MoO_4	37–39% Mo

[a] From Nikitin (4).

In recent years there has been an increase in the use of chelates to supply iron, zinc, manganese, and copper (see p. 495). These materials are especially useful on soils of high pH where mineral sources would be quickly rendered unavailable. Because of their high cost, however, these materials are often used as foliar sprays, which permits material reduction in the rate of application.

The rate of application of micronutrients should be carefully regulated since an overdose can cause severe damage. For instance, 50 pounds of borax per acre is nearing the maximum for the average soil when its pH is near 7. A few ounces of molybdenum is often all that is required. Much, of course, depends upon the crop grown. In case of doubt as to the amounts to apply, expert advice should be sought.

In some cases trace elements are placed in ordinary commercial fertilizers and their presence therein stated in the guarantee. One objection to such a practice is that not enough of any one trace element may be added to the soil to adequately meet a real deficiency should it occur. Furthermore, the amount of trace element applied varies with the dosage of the commercial fertilizer. Trace elements usually require a more careful regulation than can be attained in this way.

19:9. MIXED FERTILIZERS

For years, fertilizer manufacturers have placed on the market mixtures of materials which contain at least two of the "fertilizer elements," and usually all three. Materials such as those discussed in previous sections are mixed in proper proportion to furnish the desired amount of the nutrient elements. For example, an ammonia solution, triple superphosphate, muriate of potash, and a very small amount of an organic might be used if a complete fertilizer is desired. Such fertilizers supply about two thirds of the total fertilizer nutrient consumption in the United States, the remainder coming from separate materials. The nutrient use per acre for various regions of the United States is shown in Fig. 19:5.

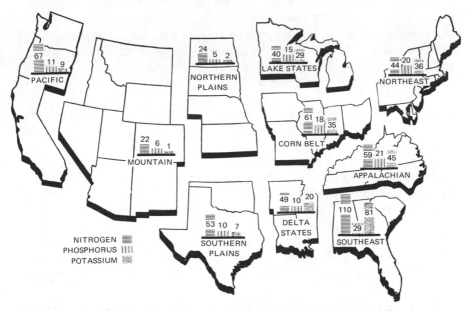

FIGURE 19:5. Nutrients added in fertilizers applied to cropland in different regions of the United States. While the Southeast uses more nutrients per acre than the other regions, fertilizer usage has increased most dramatically during the past few years in the central part of the country. [*Courtesy U. S. Department of Agriculture.*]

PHYSICAL CONDITION. In addition to supplying nitrogen, phosphorus, and potassium in desired proportions, however, a commercial fertilizer should have a good *physical condition*. It should be drillable when first purchased and should remain in this condition after storage. Mixtures of certain materials cannot be used because of their tendency to "set up" or harden. Of the fertilizer materials commonly found in mixed goods, ammonium nitrate, sulfate of ammonia, and potassium chloride are most likely to develop unsatisfactory physical conditions. The extreme deliquescence of some of the salts, especially urea and ammonium nitrate, tends to make mixed fertilizers sticky and thus reduce the drillability.

METHODS OF ENSURING FREE FLOWAGE. The free-flowing condition of a mixed fertilizer is maintained by using (a) moisture-resistant bags, (b) by including certain moisture-absorbing materials in the fertilizer mix, and (c) by granulating the mixed fertilizers. Separate materials such as sodium and ammonium nitrates have been pelleted for several years to prevent caking. In recent years the granulation of mixed fertilizers has become commonplace in this country, most of these fertilizers being sold in granular condition.

In addition to having free-flowing properties, granulated fertilizers present certain other advantages. For example, they are less disagreeable to handle since they are comparatively free of small dustlike particles. The granulation also prevents the fertilizer from being carried by the wind, thus permitting more uniform spreading. Also, granulation tends to reduce the rate of reaction of the fertilizer with the soil.

BULK BLENDING. Twenty years ago, essentially all mixed fertilizers were bagged at the manufacturing plant and then shipped to distribution points where they were sold to farmers. The farmers applied the fertilizers usually at the same time other operations, such as planting or plowing, were done. Today, bulk handling and blending of fertilizers has become more and more common, about 45 percent of the fertilizer applied in the U. S. being so handled. Granular materials are shipped in bulk to a blending plant located within a 25- to 30-mile radius of the intended use of the fertilizer (see Fig. 19:6). The fertilizers are mixed to the customer's order and move directly to the farm, where they are spread on the field. Frequently the trucks used to do the spreading are owned and operated on a custom basis by the blending plant.

The most obvious advantage of bulk blending and handling is to reduce labor. Furthermore, costs of storing, production, transportation, and spreading are all lower, at least for the medium- to large-sized farms, than the conventional means of handling fertilizers. Also, the fertilizers are generally high in analysis, and chemical incompatibility of the fertilizer materials is generally not a problem. Last, a variety of nutrient ratios can easily be formulated.

A problem of some concern in bulk blending is the segregation of the materials which could result in uneven distribution of the fertilizer nutrients in the field. Care must be taken to use materials wherein the granules are about the same size and density. Since this is not always possible, special mixing hoppers have been developed.

FIGURE 19:6. Bulk blending fertilizer plant and bulk spreading trucks. Different fertilizer analyses can be mixed to order and spread by truck on a custom basis. [*Photo courtesy Harold Sweet, Agway, Inc.,* Syracuse, N.Y.]

Fertilizer materials commonly used in the bulk blending process are urea, ammonium nitrate, ammonium sulfate, ammonium phosphates, triple superphosphate, and potassium chloride.

LIQUID FERTILIZERS. Another innovation in the formulating and handling of mixed fertilizers is the use of liquid fertilizers. These materials are used extensively in California, where the practice started, and in the North Central States. Nationally, nearly one third of the total fertilizers are applied in the liquid form. As with bulk blending, liquid fertilizers have the advantage of low labor costs since the materials are handled in tanks and pumped out for transfer or for application. However, the cost per unit of nutrient element is usually higher, more sophisticated equipment is needed for storage and handling, and only relatively low analysis mixed fertilizers can be made. (For example, 5–10–10, 7–7–7, and 6–18–6 are common analyses.) Slightly higher analyses are possible by using superphosphoric acid or ammonium polyphosphates as a source of at least part of the phosphorus. Also, "suspension fertilizers," wherein a small amount of solids is suspended in the liquid, permit higher analyses, such as 15–15–15.

19:10. Effect of Mixed Fertilizers on Soil pH

ACID-FORMING FERTILIZERS. Most complete fertilizers, unless specially treated, tend to develop an acid residue in soils. This is mainly due to the influence of certain of the nitrogen carriers, especially those which supply ammonia or produce ammonia when added to the soil. The major effect of ammonium ions is exerted when they are nitrified. Upon oxidation the ammonium compounds tend to increase acidity, as shown by the following reaction:

$$NH_4^+ + 2O_2 \rightarrow 2H^+ + NO_3^- + H_2O$$

In addition to ammonium compounds, materials such as urea and some of the organics, which upon hydrolysis yield ammonium ions, are potential sources of acidity. The phosphorus and potash fertilizers commonly used have little effect upon soil pH unless they also contain nitrogen. The approximate acidifying capacity of some fertilizer materials expressed in pounds of calcium carbonate per 20 pounds of nitrogen supplied are as follows:

Ammonium sulfate	107	Ammonium nitrate	36
Ammo-phos	100	Cottonseed meal	29
Anhydrous ammonia	36	Castor pomace	18
Urea	36	High-grade tankage	15

The basicity in the same terms for certain fertilizers is as follows:

Nitrate of soda	36	Tobacco stems	86
Calcium cyanamid	57	Cocoa meal	12

NON-ACID-FORMING FERTILIZERS. In some instances, the acid-forming tendency of nitrogen fertilizers is completely counteracted by adding dolomitic limestone to the mixture. Such fertilizers are termed *neutral* or *non-acid-forming* and exert little residual effect on soil pH. Data are available in respect to nitrogen fertilizers that make it rather easy to calculate the approximate amount of dolomitic limestone necessary in any particular case. It should be pointed out, however, that it is often economically preferable to use acid-forming fertilizers and to neutralize the soil acidity with separate bulk applications of lime.

19:11. The Fertilizer Guarantee

Every fertilizing material, whether it is a single carrier or a complete ready-to-apply mixture, must carry a guarantee as to its content of nutrient elements. The exact form is generally determined by the state in which the fertilizer is offered for sale. The *total nitrogen* is usually expressed in its *elemental form* (N). The phosphorus is quoted in terms of *available phosphorus*

(P) or *phosphoric acid* (P_2O_5); the potassium is stated as *water-soluble potassium* (K) or *potash* (K_2O).

The guarantee of a simple fertilizer such as sulfate of ammonia is easy to interpret since the name and composition of the material is printed on the bag or tag. The interpretation of an analysis of a complete fertilizer is almost as easy. The simplest form of guarantee is a mere statement of the relative amounts of N, P, and K, or N, P_2O_5, and K_2O. Thus, an 8–16–16 fertilizer contains 8 percent *total nitrogen*, 16 percent *available* P_2O_5 (7 percent available P), and 16 percent *water-soluble* K_2O (13 percent water-soluble K). In some states, figures as to *water-soluble nitrogen, water-insoluble nitrogen,* and *available insoluble nitrogen* are required by law.

TABLE 19:6. *Open Formula Guarantee of an 8–16–16 (Acid-Forming) Fertilizer*

| | | Pounds of Nutrients | | | |
| | Total N | Available | | Soluble | |
Ingredients per (lb/ton)		P_2O_5	P	K_2O	K
300 nitrogen solution	120				
100 sulfate of ammonia	20				
100 diammonium phosphate	20	52	23		
300 treble superphosphate		135	59		
666 superphosphate		133	58		
534 muriate of potash				320	266
2,000 8–16–16 (total)	160	320	140	320	266

An *open formula* guarantee is used by some companies. Such a guarantee not only gives the usual chemical analysis but also a list of the various ingredients, their composition, and the pounds of each in a ton of the mixture. Such a guarantee is given in Table 19:6 for an 8–16–16 fertilizer with the nutrients expressed as N, P_2O_5, and K_2O.

Commercial fertilizers are sometimes grouped according to their nutrient *ratio*. For instance, a 5–10–10, an 8–16–16, a 10–20–20, and a 15–30–30 all have a 1–2–2 ratio. These fertilizers should give essentially the same results when suitably applied in equivalent amounts. Thus, 1,000 pounds of 10–20–20 furnishes the same amounts of N, P, and K as does a ton of 5–10–10.

This grouping according to ratios is valuable when several analyses are offered and the comparative costs become the deciding factor as to which should be purchased. Moreover, fertilizer recommendations are sometimes made on the basis of the ratio, the particular analysis being decided on later.

19:12. FERTILIZER INSPECTION AND CONTROL

So many opportunities are open for fraud, either as to availability or the actual quantities of ingredients present, that laws controlling the sale of fertilizers are necessary. These laws apply not only to the ready-mixed goods but also to the separate carriers. Such regulations protect not only the public but also the reliable fertilizer companies since goods of unknown value are kept off the market. The manufacturer is commonly required to print the following data on the bag or on an authorized tag:

1. Number of net pounds of fertilizer to a package.
2. Name, brand, or trademark.
3. Chemical composition guaranteed.
4. Potential acidity in terms of pounds of calcium carbonate per ton.
5. Name and address of manufacturer.

An example of how this information often appears on the fertilizer bag is shown in Fig. 19:7.

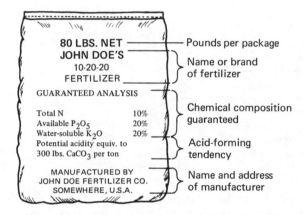

FIGURE 19:7. Fertilizer bag on which is printed the information commonly required by law in most states. Note the acid-forming tendency of this particular fertilizer.

19:13. FERTILIZER ECONOMY

HIGH VERSUS LOW ANALYSES. Whether buying mixed fertilizers or the various separate carriers such as ammonium nitrate, sulfate of ammonia, and the like, it is generally advantageous to obtain high-analysis goods. Price data indicate that the higher the grade of a mixed fertilizer, the lower is the nutrient cost. This is shown graphically in Fig. 19:8. Obviously, from the standpoint of economy, an 8–16–16 fertilizer furnishes more nutrients *per dollar* than a 4–8–8 or a 5–10–10.

In the past, certain disadvantages of high-analysis fertilizers have some-what hampered their increased usage. For example, the equipment used to place the fertilizer in the soil was too crude to distribute satisfactorily the smaller amounts of the more concentrated materials. Also, when improperly placed, the more concentrated goods often resulted in injury to young

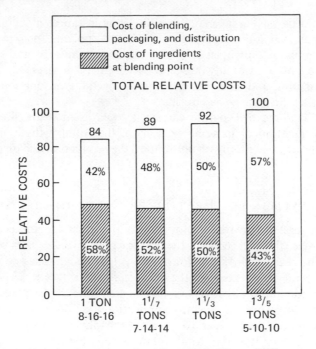

FIGURE 19:8. Relative farm costs of nutrients in fertilizer of different concentrations. Note that the same amount of *nutrients* is supplied by the indicated quantities of each fertilizer. These nutrients are furnished more cheaply by the higher analysis goods, mainly because of the lower handling costs.

seedlings. Modern fertilizer-placement machinery, plus an increasing tendency to bulk spread at least part of the fertilizer, has largely overcome these difficulties. Hence, the savings which result from the use of higher analysis materials seem to far outweigh any associated disadvantages that may now exist.

RELATIVE COSTS OF N, P_2O_5, AND K_2O. Another price factor to remember in purchasing fertilizers is the relative costs of nitrogen, phosphorus, and potassium. Nitrogen has generally been more expensive than the other two constituents. This price differential has been somewhat reduced, however, due to the production of synthetic nitrogen carriers. Phosphorus per unit of P_2O_5 is next most expensive, and potassium is the least expensive of all. However, there is considerable variation in the cost of different carriers for a given element. Thus, the manufactured cost of nitrogen in some solid carriers (for example, sodium nitrate) is about double that in anhydrous ammonia. Likewise, sulfate of potash is considerably more expensive than an equivalent amount of potassium chloride.

PURCHASING OF SEPARATE MATERIALS. In many instances a saving can be effected by purchasing the separate materials and applying them to the soil. Anhydrous ammonia, nitrogen solutions, ammonium nitrate, and sulfate of ammonia are employed to advantage in orchards, as top dressings on meadows and pastures, and as a side dressing in vegetable production and the growing of such field crops as cotton.

The bulk handling of separate carriers as well as mixtures thereof has already been mentioned. In some sections of the country these and other developments have essentially revolutionized the production and distribution of fertilizers.

19:14. MOVEMENT OF FERTILIZER SALTS IN THE SOIL

To fully understand the reasons for some of the specific methods of applying fertilizers (see the next section), the movement of fertilizer salts within the soil should be reviewed. As previously indicated, phosphorus compounds tend to move but little, except in the more sandy soils (see p. 224). Consequently, for maximum effectiveness this element should be placed in the zone of root development. Surface applications, unless worked into the soil, do not supply the deeper roots with phosphorus. Also, because of phosphorus immobility, the total quantity of this element necessary for a given season can be applied at one time without fear of loss by leaching.

On the other hand, potassium, and to an even greater extent, nitrogen, tend to move from their zones of placement. This movement is largely *vertical*, the salts moving up or down depending upon the direction of water movement. These translocations greatly influence the time and method of applying nitrogen and potassium. For example, it is often undesirable to supply nitrogen all in one annual application because of the leaching hazard. This tendency for the downward movement of nitrates is taken advantage of, however, in subsequent top dressings, where this fertilizer is applied on the surface of the ground. Water movement in the latter case is depended upon to carry the dissolved nitrate salts down to the plant roots. Top dressings of nitrogen solutions and urea may present problems in some cases because of the danger of volatilization (see p. 431).

The movement of nitrogen and, to a lesser degree, potassium must be considered in the *placement* of the fertilizer with respect to the seed. If the fertilizer salts are located in a band directly under the seeds, the upward movement of nitrates and some of the potassium salts by capillary water often results in injury to the stand. Rain immediately after planting followed by a long dry spell encourages such damage to seedlings. Placing the fertilizer immediately above the seed or on the soil surface also may result in injury, especially to row crops. The possibility of such injuries should be kept in mind in reading the following section.

19:15. METHODS OF APPLYING SOLID FERTILIZERS

Much emphasis is placed on the selection of the correct fertilizer ratio and on the adequate and economical amounts of the various fertilizers to be used. However, the method of application is equally as important and must not be overlooked. A fertilizer should be placed in the soil in such a position that it will serve the plant to the best advantage. This involves not only different zones of placement but also the time of year the fertilizer is to be applied. The methods of application will be discussed on the basis of the crops to be fertilized.

ROW CROPS. Cultivated crops such as corn, cotton, and potatoes are usually fertilized in the *hill* or the *row*, part or all of the fertilizer being applied at the time of planting. If placed in the hill, the fertilizer may be deposited slightly below and on one side, or better, on both sides of the seed. When applied to the row, the fertilizer usually is laid in as a narrow *band* on one or both sides of the row, 2 or 3 inches away and a little below the seed level (see Fig. 19:9).

FIGURE 19:9. Best fertilizer placement for row crops is to the side and slightly below the seed. This eliminates danger of fertilizer "burn" and concentrates the nutrients near the seed. [*Photo courtesy National Plant Food Institute, Washington, D.C.*]

When the amount of fertilizer is large, as is often the case with vegetable crops, it is often wise to broadcast part and thoroughly work it into the soil before the planting is done. In some cases the crop is side-dressed with an additional amount of fertilizer later in the season. This practice involves placing the fertilizer along the side of the row at a time most satisfactory to the crop. This requires experience and good judgment.

SMALL GRAINS. With small grains such as wheat the drill is equipped with a fertilizer distributor, the fertilizer entering the soil more or less in contact with the seed. As long as the fertilizer is low in analysis and the amount applied does not exceed 300 or 400 pounds per acre, germination injury is not serious. Higher rates, especially of high-analysis fertilizers, may result in serious injury if the seed and fertilizer are placed together. The more modern grain drills are equipped to place the fertilizer alongside the seed rather than in contact with it.

PASTURES AND MEADOWS. With meadows, pastures, and lawns, it is advisable to fertilize the soil well at the time of seeding. The fertilizer may be applied with the seed or, better, broadcasted, and worked thoroughly into the soil as the seedbed is prepared. The latter method is preferable, especially if the fertilization is heavy. During succeeding years it may be necessary to top-dress such crops with a suitable fertilizer mixture. This requires care. The amount of fertilizer applied and the time of treatment should be so regulated as to avoid injury to the foliage and to the root crowns of the plants.

TREES. Orchard trees usually are treated individually, the fertilizer being applied around each tree within the spread of the branches but beginning several feet from the trunk. The fertilizer is worked into the soil as much as possible. When the orchard cover crop needs fertilization, it is treated separately, the fertilizer being drilled in at the time of seeding or broadcast later.

Ornamental trees are often fertilized by what is called the *perforation* method. Numerous small holes are dug around each tree within the outer half of the branch-spread zone and extending well into the upper subsoil. Into these holes, which are afterward filled up, is placed a suitable amount of an appropriate fertilizer. This method of application places the nutrients within the root zone and avoids an undesirable stimulation of the grass that may be growing around ornamental trees.

19:16. APPLICATION OF LIQUID FERTILIZERS

The use of liquid materials as a means of fertilization is assuming considerable importance in certain areas of this country. Three primary methods of applying liquid fertilizers have been used: (a) direct application to the soil, (b) application in irrigation water, and (c) the spraying of plants with suitable fertilizer solutions.

DIRECT APPLICATION TO SOIL. The practice of making direct applications of anhydrous ammonia, nitrogen solutions, and mixed fertilizers to soils is rapidly increasing throughout the United States. In each case equipment is needed designed specifically to handle the chemical in question. Carbon steel or plastic-lined containers are used for mixed fertilizers, whereas aluminum containers are needed for most of the nitrogen solutions, and mild steel is used for anhydrous ammonia. Pumps are needed to transfer and apply the aqua ammonia, liquid mixes, and no-pressure solutions. Anhydrous ammonia and pressure solutions must be injected into the soil to prevent losses by volatilization. Depths of 6 and 2 inches, respectively, are considered adequate for these two materials.

APPLICATION IN IRRIGATION WATER. In the west, particularly in California and Arizona, there is some application of liquid fertilizers in irrigation waters. Liquid ammonia, nitrogen solutions, phosphoric acid, and even complete fertilizers are allowed to dissolve in the irrigation stream. The nutrients are thus carried into the soil in solution. This requires no added application costs and allows the utilization of relatively inexpensive nitrogen carriers. Increased usage of these materials attest to the growing popularity of the irrigation method of application. Some care must be used, however, to prevent ammonia loss by evaporation.

APPLIED AS SPRAY ON LEAVES. The direct application of micronutrients or urea to plants as a spray has been made. This type of fertilization is unique. It does not involve any extra procedures or machinery since the fertilizer is applied simultaneously with the insecticides. Apple trees seem to respond especially well to urea since much of the nitrogen is absorbed by the leaves. Moreover, that which drips or is washed off is not lost for it falls on the soil, from which it may later be absorbed by plants.

19:17. FACTORS INFLUENCING THE KIND AND AMOUNT OF FERTILIZERS TO APPLY

The agricultural value of a fertilizer is necessarily uncertain, since a material so easily subject to change is placed in contact with two wide variables, the *soil* and the *crop*. The soil and the added fertilizer react with each other chemically and biologically. The reversion of phosphoric acid is an example of the first; the microbial hydrolysis of urea illustrates the latter. The result may be an increase or, more often, a decrease in the effectiveness of the fertilizer. Allowance should be made for these reactions when deciding on the kind and amount of fertilizer to apply.

The *weather* has a tremendous effect on the soil, upon the crop, and upon the fertilizer applied. If there is either an excess or deficiency of moisture, full efficiency of the fertilizer cannot be expected. In fact, any factor which may tend to limit plant growth will necessarily reduce fertilizer efficiency and consequently the crop response to fertilization. It is only when other factors are not limiting that the amounts of fertilizers can be estimated with any degree of certainty.

In spite of the complexity of the situation, however, certain *guides* can be used in deciding the kind and amounts of fertilizers to be applied. The following are especially pertinent:

1. Kind of crop to be grown
$$\begin{cases} \text{economic value.} \\ \text{nutrient removal.} \\ \text{absorbing ability.} \end{cases}$$

2. Chemical condition of the soil in respect to
$$\begin{cases} \text{total nutrients.} \\ \text{available nutrients.} \end{cases}$$

3. Physical state of the soil, especially as to
$$\begin{cases} \text{moisture content.} \\ \text{aeration.} \end{cases}$$

The last of these factors in most cases has an indirect effect on fertilizer usage and has been discussed sufficiently elsewhere (pp. 209–10 and 264). Consequently, only the first two will be considered here.

19:18. KIND OF CROP TO BE FERTILIZED

Crops of high economic value such as vegetables justify larger fertilizer expenditures per pound of response obtained. Consequently, for these crops complete fertilizers are used in rather large amounts. As much as 2,000 pounds of analyses such as 8–16–16 are often recommended.

With crops having a low economic value per acre, much lower rates of fertilizer application are generally advisable. The extra yields obtained by applying large amounts of fertilizer are not usually sufficient to pay for the additional plant nutrients used, especially with such crops as natural pastures and meadows.

It must always be remembered that the very highest yields obtainable under fertilizer stimulation are not always the ones that give the best return on the money invested (see Fig. 19:10). In other words, the law of diminishing returns is a factor in fertilizer practice regardless of the crop being grown. Therefore, the application of moderate amounts of fertilizer is to be urged for all soils until the maximum paying quantity that may be applied for any given crop is approximately ascertained.

If the nutrient removal by a given crop is high, fertilizer applications are usually increased to compensate for this loss. The extra fertilizer may be applied directly to the particular crop under consideration or to some preceding and more responsive crop in the rotation. It should not be implied that in all cases an attempt should be made to return nutrients in amounts

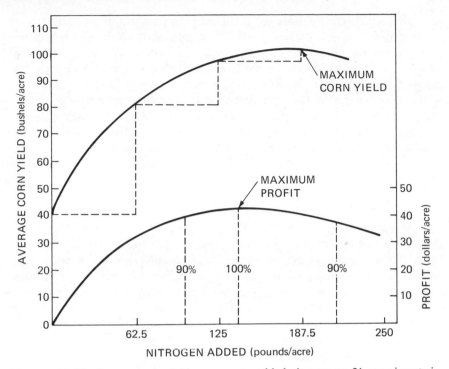

FIGURE 19:10. Average corn yield response to added nitrogen on 21 experiments in North Carolina and the calculated profit or net return to fertilizer. Note that maximum profit is obtained at a lower fertilizer rate than maximum yield. [*From Englestead* (1).]

equivalent to those that have been removed. Sometimes advantage can be taken of the nutrient-supplying power of the soil. This source is usually insufficient, however, and extra fertilizer additions must be made. Because of the vigorous reaction of phosphorus with soils, additions of this element in considerable excess of the amounts removed by crops and in other ways are usually economically sound.

Tremendous differences exist in the ability of plants to absorb nutrients from a given soil. For example, lespedeza and peanuts, both of which are legumes, can readily absorb adequate potassium under much lower soil potash levels than can alfalfa or soybeans. Consequently, responses to additions of potassium usually are much greater with the latter two crops. Obviously, crop characteristics deserve very careful study.

19:19. CHEMICAL CONDITION OF THE SOIL—TOTAL VERSUS PARTIAL ANALYSES

To determine the kind and quantity of fertilizer to add to soil, it is necessary to know what nutrient elements (or element) are deficient. Two general types of chemical analyses of soils are commonly used to obtain an estimate of nutrient deficiencies: *total* and *partial*. In a *total analysis* the entire amount of any particular constituent present in the soil is determined, regardless of its form of combination and its availability. Such data are of great value in studying soil formation and other phases of soil science. But they give little if any information on the availability of the essential elements to plants. Analyses suitable for measuring only the available portion of a given nutrient constituent must be employed. Such analyses are called *partial*, because only a portion of the total quantity of a soil constituent is determined.

19:20. TESTS FOR AVAILABLE SOIL NUTRIENTS—QUICK TESTS

Many procedures for the *partial analyses* of soils have been developed during the past half century. In general, these procedures attempt to extract from the soil amounts of certain fertility elements which are correlated with those removed by plants. The large number of extracting solutions which have been employed in trying to measure nutrient availability is mute testimony of the difficulties involved. The extraction solutions employed have varied from strong acids such as H_2SO_4 to weak solutions of CO_2 in water. Buffered salt solutions such as sodium or ammonium acetate are the extracting agents now most commonly used.

QUICK TESTS—GENERAL CONSIDERATION. The group of tests most extensively used for nutrient availability are the *rapid* or *quick tests*. As the name implies, the individual determinations are quickly made, a properly equipped laboratory being able to handle thousands of samples a month. Since these tests are the only ones having practical possibilities for widespread usage, their limitations as well as their merits will be critically reviewed.

A weak extracting solution such as buffered sodium acetate is generally employed in rapid tests and only a few minutes are allowed for the extraction. Most of the nutrients removed are those rather loosely held by the colloidal complex. The test for a given constituent may be reported either in general terms such as *low*, *medium*, or *high*, or more specifically, in pounds per acre. The constituents most commonly tested for are phosphorus, potassium, calcium, and magnesium, and aluminum, iron, and manganese are frequently included.

LIMITATIONS OF QUICK TESTS. Perhaps the first limitation is the difficulty of properly *sampling soils* in the *field*. Generally only a very small sample is taken, perhaps a pint or so from an area of land often several acres in size. The chances of error, especially if only a few borings are made, are quite high. In general, at least 15 to 20 borings are suggested for each sample in order to increase the probability that a representative portion of the soil has been obtained.

A second limitation of these tests is the fact that it is essentially impossible to extract from a soil sample in a few minutes the amount of a nutrient, or even a constant proportion thereof, that a plant will absorb from that soil in the field during the entire growing season. The test results must be correlated with crop responses before reliable fertilizer recommendations can be made. Even then, the recommendations must be made in light of a practical knowledge of the crop to be grown, the characteristics of the soil under study, and other environmental conditions. Two soils testing exactly the same in respect to a given constituent have been found to respond quite differently in terms of crop yields to identical fertilizer treatments in the field.

TRAINED PERSONNEL TO INTERPRET TESTS. Obviously, it is unwise to place any particular confidence in fertilizer recommendations when the tests are made and interpreted by an amateur. Not only might the novice make inaccurate determinations, but he is sure to lack the knowledge and judgment necessary for a rational interpretation in terms of fertilizer needs. The interpretation of quick test data is best accomplished by experienced and technically trained men, who fully understand the scientific principles underlying the common field procedures.

As already suggested, it is essential that a person have considerable related information in order to make the proper use of quick test data. Examples of supplementary data desirable are shown by the sample information sheet in Fig. 19:11. This supplementary information may be as helpful as the actual test results in making fertilizer recommendations.

MERITS OF RAPID TESTS. It must not be inferred from the preceding discussion that the limitations of rapid tests outweigh their advantages. When the precautions already described are observed, rapid tests are an invaluable tool in making fertilizer recommendations. Moreover, these tests will become of even greater importance insofar as they are correlated with the results of field fertilizer experiments. Such field trials undoubtedly will be expanded to keep pace with the increasing use of the rapid tests.

19:21. BROADER ASPECTS OF FERTILIZER PRACTICE

It is obvious that fertilizer practice involves many intricate details regarding soils, crops, and fertilizers. In fact, the interrelations of these three are so complicated and far reaching that a practical grasp of the situation requires years of experience. The unavoidable lack of exactness in fertilizer decisions should be kept in mind.

Leaving the details and viewing the situation in a broad way, it seems to be well established that any fertilizer scheme should be built around the effective use of the most expensive of the fertilizer elements, nitrogen. Applications

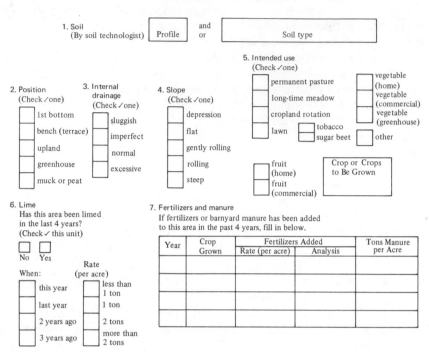

FIGURE 19:11. Satisfactory interpretation of a quick test analysis of a soil cannot be made without certain pertinent field data. The record sheet above suggests the items that are most helpful to the expert in making a fertility recommendation. [*Modified from a form used by the Ohio Agricultural Experiment Station.*]

of phosphorus and potassium should be made to balance and supplement the nitrogen supply, whether it be from the soil, crop residues, or added fertilizers.

A second aspect relates to economics. Farmers do not use fertilizers just to grow big crops or to increase the nutrient content of their soils. They do so to make a living. As a result, any fertilizer practice, no matter how sound it may be technically, which does not give a fair economic return will not long stand the test of time.

From this follows another important point. Fertilizers are added because of their effects on crops. For that reason it is inappropriate to think only in terms of soil–fertilizer interactions. The nature of the crop, its requirements, and economic value must be considered.

Not only must the soil, the crop, and the fertilizer receive careful study, but the crop rotation employed and its management should be considered. Obviously, the fertilizer applications must be correlated with the use of farm manure, crop residues, green manure, trace elements, and lime.

Moreover, the residues of previous fertilizer additions must not be forgotten. In short, fertilizer practice is only a phase, but a very important one, of the fertility management of soils.

REFERENCES

(1) Englestad, O. P., "Agronomic Response to Bulk Blended Fertilizers," in *New Frontiers in Fertilizer Technology and Use* (Muscle Shoals, Ala.: Tennessee Valley Authority, 1964).

(2) Englestad, O. P., "Phosphate Fertilizers Aren't All Alike," *Crops and Soils*, **17**:14–15, Aug.–Sept. 1965.

(3) Nelson, L. B., "Advances in Fertilizer," *Adv. in Agron.*, **17**:1–84, 1965.

(4) Nikitin, A. A., "Technological Aspects of Trace Elements Usage," *Adv. in Agron.*, **6**:183–97, 1954.

(5) Parr, J. F., "Chemical and Biochemical Considerations for Maximizing the Efficiency of Fertilizer Nitrogen," in *Effects of Intensive Fertilizer Use on the Human Environment*, Soils Bull. 16 (Rome: FAO; published by Swedish International Development Authority, 1972).

Chapter 20

ANIMAL MANURES AND GREEN MANURES

FOR centuries, farm manure has been a most important agricultural by-product. Its use has long been synonymous with a successful and stable agriculture. Not only does it supply organic matter and plant nutrients to the soil but it is associated with animal agriculture and with forage crops, which are generally soil protecting and conserving.

Unfortunately, in recent years developments relating to increased animal efficiency in the United States have drastically changed the position of manures in modern agriculture. In the past, animals were largely dispersed on the land, making possible the easy and economical application of manure to the nearby soils. Now, more than two thirds of the beef production in the United States is under concentrated confined conditions. In modern meat and poultry production factories, tens of thousands of beef and swine, and hundreds of thousands of poultry, are managed in centralized locations. In 1970 there were 184,000 feedlots with over 1,000 head of beef. Beef feedlots in the Missouri basin states and broiler- and egg-producing complexes in the Southeast and along the Atlantic Coast are examples of such large operations (see Fig. 20:1).

While these changes have provided for increased production efficiency, low labor costs, and greater ease of animal disease prevention and control, they have also provided huge mounds of animal manures, most of which are too far distant from fields to be utilized economically. Furthermore, offensive odors and runoff during wet weather have ranked animal manures so produced as one of the nation's serious air and water pollution problems. As the properties and biological values of manure are considered, this serious potential pollution problem for much of the manure in the United States must be kept in mind.

20:1. QUANTITY OF MANURE PRODUCED

The quantity of manure available for land application depends on a number of factors. The species and size of the animals and the quantity and nature of feed consumed have an influence. Likewise, the nature and amount of bedding used are factors as well as the type of housing or confinement of the animals.

It is not surprising therefore to see quite a range in values for manure production from different animal species. The data in Table 20:1 are averages of figures obtained by different investigators. To provide a basis for comparisons among the different animals, the data are expressed in terms of fresh weight per 1,000 pounds of live weight. Because of differences in moisture content, comparisons are also included on a dry-weight basis.

TABLE 20:1. *Representative Annual Rates of Manure Production Expected from Different Animals per 1,000 Pounds Live Weight Expressed as Fresh Excrement and as Dry Matter*

Note the relative constancy of the annual rate expressed in tons of dry matter.

Animal	Annual Production (tons)	
	Fresh Excrement	Dry Matter
Cattle	12.6	1.89
Poultry	5.6	2.14
Swine	13.2	1.98
Sheep	5.9	2.00
Horse	5.8	1.98

Although the comparative production of manures is of some interest, the total quantity of manure voided by farm animals is of greater significance. In the United States this figure approaches 2 billion tons annually. In terms of volume, this is ten times that produced by the human population in this country. More than 1 billion tons of this manure are produced in feedlots or giant poultry or swine complexes where manure *disposal* as a problem tends to overshadow manure *utilization* as an opportunity (see Fig. 20:1). A 50,000-head beef feedlot operation, for example, produces about 90,000 tons of manure annually after some decomposition and considerable moisture loss. At a conventional application rate of 10 tons per acre, 9,000 acres of land would be required to utilize the manure. The enormity of the disposal problem is obvious even for operations one half or one third this size (see p. 568).

[OVERLEAF]
FIGURE 20:1. Aerial view of a 320-acre feedlot in Colorado wherein 100,000 cattle are on feed at one time. Problems of manure disposal and utilization are difficult to cope with under these conditions of concentrated animal feeding. [*Photo courtesy Monfort Feed Lots Inc., Greeley, Colorado.*]

20:2. Chemical Composition

Manure as it is applied in the field is a combination of feces, urine, bedding (litter), and feed wastage which is incidentally incorporated as the animals move about their housing structures and pens. As might be expected, the chemical composition of this material varies widely from place to place depending upon factors such as (a) animal species (for example, ruminant versus nonruminant), (b) age and condition of animals, (c) nature and amount of litter, and (d) the handling and storage of the manure before it is spread on the land.

MOISTURE AND ORGANIC CONSTITUENTS. The moisture content of fresh manure is high, commonly varying from 60 to 85 percent (see Table 20:2). This excess water is a nuisance if the fresh manure is spread directly on the land. Much energy must be devoted to handling and transporting the water as well as the solids. If, on the other hand, the manure is handled and digested in a liquid form or slurry prior to application, the moisture content is not objectionable.

TABLE 20:2. *Moisture and Nutrient Content of Manure from Farm Animals*[a]

| Animal | Feces/Urine Ratio | H_2O (%) | Manure (lb/ton) | | |
			N	P_2O_5	K_2O
Dairy cattle	80:20	85	10.0	2.7	7.5
Feeder cattle	80:20	85	11.9	4.7	7.1
Poultry	100:0	62	29.9	14.3	7.0
Swine	60:40	85	12.9	7.1	10.9
Sheep	67:33	66	23.0	7.0	21.7
Horse	80:20	66	14.9	4.5	13.2

[a] Average values from a number of references.

Manures are to a considerable extent partially degraded plant materials. Animals utilize only about one half of the ingested organic matter in their feed. It is not surprising therefore that the bulk of the solid matter in manures is composed of organic compounds very similar to those found in the feed the animals consumed. Thus, although most of the cellulose, starches, and sugars are decomposed, hemicellulose and lignin are common, as are ligno-protein complexes similar to those found in soil humus. These plant materials have been only partially degraded, however, as evidenced by the ready decomposition of at least the soluble components when they are added to soil or a digestion tank.

One other important organic component of animal manures is the live component—the soil organisms. Especially in ruminant animals (for example, cattle and sheep) the manures are teeming with bacteria and other micro-organisms. Between one fourth and one half of the fecal matter of ruminants may consist of microorganism tissue. Although some of the compounds may be the same as those in the original plant materials, others have been synthesized by the microorganisms. In any case, some of these organisms continue to break down constituents in the voided feces and participate in decomposition of the manure in storage.

NITROGEN AND MINERAL ELEMENT CONTENTS. As was the case with organic matter, a fair share of the nutrient elements consumed in animal feeds are found in the voided excrement. As a generalization, it may be considered that three fourths of the nitrogen, four fifths of the phosphorus, and nine tenths of the potassium are not utilized by the animals. For this reason, animal manures are valuable sources of both macro- and micronutrients.

The quantities of nitrogen, phosphorus, and potassium that might be expected in different manures are given in Table 20:2. The range of other nutrients commonly found is as follows (2), expressed in pounds per ton:

Calcium	2.4 –74.0	Boron	0.02 –0.12
Magnesium	1.6 – 5.8	Manganese	0.01 –0.18
Sulfur	1.0 – 6.2	Copper	0.01 –0.03
Iron	0.08– 0.93	Molybdenum	0.001–0.011
Zinc	0.03– 0.18		

As shown in Table 20:2, the ratio of feces to urine in farm manure varies from 2:1 to 4:1. On the average, a little more than *one half of the nitrogen*, almost *all of the phosphorus*, and about *two fifths of the potassium* are found in the solid manure (see Fig. 20:2). Nevertheless, this apparent superiority of the solid manure is offset by the ready availability of the constituents carried by the urine. Care must be taken in handling and storing the manure to minimize the loss of the liquid portion.

The data in Table 20:2 suggest three outstanding characteristics of manures as nutrient carriers: (a) considerable variability in moisture and nutrient contents, (b) a relatively low nutrient content in comparison with commercial fertilizer, and (c) an unbalanced nutrient ratio, being considerably lower in phosphorus than in nitrogen and potassium. Modern commercial fertilizers commonly carry 20–30 times the nutrient content of manure and most have a phosphate content at least as great as the nitrogen and potassium. The bulky characteristic of manure is a distinct disadvantage, greatly increasing the handling and spreading costs per unit of fertilizer. The nutrient imbalance requires correction through supplemental soil treatment.

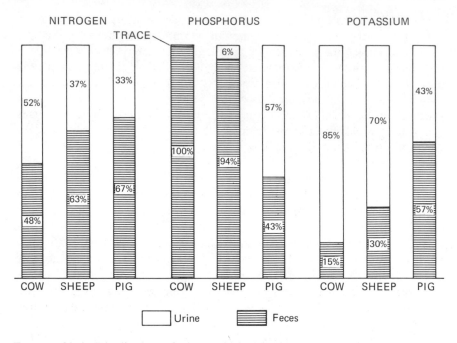

FIGURE 20:2. Distribution of nitrogen, phosphorus, and potassium between the solid and liquid portions of farm manure from cattle, sheep, and pigs. [*Calculated from data of Van Slyke* (4).]

THREE GENERALIZATIONS. From a practical point of view the characteristics shown in Table 20:2 suggest three significant generalizations. First, for purposes of calculation and discussion, 1 ton of an average farm manure is considered to supply as much N, P_2O_5, and K_2O as 100 pounds of 10–5–10 fertilizer. This envisages an average manure composition of 0.5 percent N, 0.25 percent P_2O_5, and 0.5 percent K_2O, emphasizing again the relatively low nutrient analysis of manures. However, since field rates of manure application are commonly 10–15 tons per acre, the total nutrients supplied under practical conditions are substantial and sometimes more than needed for optimum growth.

The second generalization relates to the availability of nitrogen, phosphorus, and potassium to growing crops. Again, there is considerable variation, depending on such factors as the crop, the soil, and weather conditions. In general, however, nearly one half of the nitrogen, about one fifth of the phosphorus, and about one half of the potassium may be recovered by the first crop (see Table 20:3). Thus, on the basis of readily available nutrients, 1 ton of average manure supplies 5 pounds of N, 1 pound of P_2O_5, and 5 pounds of K_2O.

TABLE 20:3. *Effect of Handling Cow Manure on the Yield and Recovery of Nitrogen, Phosphorus, and Potassium by Corn Grown on Miami Silt Loam in a Greenhouse Experiment*[a]

Type of Manure	Yield (g/pot)	Recovery by Crop (%)		
		N	P	K
No manure	11.0	—	—	—
Cow manure 15T/A				
Fresh	19.5	44.0	19.5	40.5
Fermented	19.5	42.0	22.5	49.5
Aerobic liquid	17.0	18.5	19.5	38.0
Anaerobic liquid	22.5	52.5	29.0	48.0

[a] From Hensler (3).

The third practical generalization is the need to balance the N/P/K ratio by supplying phosphorus in addition to that contained in manures. This is done in two ways: (a) by supplementing the manure by adding superphosphate to the barn floor (an example is the stantion housing of dairy animals) or to the loaded manure spreader, and (2) by adjusting commercial fertilizer analyses applied in the field to supplement the manure applications.

Before turning to methods of storing manure, one more set of figures might be helpful—that is the proportion of nutrients in a harvested crop which might be expected to survive animal removal and loss from the manure prior to its application to soil. The losses which feeds sustain during digestion and the waste of the manure in handling and storage permit the return to the soil of no more than *25 percent of the organic matter, 40 percent of the nitrogen, 50 percent of the phosphorus, and 45 percent of the potassium.* As will be seen in the next section, controlled manure digestion procedures will remove an even higher percentage of the organic matter and in some cases the nitrogen.

20:3. STORAGE, TREATMENT, AND MANAGEMENT OF ANIMAL MANURES

A generation ago, manure storage and application was a simple matter. Some farmers spread manure daily from their barns or allowed it to pile up until time and soil conditions permitted it to be spread. Animal feeding operations were small enough to permit the spreading in the spring of manure which had accumulated during the fall and winter. In any case, manure management was generally considered a private affair.

The coming of confined and concentrated animal management has drastically changed this situation. The marked concentration of animal manures has become the concern not only of the farmer but of rural nonfarm citizens as well as environmental ecologists generally. Together they have begun a revolution in the management of farm manures. Offensive odors, along with the possibilities of organic and nitrate pollution of streams and drinking water, are primary factors in stimulating the revolution.

Four general management systems are being used to handle farm manures: (a) collection and spreading of the fresh manure daily, (b) storage and packing in piles and allowing the manure to ferment before spreading, (c) aerobic liquid storage and treatment of the manure prior to application, and (d) anaerobic liquid storage and treatment prior to application. Since these methods of handling manure affect its biological value, each will be discussed briefly.

APPLYING FRESH MANURE DAILY. This system is commonly used where stantion dairy barns are employed. The manure is scraped or otherwise moved mechanically into spreaders, sometimes reinforced with superphosphate, and spread daily on the land. The obvious advantage is the prevention of loss by decomposition or volatilization, thereby maximizing the quantity of nutrients applied to the soil. One obvious drawback in northern climates is the loss of nutrients resulting when the manure is spread on frozen ground. Spring thaws carry off much potassium and nitrogen, thereby reducing crop response and increasing the probability of eutrophication[1] caused by unwanted nutrient concentration in streams, ponds, and lakes. As much as 25 percent of the applied potassium can be lost in this manner.

STORAGE OR PACKING IN PILES (FERMENTATION). Manure from dairy barns and cattle feedlots may be allowed to accumulate under the animals over a period of time or may be removed to a pile nearby. In any case, if the manure is not allowed to dry out below 40 percent moisture (as it sometimes does in the Great Plains), fermentation will occur. Depending on the moisture content and degree of compaction, both anaerobic and aerobic breakdown takes place. The most abundant products of decay are carbon dioxide and water, along with considerable heat. Of perhaps greater practical significance, however, are reactions involving elements such as nitrogen, phosphorus, and sulfur. For example, urea is hydrolyzed during the decomposition process, yielding ammonia through the following reactions:

$$CO(NH_2)_2 + 2H_2O \rightarrow (NH_4)_2CO_3$$

$$(NH_4)_2CO_3 \rightarrow 2NH_3\uparrow + CO_2\uparrow + H_2O$$

If conditions are favorable for nitrification, nitrates will appear in abundance. Either form of concentrated inorganic nitrogen can be lost and in so doing become a pollution hazard. The ammonia lost to the atmosphere may be

---

[1] Eutrophication is the process of accumulation in bodies of water of organic residues from aquatic species. The presence of inorganic nutrients, particularly nitrogen and phosphorus, encourages the growth of algae and other aquatic species which are the source of these residues.

captured by rain and snow and returned to the surface. The nitrates, being soluble and unadsorbed by the soil or the manure, are subject to leaching and to movement in runoff water. In the case of either ammonia or nitrate there are possibilities of adverse nutrient buildup in nearby streams or other bodies of water resulting from the massive organic matter breakdown.

Independent of the pollution problem, fermentation may provide a more satisfactory product for land application than fresh manure. For example, highly carbonaceous bedding or litter which may be present in the fresh manure is at least in part broken down during the fermentation process.

AEROBIC LIQUID TREATMENT. One of the more sophisticated animal-waste-treatment procedures is that of aerobic digestion of a liquid slurry containing the manure. The manure is stored in an aerated lagoon or in an oxidation ditch. Used primarily in swine operations, the manure may fall through slotted floors into the oxidation ditch or may be transported to a nearby lagoon. By vigorous stirring, oxygen is continuously incorporated into the system, bringing about continuous oxidation. Offensive odors are kept to a minimum, although some nitrogen is lost, probably as ammonia. The primary products of decomposition are carbon dioxide, water, and inorganic solids. Periodically the solids can be removed along with the resistant organic residues and applied to the land. A diagram showing an oxidation ditch and lagoon in use for swine is shown in Fig. 20:3. This method of treatment has many advantages, the most important of which is its essential freedom from pungent odors.

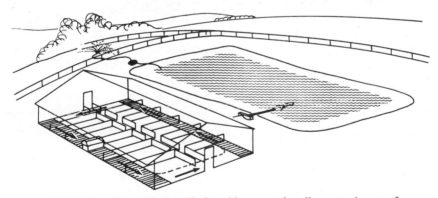

FIGURE 20:3. Use of an oxidation ditch and lagoon to handle wastes in a confinement swine building. A slotted floor around the outside of the building permits the wastes to drop into a continuous channel or ditch filled with water. A rotor on the left center of the building stirs oxygen from the air into the slurry and drives the mixture counter-clockwise around the ditch. Aerobic organisms oxidize most of the organic wastes and some of the nitrogen is volatilized. The partially purified residue is pumped or allowed to flow into a lagoon, from which it can later be applied to the land by irrigation for final purification. [*Courtesy A. J. Muehling, University of Illinois.*]

Its costs of construction and operation are high, however, and it permits some nutrient loss. This method will find its greatest use in areas where odor control must be at a maximum.

ANAEROBIC LIQUID TREATMENT. This method is similar to its aerobic counterpart except that no gaseous oxygen is added to encourage aerobiosis of the liquid slurry. Consequently, the nature of the reactions is quite different, molecular oxygen (for example, SO_4^{--} and NO_3^{-}) providing energy for much of the breakdown. The reactions are similar to those taking place in a septic tank. The principal gaseous product is methane, which makes up some 60 to 80 percent of the total. The remainder of the gas evolved is mostly carbon dioxide. Methane is thought to form through either of two types of biochemical reactions brought about by "methane bacteria":

$$CO_2 + 4H_2 \rightarrow CH_4\uparrow + 2H_2O$$

and

$$\underset{\text{Acetic acid}}{CH_3COOH} \rightarrow CH_4\uparrow + CO_2$$

By discouraging the buildup of acidity, the second of these reactions helps prevent the formation of the most disagreeable odoriferous gases. If the anaerobic process is functioning properly, it is quite satisfactory, providing a product with little nutrient loss. If the process is "poisoned" by excess acidity, heavy metals, or salts, however, the odors are most objectionable. Also at "clean -out" time, strong odors are common.

This method of treatment, along with the others, results in considerable loss of organic carbon. However, crop response to the treated product generally is as good as where fresh manure is used. Data in Table 20:3 illustrate this point. Only for the aerobic processed liquid was there evidence that nitrogen had been lost during the animal waste treatment.

Scientific developments during the past few years have markedly improved animal-waste-management technology. Continuing public pressures will probably bring about even greater changes in the coming years. Hopefully, these changes will continue to provide products which can be utilized in commercial agriculture.

20:4. UTILIZATION OF ANIMAL MANURES

Biologically, manure has many attributes. It supplies a wide variety of nutrients along with organic matter for improving the physical characteristics of soils and their water-holding capacities. Its beneficial effects on plant growth are sometimes difficult to duplicate with other materials. At the same time, its bulkiness and low analysis reduces its competitive economic value. High labor and handling costs and relatively cheap inorganic fertilizers are responsible for this unfortunate situation. Even so, manure remains a most valuable soil organic resource. Economic considerations merely make it necessary to choose more carefully the soil and crop situation wherein manure is applied.

RESPONSIVE CROPS. Manure is an effective source of nutrients for most crops. However, those with a relatively high nitrogen requirement are most likely to respond to its application. Crops such as corn, sorghum, small grains, and the grasses respond well to manure as do vegetable and ornamental crops. Care must be taken with crops such as small grains to prevent lodging (plants falling over), which may result from excessive applications. The manuring of new forage seedings is common, providing both nutrients for plant growth and a mulch to resist soil crusting. For new forage seedings and certain vegetable crops, herbicides may be needed since weed seeds commonly pass through the animal and are added back to the soil with the manure.

The rate of application of manure will depend upon the specific needs. However, rates of 10–15 tons per acre are commonly employed. In general, rates of application heavier than these give lower response per ton of manure. Rates as high as 25–30 tons per acre applied to productive soils can be justified but primarily in terms of manure disposal rather than utilization.

SPECIAL USES. There are a number of situations where manure is especially valuable. Denuded soil areas resulting from erosion or from land leveling for irrigation are good examples. Initial applications of 30–40 tons per acre may be worked into the soil in the affected areas. These rates may be justified to supply organic matter as well as nutrients.

Special cases of micronutrient deficiency can be ameliorated with manure application. Such treatments are sometimes used when there is some uncertainty as to specific nutrient deficiency. Manure applications can be made with little concern for adding toxic quantities of the micronutrients.

On soils with extreme textural makeup, high manure applications may be required. The water-holding capacity of very sandy soils can be increased with heavy manure applications. Likewise, the structural stability and tilth of heavy plastic clays is sometimes dependent upon heavy organic applications. In both cases, the physical effects of manure justify its use.

Home gardeners commonly use manures at rates far in excess of those employed in commercial agriculture. Their aim is to provide a friable, easily tilled soil, and the cost is a secondary matter. Further, in planting trees and shrubs, one-time applications are common, therefore high initial rates of application are justified.

20:5. LONG-TERM EFFECTS OF MANURES

The total benefits from manure utilization are sometimes not apparent from crop yields during the first or even second or third year following application. Manure, along with crop residues, is a primary means of replenishing soil organic matter. Although a portion of the nutrients and organic matter in manure is broken down and released during the first year or two, some is held in humuslike compounds subject to very slow decomposition. Its effect is long standing, not only on future nutrient supplies but also on the physical condition of the soil.

When manure or crop residues are added to soil, a portion of the organic carbon, nitrogen, sulfur, and perhaps other elements is converted to humus. In this form, the elements are released only very slowly, rates of 2–4 percent per year being common. Thus, the components of manure which are converted to humus will have continuing effects on soils years after their application. The same can be said for the green manures to be covered in the sections which follow.

20:6. GREEN MANURES—DEFINED

From time immemorial the turning under of a green crop to better the condition of the soil has been a common agricultural practice. Records show that the use of beans, vetches, and lupines for such a purpose was well understood by the Romans, who probably borrowed the idea from other nations. The art was lost in Europe to a great extent during the Middle Ages but was revived again as the modern era was approached.

This practice of turning into the soil undecomposed green plant tissue is referred to as *green manuring*. Material so added, if the soil is in a proper condition and well managed, brings about a number of favorable effects and may aid materially in maintaining or raising the crop-producing capacity of a soil.

20:7. BENEFITS OF GREEN MANURES

There are four major benefits from the use of green manures in a crop rotation: (a) organic matter addition, (b) nitrogen addition (if the green manure is a legume), (c) nutrient conservation, and (d) ground cover during erosion-prone periods of the year.

SUPPLY OF ORGANIC MATTER. The quantity of organic residues returned to the soil by a green manure is considerable. One to two tons of dry matter is not an unusual amount to be so added. This organic matter, especially if it is succulent and subject to ready decay, encourages microbial action not only of the heterotrophic organisms responsible for general decomposition, but for the *Azotobacter*, the "free-living" organisms which are able to fix nitrogen from the atmosphere. The organic residues from green manures also help to provide the stability of soil structure needed for optimum plant growth.

ADDITION OF NITROGEN. If the green manure is a legume, there is a good possibility that nitrogen will be added to the soil crop system. Data in Fig. 20:4 illustrate this point. Note that the effects of the rye and vetch green

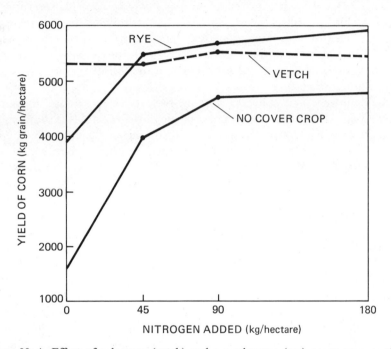

FIGURE 20:4. Effect of a legume (vetch) and a nonlegume (rye) cover crop on the 7-year average yield of corn receiving three levels of fertilizer nitrogen on a Cecil sandy loam soil. Cover crops grown each year. [*From Adams et al.* (1).]

manures on the yield of corn are essentially the same if ample fertilizer nitrogen is applied. With no fertilizer nitrogen, however, the vetch was markedly superior to the rye in increasing the yield of corn. The nitrogen-supplying ability of the vetch is obvious. Even with large amounts of nitrogen, plots on which the green manures were grown gave significantly higher corn yields than where no green manures were included. Effects other than those associated with nitrogen supply are obvious on this sandy loam soil.

Long-term effects of green manures on the soil nitrogen and organic matter contents are difficult to measure. It is probable, however, that the primary effects of these supplemental crops will be seen during the growing season following their incorporation into the soil. At the same time, improvements in the structure of some soils resulting from the use of green manures have persisted for several years after green manuring has been discontinued.

NUTRIENT AND SOIL CONSERVATION. Two other benefits of green manuring, nutrient conservation and ground cover, are closely related. Green manures are often sown in the summer or fall after a primary crop such as corn or potatoes. The green manure crop, termed a *cover crop* or *catch crop* in this case, is able to utilize excess fertilizer not assimilated by the main crop. Further, the ground is rather quickly covered by the crop, thereby protecting the soil from erosion during the fall and early spring rains. If such crops are not harvested but are plowed or disced into the soil, they are considered to be green manure crops.

Both nutrient and soil conservation have taken on added meaning in recent years. For some cash crops, particularly vegetables, the quantities of fertilizer added to produce maximum yields are far in excess of the amounts the crops take up during the growing season. If not taken up by a cover crop, these nutrients are subject to leaching and can contribute to undesirable nutrient levels in streams and lakes. Thus, a cover crop (and, when it is incorporated into the soil, a green manure crop) may conserve nutrients, reduce nutrient contamination of lakes and streams, reduce soil erosion, and supply significant quantities of nitrogen and organic matter for the succeeding crop.

20:8. PLANTS SUITABLE AS GREEN MANURES

DESIRABLE CHARACTERISTICS. An ideal green manure crop should be easily established and should grow rapidly. It should produce an abundant growth of succulent tops and roots in a short time. And its growth habit should encourage ground cover soon after its establishment. Green manure crops should have the ability to grow on poor soils because these are the ones where the beneficial effects are most needed. These include sandy soils and fine-textured soils having poor structural stability.

When other conditions are equal, it is better to make use of a leguminous green manure in preference to a nonleguminous one because of the nitrogen gained by the soil and the organic activity it promotes. A little additional nitrogen is sometimes of tremendous importance.

However, it is often difficult to obtain a catch of some legumes and they may be so valuable as animal feed that it is poor management to turn the stand under. Again, the seeds of many legumes are expensive, almost prohibiting the use of such crops as green manures. Moreover, some legumes do not fit into the common rotations in such a way as to be turned under conveniently as a green manure.

SUITABLE PLANTS. Some of the plants utilized in the various parts of the United States as green manures are listed in Table 20:4. Their value, of course, depends in part on climate, some of those mentioned being unsuited to cold regions, while others have a wide range of adaptability.

The growing of two crops together for green-manuring purposes is sometimes recommended. If properly chosen in respect to growth habits, climatic adaptation, and soil requirements, the advantages are noteworthy. Not only can larger amounts of green material be produced, but if one of the crops is a legume, nitrogen fixation may be utilized. Also one crop may offer physical support to the other, a factor of no mean importance when dealing with plants that tend to lodge. Oats and peas, and rye and vetch, are excellent examples of such green manure combinations.

TABLE 20:4. *List of Possible Green Manuring Crops*

Legumes		Nonlegumes	
Warm Regions Especially	Wide Range	Wide Range in Most Cases	
Crimson clover	Alfalfa	Rye	Ryegrass
Bur clover	Red clover	Oats	Sudan grass
Lespedeza	Sweet clover	Barley	Mustard
Crotalaria	Soybean	Millet	Rape
Vetch	Canadian field pea	Buckwheat	Weeds
Austrian winter pea	Cowpea		

20:9. PROBLEMS WITH GREEN MANURES

Two major problems with the use of green manures should be recognized. First, nonlegumes may have sufficiently high C/N ratios as to depress nitrogen uptake by the following crop. Second, green manure crops may deplete soil moisture to the point that the succeeding crop will suffer from drought.

Plowing in a heavy nonleguminous sod such as rye grass can result in nitrogen deficiency in the succeeding crop if the C/N ratio of the manure crop is sufficiently high. The microorganisms tie up inorganic soil nitrogen as they decompose the highly carbonaceous crop (see pp. 152–153). This tendency can be counteracted by adding extra fertilizer nitrogen either to the nonlegume at the time it is incorporated into the soil or at the planting of the succeeding crop. Also, turning in the manure crop before its C/N ratio becomes objectionably high will reduce the severity of the problem.

In areas of low rainfall the moisture regime dictates the judicious use of green manures. The green manure crop can deplete the soil of moisture, leaving too little for the main crop which follows. In general, green manures seem to be of little benefit to the wheat crop in regions of less than 18–20 inches of rain. This generalization is in agreement with the practice commonly accepted in these areas of fallowing or "resting" the soil from all vegetation for a year previous to the planting of wheat.

20:10. PRACTICAL UTILIZATION OF GREEN MANURES

The place of green manures in practical cropping systems will vary according to the climatic conditions and the nature of the cropping and fertilizing systems. In the United States and other economically well developed nations, the extensive use of relatively inexpensive fertilizers has lessened the need for the nitrogen supplied by green manures. In such circumstances, these crops must be justified for reasons other than nutrient supply.

In regions where nitrogen fertilizers are still quite costly and inaccessible, leguminous green manures are distinctly beneficial. The nitrogen supplied by these crops can be effectively utilized by the main crops which follow. In the humid tropics, where several crops a year can be grown, the use of such green manures should be encouraged.

Green manures as soil- and nutrient-conserving crops have application under most humid climatic conditions. Even in countries where fertilizer nutrient supplies are plentiful and cheap, the use of cover crops to help regulate the flow of nutrients from agricultural lands into streams and lakes is important and will probably become more important in the future. Also, the role of green manure crops in controlling soil erosion and concomitant silting of reservoirs and lakes cannot be disregarded.

Only in areas of low rainfall does there seem to be no place for green manure crops. In these areas, moisture conservation is the first requisite in soil and crop management and the moisture must be reserved for the main crops.

REFERENCES

(1) Adams, W. E., et al., "Effects of Cropping Systems and Nitrogen Levels on Corn (*Zea Mays*) Yields in the Southern Piedmont Region," *Agron. J.* **62**:655–59, 1970.

(2) Benne, E. J., et al., *Animal Manures*, Circ. 291, Michigan Agr. Exp. Sta., 1961.

(3) Hensler, R. F., *Cattle Manure: I. Effect of Crops and Soils: II. Retention Properties for Cu, Mu, and Zn.* Ph.D. Thesis, University of Wisconsin, 1970.

(4) Van Slyke, L. L., *Fertilizers and Crop Production* (New York: Orange Judd, 1932), pp. 216–26.

Chapter 21

SOILS AND CHEMICAL POLLUTION

IN recent years public attention has been focused increasingly on environmental pollution and its effects on man and other creatures. Among the major pollutants are wastes from an urbanized and industrialized society along with the myriad of chemicals, new and old, needed to maintain an efficient society. The soil is a primary recipient, intended or otherwise, of many of these waste products and chemicals. Furthermore, once these materials enter the soil they become part of a cycle which affects all forms of life, including man. At least a general understanding of the pollutants themselves, their reactions in soils, and means of managing, destroying, or inactivating them is essential.

Five general kinds of pollutants will receive attention. First are the thousands of *pesticide* preparations, most of which are used for agricultural purposes and all of which reach the soil. Second is a group of *inorganic pollutants*, such as mercury, cadmium, and lead, which have been discovered in toxic quantities as they move along the food chain. Third are the *organic wastes*, such as those from concentrated feedlots and food-processing plants, which will be considered along with municipal and industrial wastes, some of which may be dumped on soils. *Salts* and *radionuclides* are the fourth and fifth contaminants to be discussed.

21:1. CHEMICAL PESTICIDES — BACKGROUND

The history of man is replete with stories of his battles with pests. More than 10,000 species of insects, 600 weed species, 1,500 plant diseases, and 1,500 species of nematodes are known to be injurious at least to some degree to man, plants, and animals. Various methods have been used to tip the scales of nature in man's favor. Crop varieties and breeds of animals have been developed which resist the pests. Tillage implements are used to control unwanted plants or weeds. And man has learned to rotate his crops to prevent the buildup of pest organisms which are dependent upon a single crop species.

The use of chemicals to control pests has been practiced for centuries. For example, Greeks are said to have used sulfur to control certain plant diseases. When in the early nineteenth century Pasteur discovered that microbes were the cause of certain plant and animal diseases, chemical cures were sought and found. The use of copper-containing bordeaux mixture dates from soon after Pasteur's discoveries. Lime and sulfur mixtures and arsenic sprays have been used for half a century for the prevention of disease and insect damage in apple orchards. Naturally occurring insecticides such as rotenone and pyrethrins were also extensively used.

SYNTHETIC PESTICIDES. Even though these few chemicals were being used, it was the discovery of the insecticidal properties of DDT in 1939 and the herbicidal effects of 2-4-D in 1941 that truly began the chemical revolution in agriculture. These chemicals would kill pests and could be manufactured economically by man. While they were important themselves in controlling insects and weeds, their greatest impact was through the concept they exemplified. Man could develop and manufacture biocides to use in his war on pests. Literally tens of thousands of such chemicals and formulations have been developed, tested, and are in use. In 1970, more than 1 billion pounds of pesticides were applied in the United States, about 50 percent of which were used in agriculture (see Table 21:1). Some 900 chemicals in about 60,000 formulations are used to control pests.

TABLE 21:1. *Quantities of Organic Pesticides Used in the United States 1962–1970[a]*
Note that in recent years the greatest increase has been among the herbicides.

| Year | Organic Pesticide Sales (millions of pounds) | | | |
	Fungicides	Herbicides	Insecticides	Total
1962	97	95	442	634
1964	95	152	445	692
1966	118	221	502	841
1968	130	318	511	959
1970 (est.)	129	308	443	880

[a] Source: U. S. Tariff Commission.

BENEFITS FROM PESTICIDES. Perhaps the greatest benefit from pesticide use has been the millions of human lives saved from yellow fever, encephalitis, malaria, and other insect-borne diseases. Protection of crops and livestock has also brought economic benefits to society. Chemical weed control, for example, has in some cases virtually eliminated hand hoeing and even culti-vation. In the United States, pesticides have been a prime factor in the agri-cultural revolution which makes it possible for less than 6 percent of the population to feed the other 94 percent and export hundreds of millions of dollars worth of farm products. Pesticides have also protected food from pest damage as it moves from the farm through processing and marketing channels to the dinner table.

PROBLEMS AND DANGERS FROM PESTICIDE USE. Three major problems threaten to limit the continued usefulness of pesticides. First, some pest organisms (particularly the insects) have developed resistance to the chemicals. This necessitates higher dosages or the development of new chemicals to replace those to which the pests are resistant. Second, some pesticides are not readily biodegradable and tend to persist for years in the environment. Although this characteristic may be advantageous in controlling some pests, it is a disadvantage as the chemical moves to other parts of the environment.

This leads to the third problem, the detrimental effects of the chemicals on organisms other than the target pests. Soil flora and fauna may be adversely affected, as may be fish and other wildlife. This problem is compounded by the tendency of the chemicals to build up in organisms as movement up the food chain occurs. Birds and fish, being secondary and tertiary consumers, tend to concentrate these chemicals in their body tissue, in some cases to toxic levels. Damage to these creatures sounded the warning cry that man must know more about the ecological effects of pesticides if their use is to be continued. Even now, the use of those chemicals which are the greatest hazards are being restricted or eliminated.

21:2. KINDS OF PESTICIDES

Pesticides are commonly classified according to the target group of pest organisms: (a) insecticides, (b) fungicides, (c) herbicides (weedicides), (d) rodenticides, and (e) nematocides. Since the first three are used in largest quantities and are therefore more likely to contaminate soils, they will be given primary consideration.

INSECTICIDES. The quantity of insecticides in use today exceeds that of any other group of pesticides. Most of these chemicals are included in three general groups (Table 21:2). The *chlorinated hydrocarbons* (for example, DDT) were the most extensively used until the early 1970s. They had the advantages of low cost (especially DDT), general effectiveness, persistence, and relative low level of toxicity to man. However, their low biodegradability and excessive persistence as well as toxicity to birds and fish have made them targets for environmentalists, resulting in restrictions on their use.

The *organophosphate* pesticides are generally biodegradable, which lessens the possibility of their buildup in soils and water. At the same time, they are relatively much more toxic to man than are the chlorinated hydrocarbons, so great care must be used in handling and applying them. The *carbamates* are also popular amongst most environmentalists because of their ready biodegradability and relatively low mammal toxicity.

TABLE 21:2. *Classes of Pesticides Commonly Used in the United States and Examples of Each Group*

Chemical Group	Examples
Insecticides	
Chlorinated hyurocarbons	DDT, aldrin, dieldrin, heptachlor
Organophosphates	Diazinon, parathion, malathion
Carbamates	Sevin
Fungicides	
Thiocarbamates	Ferbam, ziram, maneb, nabam
Mercurials	Ceresan
Others	PCNB, copper sulfate
Herbicides	
Phenoxyalkyl acids	2,4-D, 2,4,5-T, silvex MCPA. 2.4-DB
Triazines	Atrazine, simazine
Phenylureas	Monuron, diuron, fenuron, linuron
Aliphatic acids	Dalapan, TCA
Carbamates	IPC, CIPC, EPTC, CDEA
Dinitroanilines	Trifluralin, dipropalin, benefin
Dipyridyls	Paraquat, diquat

FUNGICIDES. The quantity of fungicides used in the United States is far less than that of either herbicides or insecticides. Even so, large areas receive these chemicals annually. Fungicides are used primarily to control the field diseases of fruits and vegetables. They are also used to counteract seed diseases of common crops and to protect harvested fruits and vegetables from decay and rot. Fungicides are used as wood preservatives and for the protection of clothing from mildew.

HERBICIDES. Herbicides rival insecticides in total quantities used, more money being spent on them in the United States than on any other group of pesticides. Starting with 2,4-D (chlorinated phenoxyacetic acid), dozens of other chemicals in literally hundreds of formulations have been placed on the market. These include the *triazines* (rather specific for weed control in corn), *phenylureas, aliphatic acids, carbamates, dinitroanalines,* and *dipyridyls* (Table 21:2). As one might expect, this wide variation in chemical makeup provides an equally wide variation in properties. However, herbicides are biodegradable and most of them are relatively low in mammalial toxicity. Some are quite toxic to fish and perhaps to other wildlife, however, emphasizing once again the need to consider the indirect effects of these chemicals.

21:3. BEHAVIOR OF PESTICIDES IN SOILS[1]

Pesticides are commonly applied to plant foliage, on the soil surface, or are incorporated into the soil. In any case, a high proportion of the chemicals

[1] For an excellent recent review of this subject see Helling et al. (7).

eventually move into the soil, a fact which adds significance to studies of the fate of these chemicals in soil.

The wide variety of chemical structures found in pesticides (see Fig. 21:1) suggests great variability in the behavior of these chemicals in soil. Laboratory and field tests confirm just such variability. To obtain a glimpse of pesticide behavior, a few characteristics of pesticides and how they affect the behavior and reactions of chemicals in soils will first be considered. Attention will be given to five possible fates of pesticides once they are added to soils: (a) the chemicals may vaporize and be lost to the atmosphere without chemical change, (b) they may be adsorbed by soils, (c) they may move downward through the soil in liquid or solution form and be lost by leaching, (d) they may undergo chemical reactions within or on the surface of the soil, and (e) they may be broken down by soil microorganisms.

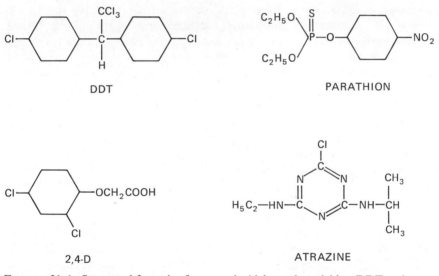

FIGURE 21:1. Structural formulas for several widely used pesticides. DDT and parathion are insecticides; 2,4-D and atrazine are herbicides. This variety of structures dictates great variability in properties and reactivity in the soil.

VOLATILITY. Pesticides vary greatly in their volatility and subsequent susceptibility to atmospheric loss. Some soil fumigants such as methyl bromide are selected because of their very high vapor pressure which permits them to penetrate soil pores to contact the target organisms. This same characteristic encourages rapid loss to the atmosphere after treatment unless the soil is covered or sealed. A few herbicides (for example, EPTA, CDEA, trifluralin) and fungicides (for example, PCNB) are sufficiently volatile to make vaporization a primary means of their loss from soil. Atmospheric analyses suggest that DDT and dieldrin are volatilized from soil in significant quantities, even though their vapor pressures are quite low in comparison with the chemicals previously mentioned. This vaporization probably accounts at least in part for the air transport of these chemicals to great distances from their point of application.

Earlier assumptions that disappearance of pesticides from soils was evidence of their breakdown are now known to be questionable. Loss of some chemicals to the atmosphere only to have them returned to the soil or surface waters in rain is now known to occur.

ADSORPTION. The tendency of pesticides to be adsorbed by soil is determined largely by the characteristics of the pesticides and of the soils to which they are added. The presence of certain functional groups, such as —OH, —NH$_2$, —NHR, —CONH$_2$, —COOR, and R$_3$N$^+$, in the molecular structure of the chemical encourages adsorption, especially on the soil humus. Hydrogen bonding (see p. 166) and protonation (adding of H$^+$ to a group such as an amino group) probably promotes some of the adsorption. In general, the larger the size of the pesticide molecules, everything else being equal, the greater will be their adsorption.

The soil characteristic with which adsorption is most closely associated is the soil organic matter content. Apparently, the complexity of the humus fraction along with its nonpolar nature encourages adsorption of the pesticides. A few pesticides, such as the herbicides diquat and paraquat, which tend to form cations (strong bases) are also adsorbed by silicate clays. Adsorption by clays of pesticides which form bases tends to be pH dependent (see Fig. 21:2), maximum adsorption occurring at the pH level where protonation occurs. Some clay-adsorbed pesticides lose their biocidal properties until they are desorbed. This has some obvious practical implications.

LEACHING. The tendency of pesticides to leach from soils is closely related to their potential for adsorption. Strongly adsorbed molecules are not likely to move down the profile. Likewise, conditions which encourage such adsorption will discourage leaching. Leaching is apt to be favored by water movement, taking place most readily in permeable sandy soils that are low in clay and organic matter. In general, herbicides seem to be more mobile than are either fungicides or insecticides. A grouping of pesticides into five classes on the basis of their probable mobility shows 58 of 61 in the four highest mobility classes to be herbicides (2). In contrast, 19 of 29 in the lowest mobility class are insecticides or fungicides.

CHEMICAL REACTIONS. Upon contacting the soil many pesticides undergo chemical modification independent of soil organisms. At the soil surface some, such as DDT and diquat, are subject to slow photodecomposition activated by solar radiation. Such degradation is relatively less important, however, than that which is catalyzed directly by the soil. Such catalysis is

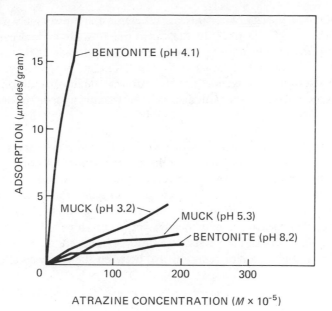

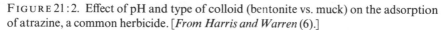

ATRAZINE CONCENTRATION ($M \times 10^{-5}$)

FIGURE 21:2. Effect of pH and type of colloid (bentonite vs. muck) on the adsorption of atrazine, a common herbicide. [*From Harris and Warren* (6).]

thought to be due largely to the silicate clay fraction, especially if the soils are acid. Chemical breakdown of DDT, endrin, heptachlor, malathion, diazinon, and atrazin are examples of this type of reaction. While the complexities of molecular structure of the pesticides suggest different mechanisms of breakdown, it is important to realize that degradation independent of soil organisms does in fact occur.

MICROBIAL METABOLISM. Biochemical degradation by soil organisms is perhaps the single most important method by which pesticides are removed from soils. Apparently the presence of certain polar groups on the pesticide molecules provide points of attack for the organisms. These include —OH, —COO⁻, —NH$_2$, and —NO$_2$ groups.

DDT can be changed by certain bacteria and fungi to the related compounds DDD under anaerobic conditions (quite rapidly) and DDE in well-aerated soils (very slowly). Further decomposition, however, takes place very slowly and the degradation routes are not well known. The other important chlorinated hydrocarbons (such as aldrin, dieldrin, and heptachlor) are subject to only slow "partial decomposition" since organisms have not adapted themselves to the rapid destruction of such compounds. This helps account for the marked persistence of these compounds in soils and elsewhere.

The organophosphate insecticides are degraded quite rapidly in soils apparently by a variety of organisms. Likewise, the most widely used herbicides, such as 2,4-D, the phenylureas, the aliphatic acids, and the carbamates, are readily attacked by a host of organisms. Exceptions are the triazines (for example, atrazine), which apparently are degraded primarily by chemical action. Most organic fungicides are also subject to microbial decomposition, although the rate of breakdown of some is slow, causing troublesome residue problems.

PERSISTENCE IN SOILS. The persistence of pesticides in soils is a summation of all the reactions, movements, and degradations affecting pesticides. Marked differences in persistence are the rule. For example, organophosphate insecticides may last only a few days in soils; the most widely used herbicide, 2,4-D, persists in soils for only two to four weeks; DDT and other chlorinated hydrocarbons may persist for from three to fifteen years or longer (see Table 21:3). The persistence times of hundreds of other herbicides, fungicides, and insecticides generally are between the extremes cited.

TABLE 21:3. *Comparative Persistence of Commonly Used Pesticides Given in Terms of the Time Needed for Half the Chemical to Disappear From the Soil[a]*

Pesticide	Approximate Half-life (years)
Lead, arsenic, copper, mercury	10–30
Dieldrin, BHC, DDT insecticides	2–4
Triazine herbicides	1–2
Benzoic acid herbicides	0.2–1
Urea herbicides	0.3–0.8
2,4-D, 2,4,5-T herbicides	0.1–0.4
Organophosphate insecticides	0.02–0.2
Carbamate insecticides	0.02–0.1

[a] Estimated from Metcalf and Pitts (13).

Among the practices suggested to reduce pesticide levels in soils is the addition of easily decomposed organic matter. The growth of high-nitrogen cover crops or the additions of large quantities of animal manures should be helpful. Apparently degradation of even the most resistant pesticides is encouraged by conditions favoring overall microbial action. Other practices suggested to reduce pesticide levels are frequent cultivation to encourage volatilization and microbial action, cropping to plants which accumulate the pesticide, and leaching the soil. Unfortunately, some of these procedures merely transfer the chemical from the soil to some other part of the environment, a process of dubious value.

This brief review of the behavior of pesticides in soils reemphasizes the complexity of the changes which take place when new and exotic chemicals are added to our environment. The wisdom is seen of evaluating as thoroughly as possible the ecological effects of new chemicals before their extensive use is permitted.

21:4. EFFECTS OF PESTICIDES ON SOIL ORGANISMS

Since the purpose of pesticides is to kill organisms, it is not surprising that some of them are toxic to specific soil organisms. At the same time, the diversity of the soil organism population is so great that except for a few fumigants, most pesticides do not kill a broad spectrum of soil organisms. It is perhaps surprising that the extensive use of pesticides in the United States has not provided more extensive evidence of damage to soil organism numbers. Even so, there is evidence that some commonly used pesticides adversely affect specific groups of organisms, some of which carry out important processes in the soil.

FUMIGANTS. These compounds have a more drastic effect on both the soil fauna and flora than do other pesticides. For example, 99 percent of the microarthropod population is usually killed by the fumigants DD and vampam, and it takes as long as 2 years for the population to recover. The recovery time for the microflora is generally much less, although the rate of recovery varies greatly among the affected organisms. Also, fumigation reduces the number of species of both flora and fauna especially if the treatment is repeated, as is often the case where nematode control is attempted. At the same time, the total number of bacteria is frequently much greater following fumigation than before. This is probably due to the relative absence of competitors and predators following fumigation.

EFFECTS OF PESTICIDES IN THE FIELD. Field pesticide treatment can also reduce the number of animal species as well as total population numbers (see Table 21:4). In general, however, this effect is probably less than the change encountered by bringing natural grasslands or forests into cultivation.

Of the various microorganisms in the soil, the nematodes, fungi, and bacteria seem to be most affected by pesticides. As with fumigation, field pesticide use tends to reduce species numbers, the effect being greatest soon after the application is made. Fungicides have decided effects on the number of viable species of soil fungi.

EFFECTS ON PLANT PATHOGENS. Insecticides and herbicides can affect the growth of fungi, as evidenced by the effects of some of these chemicals on the severity of certain plant diseases caused by fungi (7). For example,

TABLE 21:4. *Effects of Two Insecticides on the Number of Species of Soil Arthropods*[a]

While both chemicals reduced species numbers, pesticide treatment was apparently less detrimental than was cultivation (arable vs. pasture soil).

Treatment	Number of Arthropod Species	
	Arable Field	Pasture
Control	66	148
DDT	48	109
Aldrin	40	99

[a] From Edwards and Lofty (5).

aldrin, an insecticide, reduces the infestation of barley root rot, tomato wilt, and clubroot of cabbage. This fortuitous effect is apparently due to fungicidal action of the insecticide on specific organisms which cause the disease. The opposite effect, increasing disease severity, has been noted as a side effect of other pesticide applications. Increased severity of barley seedling infestation resulting from the application of heptachlor (an insecticide) and sugar beet root infection caused by the use of pebulate and pyragon (herbicides) are cases in point. Apparently the chemicals adversely affect competitors or predators of the plant pathogens, increasing opportunities for attack.

EFFECTS ON NITROGEN TRANSFORMERS. Nitrogen mineralization and nitrification are adversely affected by the application of some pesticides. Again, the most significant depressing effects are noted in fumigated soils. Retarded growth of nitrifiers results in ammonia accumulation sometimes to toxic levels, and in suboptimal growth of plants requiring nitrate sources. It is well to recognize, however, that the nitrifier organisms are not adversely affected by most pesticides applied at normal field rates.

Much is yet to be learned about the ecological effects of pesticides as they influence soil processes and life in the soil. Enough is known, however, to warn us that the addition of any new chemical to soil can adversely affect the life and processes of the soil. Even though the great diversity of soil organisms and processes have made it possible to cope with most pesticides with minimal adverse effects, the few that have been most troublesome will need to be eliminated or replaced,

21:5. CONTAMINATION WITH TOXIC INORGANIC COMPOUNDS[2]

Public attention has been called in recent years to environmental contamination by a number of inorganic compounds, including those containing

[2] For a recent review of this subject, see Lisk (10).

mercury, cadmium, lead, arsenic, nickel, copper, zinc, manganese, fluorine, and boron. To a greater or lesser degree all of these elements are toxic to man and other animals. Cadmium and arsenic are extremely poisonous; mercury, lead, nickel, and fluorine are moderately so, and boron, copper, manganese, and zinc are relatively lower in toxicity. Although the metallic elements (which exclude fluorine and boron) are not all normally included among the "heavy metals," for simplicity this term is often used in referring to them. Table 21:5 provides background information on the uses, sources, and quantity mined annually for each of these elements.

SOURCES AND ACCUMULATION. Several factors account for the emergence of inorganic chemical contamination as an important ecological problem.

TABLE 21:5. *Estimated Quantities of Certain Inorganic Elements Used Annually, Their Major Uses, and Sources of Soil Contamination*[a]

Chemical	Mined Annually (thousands of tons)	Major Uses	Major Sources of Soil Contamination
Arsenic	27	Medicines, pesticides, paints	Pesticides, industrial air pollution
Boron	100	Detergents, glass, fertilizer, gasoline additive	Gasoline combustion, irrigation water
Cadmium	10	Alloys, rust-proofing of steel, batteries, pigments	Smelting, roasting and plating minerals, fertilizer impurities
Copper	4,000	Coins, pipe, alloys, electrical wire	Industrial dusts, mine effluents, sewage treatment waters, fungicides
Fluorine	800	Refrigerants, spray-can propellants, fertilizers, pesticides	Fertilizers, pesticides, local air pollution
Lead	2,000	Gasoline additives, batteries, solders, cable covering	Combustion of leaded gasoline, smelting, fertilizers, pesticides
Manganese	6,000	Ferromanganese, batteries, chemicals, fertilizers	Mine seepage, fly ash, fertilizers
Mercury	3	Dental amalgams, drugs, fluorescent lights, electric switches, scientific instruments	Fungicides, atmospheric contamination from evaporation of metallic Hg
Nickel	300	Stainless steel and other alloys, gasoline additives	Fertilizers, gasoline combustion
Zinc	3,000	Alloys, galvanized metals, brass, paints, cosmetics	Sewage effluents, industrial waste, fertilizers, pesticides

[a] From Bowen (3).

In the first place, modern technology requires the use of these elements in much larger quantities than in the past. The burning of fossil fuels, smelting, and other processing techniques release into the atmosphere tons of these elements, which can adversely affect surrounding vegetation. These "aerosol" dust particles may be carried for miles and later deposited on the vegetation and soil.

Lead is emitted to the atmosphere as one of the products of coal burning. It is also used as an additive to gasolines, as are nickel and boron. Upon combustion these elements are released to the atmosphere and carried to the soil through rain and snow. Borax is used as a detergent and in fertilizer, both of which commonly reach the soil. Superphosphate and limestone usually contain small quantities of cadmium, copper, manganese, nickel, and zinc. Arsenic was used for many years as an insecticide on cotton, tobacco, and fruit crops. It is still being used as a defoliant or vine killer. The heavy metals are found as constituents in specific organic fungicides, herbicides, and insecticides, and tend to be concentrated in domestic and industrial sewage sludge. The process of recycling wastes further accentuates such concentration.

The quantities of most of the products in which these inorganic contaminants are used have increased notably in recent years, enhancing the opportunity for contamination. They are present in the environment in increasing amounts and are daily ingested by man either through the air or through food and water.

NEW DISCOVERIES. Another important reason for current concern over heavy metal contamination is the fact that scientists have discovered that such contamination exists. Sophisticated analytical tools have made possible measurements of contamination previously undetected. The discovery of the accumulation of one contaminant has prompted investigations of others, leading to a better understanding of the cycling of these elements in our environment. Also, in the case of mercury at least, discovery of new means of solubilizing the element has revolutionized the concept of its role as a contaminant.

CONCENTRATION IN ORGANISM TISSUE. Irrespective of their sources, toxic elements can and do reach the soil, where they become part of the life cycle of soil →plant →animal →man (see Fig. 21:3). Unfortunately, once the elements become part of this cycle they may accumulate in animal and human body tissue to toxic levels. This situation is especially critical for fish and other wildlife and for man, who resides at the end of the food chain. It has already resulted in restrictions on the use for human food of certain fish and wildlife. Also, it has become necessary to curtail the release of these toxic elements in the form of industrial wastes.

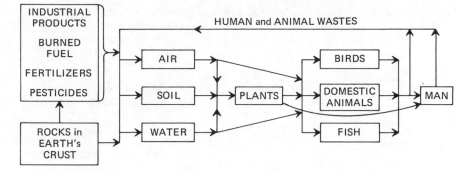

FIGURE 21:3. Sources of heavy metals and their cycling in the soil–water–air–organism ecosystem. It should be noted that the content of metals in tissue generally builds up as movement is made from left to right, indicating the vulnerability of man to heavy metal toxicity.

It is obvious that soils are only a part of the biological cycle relative to heavy metals and other inorganic toxin contamination. At the same time, soils are the ultimate depositories of large quantities of these compounds. Furthermore, the variety of chemical reactions which these elements undergo in soils controls to a considerable extent their rate of cycling if not their removal from the cycle altogether. A brief summary of these reactions follows.

21:6. BEHAVIOR OF INORGANIC CONTAMINANTS IN SOILS

There is considerable variation in the level of these elements present in soils and plants. This is borne out by the data in Table 21:6, which give the ranges commonly found. These relative concentrations are of particular significance as the behavior of each of these elements is considered.

Four of the heavy metals, zinc, copper, manganese, and nickel, have similar chemical characteristics and undergo similar reactions in soils, and so will be discussed as a group. Each of the other elements is sufficiently different in its properties to be given specific consideration.

ZINC, COPPER, MANGANESE, AND NICKEL. The reaction of these elements in soils is definitely affected by the pH, organic matter content, and the oxidation–reduction status of the soil. Ordinarily at pH values of 6.5 and above they tend to be only slowly available to plants, especially if they are present in their high-valent or oxidized forms. Consequently, most soils will tie up relatively large quantities of these elements if the soil pH is high and the drainage good.

The tendency of the cations of these elements to "chelate" in the presence of organic matter decidedly influences their behavior (see p. 493). The relative strength of chelation is generally copper > nickel > zinc > manganese.

TABLE 21:6. *Range of Concentration in Soils and Plants of Inorganic Elements which Sometimes Occur as Environmental Contaminants*[a]

Element	Common Range in Concentration (ppm)	
	Soils	Plants
Arsenic	0.1–40	0.1–5
Boron	2–100	30–75
Cadmium	0.1–7	0.2–0.8
Copper	2–100	4–15
Fluorine	30–300	2–20
Lead	2–200	0.1–10
Manganese	100–4,000	15–100
Nickel	10–1,000	1
Zinc	10–300	15–200

[a] From Allaway (2).

Since iron is more tightly adsorbed than any of them, its presence in a soluble form reduces the chelation tendency of all these elements. However, high pH and good drainage reduce the probability that soluble iron will be present in appreciable quantities.

CADMIUM. Only in recent years has this element been suspected of being toxic to human populations. About ten years ago it was reported that hypertension of laboratory animals was associated with prolonged low-level feeding of this element. There has been too little research accomplished since then to determine the soil and other factors influencing the content of cadmium in food. Likewise, there is little information available on cadmium reactions in soils. Because of its chemical similarity to zinc, however, it would be expected to behave in soils much the same as does zinc. Further research will be needed to determine how cadmium behaves in soils and how its concentration in plants might be controlled.

MERCURY. Research in Sweden and Japan as well as the United States has called attention to toxic levels of this element in certain species of fish. This situation stems from soil reactions whereby mercury is changed from insoluble inorganic forms not available to living organisms to organic forms that can be assimilated easily. Metallic mercury is first oxidized by the following chemical reaction in the sediment layer of lakes and streams:

$$Hg^0 \xrightarrow{[O]} Hg^{++}$$

The divalent mercury is then converted by microorganisms to methylmercury, which is water soluble and can be absorbed through the food chain by fish. The methylmercury can be changed to dimethylmercury through biochemical reactions such as the following:

$$Hg^{++} \rightarrow CH_3Hg^+$$
$$\text{Methylmercury}$$

$$CH_3Hg^+ \rightarrow CH_3HgCH_3$$
$$\text{Dimethylmercury}$$

Apparently the reactions will take place in either aerobic or anaerobic conditions. The methylmercury concentrates as it moves up the food chain, accumulating in some fish to levels which may be toxic to man.

Inorganic mercury compounds added to soils react quickly with the organic matter and clay minerals to form insoluble compounds. In this form the mercury is quite unavailable to growing plants. However, it can be reduced to metallic mercury, which is subject to volatilization and movement elsewhere in the environment. Mercury is not readily absorbed from soil by plants unless it is in the methylmercury form.

LEAD. Interest in the soil as a source of lead for crop plants is heightened by the concern over airborne lead from automobile exhausts. The importance of this airborne source is verified by the concentrations of lead in plants and soils along heavily traversed highways (see Fig. 21:4). The airborne particles are moved far from the point of exhaust and are an important factor in determining the lead content of foods. Just how much lead is deposited directly on the leaf surface and how much is deposited on the soil and later taken up by the plants is not known. However, behavior of this element in soil would suggest that much of the lead in food crops comes from atmospheric contamination.

Soil lead is largely unavailable to plants, as evidenced by the small increases in lead content of plants following soil applications of the element. As with the other toxic metallic cations, lead is quite insoluble in soil, especially if the soil is not too acid. Most lead is found in the surface soil, indicating little if any downward movement. As might be expected, liming reduces the availability of the element and its uptake by plants.

ARSENIC. Reasonably heavy applications of arsenical pesticides over a period of years, especially to orchard soils, have resulted in the accumulation of soil arsenic, in a few cases to toxic levels. These toxicities have in turn led to both soil and plant studies of the reaction and uptake of arsenic. Such studies suggest that arsenic behaves in soils very much like phosphate. For this reason most of the applied arsenate is relatively unavailable for plant growth and uptake. Being present in an anionic form (for example, AsO_4^{3-}), arsenic is adsorbed by hydrous iron and aluminum oxides. This adsorbed arsenate is replaceable from these oxides by phosphate through the process of anion exchange. The similarity in properties between phosphates and arsenates is important to remember.

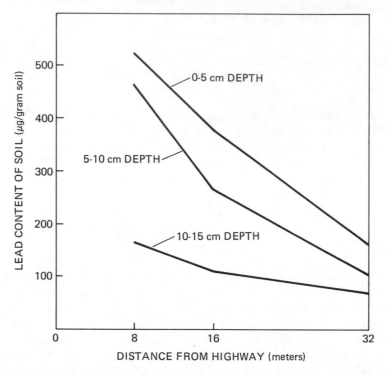

FIGURE 21:4. Lead content of soils at increasing distance from a heavily traveled highway near Beltsville, Maryland. Note that the lead is highest near the automobile traffic (source of lead) and tends to be concentrated in the upper depths. [*Adapted from Lagerwerff and Specht* (9).]

In spite of the capacity of most soils to tie up arsenates, long-term additions of arsenical sprays have in a few instances resulted in decided toxicities to some sensitive plants (see Fig. 21:5). Even though the arsenic level in the plant tissue grown on such soils generally is not toxic to animals, normal plant growth is limited by excess arsenic in the soils. The arsenic toxicity can be reduced by applying to the soil sulfates of zinc, iron, or aluminum. These probably form insoluble arsenate compounds similar to those that form with phosphates.

BORON. Soil contamination by boron can occur from irrigation water high in this element or by excess fertilizer application. The boron can be adsorbed by organic matter and clays but is still available to plants except at high soil pH. Boron is relatively soluble in soils, toxic quantities being leachable especially from acid sandy soils. Boron toxicity is usually considered a localized problem and is probably much less important than a deficiency of the element.

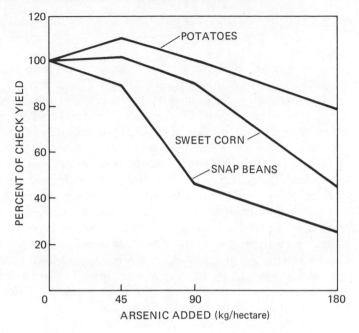

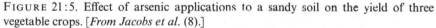

FIGURE 21:5. Effect of arsenic applications to a sandy soil on the yield of three vegetable crops. [*From Jacobs et al.* (8).]

FLUORINE. Fluorine toxicity is also generally localized; it is exhibited in drinking water for animals and in fluoride fumes from industrial processes. The fumes can be ingested directly by the animals or deposited on nearby plants. If the fluorides are adsorbed by the soil, their uptake by plants is restricted. The fluorides formed in soils are highly insoluble, the solubility being least if the soil is well supplied with lime.

21:7. PREVENTION AND ELIMINATION OF INORGANIC CHEMICAL CONTAMINATION

Two primary methods of alleviating soil contamination by toxic inorganic compounds are (a) to eliminate or drastically reduce the soil application of the toxins, and (b) to so manage the soil and crop to prevent further cycling of the toxin.

REDUCING SOIL APPLICATION. The first method requires action to reduce unintentional aerial contamination from industrial operations and from automobile exhausts. Decision makers must recognize the soil as an important natural resource which can be seriously damaged if its contamination by accidental addition of inorganic toxins is not curtailed. Also, there must be judicial reductions in intended applications to soil of the toxins through pesticides, fertilizers, irrigation water, and solid wastes.

REDUCING RECYCLING. Soil and crop management can help reduce the continued cycling of these inorganic chemicals. This is done primarily by keeping the chemicals in the soil rather than encouraging their uptake by plants. The soil becomes a "sink" for the toxins, thereby breaking the cycle of soil–plant–animal (man) through which the toxin exerts its effect. Several techniques to do this exist. For example, most of these elements are rendered less mobile and less available if the pH is kept high. Liming of acid soils should expedite this immobility. Likewise, the draining of wet soils should be beneficial since the oxidized forms of the several toxic elements are generally less soluble and less available for plant uptake than are the reduced forms.

Heavy phosphate applications reduce the availability of these cations but may have the opposite effect on arsenic, which is found in the anionic form. Leaching may be effective in removing excess boron, although moving the toxin from the soil to water may not be of any real benefit. Also advantageous are differences in the abilities of crop species or varieties to extract the toxins. "Accumulators" should be avoided if the harvests are to be fed to man or animals. Moreover, forage crops should be harvested at the maturity stage at which the concentration of the toxin is lowest. It is obvious that soil and crop management offers some potential for alleviating contamination by inorganic elements.

21:8. ORGANIC WASTES

The pollution potential of organic wastes, urban and rural, has become a national and even international problem. In the United States, nearly 1 ton of domestic organic wastes are generated per person each year. Food and fiber processing plants and other industrial operations produce millions of tons of organic wastes, all of which must be disposed of. Farm animal wastes amount to nearly 2 billion tons annually, about two thirds of which are concentrated in large animal feedlots or other confined animal production units where thousands of animals are reared (Table 21:7). Runoff from these feedlots is high in biodegradable organic matter and nitrates (see Fig. 21:6). It has been implicated in fish kills and is a source of nitrate pollution of streams and rivers. In dry climates it can become a windborne air pollutant. The odors of both ammonia and organic compounds from some of these operations are also most offensive. The loss of ammonia is serious because nitrogen is one of the prized nutrients contained in the manure. The gaseous ammonia may be readsorbed by surrounding soils or by rainwater. Much of it eventually moves into nearby lakes, reservoirs, or streams.

Environmental considerations are leading to restrictions on the disposal of organic wastes, both urban and rural, into waterways or into the

TABLE 21:7. *Number of Cattle Feedlots (Thousands) with Less Than and More Than 1,000 Head in Size and the Marketings from Them During the Period 1964–1970[a]*

Year	Under 1,000 Head		Over 1,000 Head	
	Lots	Marketings	Lots	Marketings
1964	223	11,094	1.7	7,050
1966	215	11,336	1.9	9,026
1968	197	11,775	2.0	10,461
1970	180	11,148	2.2	13,642

[a] From U. S. Department of Agriculture (16).

atmosphere by burning. The soil offers an alternative disposal sink in some instances which may be well worth considering. At the same time, care must be taken to prevent harmful contamination in the process.

Two methods of utilizing organic wastes in agriculture have been suggested (4). First, the wastes (for example, domestic sewage and animal

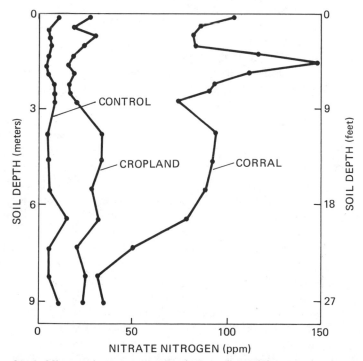

FIGURE 21:6. Nitrate nitrogen concentrations under a dairy corral as compared to those of a nearby cropped area and a control area to which no manure or irrigation water had been added. [*From Adriano et al.* (1).]

manures) can be added to currently cropped soils to increase their productivity. Second, the soils can be used exclusively for waste disposal sites without reference to crop production. Both methods have advantages. Their use in a particular situation will depend on economic as well as biological factors.

21:9. USE OF ORGANIC WASTES FOR CROP PRODUCTION

ADVANTAGES. Organic wastes have long been used to enhance crop production. Farm manure and prosperous agriculture have been closely associated for centuries. In more recent years, urban wastes such as sewage sludge have been used also in commercial agriculture. Materials such as shredded garbage (mostly paper) may have some potential for the future as a soil amendment.

There are obvious arguments in favor of using organic wastes for crop production. By so doing the wastes serve two functions. Crop yields can be increased and long-term soil productivity improved. Furthermore, in some situations the wastes can be surface-applied and thereby afford protection from erosion during critical seasons of the year. The spreading of manure on nearby cropped lands from animal feedlots and poultry farms is an example of effective organic waste utilization.

DISADVANTAGES. The primary disadvantages of using organic wastes for crop production are economic. These wastes are bulky and, compared to commercial inorganic fertilizers, low in nutrient content. Unless the field on which the manure or domestic waste is to be applied is nearby, costs of handling and applying the wastes soon exceed the cost of an equivalent amount of nutrients in commercial fertilizers. Furthermore, there is no certainty of the composition of the wastes, especially domestic sewage. In some cases high carbonaceous wastes require additional nitrogen for degradation, adding to the total cost of their use. Other organic wastes such as poultry manure may supply excess nitrogen and potassium, causing problems of both plant and animal nutrition. Magnesium deficient forage resulting from heavy poultry manure applications in the southwest is a case in point.

Another disadvantage of sewage sludge is its variable content of heavy metals and other toxic inorganic compounds (see Table 21:8). Since the sludge is often produced under anaerobic conditions, reduced forms of these elements may be present, as may be methylmercury with its pollution potential. It is obvious that care must be taken to be certain that sewage applications do not contaminate the soil. Maintaining a high soil pH is a requisite to reduce the possibilities of heavy metal toxicities.

The uninformed may make the assumption that crop production could effectively utilize all the organic wastes produced. Although this may be true biologically, the economic feasibility of such action prevents its implementation in most situations (see Fig. 21:7). Economics has prevented the

TABLE 21:8. *Contents of "Heavy Metals" in Sewage Sludge and Animal Manures in Connecticut*[a]

High level of zinc in sewage sludge is quite common.

Element	Level (ppm) of element in:	
	Sewage Sludge	Manure
Boron	211	38
Copper	758	61
Manganese	172	150
Zinc	3,205	210

[a] From Lunt (11).

widespread practice of composting of domestic wastes followed by land application. Even in Europe where this practice has been most successful in the past, only a minor part of the domestic wastes is composted.

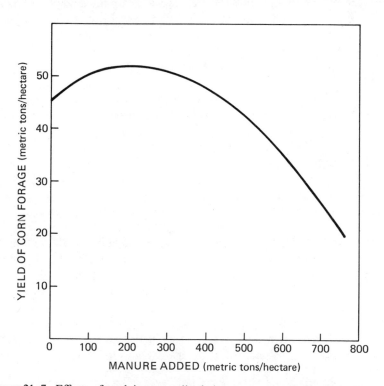

FIGURE 21:7. Effects of applying exceedingly large amounts of manure in 1969 on the forage yield of corn in 1970. Depressed yields at rates above 200 metric tons per hectare are probably due to high salt content in the manure. [*From Murphy et al.* (14).]

FUTURE PROSPECTS. The lack of attractive alternatives for organic waste disposal will likely improve the economic feasibility of the use of soil for this purpose. In the past it has always been assumed that organic wastes added to soils may pay for themselves through added crop yields. The future may well see these materials supplied to farmers at a low cost since adding them to the soil may well be the cheapest way of their disposal. Also, these wastes can be used on speciality crops and home gardens where material cost is not so important. When it is economically feasible, it is certain that land application of organic wastes for crop production will continue to be effectively utilized.

21:10. SOILS AS ORGANIC WASTE DISPOSAL SITES

Soils have long been used as disposal "sinks" for municipal refuse. "Sanitary landfills" are widely employed to dispose of a variety of wastes from our towns and cities. These wastes include paper products, garbage, and nonbiodegradable materials such as glass and metals. These sites are often located in swampy lowland areas which eventually become built up by the dumping, creating upland areas for such uses as city parks and other facilities.

Unfortunately, sanitary landfills are sometimes not so sanitary. Leaching and runoff from these sites can contaminate both surface and ground waters. Contaminants include heavy metals as well as soluble and biodegradable organic materials. The environmental hazards associated with landfills place marked restrictions on their continued use.

Soils are being used successfully for disposal of domestic sewage sludge, food processing wastes, and selected industrial wastes. The city of Chicago transports and applies liquid sewage sludge to a sandy soil area some 50 miles from the city. Organic wastes from food processing plants are commonly applied, often by sprinkling, to nearby agricultural areas. Liquid poultry manure at rates of hundreds of tons per acre has been applied on the surface and by incorporation into soil. Grassland vegetation is commonly permitted on the disposal sites and no attempt is made to harvest the areas.

ADVANTAGES. There are several advantages of soil "sinks" for organic waste disposal. In the first place, most soils have remarkable capacities to accommodate these wastes. The biodegradable components are subject to breakdown by soil organisms. Inorganic components react with soils and are effectively removed from the biological cycle. It is often more economical to use soils as waste disposal sites than to couple such disposal with crop production. Where there is no concern for the crop, rates of application are determined not be crop needs, but by the capacity of the soil to utilize the wastes.

DISADVANTAGES. The primary disadvantages of disposing of organic wastes in soils stems from exceeding the soil's capacity to accommodate the wastes. Examples include heavy metal contamination and excess nitrate leaching into ground waters. Toxicity from zinc, for instance, can prevent growth of grass or other plants used for ground cover. This in turn can reduce the liquid infiltration capacity of the site, limiting its usefulness for further waste absorption.

The formation of nitrates from high nitrogen wastes and their removal from the soil by leaching can seriously affect ground water quality. Nitrate leaching from some feedlots or from nearby storage areas is an example (see page 568). It seem incredible that animal manures, so long associated with good soil management, can become a soil pollutant. Unfortunately, high-density confined animal feeding brings together a concentration of wastes in one feedlot comparable to the human wastes from a small city (see Fig. 21:8). Unless these wastes are dispersed, the movement of nitrogen into drainage waters can be expected, the seriousness being greatest in humid and subhumid areas (12). Research by agricultural and environmental scientists will be required to develop techniques which overcome the disadvantages of applying large quantities of organic wastes to soils.

FIGURE 21:8. Mechanically loading manure from animals grown in a large feedlot. The manure to be utilized or disposed of from one of these large operations is equivalent to the human wastes from a small city. [*Photo courtesy Brookover Feed Yards, Inc., Garden City, Kansas.*]

21:11. SOIL SALINITY

Contamination of soils with salts is one form of soil pollution primarily agricultural in origin. Furthermore, it is not a new problem. Pre-Christian civilizations in both the New and Old Worlds crumbled because salts built up in their irrigated soils. The same principles govern the management of irrigated soils today and the same dangers exist of salt buildup and con- comitant soil deterioration.

Salts accumulate in soils because more of them move into the plant rooting zone than move out. This may be due to application of salt-laden irrigation waters. And it may be caused by irrigating poorly drained soils. Salts move up from the lower horizons and concentrate in the surface soil layers.

EXAMPLE. The problem of salt concentration in some of the rivers of the western United States is illustrated in Fig. 21:9. Removal of water from the river for irrigation purposes reduced total flow and probably increased the movement of salts back into the stream by drainage from irrigated lands. Consequently, in a distance of 200 miles salt concentration in the river in- creased twenty times. Similar but perhaps less dramatic changes take place moving from the headwaters to the sea in other river systems. Noted examples are the Rio Grande and the Colorado systems. It is not difficult to understand why irrigation at the lower reaches of these rivers becomes hazardous.

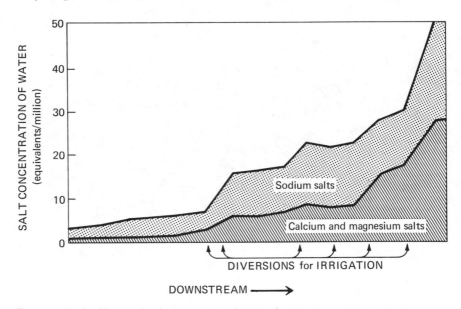

FIGURE 21:9. Changes in the amount and kind of salt in Sevier River, Utah, over a distance of 200 miles. Note the marked salt increase downstream from points of diversion of the water for irrigation purposes. [*From Thorne and Peterson (15); copyright 1967 by the American Association for the Advancement of Science.*]

SALTS IN HUMID REGIONS. One should not gain the impression that there is no salt buildup in rivers whose watersheds are in humid regions. The exact opposite is true, especially in heavily populated and industrialized areas where water is returned to streams following its domestic or industrial use. Sewage plant treatments commonly do not remove soluble inorganic salts unless they are known to be toxicants. When salts thus added are combined with salts leached from the watershed soils, the level is normally higher than that of rivers in areas of lower rainfall, where irrigation is not practiced.

The control of salinity depends almost entirely on water, its quality and management. In local areas sulfur or gypsum applications can be used to help eliminate toxic sodium carbonate (see page 401), but even in these areas, water management following chemical treatment is most vital. Water quality is determined by both public policies with respect to its use and by individual farm practices such as those which assure good soil drainage and optimum irrigation practices.

21:12. RADIONUCLIDES IN SOIL

Nuclear fission in connection with atomic weapons testing provides another source of soil contamination. To the naturally occurring radionuclides in soil (for example, ^{40}K, ^{87}Rb, and ^{14}C), a number of fission products have been added. However, only two of these are sufficiently long-lived to be of significance in soils. These are strontium-90 (half-life $= 28$ years) and cesium-137(half-life $= 30$ years). The average levels of these nuclides in soil in the United States expressed in millicuries per square mile is about 150 for ^{90}Sr and 240 for ^{137}Cs. A comparable figure for the naturally occurring ^{40}K is 20,000. It is obvious that normal soil levels of the fission radionuclides are not high enough to be hazardous. Even during periods of weapons testing, the soil did not contribute significantly to the level of these nuclides in plants. Fallout from the atmosphere directly on the vegetation was the primary source. Consequently, only in the event of a catastrophic supply of fission products could toxic soil levels of ^{90}Sr and ^{137}Cs be expected. Fortunately, considerable research has been accomplished on the soil behavior and nutrient uptake characteristics of these two nuclides.

STRONTIUM-90. Strontium-90 behaves in soil much the same as does calcium, to which it is closely related chemically. It enters soil from the atmosphere in soluble forms and is quickly adsorbed by the colloidal fraction, both organic and inorganic. It undergoes cation exchange and is available to plants much as is calcium. The possibility that strontium is involved in the same plant reactions as calcium probably accounts for the fact that high soil calcium tends to decrease the uptake of ^{90}Sr.

CESIUM-137. Although chemically related to potassium, cesium tends to be less readily available in many soils. This is apparently because ^{137}Cs is firmly fixed by vermiculite and related interstratified minerals. The fixed nuclide is nonexchangeable, much as is fixed potassium in some interlayers of clay. Plant uptake of ^{137}Cs from such soils is very limited. Where vermiculite and related clays are absent, as in some tropical soils, ^{137}Cs uptake is more rapid. In any case, the soil tends to be a damper on the movement of ^{137}Cs into the food chain of animals and man.

21:13. THREE CONCLUSIONS

Three major conclusions may be drawn about soils in relation to environmental quality. First, since soils are valuable resources, they should be protected from environmental contamination, especially that which does permanent damage. Second, because of their vastness and remarkable capacities to absorb, bind, and break down added materials, soils offer promising mechanisms for disposal and utilization of many wastes which otherwise may contaminate the environment. Third, products of soil reactions can be toxic to man and other animals if they move from the soil into the air and particularly into water.

To gain a better understanding of how soils might be used and yet protected in waste management efforts, soil scientists must devote a fair share of their research effort to environmental quality problems.

REFERENCES

(1) Adriano, D. C., et al., "Nitrate and Salt in Soil and Ground Waters from Land Disposal of Dairy Manures," *Soil Sci. Soc. Amer. Proc.*, **35**:759–62, 1971.

(2) Allaway, W. H., "Agronomic Controls over the Environmental Cycling of Trace Elements," *Adv. in Agron.*, **20**:235–74, 1970.

(3) Bowen, H. J. M., *Trace Elements in Biochemistry* (New York: Academic Press, 1966).

(4) Carlson, C. W., and Menzies, J. D., "Utilization of Urban Wastes in Crop Production," *BioScience*, **21**:561–64, 1971.

(5) Edwards, C. A., and Lofty, J. R., "Agricultural Practice and Soil Micro-Arthropods," in Sheals, J. G., Ed., *The Soil Ecosystem*, Publ. 8 (London: The Systematics Association, 1969).

(6) Harris, C. I., and Warren, G. F., "Adsorption and Desorption of Herbicides by Soil," *Weeds*, **12**:120–26, 1964.

(7) Helling, C. S., et al., "Behavior of Pesticides in Soils," *Adv. in Agron.*, **23**:147–240, 1971.

(8) Jacobs, L. W., et al., "Arsenic Residue Toxicity to Vegetable Crops Grown on Plainfield Sand," *Agron. J.*, **62**:588–91, 1970.

(9) Lagerwerff, J. V., and Specht, A. W., "Contamination of Roadside Soil and Vegetation with Cadmium, Nickel, Lead and Zinc," *Env. Sci. Tech.*, **4**:583–86, 1970.

(10) Lisk, D. J., "Trace Metals in Soils, Plants and Animals," *Adv. in Agron.*, **24**:267–325, 1972.

(11) Lunt, H. A., "*Digested Sewage Sludge for Soil Improvement*," Bull. 622, Connecticut Agr. Exp. Sta., 1959.

(12) Marriatt, L. F., and Bartlett, H. D., "Contribution of Animal Waste to Nitrate Nitrogen in Soil," in *Waste Management Research Proceedings*, Cornell Waste Management Conference, Ithaca, N.Y., 1972.

(13) Metcalf, R. L., and Pitts, J. N., "Outlines of Environmental Science," *Adv. in Env. Sci. Tech.*, **1**:1–26, 1969.

(14) Murphy, L. S., et al., "Effects of Solid Beef Feedlot Wastes on Soil Conditions and Plant Growth," in *Waste Management Research Proceedings*, Cornell Agricultural Waste Management Conference, Ithaca, N.Y., 1972.

(15) Thorne, D. W., and Peterson, H. B., "Salinity in United States Waters," in Brady, N. C., Ed., *Agriculture and the Quality of Our Environment*, (Washington, D.C.: Amer. Assoc. for the Advancement of Science, 1967).

(16) U.S. Department of Agriculture, *Number of Cattle Feedlots by Size, Groups and Number of Cattle Marketed*," SRS-14, (Washington, D.C.: USDA, 1968).

Chapter 22

SOILS AND THE WORLD'S FOOD SUPPLY

H UNGER is not new to the world. It has always been a threat to man's survival. At some place on earth through the centuries scarcity of food has brought misery, disease, and even death to man. But never in recorded history has the threat of starvation been greater than it is today. This threat is not due to the reduced capacity of the world to supply food. Indeed, this capacity is greater today than it has ever been and is continuing to grow at a reasonable rate. The problem lies in the even more rapid rate at which world population is increasing. World food production per person is at best holding its own. In selected areas it is declining.

22:1. EXPANSION OF WORLD POPULATION

Science is largely responsible for the marked expansion in world population growth, and especially that which has occurred in the developing nations. Until the near midpoint of the twentieth century, high birth rates in most of South America, Africa, and Asia were largely negated by equally high death rates. High infant mortality, poor health facilities, inadequate medical personnel, and disease-spreading insects took their toll. Population expansion was held in check.

During the past few decades advances in medical science and their application throughout the world have drastically changed this situation (see Fig. 22:1). Death rates have been drastically reduced, especially among the young. Pesticides have held in check mosquitoes and other disease-carrying pests. Medical services heretofore unheard of in remote areas of the developing nations have been initiated. The result is unprecedented population growth. The population is doubling every 18 to 27 years in the developing areas, where two thirds of the world's population now live. Experts predict the world's population will be between 6 and 7 billion by the year 2000—double that of today. Furthermore, *more than 90 percent of this increase will occur in developing nations*, where food supplies are already critical and where the technology for increased food production is wholly inadequate. It is no wonder that the world food supply is considered by some to be mankind's most serious problem.

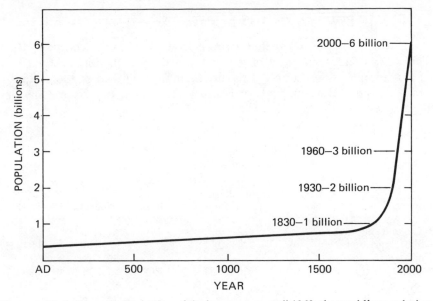

FIGURE 22:1. From the beginning of the human race until 1960, the world's population increased to 3 billion. Population estimates indicate that only 40 years will be needed to provide the second 3 billion. Furthermore, most of the population growth will occur in countries already struggling to feed themselves. [*Adapted from Brown (1).*]

22:2. FACTORS INFLUENCING WORLD FOOD SUPPLIES

The ability of a nation to produce food is determined by a multitude of variables. These include a complex of social, economic, and political factors, most of which affect the farmer's incentive to produce and his supply of production inputs. Also included are a number of physical and biological factors, such as the following:

1. The natural resources available, especially soil and water.
2. Available technology, including the knowledge of proper management of plants, animals, and soils.
3. Improved plant varieties and animal breeds which respond to proper management.
4. Supplies of production inputs such as fertilizers, insecticides, and irrigation water.

Each of these factors is affected by the quantity and quality of soils—their natural productivity and response to management. There is good reason to place satisfactory soil properties high on the list of requisites for an adequate world food supply.

22:3. THE WORLD'S LAND RESOURCES

There is a total of $32\frac{1}{2}$ billion acres of land in the major continents (see Table 22:1). However, most of it is not suited for cultivation. About half of it is completely nonarable. It is mountainous, too cold or too steep for tillage; it may be swampland; or is it desert country, too dry for any but the sparsest vegetation.

TABLE 22:1. *World Land Area in Different Climatic Zones*[a]

Area in billions of acres.

Climatic Zone	Potentially Arable	Grazing	Nonarable	Total
Polar and subpolar	0	0	1.38	1.38
Cold temperate boreal	0.12	0.47	4.28	4.87
Cool temperate	2.24	2.46	2.48	7.18
Warm-temperate subtropical	1.37	2.08	3.38	6.83
Tropical	4.13	4.02	4.08	12.23
Total	7.86	9.02	15.60	32.50

[a] From *The World Food Problem* (2), Vol. II, p. 23.

About one fourth of the land area supports enough vegetation to provide grazing for animals, but for various reasons cannot be cultivated. This leaves only about 25 percent of the land with the physical potential for cultivation (see Table 22:1). And only half of this potentially arable land is actually under cultivation. It is obvious that the kind of soils and their response to management may eventually hold the key to adequate food production, at least in some areas.

CONTINENTAL DIFFERENCES. Data in Table 22:2 suggest the role which soils may play in helping to meet the world food requirements. While the *total* potentially arable land is more than double that being cultivated today, there is great variation from continent to continent. In Asia and Europe where population pressures have been strong for years, most of the potentially arable land is under cultivation. In contrast, only 2 percent of the potentially arable land is cultivated in Australia and New Zealand. Comparable figures are 11 percent for South America and 22 percent for Africa. In these last four areas the physical potential for greater utilization of arable land is indeed great.

It is unfortunate that there is not better distribution of arable land in relation to population densities. The area of cultivated land *per person* is high in North America, the U.S.S.R., Australia, and New Zealand. It is

TABLE 22:2. *Population and Cultivated Land on Each Continent, Compared with Potentially Arable Land (1965)*[a]

| Continent | Population in 1965 (millions) | Area (billions of acres) | | | Acres of Cultivated Land per Person | Ratio of Cultivated to Potentially Arable Land (%) |
		Total	Potentially Arable	Culti-vated		
Africa	310	7.46	1.81	0.39	1.3	22
Asia	1,855	6.76	1.55	1.28	0.7	83
Australia and New Zealand	14	2.03	0.38	0.04	2.9	2
Europe	445	1.18	0.43	0.38	0.9	88
North America	255	5.21	1.15	0.59	2.3	51
South America	197	4.33	1.68	0.19	1.0	11
U.S.S.R.	234	5.52	0.88	0.56	2.4	64
Total	3,310	32.49	7.88	3.43	1.0	44

[a] From *The World Food Problem* (2), Vol. II, p. 434.

low in Asia and Europe and not much higher in South America and Africa. This does not present a serious problem in Europe or the more economically developed parts of Asia. They can readily purchase food from the countries with excess supply. Only the transportation, trade, and marketing problems must be overcome.

In developing countries of Asia, Africa, and Latin America the situation is much more critical. Their populations are increasing far more rapidly than their food production. Countries which formerly exported food crops now must import them. And their national economic growth rate is too slow to provide the resources to pay for the needed food. They must either be provided with food aids by their more fortunate neighbors or must increase dramatically their capacity to produce food.

CHOICES OF ACTION. There are two routes which nations may follow to utilize land to increase their food production: (a) they may clear and cultivate arable land which has heretofore not been tilled, or (b) they may intensify production on lands already under cultivation. Some nations, notably those in Europe and Asia, have only the latter choice. They have little opportunity to expand land under cultivation since most of their arable land is already in use. Only by increasing annual yields per acre can they produce more food.

In areas outside Asia and Europe, the physical potential for increasing land under cultivation is great. For example, in Africa and South America where current land utilization per person is not high, there are about 3 billion acres of arable land which are not now being cultivated. Unfortunately, much of this land is inaccessible to modern transportation. The cost of clearing the land, transporting the fertilizer and other needed inputs, and distributing the food produced from it is high. Also large areas on those continents have humid, tropical climates and tropical soils—the optimum management of which man has yet to learn.

For these reasons, in most areas of the world, intensification of land already under cultivation is the preferred immediate method of increasing food production. In time, however, as economic development, transportation, and knowledge of soil management progress, expansion of land under cultivation is almost certain to occur. Attention will now be directed to general kinds and acreage of soils found where food production is most critical.

22:4. POTENTIAL OF BROAD SOIL GROUPS

In Table 21:3 are shown the acreages dominated by different broad major soil groupings. Note that in terms of total area, shallow soils and sand are most prevalent (8.7 billion acres) followed closely by a variety of Latosols (Oxisols) found in tropical areas. There are vast areas of deserts soils and associated dry areas (Aridisols). Podzols (Spodosols) and podzolic soils make up the next most expansive grouping.

While the total area of different soil groupings is interesting, the potential utilization is of much greater practical importance. For example, 82 percent of the 6.73 billion acres of shallow Lithosols are considered nonarable, not even fit for grazing. In contrast, only about 9 percent of the Chernozem and related soils (Mollisols) are classed as nonarable.

More than one half of the dark-colored, base-rich soils with which the Chernozems (Mollisols) are groups are potentially arable. A similar proportion of Alluvial soils are so classed. It is interesting to note that man has long recognized the productivity of the dark-colored base-rich soils and especially the Alluvial soils (mostly Entisols). Alluvial soils are being utilized throughout the world but are especially important in lowland rice culture. Their location with respect to river and ground water make irrigation relatively easy, and their productivity is generally high. The Chernozemic and related soils (Mollisols) are the bulwark of agriculture in North America, the U.S.S.R., and parts of South America.

The arability of areas dominated by the light-colored, base-rich soils of dry areas is determined to a considerable extent by the availability of irrigation water. The soils are often quite fertile and will respond to management if supplemental water can be applied. The remarkable progress that has been made in irrigation expansion in India and West Pakistan illustrates the productivity potential of these soils. Vast acreages are physically located where they can be irrigated if water can be stored and distributed economically. There are more than 1 billion acres of these soils which can be cultivated successfully if water and other production inputs are available.

The greatest potential for increasing land under cultivation is found in tropical areas dominated by Latosols (Oxisols). Scientists estimate that 40 percent of the more than 6 billion acres of these soils are potentially arable (see Table 22:3). Another 1.75 billion acres are suitable for grazing. The potential here for increasing food production is enormous.

TABLE 22:3. *Total Acreage of Broad Soil Groups in the Major Continents and the Percentages That Are Arable, Nonarable But Suitable for Grazing, and Nonarable[a]*

Broad Soil Groupings	Total Acres (billions)	Approximate Percentage of Acreage		
		Potentially Arable	Grazing	Non-arable
Light colored, base-rich desert (Aridisols)	5.26	20.3	43.0	36.7
Dark colored, base-rich				
Chernozem, Brunizem (Mollisols)	2.03	56.7	34.5	8.8
Grumusol, Terra Rosa	0.81	53.1	33.3	13.6
Brown Forest, Rendzina	0.25	24.0	52.0	24.0
Moderately weathered, leached				
Noncalcic Brown (Alfisols)	0.72	38.0	47.9	14.1
Ando (Inceptisols)	0.06	28.6	42.8	28.6
Highly weathered, leached				
Podzol (Spodosols)	4.85	16.3	25.6	58.1
Red-yellow podzolic (Utisols)	0.96	33.0	50.5	16.5
Latosols (Oxisols)	6.18	42.4	28.5	29.1
Shallow soils and sands				
Lithosols (Inceptisols)	6.73	2.8	15.2	82.0
Regosols	1.90	7.9	18.9	73.2
Alluvial (mostly Inceptisols)	1.47	53.7	29.3	17.0
Tundra (varies)	1.28	0	0	100.0
Total	32.50	24.2	27.8	48.0

[a] Calculated from *The World Food Problem* (2), Vol. II, p. 423.

POTENTIAL BY CONTINENTS. Table 22:4 provides us with estimates of the potentially arable land by soil groupings and by continent. Again one is impressed with the total area of arable Latosols (Oxisols), the sizable arable acreages of arid lands (desert, Aridisols), Chernozems and related soils (Mollisols), and Alluvial soils (mostly Entisols).

Europe and Asia are already utilizing most of their arable land. North America, Australia, and the U.S.S.R. are either excess-food-producing areas or have the potential to become so. For these reasons, attention will be focused on Africa and South America, where food shortages are occurring and where the soil resources are far underutilized.

TABLE 22:4. *Estimates of Acreage of Potentially Arable Land by Soil Groups and Continents*[a]

Broad Soil Groupings	Arable Land[b] (millions of acres)							
	Africa	Asia	Australia	Europe	North America	South America	U.S.S.R.	Total
Light colored, base-rich desert	340	250	110	30	50	90	210	1,080
Dark colored, base-rich								
Chernozem, Brunizem	0	120	30	30	400	180	390	1,150
Grumusol, Terra Rosa	130	60	140	20	70	20	—	440
Brown Forest, Rendzina	10	10	0	20	20	10	10	80
Moderately weathered, leached								
Noncalcic Brown	20	160	20	40	20	10	—	270
Ando	0	10	0	—	0	10	—	20
Highly weathered, leached								
Podzol	—	50	2	280	250	—	220	802
Red-yellow podzolic	20	90	10	—	190	10	—	320
Latosols	1,030	250	30	—	40	1,270	—	2,620
Shallow soils and sands								
Lithosols	20	70	10	10	30	40	10	190
Regosols	120	10	10	10	10	—	10	170
Alluvials	130	480	0	10	80	60	30	790
Total	1,820	1,560	362	450	1,160	1,700	880	7,932

[a] From *The World Food Problem* (2), Vol. II, p. 430. See Table 22:3 for approximate classification of soils under the comprehensive classification system.

[b] Dash indicates soil group not present on the continent; 0 indicates essentially no potentially arable land in this group on the continent.

The light-colored, base-rich soils of dry areas provide considerable food production potential for both continents provided irrigation water is available. Potential exists for increased irrigation in southern South America and in tropical Africa. Doubling or even tripling the acreage under irrigation on these continents would appear to be economically feasible. Problems of storing and distributing the water are immediate limiting factors but in time are likely to be overcome. In both Africa and South America vast areas of Latosols (Oxisols) have yet to be exploited. Most of the unused arable Latosols of the world are found in these two continents—more than $2\frac{1}{2}$ billion acres. Furthermore, there are food population problems in tropical areas near these soil resources. For example, northeast Brazil is an economically depressed food shortage area. In that country literally millions of areas of uncleared Latosols are available. In time, total economic development of the inland areas where these soils are found will make possible realization of the production potential of that country. A similar statement might be made of central Africa, whose Latosols are largely unutilized. However, as the next section will show, there are some unique problems in tropical areas which must be solved before full use can be made of these natural resources.

22:5. PROBLEMS AND OPPORTUNITIES IN THE TROPICS

One cannot help but wonder why soils of the humid tropics have not been more widely exploited. They seemingly have many advantages over their temperate zone counterparts. The crop growing season is commonly year round. In many areas ample moisture is available throughout the growing season. And some of the soils have physical characteristics far superior to the soils of the temperate zones. The hydrous oxide and kaolinitic types of clays which dominate in these areas permit cultivation under very high rainfall conditions.

LIMITING FACTORS. There appear to be many reasons for man's failure to utilize more fully many of the humid tropical areas. In the first place, the environment of the wet tropics has not been too favorable for man and his domesticated plants and animals. Not only is the hot humid climate a source of discomfort, but it encourages diseases and pests and it results in rapid decay and breakdown of tools and equipment man needs for production. Most of those factors which favor domesticated crop and animal production also support competitors in the environment.

Another very important limiting factor to agricultural development in the tropics is the absence or inadequacy of quality transportation. Regardless of the productivity of a given soil area, if there are no highways or railroads connecting it to cities and towns where the customers are, and from whence fertilizers, pesticides, and other inputs come, agriculture will not be likely to succeed.

Coupled with transportation are all other aspects of economic and social development. The package of inputs which have been so essential in the more developed areas of the world are no less essential for agriculture in the developing countries of the tropics. Capital is needed to clear the land, build the roads, construct the irrigation dams and canals, and to build the fertilizer and pesticide plants and distributing systems. Economic development in other segments of the economy is required if money is to be available to purchase the farmer's produce. Marketing systems must be developed to move perishable foods from farms to cities and to distribute them to hungry people. Last but not least, there must be research and education related specifically to the problems of the tropics.

SPECIAL PROBLEM OF TROPICAL SOILS. Special problems are associated with soils of the tropics. In the first place, all too little is known about them and their management. Research on tropical soils in relation to their areal expanse and probable complexities is insignificant compared to that on temperate region soils. Knowledge of their characteristics makes possible identifying only the broadest categories of classification. More intensive study will undoubtedly show many different kinds of soils where now only a few can be identified.

The little we have learned about tropical soils merely reminds us of our ignorance. For example, some of the soils (Oxisols) of Hawaii and of the Philippines are excellent for pineapple and sugar cane production. They respond well to modern management and mechanization. In contrast, modern mechanized farming was a dramatic failure for the "Groundnut Scheme" carried out by the British in Tanganyika following World War II. In that case, apparently, exposing the cleared soil to tropical rains resulted in catastrophic erosion. Knowledge of the nature of the soil might well have prevented much of the project's 100 million dollar loss.

There is a good likelihood that more intensive study will identify complexities among tropical soils similar to those known for temperate regions. We already know of much variability among soils of tropical areas. Some are deep and friable, easily manipulated and tilled. At the other extreme are the lateritic soils which, when denuded of their upper horizon, expose layers which harden into a surface resembling a pavement (see Fig. 22:2). Such soils are essentially worthless from an agricultural point of view.

The chemical characteristics of Latosols (Oxisols) differ drastically from those of soils of temperate regions. The high hydrous oxide content dictates enormous phosphate-fixing capacities. The low cation exchange capacities and heavy rainfall result in removal of not only macronutrients but micronutrients as well. Indeed, the level of technology needed to manage Latosols is fully as high as that required for temperate zone soils.

SOIL AND CROP MANAGEMENT SYSTEMS IN THE TROPICS. The plantation system of agriculture has been successful in raising crops such as bananas, sugar cane, pineapples, rubber, coffee, and cacao. These crops are commonly grown for export and require considerable skill and financial inputs for their production. The plantation system generally imports the best available technology from developed countries and sometimes has associated with it sizable research staffs to gain new knowledge for improved technology. While it has been generally successful in producing and marketing crops and

FIGURE 22:2. Exposed laterite area in central India. The pavementlike surface is barren and will not permit crop production. Sizable acreages of such Latosols (Oxisols) with laterite layers are found in India, Africa, South America, and Australia.

animals, it provides little benefit for the indigenous producer. Conscquently, social and political problems have plagued this system.

At the opposite extreme from the plantation approach are indigenous systems which require little from the outside and which have evolved mostly by trial and error of the native cultivators. One of the most widespread of these systems—that of *shifting cultivation*—will be described briefly to illustrate what the natives have learned from centuries of expericnce (see Fig. 22:3).

While there are variations in the practice of shifting cultivation, in general it involves three major steps:

1. The cutting and burning of trees or other native plants, leaving their ashes on the soil. Sometimes only the vegetation in the immediate area is burned. In other cases this is supplemented with plants brought in from nearby areas.
2. Growing crops on the cleared area for a period of 1 to 5 years, thereby utilizing nutrients left from the clearing and burning of native plants. Vegetation from outside the area may be brought in and burned between plantings.
3. Fallowing the area for a period of 5 to 12 years, thereby permitting regrowth of the native trees and other plants and the consequent

FIGURE 22:3. Nigerian farmer clearing land of natural vegetation in preparation for the planting of crops. The cleared underbrush and trees will be burned and the ash derived therefrom will support crop production for a few years. The farmer will then move to another site, where the steps will be repeated. [*Photo courtesy International Institute of Tropical Agriculture, Ibadan, Nigeria.*]

"rejuvenation" of the soil. Nutrients are accumulated in the native plants and some, such as nitrogen, are released to the soil. The cutting and burning is repeated and the cycle starts again.

Shifting cultivation is primarily a system of nutrient conservation, accumulation, and recycling. The native plants absorb available nutrients from the soil and from the atmosphere. Some are nitrogen fixers. Others are deep rooted and bring nutrients from lower horizons to the surface. All help protect the soil from the devastating effects of rain and sunshine. Also, the cropped area is generally small in size and is surrounded by native vegetation. This reduces the chances of gully formation or severe sheet erosion from runoff water.

There are other benefits of the system other than those relating to nutrients and erosion control. The burning is likely to destroy some weed seeds and even some unwanted insects and disease organisms. The short period of actual cropping (1 to 5 years out of 10 to 20 years discourages the buildup of weeds, insects, and diseases harmful to the cultivated crops.

While shifting cultivation seems primitive, it deserves careful study. In some tropical areas it is more successful than the seemingly more efficient temperate zone systems. Experimentation may permit improvement and alteration of the shifting cultivation system. Perhaps the nutrients accumulated in the native plants can be supplemented with compost or fertilizers. Seeding the fallowed area with plants selected for their beneficial effects rather than allowing natural invasion may also be a forward step.

Some soils in the tropics must have continuous vegetative cover to remain productive. If they dry out, and especially if erosion removes the surface layers, the doughy laterite layers beneath harden irreversibly, making plant growth impossible. This turn of events may be prevented by planting the crop desired among native or other crop plants without completely removing the latter. Since the plants involved are in most cases trees, this system has been termed the *mixed tree crop* system. The desired crop or crops are introduced by removing some of the existing plants and replacing them with crop plants. In time a given area may be planted entirely to a number of crop plants. The essential feature of system, however, is that at no time should the soil be free of vegetation. As primitive as this system may appear, up to now science has not been able to develop more successful alternatives for the management of laterite-containing soils (see Fig. 22:2).

POTENTIAL OF TROPICAL AGRICULTURE. In spite of our inadequate knowledge of tropical agriculture in general, one cannot help but be optimistic about the future of agriculture in the tropics. The basic requirements for maximum year-round production appear to be higher in the humid tropics than anywhere else. Total annual solar radiation provides unmatched photosynthetic potential. The unused soil resources are plentiful (see Table 22:5). There is an increasing tendency to grow several crops a year in a

TABLE 22:5. *Estimated Acreage of Potentially Arable Land of Different Soil Groups in the Tropical Zone*[a]

(In millions of acres)

Soil Groups	Africa	Asia	Latin America	Australia and New Zealand	Total
Light-colored, base-rich soils	160	80	50	40	330
Dark-colored, base-rich soils	267	60	125	20	345
Moderately weathered and leached soils	20	135	15	20	190
Highly weathered and leached soils	1,200	270	1,135	40	2,645
Shallow soils and sands	90	30	40	30	190
Alluvial soils	105	285	40	—	430
Total	1,715	260	1,405	150	4,130

[a] From *The World Food Problem* (2), Vol. II, p. 483. Assumes application of irrigation water.

given land area of the tropics. This is of special importance to food production since the practice can be followed by small land holders whose production almost invariably involves some food crops.

Already researchers have developed new crop varieties especially adapted to the tropics. Adaptive research is identifying means of controlling pests. And fertilizer usage is becoming more common in tropical areas. The potential for food production there is enormous. Its realization depends only on man's ability to exploit this potential.

22:6. REQUISITES FOR THE FUTURE

The world's ability to feed itself depends upon many factors not the least of which is improved agricultural technology in the developing nations of the world. This technology is in turn dependent largely on science and more specifically on research and education. Furthermore, the research and education must have direct relevance to the developing countries and not be a mere transplant of what is available in the more developed nations. Too often the mistake is made of assuming that the technology of western Europe or of the United States can be transferred directly to the underdeveloped countries. Disastrous failures have shown the fallacy of this concept.

PACKAGE APPROACH. Most importantly, improved technologies must provide a package of all the inputs, cultural techniques, incentives, etc., upon which a successful food production system depends. The law of limiting factors described briefly in Chapter 2 is applicable here. Each of the dozens of economic, social, political, and biological factors which affect a successful agricultural system must be considered.

Assuming for the moment that the social, economic, and political factors can be made reasonably favorable, those components of the biological package which have some bearing on soil utilization are now considered.

CROP VARIETIES. New crop varieties adapted to conditions in the developing areas along with methods of pest control are among the first elements of the package. For example, the new dwarf Mexican wheats, which are highly responsive to fertilizer applications, have set a new plateau for wheat yields over a wide geographic area. The average yield of wheat in Mexico has doubled as a result of introducing these new varieties, especially in irrigated areas. Luckily, these new varieties are adapted to food-deficit countries such as India and Pakistan and have been widely adopted there with unprecedented speed. Disease and insect problems are being attacked through the use of pesticides and through the development of resistant varieties.

IRRIGATION AND DRAINAGE. Water supply is a critical factor in crop production in most areas of the world. In some cases, only supplemented moisture is needed to meet temporary deficits or to lengthen the growing season. At other locations, irrigation must be looked to for the bulk of the growing season moisture. Such locations may be in a year-round dry climatic area which is traversed by rivers flowing from areas of higher precipitation. Or there may be a dry period during part of an otherwise wet year. There is a growing recognition that, coupled with irrigation, soil drainage systems are also often essential.

Remarkable progress has been made in expanding irrigation. World-wide acreage under irrigation in 1968 was about four times that in 1900. In India, a country with the most serious of food problems, 69 million acres are under irrigation. Much of that increase has come in the past 15 years. This acreage is being expanded annually, not only with major dam and reservoir projects but with "tube" wells which utilize ground water and which require only the simplest of distribution systems. The potentially irrigable land, in comparison to that currently irrigated, is near triple for India. Similar potentials and plans for their realization exist for parts of South America, Africa, and Australia.

FERTILIZER. It has been estimated (2, Vol. II, p. 379) that "less than 15 percent of the world's fertilizer is used in areas whose agriculture must feed half the world's people." To move from yields common in subsistence agriculture to those dictated by today's food requirements demands dramatic increases in supplies of fertilizer nutrients. It also demands widespread information on soil characteristics to determine the kinds of fertilizers that are needed. In highly leached areas, evaluations must be made of micro-nutrient deficiencies as well as those for nitrogen, phosphorus, potassium, and lime. And attention must be given to fertilizer which resists rapid reaction with the soil or volatilization by microbial action.

While the use of manufactured fertilizers is to be encouraged, economic considerations may dictate alternative sources, especially of nitrogen. Native and improved legumes can and should be used. And animal manures will help supplement the manufactured fertilizers.

SOIL MANAGEMENT. In some areas there are serious soil management problems which need attention. For example, the "black cotton" soils of India and of the Sudan (classified as Vertisols) are heavy textured and probably contain a high proportion of 2:1-type silicate clays. These soils are sticky when wet and hard when dry. The primitive implements and small draft animals used to plow and cultivate them do not permit timely soil manipulation. This means that much of the potential productive capacity of these soils is lost. Methods of incorporating organic matter into the surface of these soils must be sought. And tractors and other machines must be used, perhaps by cooperatives of small landowners, to permit timely tillage of these soils.

Agricultural development activities have emphasized the critical need for the characterization of soils. For example, the salinity and alkalinity status can well determine the probable success of an irrigation project. Nutrient deficiencies can be identified, as can the potential for erosion and drainage problems. The time and effort being devoted to soil characterization is insignificant compared to the needs for this kind of information.

SOIL SURVEYS. Reliable soil survey information is inadequate or unavailable in most of the developing areas. In part this is due to lack of information of the soil characteristics upon which a classification scheme can be based. More frequently it is due to ignorance of the significance and value of the soil survey. Planners sometimes look upon soils as soils, without regard to the vast differences that exist among them—differences that could affect markedly the plans that are made.

Soil surveys will be of special significance in two ways. First, they will make possible the extrapolation of research results from a given area to other areas where the same kinds of soil are found. Second, they will provide one of the criteria to determine the economic feasibility of clearing and preparing for tillage lands which have as yet been unexploited.

MANPOWER. A final requisite for increased food production is trained manpower. The range needed goes from basic scientists (from whose test tubes and field plots new technologies and perhaps new food products are to come) to the cultivators and their assistants. We must have researchers whose interest relates directly to the solution of the world food problem. We must have technicians, farm managers, field service men, and individuals trained in processing and marketing trades. And we must have educators to teach not only the students but the farmers as well.

The fight to feed the world is not yet lost. But to win it will require technological and scientific inputs of a magnitude not yet realized. And among the most important of these inputs are those relating to soils and soil science.

REFERENCES

(1) Brown, L. R., *Man, Land and Food*, Foreign Agr. Econ. Rept. 11 (Washington, D.C.: U. S. Department of Agriculture, 1963).

(2) The President's Science Advisory Committee Panel on World Food Supply, *The World Food Problem*, Vols. I, II, and III (Washington, D.C.: The White House, 1967).

GLOSSARY OF SOIL SCIENCE TERMS[1]

A horizon. The surface horizon of a mineral soil having maximum organic matter accumulation, maximum biological activity and/or eluviation of materials such as iron and aluminum oxides and silicate clays.

ABC soil. A soil with a distinctly developed profile, including A, B, and C horizons.

AC soil. A soil having a profile containing only A and C horizons with no clearly developed B horizon.

accelerated erosion. SEE *erosion.*

acid soil. A soil with a preponderance of hydrogen and aluminum ions in proportion to hydroxyl ions. Specifically, soil with a pH value < 7.0. For most practical purposes a soil with a pH value < 6.6. (The term is usually applied to the surface layer or to the root zone unless specified otherwise.)

acidity, active. The activity of hydrogen ion in the aqueous phase of a soil. It is measured and expressed as a pH value.

acidity, potential. The amount of acidity that must be neutralized to bring an acid soil to neutrality or to some predetermined higher pH value. It is approximated by the sum of the absorbed hydrogen and aluminum. Usually expressed in milliequivalents per unit mass of soil.

actinomycetes. A group of organisms intermediate between the bacteria and the true fungi that usually produce a characteristic branched mycelium. Any organism belonging to the order of Actinomycetales.

adhesion. Molecular attraction which holds the surfaces of two substances in contact, that is, water and sand particles.

adsorption. The attraction of ions or compounds to the surface of a solid. Soil colloids adsorb large amounts of ions and water.

adsorption complex. The group of substances in soil capable of adsorbing other materials. Colloidal particles account for most of this adsorption.

aerate. To impregnate with gas, usually air.

aeration, soil. The process by which air in the soil is replaced by air from the atmosphere. In a well-aerated soil, the soil air is very similar in composition to the atmosphere above the soil. Poorly aerated soils usually contain a much higher percentage of carbon dioxide and a correspondingly lower percentage of oxygen than the atmosphere above the soil. The rate of aeration depends largely on the volume and continuity of pores within the soil.

---

[1] This glossary was compiled from several sources including the following: (1) J. F. Lutz (Committee Chairman) "Glossary of Soil Science Terms," *Soil Sci. Soc. Amer. Proc.* **29**:330–351, 1965; (2) *Soil, The 1957 Yearbook of Agriculture*, U. S. Department of Agriculture, Washington, D.C. 1957; and (3) Soil Survey Staff, *Soil Classification, A Comprehensive System—7th Approximation*, U. S. Department of Agriculture, 1960 and *Supplement to Soil Classification System (7th Approximation)*, 1967; and (4) *Resource Conservation Glossary*, Soil Conservation Society of America, 1970.

aerobic. (i) Having molecular oxygen as a part of the environment. (ii) Growing only in the presence of molecular oxygen, as aerobic organisms. (iii) Occurring only in the presence of molecular oxygen (said of certain chemical or biochemical processes such as aerobic decomposition).

aggregate (*soil*). Many soil particles held in a single mass or cluster such as a clod, crumb, block, or prism.

agric horizon. Horizon immediately below the plow layer of cultivated soils containing accumulated clay and humus to the extent of at least 15 percent of the horizon volume. (New comprehensive classification system.)

agronomy. A specialization of agriculture concerned with the theory and practice of field-crop production and soil management. The scientific management of land.

air-dry. (i) The state of dryness (of a soil) at equilibrium with the moisture content in the surrounding atmosphere. The actual moisture content will depend upon the relative humidity and the temperature of the surrounding atmosphere. (ii) To allow to reach equilibrium in moisture content with the surrounding atmosphere.

air porosity. The proportion of the bulk volume of soil that is filled with air at any given time or under a given condition, such as a specified moisture tension. Usually the large pores; that is, those drained by a tension of less than approximately 100 cm of water. SEE *moisture tension.*

albic horizon. A light-colored surface or lower horizon from which clay and free iron oxides have been removed or so segregated as to permit the color to be determined primarily by the primary sand and slit particles. (New comprehensive classification system.)

Alfisols. Mineral soils that have no mollic epipedon, or oxic, or spodic horizon, but do have an argillic or natric horizon which is at least 35 percent base saturated. Most soils classified as Gray-Brown Podzolic, Noncalcic Brown, and Gray Wooded in the old classification system belong in this soil order. (New comprehensive classification system.)

alkali soil. (i) A soil with a high degree of alkalinity (pH of 8.5 or higher) or with a high exchangeable sodium content (15 percent or more of the exchange capacity), or both. (ii) A soil that contains sufficient alkali (sodium) to interfere with the growth of most crop plants. SEE *saline-sodic soil* and *sodic soil.*

alkaline soil. Precisely any soil that has a pH value > 7. Practically, a soil with a pH of > 7.3. The term is usually applied to surface layer or root zone but may be used to characterize a horizon or a sample thereof.

alkalization. The process whereby the exchangeable sodium content of a soil is increased.

alluvial soil. A soil developing from recently deposited alluvium and exhibiting essentially no horizon development or modification of the recently deposited materials.

alluvium. A general term for all detrital material deposited or in transit by streams, including gravel, sand, silt, clay, and all variations and mixtures of these. Unless otherwise noted, alluvium is unconsolidated.

Alpine Meadow soils. A great soil group of the intrazonal order, comprised of dark soils of grassy meadows at altitudes above the timberline. (1949 classification system.)

alumino-silicates. Compounds containing aluminum, silicon, and oxygen as main constituents. An example is microcline, $KAlSi_3O_8$.

amendment, soil. Any substance such as lime, sulfur, gypsum, and sawdust used to alter the properties of a soil, generally to make it more productive. Strictly speaking, fertilizers are soil amendments, but the term is used most commonly for materials other than fertilizers.

amino acids. Nitrogen-containing organic acids which couple together to form proteins. Each acid molecule contains one or more amino groups ($-NH_2$) and at least one carboxyl group ($-COOH$). In addition, some amino acids contain sulfur.

ammonification. The biochemical process whereby ammoniacal nitrogen is released from nitrogen-containing organic compounds.

ammonium fixation. The adsorption or absorption of ammonium ions by the mineral or organic fractions of the soil in such a manner that they are relatively insoluble in water and relatively nonexchangeable by the usual methods of cation exchange.

anaerobic. (i) The absence of molecular oxygen. (ii) Living or functioning in the absence of air or free oxygen.

Ando soils. A zonal group of dark colored soils high in organic matter developed in volcanic ash deposits.

anion. Negatively charged ion; ion which during electrolysis is attracted to the anode.

anion exchange capacity. The sum total of exchangeable anions that a soil can adsorb. Expressed as milliequivalents per 100 grams of soil (or of other adsorbing material such as clay).

anthropic epipedon. A thick, dark surface horizon, which is more than 50 percent saturated with bases, has a narrow C/N ratio and more than 250 ppm of P_2O_5 soluble in citric acid. It is formed under long continued cultivation where large amounts of organic matter and fertilizers have been added.

antibiotic. A substance produced by one species of organism that, in low concentrations, will kill or inhibit growth of certain other organisms.

Ap. The surface layer of a soil disturbed by cultivation or pasturing.

apatite. A naturally occurring, complex calcium phosphate which is the original source of most of the phosphate fertilizers. Formulas such as $3Ca_3(PO_4)_2 \cdot CaF_2$ illustrate the complex compounds which make up apatite.

argillic horizon. A diagnostic illuvial subsurface horizon characterized by an accumulation of silicate clays. (New comprehensive classification system.)

Aridisols. Soils characteristic of dry places. Includes soils such as Desert, Red Desert, Sierozems and Solochak in the old classification system. (New comprehensive classification system.)

artificial manure. SEE *compost.* (In European usage may denote commercial fertilizers.)

association, soil. SEE *soil association.*

Atterberg limits. Atterberg limits are measured for soil materials passing the No. 40 sieve:

shrinkage limit (SL). The shrinkage limit is the maximum water content at which a reduction in water content will not cause a decrease in the volume of the soil mass. This defines the arbitrary limit between the solid and semisolid states.

plastic limit (PL). The plastic limit is the water content corresponding to an arbitrary limit between the plastic and semisolid states of consistency of a soil.

liquid limit (LL). The liquid limit is the water content corresponding to the arbitrary limit between the liquid and plastic states of consistency of a soil.

autotrophic. Capable of utilizing carbon dioxide or carbonates as the sole source of carbon and obtaining energy for life processes from the oxidation of inorganic elements or compounds such as iron, sulfur, hydrogen, ammonium, and nitrites, or from radiant energy. CONTRAST WITH *heterotrophic*.

available nutrient. That portion of any element or compound in the soil that can be readily absorbed and assimilated by growing plants. ("Available" should not be confused with "exchangeable.")

available water. The portion of water in a soil that can be readily absorbed by plant roots. Considered by most workers to be that water held in the soil against a pressure of up to approximately 15 bars. SEE *field capacity* and *moisture tension*.

azonal soils. Soils without distinct genetic horizons. A soil order under the 1949 classification system.

B horizon. A soil horizon usually beneath the A which is characterized by one or both of the following: (1) an accumulation of silicate clays, iron and aluminum oxides, and humus, alone or in combination; (2) a blocky or prismatic structure.

bar. A unit of pressure equal to one million dynes per square centimeter.

base-saturation percentage. The extent to which the adsorption complex of a soil is saturated with exchangeable cations other than hydrogen and aluminum. It is expressed as a percentage of the total cation exchange capacity.

BC soil. A soil profile with B and C horizons but with little or no A horizon. Most BC soils have lost their A horizons by erosion.

bedding (soil). Arranging the surface of fields by plowing and grading into a series of elevated beds separated by shallow depressions or ditches for drainage.

bedrock. The solid rock underlying soils

and the regolith in depths ranging from zero (where exposed by erosion) to several hundred feet.

bench terrace. An embankment constructed across sloping fields with a steep drop on the down slope side.

biomass. The amount of living matter in a given area.

Black Earth. A term used by some as synonymous with "Chernozem"; by others (in Australia) to describe self-mulching black clays.

Black Soils. A term used in Canada to describe soils with dark surface horizons of the black (Chernozem) zone; includes Black Earth or Chernozem, Wiesenboden, Solonetz, etc.

bleicherde. The light-colored, leached A2 horizon of Podzol soils.

blown-out land. Areas from which all or almost all of the soil and soil material has been removed by wind erosion. Usually unfit for crop production. A miscellaneous land type.

bog iron ore. Impure ferruginous deposits developed in bogs or swamps by the chemical or biochemical oxidation of iron carried in solution.

Bog soil. A great soil group of the intrazonal order and hydromorphic suborder. Includes muck and peat. (1949 classification system.)

border-strip irrigation. SEE *irrigation methods*.

bottomland. SEE *flood plain*.

breccia. A rock composed of coarse angular fragments cemented together.

broad-base terrace. A low embankment with such gentle slopes that it can be farmed, constructed across sloping fields to reduce erosion and runoff.

Brown Earths. Soils with a mull horizon but having no horizon of accumulation of clay or sesquioxides. (Generally used as a synonym for "Brown Forest soils" but sometimes for similar soils acid in reaction.)

Brown Forest soils. A great soil group

of the intrazonal order and calcimorphic suborder, formed on calcium-rich parent materials under deciduous forest, and possessing a high base status but lacking a pronounced illuvial horizon. (1949 classification system.) (A much narrower group than the European Brown Forest or Braunerde.)

Brown Podzolic soils. A zonal great soil group similar to Podzols but lacking the distinct A_2 horizon characteristic of the Podzol group. (1949 classification system.) (Some American soil taxonomists prefer to class this soil as a kind of Podzol and not as a distinct great soil group.)

Brown soils. A great soil group of the temperate to cool arid regions, composed of soils with a brown surface and a light-colored transitional subsurface horizon over calcium carbonate accumulation. (1949 classification system.) They develop under short grasses.

Brunizem. Synonymous with *Prairie soils.* (1949 classification system.)

buffer compounds, soil. The clay, organic matter, and compounds such as carbonates and phosphates that enable the soil to resist appreciable change in pH.

bulk density, soil. The mass of dry soil per unit bulk volume including the air space. The bulk volume is determined before drying to constant weight at 105°C.

buried soil. Soil covered by an alluvial, loessal, or other deposit, usually to a depth greater than the thickness of the solum.

C horizon. A horizon generally beneath the solum which is relatively little affected by biological activity and pedogenesis and is lacking properties diagnostic of an A or B horizon. It may or may not be like the material from which the A and B have formed.

calcareous soil. Soil containing sufficient calcium carbonate (often with magnesium carbonate) to effervesce visibly when treated with cold $O.1N$ hydrochloric acid.

calcic horizon. A horizon of secondary carbonate accumulation more than 6 inches in thickness, with a $CaCO_3$ equivalence of more than 15 percent and at least 5 percent more $CaCO_3$ than the C horizon. (New comprehensive classification system.)

caliche. A layer near the surface, more or less cemented by secondary carbonates of calcium or magnesium precipitated from the soil solution. It may occur as a soft thin soil horizon, as a hard thick bed just beneath the solum, or as a surface layer exposed by erosion.

cambic horizon. A horizon which has been altered or changed by soil forming processes. It usually occurs below a diagnostic surface horizon (epipedon). (New comprehensive classification system.)

capillary conductivity. SEE *hydraulic conductivity.*

capillary porosity. The small pores, or the bulk volume of small pores, which hold water in soils against a tension usually > 60 cm of water. SEE *moisture tension.*

capillary water. (Obsolete) The water held in the "capillary" or *small* pores of a soil, usually with a tension > 60 cm of water. SEE *moisture tension.*

carbon cycle. The sequence of transformations whereby carbon dioxide is fixed in living organisms by photosynthesis or by chemosynthesis, liberated by respiration and by the death and decomposition of the fixing organism, used by heterotrophic species, and ultimately returned to its original state.

carbon–nitrogen ratio. The ratio of the

weight of organic carbon (C) to the weight of total nitrogen (N) in a soil or in organic material.

"Cat" clays. Wet clay soils high in reduced forms of sulfur which upon being drained become extremely acid due to the oxidation of the sulfur compounds.

catena. A sequence of soils of about the same age, derived from similar parent material, and occurring under similar climatic conditions, but having different characteristics due to variation in *relief* and in *drainage.*

cation exchange. The interchange between a cation in solution and another cation on the surface of any surface-active material such as clay or organic matter.

cation exchange capacity. The sum total of exchangeable cations that a soil can adsorb. Sometimes called "total-exchange capacity," "base-exchange capacity," or "cation-adsorption capacity." Expressed in milliequivalents per 100 grams of soil (or of other adsorbing material such as clay).

cemented. Indurated; having a hard, brittle consistency because the particles are held together by cementing substances such as humus, calcium carbonate, or the oxides of silicon, iron, and aluminum.

channery. Thin, flat fragments of limestone, sandstone, or schist up to 6 inches in major diameter.

chelate. (Greek, claw) A type of chemical compound in which a metallic ion is firmly combined with a molecule by means of multiple chemical bonds.

Chernozem. A zonal great soil group consisting of soils with a thick, nearly black or black, organic matter-rich A horizon high in exchangeable calcium, underlain by a lighter colored transitional horizon above a zone of calcium carbonate accumulation; occurs in a cool subhumid climate

under a vegetation of tall and midgrass prairie. (1949 classification system.)

chert. A structureless form of silica, closely related to flint which breaks into angular fragments.

Chestnut soil. A zonal great soil group consisting of soils with a moderately thick, dark-brown A horizon over a lighter colored horizon that is above a zone of calcium carbonate accumulation. They develop under mixed tall and short grasses in a temperate to cool and subhumid to semiarid climate. (1949 classification system.)

chisel, subsoil. A tillage implement with one or more cultivator-type feet to which are attached strong knifelike units used to shatter or loosen hard, compact layers, usually in the subsoil, to depths below normal plow depth. SEE *subsoiling.*

chlorosis. A condition in plants relating to the failure of chlorophyll (the green coloring matter) to develop. Chlorotic leaves range from light green through yellow to almost white.

chroma. The relative purity, strength, or saturation of a color; directly related to the dominance of the determining wavelength of the light and inversely related to grayness; one of the three variables of color. SEE *Munsell color system, hue,* and *value, color.*

class, soil. A group of soils having a definite range in a particular property such as acidity, degree of slope, texture, structure, land-use capability, degree of erosion, or drainage. SEE *Soil texture* and *soil structure.*

classification, soil. The systematic arrangement of soils into groups or categories on the basis of their characteristics. Broad groupings are made on the basis of general characteristics and subdivisions on the basis of more detailed differences in specific properties.

clastic. Composed of broken fragments of rocks and minerals.

clay. (i) A soil separate consisting of particles <0.002 mm in equivalent diameter. (ii) Soil material containing more than 40 percent clay, less than 45 percent sand, and less than 40 percent silt.

clay mineral. Naturally occurring inorganic material (usually crystalline) found in soils and other earthy deposits, the particles being of clay size, that is, <0.002 mm in diameter.

clayey. Containing large amounts of clay or having properties similar to those of clay.

claypan. A compact slowly permeable layer in the subsoil having a much higher clay content than the overlying material, from which it is separated by a sharply defined boundary. Claypans are usually hard when dry, and plastic and sticky when wet.

clod. A compact, coherent mass of soil produced artificially, usually by the activity of man by plowing, digging, and so on, especially when these operations are performed on soils that are either too wet or too dry for normal tillage operations.

coarse texture. The texture exhibited by sands, loamy sands, and sandy loams except very fine sandy loam.

cobblestone. Rounded or partially rounded rock or mineral fragments between 3 and 10 inches in diameter.

cohesion. Holding together: force holding a solid or liquid together, owing to attraction between like molecules. Decreases with rise in temperature.

colloid, soil. (Greek, gluelike) Organic and inorganic matter with very small particle size and a correspondingly large surface area per unit of mass.

colluvium. A deposit of rock fragments and soil material accumulated at the base of steep slopes as a result of gravitational action.

compost. Organic residues, or a mixture of organic residues and soil, that have been piled, moistened, and allowed to undergo biological decomposition. Mineral fertilizers are sometimes added. Often called "artificial manure" or "synthetic manure" if produced primarily from plant residues.

concretion. A local concentration of a chemical compound, such as calcium carbonate or iron oxide, in the form of a grain or nodule of varying size, shape, hardness, and color.

conductivity, hydraulic. SEE *hydraulic conductivity.*

conifer. A tree belonging to the order *Coniferae,* usually evergreen, with cones and needle-shaped or scale-like leaves and producing wood known commercially as "soft wood."

consistence. The combination of properties of soil material that determine its resistance to crushing and its ability to be molded or changed in shape. Such terms as loose, friable, firm, soft, plastic, and sticky describe soil consistence.

consumptive use. The water used by plants in transpiration and growth, plus water vapor loss from adjacent soil or snow, or from intercepted precipitation in any specified time. Usually expressed as equivalent depth of free water per unit of time.

contour. An imaginary line connecting points of equal elevation on the surface of the soil. A contour terrace is laid out on a sloping soil at right angles to the direction of the slope and nearly level throughout its course.

cover crop. A close-growing crop grown primarily for the purpose of protecting and improving soil between periods of regular crop production or between trees and vines in orchards and vineyards.

creep. Slow mass movement of soil

and soil material down relatively steep slopes primarily under the influence of gravity, but facilitated by saturation with water and by alternate freezing and thawing.

crotovina. A former animal burrow in one soil horizon that has been filled with organic matter or material from another horizon (also spelled "krotovina").

crumb. A soft, porous, more or less rounded natural unit of structure from 1 to 5 mm in diameter.

required crushing strength. The force required to crush a mass of dry soil or, conversely, the resistance of the dry soil mass to crushing. Expressed in units of force per unit area (pressure).

crust. A surface layer on soils, ranging in thickness from a few millimeters to perhaps as much as an inch, that is much more compact, hard, and brittle, when dry, than the material immediately beneath it.

crystal. A homogeneous inorganic substance of definite chemical composition bounded by plane surfaces that form definite angles with each other, thus giving the substance a regular geometrical form.

crystal lattice. SEE *lattice structure.*

crystalline rock. A rock consisting of various minerals that have crystallized in place from magma. SEE *igneous rock* and *sedimentary rock.*

deflocculate. (i) To separate the individual components of compound particles by chemical and/or physical means. (ii) To cause the particles of the *disperse phase* of a colloidal system to become suspended in the *dispersion medium.*

Degraded Chernozem. A zonal great soil group consisting of soils with a very dark brown or black A1 horizon underlain by a dark gray, weakly expressed A2 horizon and a brown B (?) horizon; formed in the forest-prairie transition of cool climates. (1949 classification system.)

denitrification. The biochemical reduction of nitrate or nitrite to gaseous nitrogen either as molecular nitrogen or as an oxide of nitrogen.

desalinization. Removal of salts from saline soil, usually by leaching.

desert crust. A hard layer, containing calcium carbonate, gypsum, or other binding material, exposed at the surface in desert regions.

Desert soil. A zonal great soil group consisting of soils with a very thin, light-colored surface horizon, which may be vesicular and is ordinarily underlain by calcareous material; formed in arid regions under sparse shrub vegetation. (1949 classification system.)

desorption. The removal of sorbed material from surfaces.

diagnostic horizons. (As used in the Soil Classification System of the National Cooperative Soil Survey in the United States): Combinations of specific soil characteristics that are indicative of certain classes of soils. Those which occur at the soil surface are called epipedons, those below the surface, diagnostic subsurface horizons.

diatoms. Algae having siliceous cell walls that persist as a skeleton after death. Any of the microscopic unicellular or colonial algae constituting the class Bacillariaceae. They occur abundantly in fresh and salt waters and their remains are widely distributed in soils.

diatomaceous earth. A geologic deposit of fine, grayish, siliceous material composed chiefly or wholly of the remains of diatoms. It may occur as a powder or as a porous, rigid material.

diffusion. The transport of matter as a result of the movement of the constituent particles. The intermingling of two gases or liquids in contact with each other takes place by diffusion.

disintegration. The breakdown of rock and mineral particles into smaller particles by physical forces such as frost action.

disperse. (i) To break up compound particles, such as aggregates, into the individual component particles. (ii) To distribute or suspend fine particles, such as clay, in or throughout a dispersion medium, such as water.

diversion dam. A structure or barrier built to divert part or all of the water of a stream to a different course.

double layer. In colloid chemistry, the electric charges on the surface of the disperse phase (usually negative), and the adjacent diffuse layer (usually positive) of ions in solution.

drainage, soil. As a natural condition of the soil, soil drainage refers to the frequency and duration of periods when the soil is free of saturation; for example, in well-drained soils the water is removed readily but not rapidly; in poorly drained soils the root zone is waterlogged for long periods unless artificially drained, and the roots of ordinary crop plants cannot get enough oxygen; in excessively drained soils the water is removed so completely that most crop plants suffer from lack of water.

drift. Material of any sort deposited by geological processes in one place after having been removed from another. Glacial drift includes material moved by the glaciers and by the streams and lakes associated with them.

drumlin. Long, smooth cigar-shaped low hills of glacial till, with their long axes parallel to the direction of ice movement.

dryland farming. The practice of crop production in low-rainfall areas without irrigation.

duff. The matted, partly decomposed organic surface layer of forest soils.

duripan (*hardpan*). An indurated horizon cemented by materials such as aluminum silicate, silica, $CaCO_3$, and iron.

dust mulch. A loose, finely granular, or powdery condition on the surface of the soil, usually produced by shallow cultivation.

ectotrophic mycorrhiza. A mycorrhizal association in which the fungal hyphae form a compact mantle on the surface of the roots. Mycelial strands extend inward between cortical cells and outward from the mantle to the surface soil.

edaphology. The science that deals with the influence of soils on living things, particularly plants, including man's use of land for plant growth.

electrokinetic potential. In a colloidal system, the difference in potential between the immovable layer attached to the surface of the dispersed phase and the dispersion medium.

eluviation. The removal of soil material in suspension (or in solution) from a layer or layers of a soil. (Usually, the loss of material in *solution* is described by the term "leaching.") See *leaching.*

endotrophic. Nourished or receiving nourishment from within, as fungi or their hyphae receiving nourishment from plant roots in a mycorrhizal association.

endotrophic mycorrhiza. A mycorrhizal association in which the fungal hyphae are present on root surfaces only as individual threads that may penetrate directly into root hairs, other epidermal cells, and occasionally into cortical cells. Individual threads extend from the root surface outward into the surrounding soil.

Entisols. Soils which have no natural genetic horizons or only the beginning of such horizons. Typified by Grumusols in the old classification system. (New comprehensive classification system.)

epipedon. A diagnostic surface horizon which includes the upper part of the soil that is darkened by organic matter, or the upper eluvial horizons or both. (New comprehensive classification system.)

erosion. (i) The wearing away of the land surface by running water, wind, ice, or other geological agents, including such processes as gravitational creep. (ii) Detachment and movement of soil or rock by water, wind, ice, or gravity. The following terms are used to describe different types of water erosion:

accelerated erosion. Erosion much more rapid than normal, natural, geological erosion; primarily as a result of the influence of the activities of man or, in some cases, of animals.

gully erosion. The erosion process whereby water accumulates in narrow channels and, over short periods, removes the soil from this narrow area to considerable depths, ranging from 1 or 2 feet to as much as 75 to 100 feet.

natural erosion. Wearing away of the earth's surface by water, ice or other natural agents under natural environmental conditions of climate, vegetation, and so on, undisturbed by man. SYNONYMOUS WITH *geological erosion.*

normal erosion. The gradual erosion of land used by man which does not greatly exceed natural erosion. SEE *natural erosion.*

rill erosion. An erosion process in which numerous small channels of only several inches in depth are formed; occurs mainly on recently cultivated soils. SEE *rill.*

sheet erosion. The removal of a fairly uniform layer of soil from the land surface by runoff water.

splash erosion. The spattering of small soil particles caused by the impact of raindrops on very wet soils. The loosened and separated particles may or may not be subsequently removed by surface runoff.

eutrophic. Having concentrations of nutrients optimal (or nearly so) for plant or animal growth. (Said of nutrient solutions or of soil solutions.)

eutrophication. A means of aging of lakes whereby aquatic plants are abundant and waters are deficient in oxygen. The process is usually accelerated by enrichment of waters with surface runoff containing nitrogen and phosphorus.

evapotranspiration. The combined loss of water from a given area, and during a specified period of time, by evaporation from the soil surface and by transpiration from plants.

exchange acidity. The titratable hydrogen and aluminum that can be replaced from the adsorption complex by a neutral salt solution. Usually expressed as milliequivalents per 100 grams of soil.

exchange capacity. The total ionic charge of the adsorption complex active in the adsorption of ions. SEE *anion exchange capacity* and *cation exchange capacity.*

exchangeable-sodium percentage. The extent to which the adsorption complex of a soil is occupied by sodium. It is expressed as follows:

$$ESP = \frac{\text{Exchangeable sodium (meq/100 g soil)}}{\text{Cation-exchange capacity (meq/100 g soil)}} \times 100$$

fallow. Cropland left idle in order to restore productivity, mainly through accumulation of water, nutrients, or both. Summer fallow is a common stage before cereal grain in regions of limited rainfall. The soil is kept free of weeds and other vegetation thereby conserving nutrients and water for the next year's crop.

family, soil. In soil classification one of the categories intermediate between the great soil group and the soil series. Families are defined largely on the basis of physical and mineralogical properties of importance to plant growth.

fertility, soil. The status of a soil with respect to the amount and availability to plants of elements necessary for plant growth.

fertilizer. Any organic or inorganic material of natural or synthetic origin which is added to a soil to supply certain elements essential to the growth of plants.

fertilizer grade. The guaranteed minimum analysis, in percent, of the major plant nutrient elements contained in a fertilizer material or in a mixed fertilizer. (Usually refers to the percentage of $N–P_2O_5–K_2O$ but proposals are pending to change the designation to the percentage of $N–P–K$.)

fertilizer requirement. The quantity of certain plant nutrient elements needed, in addition to the amount supplied by the soil, to increase plant growth to a designated optimum.

fibric materials. SEE *organic soil materials.*

field capacity (field moisture capacity). The percentage of water remaining in a soil two or three days after having been saturated and after free drainage has practically ceased.

fine texture. Consisting of or containing large quantities of the fine fractions, particularly of silt and clay. (Includes all clay loams and clays; that is, clay loam, sandy clay loam, silty clay loam, sandy clay, silty clay, and clay textural classes.)

first bottom. The normal flood plain of a stream.

fixation. The process or processes in a soil by which certain chemical elements essential for plant growth are converted from a soluble or exchangeable form to a much less soluble or to a nonexchangeable form; for example, phosphate "fixation." CONTRAST WITH *nitrogen fixation.*

fixed phosphorus. That phosphorus which has been changed to a less-soluble form as a result of reaction with the soil; moderately available phosphorus.

flocculate. To aggregate or clump together individual, tiny soil particles, especially fine clay, into small clumps or granules. Opposite of deflocculate or disperse.

flood plain. The land bordering a stream, built up of sediments from overflow of the stream and subject to inundation when the stream is at flood stage.

fluorapatite. A member of the apatite group of minerals rich in fluorine. Most common mineral in rock phosphate.

fragipan. Dense and brittle pan or layer in soils that owe their hardness mainly to extreme density or compactness rather than high clay content or cementation. Removed fragments are friable, but the material in place is so dense that roots penetrate and water moves through it very slowly.

friable. A soil consistency term pertaining to the ease of crumbling of soils.

fulvic acid. A term of varied usage but usually referring to the mixture of organic substances remaining in solution upon acidification of a dilute alkali extract from the soil.

furrow irrigation. SEE *irrigation methods.*

genesis, soil. The mode of origin of the soil, with special reference to the processes responsible for the development of the solum, or true soil, from the unconsolidated parent material.

geological erosion. SEE *erosion.*

gilgai. The microrelief of soils produced by expansion and contraction with changes in moisture. Found in soils that contain large amounts of clay that swells and shrinks considerably with wetting and drying. Usually a succession of microbasins and microknolls in nearly level areas or of microvalleys and microridges parallel to the direction of the slope.

glacial drift. Rock debris that has been transported by glaciers and deposited, either directly from the ice or from the melt-water. The debris may or may not be heterogeneous.

glacial till. SEE *till.*

glaciofluvial deposits. Material moved by glaciers and subsequently sorted and deposited by streams flowing from the melting ice. The deposits are stratified and may occur in the form of outwash plains, deltas, kames, eskers, and kame terraces.

gley soil. Soil developed under conditions of poor drainage resulting in reduction of iron and other elements and in gray colors and mottles.

granular structure. Soil structure in which the individual grains are grouped into spherical aggregates with indistinct sides. Highly porous granules are commonly called crumbs. A well-granulated soil has the best structure for most ordinary crop plants.

gravitational water. Water which moves into, through, or out of the soil under the influence of gravity.

Gray-Brown Podzolic soil. A zonal great soil group consisting of soils with a thin, moderately dark A1 horizon and with a grayish-brown A2 horizon underlain by a B horizon containing a high percentage of bases and an appreciable quantity of illuviated silicate clay; formed on relatively young land surfaces, mostly glacial deposits, from material relatively rich in calcium. under deciduous forests in humid temperate regions. (1949 classification system.)

Gray Desert soil. A term used in Russia, and frequently in the United States, synonymous with Desert soil. SEE *Desert soil.*

Gray Wooded soils. A zonal great soil group consisting of soils with a thin A1 horizon over a light-colored, bleached (A2) horizon underlain by a B horizon containing a high percentage of bases and an appreciable quantity of illuviated silicate clay. These soils occur in subhumid to semiarid, cool climatic regions under coniferous, deciduous, or mixed forest cover.

Great Group. A category in the new comprehensive classification system between that of the suborder and the subgroup.

Great soil group. Any one of several broad groups of soils with fundamental characteristics in common. Examples are Chernozems, Gray-Brown Podzolic, and Podzol. (1949 classification system.)

green manure. Plant material incorporated with the soil while green, or soon after maturity, for improving the soil.

groundwater. Water that fills all the unblocked pores of underlying material below the water table, which is the upper limit of saturation.

Ground-Water Laterite soil. A great soil group of the intrazonal order and hydromorphic suborder, consisting of soils characterized by hard-pans or concretional horizons rich in iron and aluminum (and sometimes manganese) that have formed immediately above the water table. (1949 classification system.)

Ground-Water Podzol soil. A great soil group of the intrazonal order and hydromorphic suborder, consisting of soils with an organic mat on the surface over a very thin layer of acid humus material underlain by a whitish-gray leached layer, which may be as much as 2 or 3 feet in thickness, and is underlain by a brown, or very dark-brown, cemented hardpan layer; formed under various types of forest vegetation in cool to tropical humid climates under conditions of poor drainage. (1949 classification system.)

gully erosion, SEE *erosion.*

gypsic horizon. A horizon of accumulation of secondary $CaSO_4$, more than 6 inches thick which has at least 5 percent more gypsum than the C or underlying stratum, and in which the product of thickness in inches and the percent gypsum is at least 60 percent-inches. (New comprehensive classification system.)

Half-Bog soil. A great soil group, of the intrazonal order and hydromorphic suborder consisting of soil with dark-brown or black peaty material over grayish and rust mottled mineral soil; formed under conditions of poor drainage under forest, sedge, or grass vegetation in cool to tropical humid climates. (1949 classification system.)

halomorphic soil. A suborder of the intrazonal soil order, consisting of saline and alkali soils formed under imperfect drainage in arid regions and including the great soil group Solonchak or Saline soils, Solonetz soils, and Soloth soils. (1949 classification system.)

hardpan. A hardened soil layer, in the lower A or in the B horizon, caused by cementation of soil particles with organic matter or with materials such as silica, sesquioxides, or calcium carbonate. The hardness does not change appreciably with changes in moisture content and pieces of the hard layer do not slake in water. SEE *caliche, claypan, and duripan.*

heaving. The partial lifting of plants out of the ground, frequently breaking their roots, as a result of freezing and thawing of the surface soil during the winter.

heavy soil. (Obsolete in scientific use) A soil with a high content of the fine separates, particularly clay, or one with a high drawbar pull and hence difficulty to cultivate.

hemic materials. SEE *organic soil materials.*

heterotrophic. Capable of deriving energy for life processes only from the decomposition of organic compounds and incapable of using inorganic compounds as sole sources of energy or for organic synthesis. CONTRAST WITH *autotrophic.*

histic epipedon. A horizon at or near the surface, saturated with water at some season and containing a minimum of 20 percent organic matter if no clay is present and at least 30 percent organic matter if it has 50 percent or more of clay. (New comprehensive classification system.)

Histosols. Soils characterized by their high organic matter content. Bog soils and half-bog soils are included in this soil order. (New comprehensive classification system.)

horizon, soil. A layer of soil, approximately parallel to the soil surface, with distinct characteristics produced by soil-forming processes.

hue. One of the three variables of color. It is caused by light of certain wavelengths and changes with the wavelength. SEE *Munsell color system, chrome,* and *value, color.*

humic acid. A mixture of variable or

indefinite composition of dark organic substances, precipitated upon acidification of a dilute-alkali extract from soil.

Humic Gley soil. Soil of the intrazonal order and hydromorphis suborder that includes Wiesenboden and related soils, such as Half-Bog soils, which have a thin muck or peat O2 horizon and an A1 horizon. Developed in wet meadow and in forested swamps. (1949 classification system.)

humification. The processes involved in the decomposition of organic matter and leading to the formation of humus.

humus. That more or less stable fraction of the soil organic matter remaining after the major portion of added plant and animal residues have decomposed. Usually it is dark in color.

hydraulic conductivity. An expression of the readiness with which a liquid such as water flows through a soil in response to a given potential gradient.

hydrologic cycle. The circuit of water movement from the atmosphere to the earth and return to the atmosphere through various stages or processes, as precipitation, interception, runoff, infiltration, percolation, storage, evaporation, and transpiration.

hydromorphic soils. A suborder of intrazonal soils, all formed under conditions of poor drainage in marshes, swamps, seepage areas, or flats. (1949 classification system.)

hydrous mica. A silicate clay with 2:1 lattice structure, but of indefinite chemical composition since usually part of the silicon in the silica tetrahedral layer has been replaced by aluminum and containing a considerable amount of potassium which serves as an additional bonding between the crystal units, resulting in particles larger than normal in montmorillonite and, consequently, in a lower cation-exchange capacity. Sometimes referred to as illite.

hydroxyapatite. A member of the apatite group of minerals rich in hydroxyl groups. A nearly insoluble calcium phosphate.

hygroscopic coefficient. The amount of moisture in a dry soil when it is in equilibrium with some standard relative humidity near a saturated atmosphere (about 98 percent), expressed in terms of percentage on the basis of oven-dry soil.

igneous rock. Rock formed from the cooling and solidification of magma, and that has not been changed appreciably since its formation.

illite. A hydrous mica. SEE *hydrous mica.*

illuvial horizon. A soil layer or horizon in which material carried from an overlying layer has been precipitated from solution or deposited from suspension. The layer of accumulation.

immature soil. A soil with indistinct or only slightly developed horizons because of the relatively short time it has been subjected to the various soil-forming processes. A soil that has not reached equilibrium with its environment.

immobilization. The conversion of an element from the inorganic to the organic form in microbial tissues or in plant tissues, thus rendering the element not readily available to other organisms or to plants.

impervious. Resistant to penetration by fluids or by roots.

Inceptisols. Soils with one or more diagnostic horizons that are thought to form rather quickly and that do not represent significant illuviation or eluviation or extreme weathering. Soils classified as Brown Forest, Subarctic Brown Forest, Ando, Sols Bruns Acides, and associated Humic Gley and Low-Humic Gley soils are included in this order. (New comprehensive classification system.)

indurated (*soil*). Soil material cemented into a hard mass that will not soften on wetting. See hardpan; consistence.

infiltration. The downward entry of water into the soil.

infiltration rate. A soil characteristic determining or describing the *maximum* rate at which water *can* enter the soil under specified conditions, including the presence of an excess of water.

intergrade. A soil that possesses moderately well-developed distinguishing characteristics of two or more genetically related great soil groups.

intrazonal soils. A soil with more or less well-developed soil characteristics that reflect the dominating influence of some local factor of relief, parent material, or age, over the normal effect of climate and vegetation. (1949 classification system.)

ions. Atoms, groups of atoms, or compounds, which are electrically charged as a result of the loss of electrons (cations) or the gain of electrons (anions).

iron-pan. An indurated soil horizon in which iron oxide is the principal cementing agent.

irrigation efficiency. The ratio of the water actually consumed by crops on an irrigated area to the amount of water diverted from the source onto the area.

irrigation methods. The manner in which water is artificially applied to an area. The methods and the manner of applying the water are as follows:

border-strip. The water is applied at the upper end of a strip with earth borders to confine the water to the strip.

check-basin. The water is applied rapidly to relatively level plots surrounded by levees. The basin is a small check.

corrugation. The water is applied to small, closely-spaced furrows, frequently in grain and forage crops, to confine the flow of irrigation water to one direction.

flooding. The water is released from field ditches and allowed to flood over the land.

furrow. The water is applied to row crops in ditches made by tillage implements.

sprinkler. The water is sprayed over the soil surface through nozzles from a pressure system.

subirrigation. The water is applied in open ditches or tile lines until the water table is raised sufficiently to wet the soil.

wild-flooding. The water is released at high points in the field and distribution is uncontrolled.

isomorphous substitution. The replacement of one atom by another of similar size in a crystal lattice without disrupting or changing the crystal structure of the mineral.

kame. An irregular ridge or hill of stratified glacial drift.

kaolinite. An aluminosilicate mineral of the 1:1 crystal lattice group; that is, consisting of one silicon tetrahedral layer and one aluminum oxide-hydroxide octahedral layer.

lacustrine deposit. Material deposited in lake water and later exposed either by lowering of the water level or by the elevation of the land.

land. Land is a broader term than soil. In addition to soil, its attributes include other physical conditions such as mineral deposits and water supply; location in relation to centers of commerce, populations, and other land; the size of the individual tracts or holdings; and existing plant cover, works of improvement, and the like.

land capability classification. A grouping of kinds of soil into special units, subclasses, and classes according to their capability for intensive use and the treatments required for sustained use, prepared by the Soil Conservation Service, USDA.

land classification. The arrangement of land units into various categories based upon the properties of the land or its suitability for some particular purpose.

land-use planning. The development of plans for the uses of land that, over long periods, will best serve the general welfare, together with the formulation of ways and means for achieving such uses.

Latosol. A suborder of zonal soils including soils formed under forested, tropical, humid conditions and characterized by low silica sesquioxide ratios of the clay fractions, low base exchange capacity, low activity of the clay, low content of most primary minerals, low content of soluble constituents, a high degree of aggregate stability, and usually having a red color. (1949 classification system.)

lattice structure. The orderly arrangement of atoms in a crystalline material.

leaching. The removal of materials in solution from the soil. SEE *eluviation.*

light soil. (Obsolete in scientific use) A coarse-textured soil; a soil with a low drawbar pull and hence easy to cultivate. SEE *coarse texture* and *soil texture.*

lime (agricultural). In strict chemical terms, calcium oxide. In practical terms, it is a material containing the carbonates, oxides and/or hydroxides of calcium and/or magnesium used to neutralize soil acidity.

lime requirement. The mass of agricultural limestone, or the equivalent of other specified liming material, required per acre to a soil depth of 6 inches (or on 2 million pounds of soil) to raise the pH of the soil to a desired value under field conditions.

limestone. A sedimentary rock composed primarily of calcite ($CaCO_3$). If dolomite ($CaCO_3 \cdot MgCO_3$) is present in appreciable quantities it is called a dolomitic limestone.

liquid limit (LL). SEE *Atterberg limits.*

Lithosols. A great soil group of azonal soils characterized by an incomplete solum or no clearly expressed soil morphology and consisting of freshly and imperfectly weathered rock or rock fragments. (1949 classification system.)

loam. The textural class name for soil having a moderate amount of sand, silt, and clay. Loam soils contain 7 to 27 percent of clay, 28 to 50 percent of silt, and less than 52 percent of sand.

loamy. Intermediate in texture and properties between fine-textured and coarse-textured soils. Includes all textural classes with the words "loam" or "loamy" as a part of the class name, such as clay loam or loamy sand. SEE *loam* and *soil texture.*

loess. Material transported and deposited by wind and consisting of predominantly silt-sized particles.

Low Humic Gley soils. An intrazonal group of somewhat poorly to poorly drained soils with very thin surface horizons moderately high in organic matter over gray and brown mineral horizons, which are developed under wet conditions. (1949 classification system.)

luxury consumption. The intake by a plant of an essential nutrient in amounts exceeding what it needs. Thus if potassium is abundant in the soil, alfalfa may take in more than is required.

lysimeter. A device for measuring percolation and leaching losses from a column of soil under controlled conditions.

macronutrient. A chemical element necessary in large amounts (usually > 1 ppm in the plant) for the growth of plants and usually applied artificially in fertilizer or liming materials ("macro" refers to quantity and not to the essentiality of the element). SEE *micronutrient.*

marl. Soft and unconsolidated calcium carbonate, usually mixed with varying amounts of clay or other impurities.

marsh. Periodically wet or continually flooded areas with the surface not deeply submerged. Covered dominantly with sedges, cattails, rushes, or other hydrophytic plants. Subclasses include freshwater and salt-water marshes.

mature soil. A soil with well-developed soil horizons produced by the natural processes of soil formation and essentially in equilibrium with its present environment.

maximum water-holding capacity. The average moisture content of a disturbed sample of soil, 1 cm high, which is at equilibrium with a water table at its lower surface.

mechanical analysis. (Obsolete.) SEE *particle size analysis* and *particle size distribution.*

medium-texture. Intermediate between fine-textured and coarse-textured (soils). (It includes the following textural classes: very fine sandy loam, loam, silt loam, and silt.)

mellow soil. A very soft, very friable, porous soil without any tendency toward hardness or harshness. SEE *consistence.*

metamorphic rock. A rock that has been greatly altered from its previous condition through the combined action of heat and pressure. For example, marble is a metamorphic rock produced from limestone, gneiss is produced from granite, and slate is produced from shale.

micas. Primary alumino-silicate minerals in which two silica layers alternate with one alumina layer. They separate readily into thin sheets or flakes.

microfauna. That part of the animal population which consists of individuals too small to be clearly distinguished without the use of a microscope. Includes protozoa and nematodes.

microflora. That part of the plant population which consists of individuals too small to be clearly distinguished without the use of a microscope. Includes actinomycetes, algae, bacteria, and fungi.

micronutrient. A chemical element necessary in only extremely small amounts (< 1 ppm in the plant) for the growth of plants. Examples are B, Cl, Cu, Fe, Mn, and Zn.) ("Micro" refers to the amount used rather than to its essentiality.) SEE *macronutrient.*

microrelief. Small-scale, local differences in topography, including mounds, swales, or pits that are only a few feet in diameter and with elevation differences of up to 6 feet. SEE *gilgai.*

mineralization. The conversion of an element from an organic form to an inorganic state as a result of microbial decomposition.

mineral soil. A soil consisting predominantly of, and having its properties determined predominantly by, mineral matter. Usually contains < 20% organic matter, but may contain an organic surface layer up to 30 cm thick.

minor element. (Obsolete) SEE *micronutrient.*

moderately coarse texture. Consisting predominantly of coarse particles. (In soil textural classification, it includes all the sandy loams except the very fine sandy loam.) SEE *coarse texture*.

moderately fine texture. Consisting predominantly of intermediate-sized (soil) particles or with relatively small amounts of fine or coarse particles. (In soil textural classification, it includes clay loam, sandy loam, sandy clay loam, and silty clay loam.) SEE *fine texture*.

moisture equivalent. The weight percentage of water retained by a previously saturated sample of soil 1 cm in thickness after it has been subjected to a centrifugal force of one thousand times gravity for 30 minutes.

moisture tension (or *pressure*). The equivalent negative pressure in the soil water. It is equal to the equivalent pressure that must be applied to the soil water to bring it to hydraulic equilibrium, through a porous permeable wall or membrane, with a pool of water of the same composition.

mollic epipedon. A thick, dark surface layer, more than 50 percent base saturated (dominantly with bivalent cations), having a narrow C/N ratio (17/1 or less in virgin state and 13/1 and less with cultivated soils), a strong soil structure, a relatively soft consistence when dry, and less than 250 ppm of P_2O_5 soluble in citric acid. This horizon characterizes the Mollisol Soil Order, but is found in other orders as well.

Mollisols. Soils characterized by a thick, dark mineral surface horizon which is dominantly saturated with bivalent cations and has moderate to strong structure. Includes soils such as Chernozem, Prairie, Chestnut in the old classification system. (New comprehensive classification system.)

montmorillonite. An aluminosilicate clay mineral with a 2:1 expanding crystal lattice; that is, with two silicon tetrahedral layers enclosing an aluminum octahedral layer. Considerable expansion may be caused along the C axis by water moving between silica layers of contiguous units.

mor. Raw humus; a type of forest humus layer of unincorporated organic material, usually matted or compacted or both; distinct from the mineral soil, unless the latter has been blackened by washing in organic matter.

moraine. An accumulation of drift, with an initial topographic expression of its own, built within a glaciated region chiefly by the direct action of glacial ice. Examples are ground, lateral, recessional, and terminal moraines.

morphology, soil. The constitution of the soil including the texture, structure, consistence, color, and other physical, chemical, and biological properties of the various soil horizons that make up the soil profile.

mottling. Spots or blotches of different color or shades of color interspersed with the dominant color.

muck. Highly decomposed organic material in which the original plant parts are not recognizable. Contains more mineral matter and is usually darker in color than peat. SEE *muck soil*, *peat*.

muck soil. An organic soil in which the organic matter is well decomposed (U. S. usage).

mulch. Any material such as straw, sawdust, leaves, plastic film, and loose soil that is spread upon the surface of the soil to protect the soil and plant roots from the effects of raindrops, soil crusting, freezing, evaporation, etc.

mulch farming. A system of farming in which the organic residues are not plowed into or otherwise mixed with the soil but are left on the surface as a mulch.

mull. A humus-rich layer of forested soils consisting of mixed organic and mineral matter. A mull blends into the upper mineral-layers without an abrupt change in soil characteristics.

Munsell color system. A color designation system that specifies the relative degree of the three simple variables of color: hue, value, and chroma. For example: 10YR 6/4 is a color (of soil) with a hue = 10YR, value = 6, and chroma = 4. These notations can be translated into several different systems of color names as desired. SEE *chroma hue*, and *value, color*.

mycorrhiza. The association, usually symbiotic, of fungi with the roots of seed plants. SEE *ectotrophic mycorrhiza* and *endotrophic mycorrhiza.*

natric horizon. A special argillic horizon which has prismatic or columnar structure and is more than 15 percent saturated with exchangeable sodium. This is common in Solonctz and Solodized-Solonetz soils. (New comprehensive classification system.)

necrosis. Death associated with discoloration and dehydration of all or parts of plant organs, such as leaves.

nematodes. Very small worms abundant in many soils and important because many of them attack and destroy plant roots.

neutral soil. A soil in which the surface layer, at least to normal plow depth, is neither acid nor alkaline in reaction. SEE *acid soil, alkaline soil, pH,* and *reaction, soil.*

nitrification. The biochemical oxidation of ammonium to nitrate.

nitrogen assimilation. The incorporation of nitrogen into organic cell substances by living organisms.

nitrogen fixation. The conversion of elemental nitrogen (N_2) to organic combinations or to forms readily utilizable in biological processes.

nodule bacteria. SEE *rhizobia.*

noncalcic brown soils. The zonal group

of soils with slightly acid, light pinkish or light reddish brown A horizons over light reddish brown or dull red B horizons developed under mixed grass and forest vegetation in a subhumid, wet-dry climate. See horizon, soil; zonal soil.

nucleic acids. Complex compounds found in the nuclei of plant and animal cells and usually combined with proteins as nucleoproteins.

O horizon. Organic horizon of mineral soils.

ochric epipedon. A light-colored surface horizon generally low in organic matter which includes the eluvial layers near the surface. It is often hard and massive when dry. Although characteristic of Aridisols, it is found in several of the other soil orders. (New comprehensive classification system.)

order. The highest category in soil classification. The three orders are zonal soils, intrazonal soils, and azonal soils (1949 classification system). Ten orders are recognized in the new comprehensive classification system: Entisol, Vertisol, Inceptisol, Aridisol, Mollisol, Spodisol, Alfisol, Ultisol, Oxisol, and Histisol.

organic soil. A soil which contains a high percentage (> 20%) of organic matter throughout the solum.

organic soil materials. (As used in the Soil Classification System of the National Cooperative Soil Survey in the United States): (i) Saturated with water for prolonged periods unless artificially drained and having more than 30 percent organic matter if the mineral fraction is more than 50 percent clay, or more than 20 percent organic matter if the mineral fraction has no clay. (ii) Never saturated with water for more than a few days and having more than 34 percent organic matter. SEE *Histosols.* Kinds of organic materials:

fibric materials. The least decomposed of all the organic soil materials, containing very high amounts of fiber that are well preserved and readily identifiable as to botanical origin.

hemic materials. Intermediate in degree of decomposition of organic materials between the less decomposed fibric and the more decomposed sapric materials.

sapric materials. The most highly decomposed of the organic materials, having the highest bulk density, least amount of plant fiber, and lowest water content at saturation.

ortstein. An indurated layer in the B horizon of Podzols in which the cementing material consists of illuviated sesquioxides (mostly iron) and organic matter.

osmotic. A type of pressure exerted in living bodies as a result of unequal concentration of salts on both sides of a cell wall or membrane. Water will move from the area having the least salt concentration through the membrane into the area having the highest salt concentration and, therefore, exerts additional pressure on this side of the membrane.

oven-dry soil. Soil which has been dried at 150°C until it reaches constant weight.

oxic horizon. A highly-weathered diagnostic subsurface horizon from which most of the combined silica has been removed leaving a mixture dominated by hydrous oxide clays with some 1:1 type silicate minerals and quartz present. (New comprehensive classification system.)

Oxisols. Soils of tropical and subtropical regions characterized by the presence of a horizon (oxic) from which most of the combined silica has been removed by weathering, leaving oxides of iron and aluminum and some quartz. Includes soils referred to as Latosols and some called Ground-Water Laterites. (New comprehensive classification system.)

pans. Horizons or layers, in soils, that are strongly compacted, indurated, or very high in clay content. SEE *caliche, claypan, duripan, fragipan,* and *hardpan.*

parent material. The unconsolidated and more or less chemically weathered mineral or organic matter from which the solum of soils is developed by pedogenic processes.

particle density. The mass per unit volume of the soil particles. In technical work, usually expressed as grams per cubic centimeter.

particle size. The effective diameter of a particle measured by sedimentation, sieving, or micrometric methods.

particle-size analysis. Determination of the various amounts of the different separates in a soil sample, usually by sedimentation, sieving, micrometry, or combinations of these methods.

particle-size distribution. The amounts of the various soil separates in a soil sample, usually expressed as weight percentages.

parts per million (ppm). Weight units of any given substance per one million equivalent weight units of oven-dry soil; or, in the case of soil solution or other solution, the weight units of solute per million weight units of solution.

peat. Unconsolidated soil material consisting largely of undecomposed, or only slightly decomposed, organic matter accumulated under conditions of excessive moisture. (SEE *organic soil materials.*)

ped. A unit of soil structure such as an aggregate, crumb, prism, block, or granule, formed by natural processes (in contrast with a clod, which is formed artificially).

pedalfer. (Obsolete) A subdivision of a soil order comprising a large group of soils in which sesquioxides increased relative to silica during soil formation.

pedocal. (Obsolete) A subdivision of a soil order comprising a large group of soils in which calcium accumulated during soil formation.

peneplain. A once high, rugged area which has been reduced by erosion to a low, gently rolling surface resembling a plain.

penetrability. The ease with which a probe can be pushed into the soil. (May be expressed in units of distance, speed, force, or work depending on the type of penetrometer used.)

percolation, soil water. The downward movement of water through soil. Especially, the downward flow of water in saturated or nearly saturated soil at hydraulic gradients of the order of 1.0 or less.

permafrost. (i) Permanently frozen material underlying the solum. (ii) A perennially frozen soil horizon.

permanent charge. The net negative (or positive) charge of clay particles inherent in the crystal lattice of the particle; not affected by changes in pH or by ion-exchange reactions.

permeability, soil. The ease with which gases, liquids, or plant roots penetrate or pass through a bulk mass of soil or a layer of soil.

pF. (Obsolete) The logarithm of the soil moisture tension expressed in centimeters height of a column of water.

pH, soil. The negative logarithm of the hydrogen-ion activity of a soil. The degree of acidity (or alkalinity) of a soil as determined by means of a glass, quinhydrone, or other suitable electrode or indicator at a specified moisture content or soil–water ratio, and expressed in terms of the pH scale.

pH-dependent charge. That portion of the total charge of the soil particles which is affected by, and varies with, changes in pH.

phase, soil. A subdivision of a soil series or other unit of classification having characteristics that affect the use and management of the soil but which do not vary sufficiently to differentiate it as a separate series. A variation in a property or characteristic such as degree of slope, degree of erosion, and content of stones.

photomap. A mosaic map made from aerial photographs with physical and cultural features shown as on a planimetric map.

physical properties (of soils). Those characteristics, processes, or reactions of a soil which are caused by physical forces and which can be described by, or expressed in, physical terms or equations. Examples of physical properties are bulk density, water-holding capacity, hydraulic conductivity, porosity, pore-size distribution, and so on.

plaggen epipedon. Man-made surface layer more than 20 inches thick and produced by long continued manuring. Common in Europe but unknown in the U. S.

Planosol. A great soil group of the intrazonal order and hydromorphic suborder consisting of soils with eluviated surface horizons underlain by B horizons more strongly eluviated, cemented, or compacted than associated normal soil. (1949 classification system.)

plastic limit (PL). SEE *Atterberg limits.*

plastic soil. A soil capable of being molded or deformed continuously and permanently, by relatively moderate pressure, into various shapes. SEE *consistence.*

platy. Consisting of soil aggregates that are developed predominately along the horizontal axes; laminated; flaky.

plinthite (brick). A highly weathered mixture of sesquioxides or iron and aluminum with quartz and other diluents which occurs as red mottles and which changes irreversibly to hardpan upon alternate wetting and drying.

plow layer. The soil ordinarily moved in tillage; equivalent to surface soil.

plow-plant. Plowing and planting a crop in one operation, with no additional seedbed preparation.

Podzol. A great soil group of the zonal order consisting of soils formed in cool-temperate to temperate, humid climates, under coniferous or mixed coniferous and deciduous forest, and characterized particularly by a highly-leached, whitish-gray A2 horizon. (1949 classification system.)

podzolization. A process of soil formation resulting in the genesis of Podzols and podzolic soils.

polypedon. (As used in the Soil Classification System of the National Cooperative Soil Survey in the United States): Two or more contiguous pedons, all of which are within the defined limits of a single soil series. In early stages of the development of the classification scheme this was called a soil individual.

pore size distribution. The volume of the various sizes of pores in a soil. Expressed as percentages of the bulk volume (soil plus pore space).

porosity. The volume percentage of the total bulk not occupied by solid particles.

potassium fixation. The process of converting exchangeable or water-soluble potassium to a form not easily exchanged from the adsorption complex with a cation of a neutral salt solution.

Prairie soils. A zonal great soil group consisting of soils formed under temperate to cool-temperate, humid regions under tall grass vegetation. (1949 classification system.)

primary mineral. A mineral that has not been altered chemically since depo-

sition and crystallization from molten lava.

prismatic soil structure. A soil structure type with prismlike aggregates that have a vertical axis much longer than the horizontal axes.

productivity, soil. The capacity of a soil for producing a specified plant or sequence of plants under a specified system of management. Productivity emphasizes the capacity of soil to produce crops and should be expressed in terms of yields.

profile, soil. A vertical section of the soil through all its horizons and extending into the parent material.

protein. Any of a group of nitrogen-containing compounds that yield amino acids on hydrolysis and have high molecular weights. They are essential parts of living matter and are one of the essential food substances of animals.

puddled soil. Dense, massive soil artificially compacted when wet and having no regular structure. The condition commonly results from the tillage of a clayey soil when it is wet.

reaction, soil. The degree of acidity or alkalinity of a soil, usually expressed as a pH value.

Extremely acid	Below 4.5
Very strongly acid	4.5–5.0
Strongly acid	5.1–5.5
Medium acid	5.6–6.0
Slightly acid	6.1–6.5
Neutral	6.6–7.3
Mildly alkaline	7.4–7.8
Moderately alkaline	7.9–8.4
Strongly alkaline	8.5–9.0
Very strongly alkaline	9.1 and higher

Red Desert soil. A zonal great soil group consisting of soils formed under warm-temperate to hot, dry regions under desert-type vegetation, mostly shrubs. (1949 classification system.)

Reddish Brown soils. A zonal group of soils with a light brown surface horizon of a slightly reddish cast which grades into dull reddish brown or red material heavier than the surface soil, then into a horizon of whitish or pinkish lime accumulation; developed under shrub and short-grass vegetation of warm-temperate to tropical regions of semiarid climate.

Reddish Brown Lateritic soils (of U. S.). A zonal group of soils with dark reddish brown granular surface soils, red friable clay B horizons, and red or reticulately mottled lateritic parent material, developed under humid tropical climate with wet-dry seasons and tropical forest vegetation.

Red-Yellow Podzolic soils. A zonal great soil group consisting of soils formed under warm-temperate to tropical, humid climates, under deciduous or coniferous forest vegetation and usually under conditions of good drainage. (1949 classification system.)

regolith. The unconsolidated mantle of weathered rock and soil material on the earth's surface; loose earth materials above solid rock. (Approximately equivalent to the term "soil" as used by many engineers.)

Regosol. Any soil of the azonal order without definite genetic horizons and developing from or on deep, unconsolidated, soft mineral deposits such as sands. loess, or glacial drift. (1949 classification system.)

Rendzina. A great soil group of the intrazonal order and calcimorphic suborder consisting of soils with brown or black friable surface horizons underlain by light-gray to pale-yellow calcareous material; developed from soft, highly calcareous parent material under grass vegetation or mixed grasses and forest in humid and semiarid climates. (1949 classification system.)

residual material. Unconsolidated and partly weathered mineral materials accumulated by disintegration of consolidated rock in place.

reticulate mottling. A network of streaks of different color; most commonly found in the deeper profiles of Lateritic soils.

rhizobia. Bacteria capable of living symbiotically with higher plants, usually legumes, from which they receive their energy, and capable of using atmospheric nitrogen; hence, the term symbiotic nitrogen-fixing bacteria. (Derived from the generic name *Rhizobium.*)

rhizosphere. That portion of the soil in the immediate vicinity of plant roots in which the abundance and composition of the microbial population are influenced by the presence of roots.

rill. A small, intermittent water course with steep sides; usually only a few inches deep and, hence, no obstacle to tillage operations.

rill erosion. SEE *erosion.*

riprap. Broken rock, cobbles, or boulders placed on earth surfaces, such as the face of a dam or the bank of a stream, for protection against the action of water (waves); also applied to brush or pole mattresses, or brush and stone, or other similar materials used for soil erosion control.

rock. The material that forms the essential part of the earth's solid crust, including loose incoherent masses such as sand and gravel, as well as solid masses of granite and limestone.

rough broken land. Land with very steep topography and numerous intermittent drainage channels but usually covered with vegetation.

runoff. That portion of the precipitation on an area which is discharged from the area through stream channels. That which is lost without entering the soil is called *surface runoff* and that which enters the soil before reaching the stream is called *ground water run-off* or *seepage flow* from ground water. (In soil science "runoff" usually refers to the water lost by surface flow; in geology and hydraulics "runoff" usually includes both surface and sub-surface flow.)

salic horizon. A horizon at least 6 inches thick with secondary enrichment of salts more soluble in cold water than gypsum. (New comprehensive classification system.)

saline–sodic soil. A soil containing sufficient exchangeable sodium to interfere with the growth of most crop plants and containing appreciable quantities of soluble salts. The exchangeable-sodium percentage is >15, the conductivity of the saturation extract > 4 millimhos per centimeter (at 25°C), and the pH is usually 8.5 or less in the saturated soil.

saline soil. A nonsodic soil containing sufficient soluble salts to impair its productivity.

salinization. The process of accumulation of salts in soil.

saltation. Particle movement in water or wind where particles skip or bounce along the stream bed or soil surface.

sand. A soil particle between 0.05 and 2.0 mm in diameter.

sapric materials. SEE *organic soil materials.*

savanna (savannah). A grassland with scattered trees, either as individuals or clumps. Often a transitional type between true grassland and forest.

second bottom. The first terrace above the normal flood plain of a stream.

secondary mineral. A mineral resulting from the decomposition of a primary mineral or from the reprecipitation of the products of decomposition of a primary mineral. SEE *primary mineral.*

sedimentary rock. A rock formed from materials deposited from suspension or precipitated from solution and usually being more or less consolidated. The principal sedimentary rocks are sandstones, shales, limestones, and conglomerates.

self-mulching soil. A soil in which the surface layer becomes so well aggregated that it does not crust and seal under the impact of rain but instead serves as a surface mulch upon drying.

separate, soil. One of the individual-sized groups of mineral soil particles— sand, silt, or clay.

series, soil. SEE *soil series.*

shear. Force, as of a tillage implement, acting at right angles to the direction of movement.

sheet erosion. SEE *erosion.*

shelterbelt. A wind barrier of living trees and shrubs established and maintained for protection of farm fields. Syn. windbreak.

shrinkage limit (SL). SEE *Atterberg limits.*

Sierozem. A zonal great soil group consisting of soils with pale-grayish A horizons grading into calcareous material at a depth of 1 foot or less, and formed in temperate to cool, arid climates under a vegetation of desert plants, short grass, and scattered brush. (1949 classification system.)

silica/alumina ratio. The molecules of silicon dioxide (SiO_2) per molecule of aluminum oxide (Al_2O_3) in clay minerals or in soils.

silica sesquioxide ratio. The molecules of silicon dioxide (SiO_2) per molecule of aluminum oxide (Al_2O_3) plus ferric oxide (Fe_2O_3) in clay minerals or in soils.

silt. (i) A soil separate consisting of particles between 0.05 and 0.002 mm in equivalent diameter. (ii) A soil textural class.

silting. The deposition of water-borne sediments in stream channels, lakes, reservoirs, or on flood plains, usually resulting from a decrease in the velocity of the water.

site index. A quantitative evaluation of the productivity of a soil for forest growth under the existing or specified environment.

slick spots. Small areas in a field that are slick when wet, due to a high content of alkali or of exchangeable sodium.

sodic soil. A soil that contains sufficient sodium to interfere with the growth of most crop plants, and in which the exchangeable-sodium percentage is 15 or more.

sodium adsorption ratio (SAR).

$$\text{SAR} = \frac{1}{2} \frac{\text{Na}^+}{\sqrt{\text{Ca}^{2+} + \text{Mg}^{2+}}}$$

where the cation concentrations are in milliequivalents per liter.

soil. (i) A dynamic natural body on the surface of the earth in which plants grow, composed of mineral and organic materials and living forms. (ii) The collection of natural bodies occupying parts of the earth's surface that support plants and that have properties due to the integrated effect of climate and living matter acting upon parent material, as conditioned by relief, over periods of time.

soil air. The soil atmosphere; the gaseous phase of the soil, being that volume not occupied by solid or liquid.

soil alkalinity. The degree or intensity of alkalinity of a soil, expressed by a value >7.0 on the pH scale.

soil association. A group of defined and named taxonomic soil units occurring together in an individual and characteristic pattern over a geographic region, comparable to plant associations in many ways.

soil complex. A mapping unit used in detailed soil surveys where two or more defined taxonomic units are so intimately intermixed geographically that it is undesirable or impractical, because of the scale being used, to separate them. A more intimate mixing of smaller areas of individual taxonomic units than that described under *soil association.*

soil conservation. A combination of all management and land use methods which safeguard the soil against depletion or deterioration by natural or by man-induced factors.

soil erosion. SEE *erosion.*

soil fertility. SEE *fertility, soil.*

soil genesis. The mode of origin of the soil with special reference to the processes or soil-forming factors responsible for the development of the solum, or true soil, from the unconsolidated parent material.

soil geography. A subspecialization of physical geography concerned with the areal distributions of soil types.

soil horizon. SEE *horizon, soil.*

soil management. The sum total of all tillage operations, cropping practices, fertilizer, lime, and other treatments conducted on or applied to a soil for the production of plants.

soil map. A map showing the distribution of soil types or other soil mapping units in relation to the prominent physical and cultural features of the earth's surface.

soil mechanics and engineering. A subspecialization of soil science concerned with the effect of forces on the soil and the application of engineering principles to problems involving the soil.

soil microbiology. A subspecialization of soil science concerned with soil-inhabiting microorganisms and with their relation to agriculture, including both plant and animal growth.

soil monolith. A vertical section of a soil profile removed from the soil and mounted for display or study.

soil morphology. The physical constitution, particularly the structural properties, of a soil profile as exhibited by the kinds, thickness, and arrangement of the horizons in the profile, and by the texture, structure, consistency, and porosity of each horizon.

soil organic matter. The organic fraction of the soil that includes plant and animal residues at various stages of decomposition, cells and tissues of soil organisms, and substances synthesized by the soil population. Commonly determined as the amount of organic material contained in a soil sample passed through a 2-millimeter sieve.

soil porosity. SEE *porosity*.

soil productivity. SEE *productivity, soil*.

soil reaction. SEE *reaction, soil* and *pH, soil*.

soil salinity. The amount of soluble salts in a soil, expressed in terms of percentage parts per million, or other convenient ratios.

soil science. That science dealing with soils as a natural resource on the surface of the earth including soil formation, classification and mapping, and the physical, chemical, biological, and fertility properties of soils *per se*; and these properties in relation to their management for crop production.

soil separates. SEE *separate, soil*.

soil series. The basic unit of soil classification, being a subdivision of a family and consisting of soils that are essentially alike in all major profile characteristics.

soil solution. The aqueous liquid phase of the soil and its solutes consisting of ions dissociated from the surfaces of the soil particles and of other soluble materials.

soil structure. The combination or arrangement of primary soil particles into secondary particles, units, or peds. These secondary units may be, but usually are not, arranged in the profile in such a manner as to give a distinctive characteristic pattern. The secondary units are characterized and classified on the basis of size, shape, and degree of distinctness into classes, types, and grades, respectively.

soil survey. The systematic examination, description, classification, and mapping of soils in an area. Soil surveys are classified according to the kind and intensity of field examination.

soil texture. The relative proportions of the various soil separates in a soil.

soil type. The lowest unit in the natural system of soil classification; a subdivision of a soil series and consisting of or describing soils that are alike in all characteristics including the texture of the A horizon. (1949 classification system.)

Sol Brun Acide. A zonal group of soils developed under forest vegetation with thin A1 horizon, a paler A2 horizon which is poorly differentiated from the B2 horizon, a B2 horizon with uniform color from top to bottom, weak subangular blocky structure, and lacking evidence of silicate clay accumulation. The sola are strongly to very strongly acid and have low base status.

solodized soil. A soil that has been subjected to the processes responsible for the development of a Soloth and having at least some of the characteristics of a Soloth. (1949 classification system.)

Solonchak soils. An intrazonal group of soils with high concentrations of soluble salts in relation to those in other soils, usually light-colored, without characteristic structural form, developed under salt-loving plants, and occurring mostly in a subhumid or semiarid climate. (1949 classification system.)

Solonetz soils. An intrazonal group of soils having surface horizons of varying degrees of friability underlain by dark hard soil, ordinarily with columnar structure (prismatic structure with rounded tops). This hard layer is usually highly alkaline. Such soils are developed under grass or shrub vegetation, mostly in subhumid or semiarid climates.

soluble-sodium percentage (SSP). The proportion of sodium ions in solution in relation to the total cation concentration, defined as follows:

$$SSP = \frac{\text{Soluble sodium conc. (meq/liter)}}{\text{Total cation conc. (meq/liter)}} \times 100$$

solum (plural: sola). The upper and most weathered part of the soil profile; the A and B horizons.

splash erosion. SEE erosion.

spodic horizon. A subsurface diagnostic horizon containing an illuvial accumulation of free sesquioxides of iron and aluminum and of organic matter. (New comprehensive classification system.)

Spodosols. Soils characterized by the presence of a spodic horizon, an eluvial horizon in which active organic matter and amorphous oxides of aluminum and iron have precipitated. These soils include most Podzols, Brown Podzolics, and Ground Water Podzols of the old classification system. (New comprehensive classification system.)

sprinkler irrigation. SEE irrigation methods.

stratified. Arranged in or composed of strats or layers.

strip cropping. The practice of growing crops which require different types of tillage, such as row and sod, in alternate strips along contours or across the prevailing direction of wind.

structure, soil. SEE soil structure.

stubble mulch. The stubble of crops or crop residues left essentially in place on the land as a surface cover before and during the preparation of the seedbed and at least partly during the growing of a succeeding crop.

subirrigation. SEE irrigation methods.

subsoil. That part of the soil below the plow layer.

subsoiling. Breaking of compact subsoils, without inverting them, with a special knife-like instrument (chisel) which is pulled through the soil at depths usually of 12 to 24 inches and at spacings usually of 2 to 5 feet.

summer fallow. The tillage of uncropped land during the summer in order to control weeds and store moisture in the soil for the growth of a later crop.

surface runoff. SEE runoff.

surface soil. The uppermost part of the soil, ordinarily moved in tillage, or its equivalent in uncultivated soils and ranging in depth from 3 to 4 inches to 8 or 10. Frequently designated as the "plow layer," the "Ap layer," or the "Ap horizon."

symbiosis. The living together in intimate association of two dissimilar organisms, the cohabitation being mutually beneficial.

talus. Fragments of rock and other soil material accumulated by gravity at the foot of cliffs or steep slopes.

tensiometer. A device for measuring the negative pressure (or tension) of water in soil in situ; a porous, permeable ceramic cup connected through a tube to a manometer or vacuum gauge.

tension, soil-moisture. The equivalent negative pressure of suction of water in soil.

terrace. (i) A level, usually narrow, plain bordering a river, lake, or the sea. Rivers sometimes are bordered by terraces at different levels. (ii) A raised, more or less level or horizontal strip of earth usually constructed on or nearly on a contour and designed to make the land suitable for tillage and to prevent accelerated erosion.

texture. SEE *soil texture.*

thermal analysis (differential thermal analysis). A method of analyzing a soil sample for constituents, based on a differential rate of heating of the unknown and standard samples when a uniform source of heat is applied.

thermophilic organisms. Organisms which grow readily at temperatures above 45°C.

tile, drain. Pipe made of burned clay, concrete, or similar material, in short lengths, usually laid with open joints to collect and carry excess water from the soil.

till. (i) Unstratified glacial drift deposited directly by the ice and consisting of clay, sand, gravel, and boulders intermingled in any proportion. (ii) To plow and prepare for seeding; to seed or cultivate the soil.

tilth. The physical condition of soil as related to its ease of tillage, fitness as a seedbed, and its impedance to seedling emergence and root penetration.

toposaic. A photomap on which topographic or terrain-form lines are shown, as on standard topographic quadrangles. SEE *photomap.*

toposequence. A sequence of related soils that differ, one from the other, primarily because of *topography* as a soil-formation factor.

topsoil. (i) The layer of soil moved in cultivation. SEE *surface soil.* (ii) Presumably fertile soil material used to topdress roadbanks, gardens, and lawns.

trace elements. (Obsolete.) SEE *micronutrient.*

truncated. Having lost all or part of the upper soil horizon or horizons.

tuff. Volcanic ash usually more or less stratified and in various states of consolidation.

tundra. A level or undulating treeless plain characteristic of arctic regions.

Tundra soils. (i) Soils characteristic of tundra regions. (ii) A zonal great soil group consisting of soils with dark-brown peaty layers over grayish horizons mottled with rust and having continually frozen substrata; formed under frigid, humid climates, with poor drainage, and native vegetation of lichens, moss, flowering plants, and shrubs. (1949 classification system.)

type, soil. SEE *soil type.*

Ultisol. Soils of humid areas characterized by the presence of either an argillic horizon or a fragipan each of which is less than 35 percent saturated with bases. They have no spodic horizon and no oxic or natric horizons. They include soils formerly classified as Red-Yellow Podzolics, Reddish-Brown Laterites, and Rubrozems. (New comprehensive classification system.)

umbric epipedon. A thick, dark surface layer which is less than 50 percent saturated with bases. Ando soils have umbric epipedons. (New comprehensive classification system.)

unsaturated flow. The movement of water in a soil which is not filled to capacity with water.

value, color. The relative lightness or intensity of color and approximately a function of the square root of the total amount of light. One of the three variables of color. SEE *Munsell color system, hue,* and *chroma.*

varnish, desert. A glossy sheen or coating on stones and gravel in arid regions.

varve. A distinct band representing the annual deposit in sedimentary materials regardless of origin and usually consisting of two layers, one a thick, light-colored layer of silt and fine sand, the other a thin, dark layer of clay.

Vertisols. Soils high in swelling clays which crack widely upon drying resulting in shrinking, shearing and soil mass movement. (New comprehensive classification system.)

virgin soil. A soil that has not been significantly disturbed from its natural environment.

volume weight. (Obsolete) SEE *bulk density, soil.*

waterlogged. Saturated with water.

water-stable aggregate. A soil aggregate stable to the action of water such as falling drops, or agitation as in wet-sieving analysis.

water table. The upper surface of ground water or that level below which the soil is saturated with water; locus of points in soil water at which the hydraulic pressure is equal to atmospheric pressure.

water table, perched. The surface of a local zone of saturation held above the main body of groundwater by an impermeable layer or stratum, usually clay, and separated from the main body of groundwater by an unsaturated zone.

weathering. All physical and chemical changes produced in rocks, at or near the earth's surface, by atmospheric agents.

wilting point (or *permanent wilting point*). The moisture content of soil, on an oven-dry basis, at which plants (specifically sunflower plants) wilt and fail to recover their turgidity when placed in a dark humid atmosphere.

windbreak. A planting of trees, shrubs, or other vegetation, usually perpendicular or nearly so to the principal wind direction, to protect soil, crops, homesteads, roads, and so on, against the effects of winds, such as wind erosion, and the drifting of soil and snow.

xerophytes. Plants that grow in or on extremely dry soils or soil materials.

zeta potential. SEE *electrokinetic potential.*

zonal soil. A soil characteristic of a large area or zone. (1949 classification system.)

INDEX

Absorption by plants
  form in which nutrients are taken up, 35, 36
  influenced by cation exchange, 105
  influenced by soil aeration, 263
  luxury consumption of potassium, 474
  of ammonium, 36
  of calcium ions, 35
  of carbonate ions, 35
  of nitrate nitrogen, 36
  of organic materials, 150
  of organic nitrogen compounds, 150
  of organic phosphorus compounds, 146, 458, 469
  of phosphate ions, 36
  of potassium ions, 35
  of sulfate ions, 36
  of water, 197, 263
Acidity (*see* Soil acidity)
Acre–furrow slice
  energy of, 143
  organic and mineral soils compared, 11, 361, 365
  pounds of nutrients, 23, 24
  pounds of organic matter in, 24, 25
  relation to profile, 10
  weight of, mineral soils, 54
  weight of, peat soils, 361
Actinomycetes of soil, 127, 130, 390, 395
Adhesion, water retention by, 166
Adsorption of anions by soils
  mechanism of, 465
  of phosphorus, 465
Adsorption of cations by soils
  capacity of colloidal clays, 81, 99
  capacity of humus, 148
  capacity of mineral soils, 102
  capacity of peat, 363, 364
  influence of organic matter, 103
  mechanism of, 97
  of calcium, 26–28
  of hydrogen, 97
  of magnesium, 26–28
  of metallic cations, 75
  of potassium, 26–28
Adsorption of pesticides, 556
Adsorption of water by soils, 75, 166
Adsorption vs. absorption, 31
Aeolian deposits (*see also* Glacial-aeolian deposits)
  loessial, 299–300
  other than loessial, 288, 289, 300
Aeration of soil
  importance, 260–264
  influence of cropping on, 258–260
  influence of drainage on, 233, 254, 265
  influence of organic matter on, 55–56, 258
  influence of soil moisture on, 189, 253

Aeration of soil [*cont.*]
  influence of soil texture on, 43, 62, 254, 258-259
  influence of structure on, 258–259
  influence of tillage on, 68, 265–266
  influence on crop management, 265–266
  influence on nitrification, 145, 260, 264, 429
  influence on nitrogen loss, 431
  influence on nutrient absorption, 263
  influence on oxidation, 260–262
  influence on plant root development, 261
  influence on reduction, 264, 432
  influence on soil color, 265
  influence on soil organisms, 126, 129, 140, 260, 429
  influence on solubility of nutrients, 263
  influence on water absorption by plants, 263
  measurement of, 256
  subsoil vs. topsoil, 258–259
Age of soil, fertility relations, 303, 309
Aggregation of soil particles (*see also* Granulation of soil)
  formation of aggregates, 58
  influenced by cropping, 63
  influenced by liming, 60, 413
  influenced by organic matter, 58
  influenced by organisms, 58
  influenced by tillage, 60
  influence on aeration, 55
  influence on crop yields, 68
  influence on erosion losses, 238
  stability of aggregates, 60, 148
Air of soil (*see also* Aeration of soil)
  a general consideration, 16
  composition of, 256
  factors affecting composition of, 256–258
  gases present, 16, 253
  importance of carbon dioxide, 31, 98
  importance of oxygen, 255
  movement of, 254–256
  relative humidity of, 188, 256
  variability of, 14
Air space of soil
  amount, 12, 67
  as related to structure, 55–77
  as related to texture, 55–77
  fluctuation of, 12
  influence of cropping on, 55
  influence of tillage on, 60, 68
  regulation, 62, 67–68
Alfisol soil order, 323, 330, 333
Algae of soil, 123, 457
Alkaline and saline soils
  chemical treatment of, 401
  classification of, 326, 396
  effect of leaching, 342, 397, 400
  irrigation of, 216–217, 400
  management of, 400

Alkaline and saline soils [*cont.*]
 origin of, 396
 pH of, 342, 397
 plant tolerance of, 399
 problems of, 217–218, 397
 reclamation of, 400
 saline-sodic, 397
 saline soils, 396
 salts, concentration in, 218, 326
 sodic soils, 397
 sodium in, 396
 Solonchak, 323, 341
 Solonetz, 323, 341
 spots, 398
 white, 326, 397
Alluvial soil materials, 582
 deltas, 292
 fans, 291
 flood plains, 290
 importance as soil, 290–292, 324
 of Mississippi River, 291
Aluminum
 accumulation in B horizon, 313
 amount in soils, 24
 hydroxyl ions, 373–375
 influence on phosphates, 460–466
 in silicate clays, 76–78
 role in acid soils, 373–375
 solubility and pH, 389, 414
 toxicity to plants, 389
Amino compounds
 carriers of nitrogen, 27, 141
 decomposition of, 29–30, 141, 427
 produced by nodule bacteria, 434
 reaction with clay, 146–147
 reduction of, 432
 used by microorganisms, 141, 426
Ammonia
 ammonia liquor, 506
 as source of fertilizer salts, 504, 505
 economic source of nitrogen, 505, 523
 fixation, 427, 477
 used directly as fertilizer, 506, 527
Ammoniated superphosphate
 as a fertilizer, 510
 manufacture of, 511
Ammoniates, 504
Ammonification
 explained, 29–30
 organisms involved, 145
Ammonium carbonate
 formation as organic matter decomposes,
  426
 formation from urea, 508
 influence on soil pH, 508
Ammonium nitrate
 as a fertilizer, 505, 507
 deliquescence of, 507
 effect on soil pH, 520
Ammonium salts
 in rain water, 441
 used as fertilizers, 505
Ammonium sulfate
 as a fertilizer, 505, 506
 cause of acidity, 379
Amoeba, 119

Analysis (*see* Chemical analysis; Mechanical
  analysis)
Anhydrous ammonia, as a separate fertilizer,
  505, 527
Animal manure
 aerobic treatment of, 543
 anaerobic treatment of, 544
 annual rates of production of, 535
 artificial, 161
 as pollutant, 534, 544, 568–573
 chemical composition of, 538–541
 effects of, 546
 influence on soil organic matter, 161
 mineral content of, 539
 moisture in, 538
 N/P/K ratio in, 541–543
 nutrients in, 539, 540
 organic constituents in, 538
 quantity of, 534–535
 storage of, 541
 utilization of, 544–545
Animals in soil
 macro, 111–113
 micro, 118
Anion exchange and phosphate fixation, 465
Antibiotics in soil, 132
Apatite
 as a soil mineral, 279
 reversion product, 466
 source of phosphorus to plants, 458, 512
Aridisol soil order, 323, 326, 333, 582, 583
Arid region soils
 classification, wet and dry as, 333, 334
 compared to humid region soils, 24
 general composition of, 24
 management of, 215, 216, 396
 nutrient content of, 24
 pH of, 35, 396–397
 soluble salts of, 218, 396–399
 water relations of, 215–219
Artificial farm manure, 161
Association of soils, 338–340
Atmospheres of tension of soil water, 173
Autotrophic bacteria
 nature of, 129
 types of, 129, 428–429
Availability of nutrient elements
 influenced by aeration, 260
 influenced by pH, 36–37, 394–398,
  413–414, 462
Azonal soil groups, 322, 340–343
Azotobacter in soil, 439

Bacteria in soil (*see also* Soil organisms)
 aerobic vs. anaerobic, 129
 autotrophic vs. heterotrophic, 129, 428
 Azotobacter, 439
 cause of plant disease, 130
 Clostridium group, 457
 colonies of, 128
 competition with plants, 132
 conditions affecting, 129
 nature of, 128
 nitrifiers, 429
 nitrogen fixers, 131, 434, 438

Bacteria in soil [*cont.*]
  nitrogen reducers, 431
  nodule organisms, 434
  numbers and weight, 128
  rate of multiplication, 128
  relations to pH, 398, 414
  size of, 128
  sulfur oxidizers, 129, 146
  surface exposed by, 128
Bactericidal substances in soils, 132
Bars of tension of soil water, 173
Base saturation of soils
  meaning of, 103, 385
  of representative soils, 105
  relation to pH, 104, 378
Basic exchange (*see* Cation exchange)
Basic slag as a fertilizer, 510, 512
Bedding in barnyard manure, 538, 542, 544
Beijerinckia in soils, 439
Biological cycles of soil
  carbon, 143–144
  nitrogen, 424
Biological nature of soil, 17
Black alkali (*see* Alkali and saline soils)
Bog lime, 367, 405
Bog soils, 323, 341
Bone meal, 510, 512
Boron in soil
  amounts in soils, 23, 489
  an essential element, 22, 484
  as fertilizer, 489, 516
  as pollutant, 566
  availability of, 496
  forms in nature, 489
  influence of pH on, 388, 389, 415, 496
  ionic form in soil, 36, 496
  role of, 487
Brown forest soils, 323, 341
Brown podzolic soils, 323, 341
Brunigra soils (*see* Prairie soils)
Buffering (*see* Soil buffering)
Bulk density
  contrasted with particle density, 50
  data of, 52, 63
  determination of, 51
  explained, 50
  factors affecting, 50
  influence of cropping, 51
  of peat soil, 361
  range of, 51
  sod vs. cultivated soil, 53
Burned lime
  guarantee, 407
  source and nature of, 404

Calcimorphic soils, 341
Calcite, 279, 404–405
Calcium
  adsorbed, prominence of, 92
  amount in soils, 24, 74, 365
  amounts exchangeable in mineral soils, 28, 105
  amounts exchangeable in peats, 364
  an essential element, 21
  as pollutant, 564

Calcium [*cont.*]
  compounds used as lime, 404–405
  diagram, gains and losses of, 421
  flocculating effects, 60, 412
  forms in soils, 27
  influence on boron absorption, 389, 415, 496
  influence on microbial activity, 127, 390, 416
  influence on nitrification, 414, 446
  influence on phosphate availability, 389, 413, 415, 462
  influence on potassium availability, 413, 479
  ionic form in soil, 36
  loss by erosion, 412
  loss in drainage, 226, 227, 424
  pH and exchangeable calcium, 98, 105, 365, 387, 415
  removal by crops, 412
  transfer to available forms, 31
  why compounds used as lime, 404–405
Calcium carbonate
  in liming materials, 405
  solubility influenced by CO_2, 411
Calcium cyanamid, as a fertilizer, 505, 508
Calcium metaphosphate, as a fertilizer, 510, 512
Calcium nitrate, as a fertilizer, 505
Calcium oxide equivalent of lime, 406–407
Cal-nitro as a fertilizer, 505, 507
Capillary movement of soil water
  capillary conductivity, 182
  explained, 178–180
  factors influencing, 180–185
  formula for height, 179
  supplying plants with water, 197
  type most common, 184
Capillary water in soil
  amounts in soils, 192
  as soil solution, 178
  factors governing amount, 192
  tension limits, 172
Carbon
  an essential element, 21
  carbonate source for plants, 36, 144
  ratio to nitrogen in plants, 151
  ratio to nitrogen in soil, 151, 365, 372
Carbon cycle
  diagram of, 144
  importance of, 143
Carbon dioxide (*see also* Carbonic acid)
  diffusion of, 254–256
  generation in farm manure, 565
  generation in soil, 143
  in soil air, 16, 257, 258
  place in the carbon cycle, 144
  source of carbon for plants, 21
  used by autotrophic bacteria, 429
Carbon–nitrogen ratio
  conservation of nitrogen, 151–152
  constancy of, 153
  explanation of, 153
  influence of climate on, 151
  influence on nitrates, 153, 365, 430
  maintenance of soil humus, 154
  of green manures, 549

Carbon-nitrogen ratio [*cont.*]
  of mineral soils, 151, 365
  of peat soils, 365
  of plants, 151
  practical importance, 152
Carbonate of lime (*see* Lime)
Carbonates, absorbed by plants, 36, 143–144
Carbonation, a process of chemical
  weathering, 283
Carbonic acid
  cation exchange reaction, 31, 97
  developed in soil solution, 144
  influence on soil pH, 99, 379
  place in the carbon cycle, 144
  reaction with lime, 411
  release of calcium, 31, 99
  solubility of phosphates, 30
  solubility of potassium, 31
"Cat" clays, 453
Catenas, soil, 338–340
Cation exchange
  availability of nutrients, 32, 38, 95
  due to H_2CO_3, 31, 98
  due to KCl, 95
  function of roots, 37
  graphically expressed, 97–99
  loss of lime by, 99
  mechanism of, 97–98
  reversibility of, 98
Cation exchange capacity
  calculation of, 101
  factors influencing, 99, 103
  of hydrous oxide clays, 91
  of peats, 363, 364
  of representative mineral soils, 102
  of silicate clays, 99–102
  of soil humus, 102
  of various mineral soils, 102
Cations
  adsorbed by colloids, 75
  flocculating capacity, 60
  hydration, 75
  present in soil solution, 35, 36
Cesium, 137, 575
Chemical analysis
  cation exchange capacity, 102, 105, 364
  complete analysis of soil, 24, 284, 370
  for fertilizer needs, 529-530
  for organic carbon, 155
  of caustic limes, 407
  of farm manure, 562
  of fertilizers, 505, 510, 514, 516, 521, 522
  of granite and its clay, 284
  of ground limestone, 408
  of limestone and its clay, 284
  of peat soils, 364–366
  of plant tissue, 138
  of silicate clays, 86
  of soil air, 256
  of soil separates, 46
  peat vs. mineral soil, 364–365
  rapid tests for soil, 530
  representative arid region soil, 24
  representative humid region soil, 24
Chernozems (*see also* Mollisol)
  cation exchange data of, 105

Chernozems [*cont.*]
  described, 317, 323, 341
Chestnut soils, 323, 341
Chlorine
  added in rain, 489, 496
  amounts in soil, 23, 489
  an essential element, 21, 22, 489, 496
  forms in nature, 489
  ionic form in soil, 36
Chlorite, a soil mineral, 83, 89
Classification of soil (*see* Soil classification)
Clays (*see* Hydrous oxide clays; Silicate
  clays)
Climate
  a factor in soil classification, 302, 340
  cause of variation in loess soils, 299
  genesis of soils, 302, 340
  influence on soil nitrogen, 156
  influence on soil organic matter, 156, 340
  influence on transpiration ratio, 209
  influence on vegetation, 302, 340
  influence on weathering, 285, 298, 310
Clostridium in soils, 439
Cohesion in soil
  of clay, 43
  of peat, 362
  of water, 166
Colloido-biological concept of soils, 18
Colloids, soil (*see also* Humus; Hydrous
    oxide clays; Silicate clays)
  acid-salt-like nature, 96
  availability of nutrients from, 37, 105
  cation exchange capacity of, 99
  cation exchange of, 97
  control of soil pH by, 372
  dispersion of, 108, 342, 397
  flocculation of, 60, 108
  general organization of, 71
  humus-clay complex of, 18, 149
  kinds of clay, 78–83
  organic colloids, 94, 362
  reaction with lime, 411
  seat of soil activity, 18
  shape of particles, 43, 71, 78, 81
  size of particles, 81, 94
  surface area of, 74
  swelling of, 43, 107
Colluvial parent materials, 290
Color of soils, 14–15, 148, 323, 341
  influence of aeration on, 265
  influence of iron oxides on, 91, 283
  influence of organic matter on, 148
  influence on soil temperature, 272
  mottling, 265
  of parent material, 283
  of peat, 361
Commercial fertilizers (*see* Fertilizers)
Composition of soil by volume, 12
Conduction of heat, 272
Consumptive use of water, 207–208
Conversion factors for lime, 407
Copper
  amounts in soil, 23, 489
  an essential element, 21–22, 484
  as a fertilizer, 388, 389, 415, 490, 491,
    501, 516

Copper [*cont.*]
  as pollutant, 563
  availability of, 388, 415, 490
  forms in nature, 489
  ionic forms in soil, 36
  role of, 487
Critical moisture (*see* Wilting coefficient)
Crop residues (*see also* Soil organic matter)
  C/N ratio of, 151–152, 549
  importance of, 161, 549
  influence on organic matter and nitrogen
    maintenance, 161
Cultivation of soil
  benefits of, 68, 210–214
  influence on granulation, 60
  influence on nitrates, 430
  summer fallow, 216

DNA molecule, 166
Decay of organic matter, 134, (*see also*
    Organic matter)
  energy released by, 143
  of farm manure, 542–543
  of green manure, 550
Decomposition of minerals (*see* Weathering)
Delta deposits, 292
Denitrification, 431
Density, bulk (*see* Bulk density)
Density of particles, 49
Desert soils, 341
Diagnostic horizons, 319
Diffusion of soil gases, 254–256
Disease organisms
  control of, 131
  influence by pH, 131, 404
Disintegration of minerals (*see* Weathering)
Dolomite
  a soil-forming mineral, 279
  in liming materials, 404–405
Dolomitic limestone
  contrasted with calcic limestone, 405, 417
  rate of reaction with soil, 417
  used in mixed fertilizers, 520
Drainage of soils (*see also* Percolation)
  benefits of, 232–233
  influence on aeration, 233, 254, 265
  influence on soil organic matter, 158
  influence on temperature, 274–275
  mole drains, 229
  layout of tile drains, 229–232
  loss of nutrients by, 223–227
  open ditch drainage, 228
  operation of tile drainage, 229
  percentage of rainfall, 222
  percolation, mechanism of, 182
  types of drainage, 228, 232
Drainage of water
  loss of, from soils, 221
  loss of nutrients in, 223–227
  lysimeter studies of, 220
  nutrient losses compared to crop
    removals, 226
Drumlins, 296
Dry-land farming, 215

Earthworms in soil, 116–118
Edaphological approach, 6
Edaphology defined, 6
Elements, essential (*see* Essential elements)
Eluvial horizons, 312
Energy in soil (*see also* Soil water tension)
  concept of soil moisture, 167, 175
  of soil organic matter, 142
  release of, 143
Entisol soil order, 322, 323–324, 582
Epipedon, 319
Erosion (*see* Soil erosion)
Essential elements
  amounts in soil, 23–26, 459
  amounts removed by crops, 226
  available by cation exchange, 98, 105
  complex vs. simple forms, 27
  forms absorbed by plants, 27, 37, 426, 431,
    434, 461, 469
  forms of, in soil, 26–28
  from air and water, 20–21
  from soil solids, 21–23
  ionic forms in soil, 36–38
  list of, 21
  losses by erosion, 240
  losses by leaching, 223–227
  macronutrients, 21–22, 23–28, 404–482,
    503–515
  micronutrients, 22, 484–502, 515–516
  oxidized forms of, 264
  primary elements, 21
  reduced forms of, 264
  sodium, case for, 20
  solute forms, 27, 36
  trace elements, 22, 484–502
  transfer to available forms, 29–32, 424–426,
    469–470, 475–478
Eutrophication, 542
Evaporation of water from soil
  competition with crops, 208–209
  factors affecting, 204–207
  importance, 207
  magnitude of loss, 207
  methods of control, 210–214
Exchange capacity (*see* Cation exchange
    capacity)
Exfoliation, a weathering process, 281

Fallow, use of, to conserve moisture, 216
Family of soils, 320, 321, 337
Farm manure (*see* Animal manure)
Ferric compounds (*see also* Iron)
  in soils, 36, 264, 284, 490, 494
Fertility management of soil
  fertility defined, 10
  relation of animal manure to, 546
  relation of calcium to, 421
  relation of green manure to, 550
  relation of lime to, 415–421
  relation of magnesium to, 421
  relation of nitrogen to, 443
  relation of phosphorus to, 470–471
  relation of potassium to, 480–482
  relation of sulfur to, 453, 514

Fertility management [*cont.*]
  relation of trace elements to, 501
  and world food supply, 582–592
Fertilizer elements (*see also* Essential
    elements)
  fertilizers, 503–533
  list of, 21
  micronutrients, 484–502
  nitrogen and sulfur, 422–454
  phophorus, 456–471
  potassium, 472–482
Fertilizers
  acidity or basicity of, 468, 520, 521
  ammonia, source of fertilizer salts, 505
  ammonia, use as a fertilizer, 505, 527
  amounts to apply, 527
  analysis of, 521
  brands of, 522
  classification of phosphate, 513
  complete, 517
  economy buying, 519, 522
  factors influencing use of, 527
  guarantee of, 520
  high vs. low analysis, 522
  incomplete, 517
  influence on nitrification, 430
  influence on pH of soils, 520
  inorganic carriers, 505–509
  inspection and control of, 521
  methods of applying liquid, 526
  methods of applying solid, 525
  micronutrients, 516
  mixed, 517
  neutral, 521
  nitrogen carriers, 504–509
  open formula, 521
  organic carriers, 504
  phosphatic, 510–513
  physical condition of, 517
  potassium carriers, 513
  "quick" tests and use of, 530
  ratio of constituents, 521
  sulfur, 515
  trace elements, 516
  use on various crops, 525–527
  and world food supply, 591
Field capacity of soil
  data of, 192
  defined, 189
Fixation of elemental nitrogen
  amount by nodule bacteria, 436
  amount by nonsymbiosis, 439–440
  factors affecting, 390, 414, 436, 439
  fate of nitrogen fixed, 437
  fertility relations, 443
  importance of, 131, 436–439
  influence on pH on, 390, 414, 440
  mechanism of nodule fixation, 434
Fixation of phosphorus
  as calcium compounds, 390, 415, 462
  by iron and aluminum compounds, 390, 462
  by silicate minerals, 465
Fixation of potassium
  factors affecting, 479
  importance of, 478
  minerals responsible for, 479

Flocculation of colloids
  by various cations, 60, 108
  influence of lime on, 60, 108
  influence on granulation, 60, 413
  mechanism of, 108
Flood plains, 290–291
Food supply, and soil factors, 578–592
Formation of soil (*see* Soil formation)
Freezing and thawing
  force exerted by freezing, 281
  granulating effects, 58
  importance in soil formation, 280
Fungi in soils (*see also* Soil organisms),
    123–126, 426
Fungicides, 553–554

Gaseous loss of nitrogen, 431
Genesis of soil (*see* Soil formation)
Gibbsite, 91
Glacial-aeolian deposits
  deposits other than loess, 300
  distribution of loess in U.S., map, 301
  importance of soils from, 300
  nature of loess, 300
  origin of loess, 299
Glacial deposits (*see* Glaciation)
Glacial-lacustrine deposits
  formation of lakes, 298
  importance of soils from, 299
  Lake Agassiz, 298–299
  location on map of U.S., 297
  nature of deposits, 299
Glacial periods in America, 295
Glacial till and associated deposits, 295–298
  drumlins, 296
  ground moraines, 296
  terminal moraines, 295
Glaciation
  aeolian deposits of, 299
  agricultural importance, 300–302
  area covered in U.S., map, 297
  centers of ice accumulation, map, 294
  glacial till, 295
  importance, 300
  interglacial periods, 294
  length of ice age, 295
  loess of, 299–300
  movement of ice, 294
  outwash materials, 298
  periods of, in U.S., 295
  relation to peat accumulation, 355
Gley horizons, 323
Granite
  as a source of soil minerals, 277
  chemical composition of, 284
  chemical composition of clay from, 284
Granulation of soil (*see also* Aggregation of
    soil particles)
  conditions encouraging, 58–62, 148, 232
  destruction by raindrops, 236
  effect of tillage on, 60
  effects of various crops on, 63
  genesis of granules, 58
  influence of lime on, 60, 413

Granulation of soil [*cont.*]
  influence on aeration, 55
  influence on bulk density, 50–52
  influence on pore space, 55
  stability of granules, 60
Gravitational (free) water (*see also* Drainage of soils), 169–170, 192
Gray-Brown Podzolic soils, 323, 341
Gray wooded soils, 323, 341
Great soil groups, 320, 321, 335–336
Green manures
  benefits from, 546
  defined, 546
  desirable characteristics, 548
  influence on soil nitrogen, 547
  list of crops suitable, 549
  maintenance of soil fertility by, 550
  maintenance of soil humus by, 161
  practical utilization, 550
Groundwater Podzol, 323, 341
Grumusols, 323, 324
Guarantee, chemical
  of a commercial fertilizer, 520
  of caustic limes, 406
  of ground limestone, 408
Guarantee, fineness of lime, 409
Gullying and its control, 245
Gypsum
  as a soil mineral, 279, 286, 337
  as a source of calcium, 405
  as a source of sulfur, 28, 446
  in management of alkali soils, 401
  in superphosphate, 527
  influence on pH, 405

Half-bog soils, 323, 341
Halomorphic soils, 42, 43, 269, 270, 271, 341, 399
Hematite
  a soil mineral, 279
  color of, 283
  development of, 283
Herbicides, 553–554
Heterotrophic bacteria, 129, 152
Histisol soil order, 323, 332, 359
Horizons, diagnostic, 319
Horizons of soil (*see* Soil profile)
Humic Gley soils, 323, 341
Humid vs. arid region soils, compared, 24, 102, 104, 333, 334
Humus, soil (*see also* Soil organic matter)
  adsorption of cations, 94, 149
  carbon–nitrogen ratio, 151
  cation exchange by, 94, 149
  cation exchange capacity, 12, 102, 148
  characteristics of, 90–91, 94–95, 148–149
  chemical composition of, 94
  colloidal organization, 94, 148
  contrasted with clay, 90, 94, 148
  definition of, 141, 146–147
  determination of, 155
  direct influence on plants, 150
  energy of, 142
  genesis of, 146
  importance of, 14–15, 60, 147

Humus, soil [*cont.*]
  influence of farm manures on, 161, 546
  influence of green manures on, 161
  lignin (modified), 145
  micelle of, 90, 145
  physical properties of, 148
  polyuronides in, 147
  ratio to nitrogen, 152, 154
  reaction with clay, 147
  source of change, 90, 148
  stability of, 149
Hydrated lime
  changes in soil, 411
  guarantee, 407
  source and nature of, 405
Hydration
  in weathering, 282–283
  of cations, 73
  of clays, 75
Hydrogen, an essential element, 20
Hydrogen bonding, in soil water, 166
Hydrogen ion
  adsorbed by colloidal complex, 75, 378
  as active acidity, 382
  as exchangeable acidity, 382
  cation exchange by, 97, 382
  effect on plants, 36, 390, 391
  flocculating effect by, 108
  pH expression of, 33–34
  source of, 372
  strength of adsorption, 109
Hydrolysis
  alkalinity developed by, 399
  in genesis of clay, 282
  in weathering, 280
Hydromorphic soils, 341
Hydrous micas (*see* Illite)
Hydrous oxide clays
  an important type of clay, 71, 91
  cation exchange capacity, 91, 102
  contrasted with silicate clay, 91
  general organization, 91
  origin of, 90
  phosphorus fixation by, 463
Hydroxide of lime, 405
Hygroscopic coefficient
  data of, 192
  defined, 191
Hygroscopic water
  amount in soils, 192
  nature of, 192–193
  plant relations of, 193
Hysteresis in soil water, 175

Illite
  cation exchange capacity, 102
  genesis, 90
  geographic distribution in U.S., 93
  mineralogical organization, 82
  physical properties of, 45, 107
Illuvial horizon in soil, 313
Inceptisol soil order, 323, 325–326
Indicators, determination of pH, 393
Insecticides, 452, 553–554
Insects in soils, 112, 113, 114

Interglacial periods, 295
Intrazonal soils, 340–343
Ionic exchange (*see* Cation exchange)
Ions present in soil solution, 35–36
Iron
    accumulation of, in B horizon, 313
    amount in soils, 22, 23, 24
    an essential element, 21, 22
    availability of, 264, 390, 414, 491, 494
    cause of mottling, 265
    chelates of, 493
    compounds of, 264
    fertilizers of, 489, 516
    influence on phosphate availability, 462
    ionic forms in soil, 36
    oxidation and color, 265, 390, 412, 413
    pH and active iron, 491
    reduction of, 264
    role of, 487
Irrigation
    alkali control under, 218
    need for moisture control, 217

Kainit as a fertilizer, 513
Kalinite in soils
    cation exchange capacity, 97
    genesis, 88
    geographical distribution in U.S., 93
    mineralogical organization, 78
    physical properties, 107

Lacustrine deposits (*see* Glacial-lacustrine deposits)
Land-capability classification
    and classes of soil, 347–352
    description of subclasses, 352
    use of soil survey in, 347
Land resources, world, 580
Latosols, 323, 341
Leaching, nutrient losses from, 223–227, 412, 556
Legumes
    as a green manure, 546, 547
    bacteria in root nodules, 434
    cross inoculation, 435
    effect on soil nitrogen, 438
    fertility maintenance by, 438, 443
    fixation of soil nitrogen by, 434–437
    importance of, 438, 443
    influence of soil reaction on, 392, 414
    mechanism of nitrogen fixation, 434–435
Lignin
    protection of soil nitrogen, 147
    rate of decomposition, 140
    role in humus formation, 147
    stability of, 140
Lime
    amounts in soil, 25
    amounts to apply, 417, 419
    as a cure-all, 415
    burned lime, 404
    calcium carbonate equivalent of, 406–407

Lime [*cont.*]
    calcium oxide equivalent of, 406, 407
    carbonate of, 405
    changes in soil, 410
    chemical effects of, 413
    chemical guarantees, 409
    conversion factors for, 407
    crop response to, 414
    diagram of gain and loss, 421
    effects of excess, 389, 415
    effects on microorganisms, 130, 414
    effects on nitrification, 414, 431
    effects on nitrogen fixation, 414, 436, 439
    effects on phosphate availability, 389–390, 413, 415, 466
    effects on potassium, 413, 479
    effects on soil, 413
    effects on soil structure, 413
    fineness guarantee of limestone, 407
    forms of, 404
    forms to apply, 416
    hydrated lime, 405
    hydroxide of, 405
    interpretation of fineness guarantee, 409
    losses from soil, 412
    methods of application, 420
    neutralizing power, 406, 407, 409
    overliming, 389, 415
    oxide of lime, 404
    physical effects of, 413
    practical problems of, 415
    reactions with the soil, 411
    removal of, by crops, 412
    slaked lime, 405
    total carbonates of, 407
    when to apply, 415
Limestone
    as a source of minerals, 25, 278
    calcic, 408, 409, 417
    chemical composition of, 283, 407
    chemical composition of clay from, 284
    chemical guarantee, 406
    dolomitic, 408, 409, 416–417
    fineness of, 409–410
    rate of reaction with soil, 410–412, 416, 417
Liming
    crop response to, 415
    effects on soil, 413
    influence of overliming, 415
    methods of applying lime, 420
    problems of, 415, 416
Limiting factors, principle of, 19
Limonite
    a soil mineral, 279
    produced by weathering, 283
    soil color, 282
Lithosols, 341, 582
Loam, a soil class, 47
Loess (*see also* Glacial-aeolian deposits)
    location of in U.S., 299, 301
    nature of, 300
    origin of, 299
Loss of nutrients from soils
    by cropping, 226, 412, 443
    by drainage, 223–227, 412
    by erosion, 240, 412, 443

Loss of nutrients [*cont.*]
  by volatilization, 431, 443
Low-humic Gley soils, 323, 341
Lysimeters
  description of types, 220
  loss of nutrients through, 223–227

Macronutrients (*see also* Calcium;
    Magnesium; Nitrogen; Phosphorus;
    Potassium; Sulfur), 21–22, 23–28
Macroorganisms, 111–116
Macro pore spaces
  importance of, in aeration and percolation,
    54–55, 265
  influence of cultivation on, 54, 55, 60, 265
  influence of rotation on, 62
Magnesium
  as a nutrient, 20, 21, 417
  content of mineral soils, 24, 25
  content of peat soils, 365, 366
  diagram, gains and losses of, 421,
  exchangeable in mineral soils, 28
  exchangeable in peat soils, 364
  forms in soil, 27, 28
  ionic form in soil solution, 36
  in dolomite, 404, 405, 406, 407, 417
  loss by erosion, 243
  loss by leaching, 226, 227, 412
  pH and availability of, 382, 388, 413
  transfer to available form, 31, 32
Manganese
  active in acid soil, 264, 265
  amount in soils, 21, 22, 24, 489
  as a nutrient, 20, 21, 484, 516
  as pollutant, 563
  chelates of, 493
  influence on phosphorus availability, 397,
    460
  ionic form in soil, 36, 264, 491
  reduction-oxidation of, 264, 491, 492
  relationship to pH, 413, 415
  role of, 486
  suppression by overliming, 415, 491
Manure (*see* Animal manure; Green manure)
Manure salts, 514
Marine soil materials
  area of in U.S., map, 289
  importance of soils from, 293
  nature of, 292
Marl, bog lime
  under peat, 367
  use of, 367, 405
Maximum retentive capacity of soil for water,
    189
Mechanical analysis
  data of, 41, 49
  explained, 41
  used in naming soils, 45–47
Meta-phos, a fertilizer, 512
Micelle
  meaning of term, 74
  nature of and soil pH, 378
  of humus, 94, 149
  of hydrous oxides, 91

Micelle [*cont.*]
  of peat, 362
  of silicate clays, 75–83
Microcline
  a soil mineral, 279
  as source of clay, 83, 282
  liberation of potash from, 30, 282
  reaction of acid humus on, 149
Micronutrients, 22, 484–502, 515–516
  availability of anions, 496
  availability of cations, 490
  chelates of, 493
  essential elements, 22, 484
  forms in soil, 489
  in fertilizers, 501, 515–516
  ionic forms in soils, 36
  management needs, 499
Milliequivalents, 101
Minerals (*see* Soil minerals)
Minor elements (*see* Trace elements)
Mixed layer silicate clays, 83
Microorganisms (*see* Soil organisms)
Moisture (*see* Soil water)
Moisture equivalent
  compared to field capacity, 190
  data of, 190
  explained, 189
Mole drain, 229
Mollisol soil order, 323, 326–327, 329, 338,
    582, 583
Molybdenum
  amount in soils, 21, 24, 489
  essential element, 20, 21, 484
  in fertilizers, 501, 516
  ionic form in soils, 36
  relationship to pH, 490
  role of, 486
  source of, 489
Montmorillonite clay
  cation exchange capacity, 100
  genesis, 88
  geographic distribution, 93
  mineralogical organization, 80
  physical properties of, 46, 106–107
Moraines
  soils from, 295–296
  various types, 295–298
Mottling, 265
Muck (*see also* Peat deposits)
  meaning of term, 359
  peat vs. muck, 359
Mulches, 210–214
  and soil temperature, 274
  artificial, 211
  soil, 214
  stubble mulch, 67, 213, 215, 216
Mycorrhiza of soil, 125

Naming of soils texturally
  by feel, 48
  by mechanical analysis, 49
  names used, 47, 48
Nematodes in soil, 119
Nitrate accumulation in soil (*see*
    Nitrification)

Nitrate of soda
  as a fertilizer, 505, 507
  influence on soil pH, 521
Nitrates
  in rain and snow, 441
  influence of carbon–nitrogen ratio of
    residues on, 151–154, 430, 549
  reduction in soil, 431
  sources of, 145, 505
  uptake by plants, 423–425, 431
  used as top dressings, 507, 526
Nitrification
  a bacterial monopoly, 127, 428
  briefly explained, 145
  fate of nitrogen oxidized, 431
  influence of aeration on, 145, 260, 264,
    429
  influence of fertilizers on, 430
  influence of moisture on, 430
  nature of transformation, 145, 428
  organisms involved, 145, 428
  relation to carbon–nitrogen ratio, 151,
    365, 430
  relation to pH and lime, 390, 414, 430
  soil conditions affecting, 429–430
  temperature range of, 429
  vigor in peat soils, 365, 430
Nitrogen, 21, 422–444
  added to soil in rain, 441
  ammonification, 29
  amounts in mineral soils, 24, 155
  amounts in peat soils, 365
  an essential element, 20, 21, 422
  balance sheet for soils, 443
  carbon–nitrogen ratio, 151, 430
  control in soil, 161, 443
  control of soil humus by, 154
  cycle in soil, 424
  diagram, gains and losses of, 443
  fertilizers carrying, 504–509
  fixation by legume bacteria, 135, 434–437
  fixation by nonsymbiosis, 135, 440
  forms in soil, 26, 27, 138, 423
  immobilization of, 425
  in farm manure, 539
  increase due to legumes, 438, 547
  influence of annual temperature on, 155
  influence of rainfall on, 156
  influence on plants, 422
  ionic forms in soil, 36, 143, 423
  losses by erosion, 243
  losses from farm manure, 542, 544
  losses from volatilization, 431–433
  losses in drainage, 226, 227
  mineralization of, 426
  nitrification, 29, 145, 428–429
  simple forms in farm manure, 542
  transfer to available forms, 30, 145,
    426–429
  volatilization in farm manure, 542
  volatilization in soil, 145, 431
Nitrogen cycle
  control of nitrogen in cycle, 443
  diagram of, 424
  farm manure a part of, 424
Nitrogen fertilizers (see Fertilizers)

Nitrogen fixation in soils (see Fixation of
  elemental nitrogen)
Nodule bacteria, 434–436
Nutrient elements (see Essential elements)

Olivine, weathering of, 284
Open ditch drainage, 228
Optimum water, control of in soil, 194–195
Orders, soil (see Soil orders)
Organic acids
  amino, 29, 145, 426
  cause of soil acidity, 380
Organic carbon
  carbon cycle, 143, 144
  carbon–humus ratio, 156
  carbon–nitrogen ratio, 151
  determination of, 154
Organic deposits (see Peat deposits)
Organic matter (see Humus; Soil organic
  matter)
Organic soils (see Peat soils)
Organic wastes, as pollutants, 534, 544,
  568–573
Organic vs. mineral soils, 11, 370
Organisms of soils (see Soil organisms)
Osmosis in soil water, 168, 170–172
Overliming, effects of, 415
Oxidation
  influenced by aeration, 262
  of organic matter, 140
  in weathering, 283–284
Oxide of lime (see Lime)
Oxisol soil order, 323, 331–332, 582, 583
Oxygen
  an essential element, 21
  diffusion of, 254–256
  influence on bacteria, 129
  influence on nitrification, 260, 264, 429
  influenced by aeration, 255
  of the soil air, 256
  role in weathering, 284

Parent material (see Soil parent materials)
Particle density, 49
Particles, soil (see Soil particles)
Peat deposits
  classification of, 356
  distribution and extent, 355
  fibrous types, 358
  genesis of, 353
  muck vs. peat, 359
  relation to glaciation, 355
  sedimentary types, 356
  types of, 356
  various uses, 358
  woody types, 358
Peat soils
  anomalous features of, 366
  bog lime, 367, 416
  buffering of, 363
  bulk density of, 361
  cation exchange capacity, 364
  cationic conditions of, 363
  chemical composition, 364–366

Peat soils [*cont.*]
    classification of, 356
    colloidal nature, 362–363
    exchange data vs. mineral soil, 364
    factors that determine value of, 367
    management of, 368
    nitrification in, 365
    physical characteristics of, 361
    preparation for cropping, 368
    profiles, 360
    trace elements in, 370, 510
    vs. mineral soils, 11, 370
    water-holding capacity of, 361
    weight of, 361
Pedology, defined, 6
Pedon, concept of, 316–317
Percentage base saturation
    explained, 103
    of representative soils, 105
    relation to leaching loss of potassium, 480
    relation to pH, 104, 378, 386, 387
Percolation
    bare vs. cropped soil, 221–223
    factors affecting, 183
    hastened by tile drains, 229
    importance, 220
    influence on aeration, 233
    influence on temperature, 232
    losses of nutrients by, 221–223
    losses of water by, 201, 221
    mechanics of, 182
    methods of study, 220
    percentage of rainfall, 222
Pesticides
    background of, 551–553
    behavior of, 554–559
    effects of, 559–560
    kinds of, 553–554
pF of soil water, 173
pH of soil (*see also* Lime; Soil reaction)
    accuracy of determination, 393
    buffer relations, 384–387
    colloidal control of, 378
    comparative terms, diagram, 33
    correlation with soil nutrients, 146,
        365–366, 387–390, 460–461
    correlation with soil organisms, 388,
        390–391
    diagrams of correlations, 388, 392, 462
    dye method of determination, 393
    electrometric determination of, 393
    explanation of, and diagram, 32–35, 383
    for arid region soils, 34, 396
    for humid region soils, 34, 373
    heterogeneity of hydrogen ions, 381
    influence of fertilizers on, 520
    limitations in use, 393
    major changes in, 379
    minor changes in, 381
    nutrient significance of, 35, 387–390, 414
    of latosols, 335, 379
    of peats, 363, 365–366, 379
    range of, diagram, 34
    relation to disease organisms, 131, 396
    relation to percentage base saturation, 104,
        378, 386, 387

pH of soil [*cont.*]
    stabilization by buffering, 363, 384–387
    use of in liming decisions, 387–393, 415
Phase of soils, 337
Phosphorus, 456–471
    accumulation of in soil, 457–458
    amount in mineral soils, 24, 457, 458, 459
    amount in peat soils, 365
    an essential element, 21, 456
    anion exchange of, 465
    available phosphorus, 389, 415, 462, 466
    correlation with pH, diagram, 388, 462
    diagram of gains and losses of, 472
    exchangeable form, 465
    factors affecting availability of, 460–470
    fertilizer problem of, 457
    fertilizers carrying, 510
    fixation of in soils, 388–390, 415, 462–467
    forms in soil, 29, 458
    importance of, 456
    in farm manure, 540
    influence on plants, 457
    inorganic forms of, 458, 460–468
    ionic forms and pH, 146, 389, 460–461
    ionic forms in soil, 36, 146, 389, 461
    loss by erosion, 243
    loss by leaching, 226, 227
    losses from farm manure, 544
    management of, 470–471
    mineralization of organic forms, 146
    organic forms, 27, 460, 469
    pH and available forms, 146, 389, 461
    recovery of from fertilizers, 471
    transfer to available forms, 29, 146
    unavailable phosphorus, 389–390, 463, 513
Phosphorus fertilizers (*see* Fertilizers)
Planosols, 323, 341
Plant growth
    factors affecting, 19
    influence of aeration on, 260
    influence of drainage on, 232
    influence of nitrogen on, 422
    influence of phosphorus on, 457
    influence of potassium on, 472
    influence of temperature on, 266
    optimum water for, 194–195
    relation to soil reaction, 392, 414
    temperature relations, 266
    tolerance to alkali, 400
Plant nutrients (*see* Essential elements)
Plant roots
    acquisition of water by, 197
    as true soil organisms, 121
    contact relations with soil, 37
    extension of, 198, 232, 261
    influence of drainage on depth of, 232, 260
    influence of phosphorus on, 457
    mycorrhiza, 125
    organisms in rhizosphere, 122
    pruning of by cultivation, 214
    source of organic matter, 122
Plant tissue
    carbon–nitrogen ratio of, 151
    composition of, 137–138
    decomposition in soil, 139–142, 424
    energy from decay of, 142

Plants
    acid-tolerant, 391, 414
    competition with organisms, 132
    influence on erosion, 239, 241
    interception of precipitation by, 200
    lime-loving, 392, 414
    nutrients removed by, 226, 227
    pH correlations, diagram, 226, 227, 388
    response to liming, 414
    tolerance to alkali, 400
    transpiration ratio of, 207
Plasmolysis, 399
Plasticity
    of clay, 43, 62
    of peat, 362
Pleistocene ice age, 293
Plowing
    criticism of, 67
    influence on soil properties, 66–68
Podzolic soils, 317, 323, 341, 582, 583
Pollution of soil (see Soil pollution),
Polypedon, 316–317
Population expansion, 578
Pore space in soil (see also Aeration of soil)
    amount, 12, 54
    calculation of, 53–54
    factors affecting amount, 54–55
    influence of cropping, 55
    influence of granulation on, 54
    size and shape of pores, 54
    sod vs. cultivated soil, 55
    subsoil vs. topsoil, 258
Potassium, 472–482
    amount in mineral soils, 24, 473
    amount in peat soils, 364, 365
    an essential element, 21
    available forms, 27, 28, 475
    cation exchange of, 99
    diagram of gains and losses of, 481
    exchangeable form, 28, 475, 476
    fertilizer problems, 473
    fertilizers carrying, 513–515
    fixation in soil, 479
    forms in soil, 27, 475, 476
    influence of lime on, 414, 479
    influence on plants, 472
    ionic form in soil, 36
    liberation by acid complex, 149, 282
    liberation by exchange, 99
    liberation by hydrolysis, 31, 282
    loss by erosion, 243
    loss by leaching, 223, 226, 227
    loss from farm manure, 544
    luxury consumption by crops, 474
    management of, 480
    nonexchangeable, 477–478
    pH correlation, 388, 480
    removal by crops, 227, 474
    transfer to available forms, 31, 475–479
    unavailable, 477–478
Potassium fertilizers (see Fertilizers)
Potential of soil water, 168–172
Potentiometer, determination of pH, 393
Prairie soils
    description of, 323, 341
    distribution of organic matter in, 155

Profile (see Soil profile)
Proteins
    amounts in plants, 139, 140
    decomposition of, 140
    reaction with clays, 147
    reaction with lignin, 147
Protozoa in soil, 119
Puddling of soil, 62

Quartz
    a soil mineral, 13
    in sand and silt, 43, 44
"Quick" tests
    compared to total anlyses, 530
    interpretation of, 531
    limitations of, 530
    merits of, 531
    principles of, 530
    supplementary information sheet for, 532

Radiation
    of heat from soils, 269–270
    solar, 268–270
Radionuclides, 575–576
Rain (see also Climate)
    and erosion, 234–238
    influence on soil temperature, 275
    nitrogen in, 441
    sulfur in, 515
Rapid tests for soils (see "Quick" tests)
Reaction of soil (see pH; Soil reaction)
Red and yellow Podzolic soils, 323, 341
Red desert soils, 323, 341
Reddish brown soils, 323, 341
Reddish chestnut soils, 323, 341
Reddish prairie soils, 323, 341
Regolith, nature of, 7, 277, 310
Regosols, 431
Relative humidity
    influence on evaporation, 206
    of soil air, 188
Rendzina soils, 323, 341
Residual parent materials, 288
Reversion of soluble phosphates, 466
Rhizosphere, organisms of, 37
Rock phosphate
    compared to superphosphate, 512
    nature and composition, 510, 512
    used to make superphosphate, 510
Rocks, common in soil formation, 277–278
Rodents, soil, 112
Roots (see Plant roots)
Rotifers in soil, 121
Runoff water
    factors influencing amount of, 237–240,
      243
    proportion of total precipitation, 234
    relation to erosion losses, 234–236

Saline soils (see also Alkaline and saline
      soils)
    characteristics of, 396–397

Saline soils [*cont.*]
  effect on plants, 399
  management of, 400
  pH of, 397
Saline-sodic soils (*see also* Alkaline and
    saline soils)
  characteristics of, 397
  effect on plants, 399
  management of, 400
  pH of, 397
Sand (*see* Soil particles)
Separates of soil (*see* Soil particles)
Series, soil (*see* Soil series)
Serpentine, a soil mineral, 284
Sheet erosion (*see* Soil erosion)
Shifting cultivation, 587
Sierozem soils, 323, 341
Silicate clays (*see also* Colloids)
  acid-salt nature of, 96
  adsorbed cations, 75
  anion exchange, mechanism of, 465
  cation exchange, mechanism, 97
  cation exchange capacity of, 99, 148
  chemical composition of, 86
  contrasted with humus, 148
  double layer phenomenon, 71, 74
  electronegative charge on, 74
  exchangeable ions of, 75
  external and internal surfaces, 74
  fixation of phosphorus by, 465
  flocculation of, 60, 108
  formula for, 96
  genesis of, 88, 283
  hydration of, 75
  hydrous mica (illite) type, 78, 477, 479
  micelle, nucleus of, 74
  mineralogical nature of, 78–83
  mixed layer, 83
  plasticity and cohesion, 43, 107
  protein combinations with, 147, 148
  regional distribution, 92
  size and shape of particles, 71–74
  source of negative charges, 83
  surface area, 74, 79, 80
  swelling of, 107
  types in soil, 78–83
Silicon
  amounts in soils, 24
  in silicate clays, 78–83
Silt (*see* Soil particles)
Slaked lime, 405
Slope
  influence on erosion, 238–239
  influence on soil temperature, 269
Sod
  check on erosion, 240
  influence on drainage losses, 226, 227
  influence on organic matter accumulation,
    161
Sodic soils, 397, 399
Sodium
  adsorbed by colloids, 73, 397, 399
  as a nutrient, 21
  detrimental effects of salts, 397
  in alkali soils, 397–399
  in saline soils, 398

Sodium [*cont.*]
  of black alkali, 399
  of white alkali, 397
Sodium nitrate
  as a fertilizer, 505, 507
  influence on soil reaction, 520
Soil(s) (*see also* following entries and
    specific soil names)
  a soil vs. the soil, 8
  active portion of, 18
  chemical composition of, 24, 264
  colloido-biological concept, 18
  edaphological definition, 6
  four components of, 12
  general biological nature, 17
  individual soils concept, 315–317
  light vs. heavy, 41–42
  minerals of, 13
  mineral vs. organic, 11, 370
  parent materials of, 277–302, 306
  pedological definition, 6
  profile of, 9, 312–315
  specific, tables of, 323, 333, 336, 341
  specific heat of, 270
  subsoil vs. surface soil, 10, 314
  vs. land, 347
  volume composition of, 12
  weight of, 54, 256
Soil acidity (*see also* Lime; pH of soil)
  compounds suitable for reduction of, 404
  exchangeable vs. active, 382
  important correlations with, 387–391
  influence on nitrification, 390, 414, 430
  influence on phosphate fixation, 460–466
  magnitude of active, 382
  magnitude of exchangeable, 382
  methods of increasing, 395
  problems of, 394
  relation of carbon dioxide to, 379
  relation to higher plants, 35, 391
Soil age, 303, 309
Soil aggregates (*see* Aggregation of soil
    particles)
Soil air (*see* Aeration of soil; Air of soil)
Soil alkali (*see* Alkaline and saline soils)
Soil analysis (*see* Chemical analysis;
    Mechanical analysis)
Soil association, 338–340
Soil bacteria (*see* Bacteria in soil; Soil
    organisms)
Soil buffering
  capacity of, 385
  curves of, 386, 387
  explained, 384–385
  importance of, 386–387
  magnitude of, 363, 387
  of peat soils, 363
Soil catenas, 338–340
Soil class
  determination of, 48
  explained, 45–47
  names used, 46–47
Soil classification, 318–343
  Alfisols, 322, 323, 329, 330
  Aridisols, 322, 323, 326
  azonal groups, 340–343

Soil classification [*cont.*]
  bases of, 319
  categories used, 320–321
  Chernozems, 323, 327
  climatic relations, 302–306
  diagnostic horizons, 319
  Entisols, 322, 323, 324
  great soil groups, 320, 321, 335–336
  Histisols, 323, 332
  Inceptisols, 322, 323, 325, 326
  intrazonal groups, 340–343
  Mollisols, 322, 323, 326–327
  new comprehensive system, 318–340
  nomenclature of, 322, 323
  old system, 340–343
  orders of, 322–323
  Oxisols, 323, 331–332
  parent material relations, 306
  peat soils, 323, 341, 356
  phases of, 337, 343
  soil series, 320, 321, 337, 343
  soil survey maps, nature and use, 343–347
  Spodosols, 323, 327–330
  suborders, 321, 333–334
  tables of soil clasification, 323, 333, 336,
    341
  Ultisols, 323, 331
  Vertisols, 323, 324, 325
  "wet and dry," 333, 334
  zonal soil groups, 340–343
Soil colloidal matter (*see* Colloids, soil)
Soil color (*see* Color of soils)
Soil development (*see* Soil formation)
Soil erosion
  accelerated type, 236
  by water, 234–245
  by wind, 245–250
  contour strip cropping, 242, 244
  effects of crops on, 239, 241
  factors governing rate, 237–240
  gully control, 245
  in U.S., map of, 235
  loss of nutrients by, 241–243
  loss of soil by, 240
  magnitude of, 234–236
  mechanics of, 236–237
  normal erosion, 236
  practical control, 240
  raindrop impact, 236–237
  raindrop splash, 237
  sheet and rill, 240–245
  strip cropping control, 244
Soil family, 320, 321, 337
Soil fertility (*see* Fertility management
    of soil)
Soil formation, 302–312
  chemical forces of, 282
  described, 303–309
  development of parent material, 281
  factors influencing, 303
  influence of climate on, 303
  influence of parent material on, 306
  influence of time on, 309
  influence of topography on, 309
  influence of living organisms on, 306
  mechanical forces of, 285

Soil granulation (*see* Granulation of soil)
Soil groups, 320, 321, 335–336, 582–585
Soil horizons (*see* Soil profile)
Soil humus (*see* Humus, soil)
Soil minerals
  as pollutants, 560–567, 573–575
  in soil separates, 44
  of clays, 78
  primary and secondary, 13–14
Soil moisture (*see* Soil water)
Soil mulch, 214 (*see also* Mulches)
Soil names, tables of, 323, 333, 336, 341
Soil orders, 320, 322–332
  U.S., map, 304
  world, map, facing, 305
Soil organic matter (*see also* Humus, soil)
  ammonium and nitrates from, 145, 426
  amounts in soil, 24, 154
  calculation from nitrogen, 155
  carbon cycle of, 144
  carbon–nitrogen ratio of, 151
  colloidal nature of peat, 362
  composition of original tissue, 138
  conditions for rapid turnover, 162
  controlled by soil nitrogen, 154
  decomposition of, 139, 424
  determination of amount, 154
  direct influence on plants, 150
  distribution of in profile, 156, 307, 315,
    329, 338, 339
  economy of maintenance, 163
  encouragement of granulation by, 61, 148
  energy, 142
  humus proportion of, 147
  influence of carbon–nitrogen ratio on, 154
  influence of drainage on, 158
  influence of farm manure on, 161, 546
  influence of green manure on, 161, 546
  influence of precipitation on, 156
  influence of rotation on, 161
  influence of temperature on, 156
  influence of texture on, 158
  influence of tillage on, 60, 162
  influence on capillary water, 169
  influence on soil properties, 159
  level of maintenance, 154, 163
  management of, 161–163
  nitrogen cycle relations, 424–426
  phosphorus compounds from, 146, 460
  products of decomposition, 139, 423–429
  ratio of, to carbon, 156
  ratio of, to nitrogen, 156
  source of, 137, 161
  sulfates and other sulfur products, 143, 146
Soil organisms (*see also* specific names:
    Actinomycetes, Bacteria, Earthworms,
    etc.)
  antibiotics, 132
  beneficial to plants, 132, 135
  classification of, 112
  competition between groups, 132
  competition with higher plants, 132, 153
  correlation with pH, 390
  in rhizospheres, 37, 123, 439
  influence of aeration on, 126, 129, 140,
    260, 429

Soil organisms [*cont.*]
  influence of temperature on, 130, 266
  injurious to plants, 130, 412
  macro animals, 112, 130
  micro animals, 118, 130
  mycorrhiza, 125, 426
  nitrogen cycle, 424
  roots of plants, 121, 233, 262
Soil parent materials
  alluvial, 312–313
  characteristics of, 306–309, 356–358
  classification of, 287, 306
  colluvial, 290
  defined, 277
  genesis described, 287
  glacial-aeolian, 299
  glacial deposits, 295–300
  glacial-lacustrine, 298
  glacial outwash, 298
  glacial till, 295
  map of U.S., 289
  marine sediments, 292
  minerals of, 279
  product of weathering, 8, 277, 287, 310
  regolith, 7, 277, 310
  relation, 9, 310
  residual, 288
  rocks of, 278
Soil particles (*see also* specific names:
    Clay, Humus, Sand, etc.)
  chemical composition of, 45
  classification of, 40
  density of, 49
  feel of, 48
  general description, 14
  mineralogical nature of, 44
  names of, 41
  physical nature of, 42
Soil pH (*see* pH of soil; Soil reaction)
Soil phase, 337
Soil pollution
  effects of, 559–560
  history of, 551–553
  inorganic compounds, 560–567
  minerals, 560–567
  organic wastes, 534, 544, 569–573
  pesticides, 551–560
  prevention of, 567–568
  radionuclides, 575–576
  salts, 573–575
Soil pores
  adjustment of water in, 180–181
  calculation of pore space, 53–54
  macro and micro, 54, 189–190, 192–193
  size and shape, 54–55
Soil profile
  acquired characteristics, 312
  and weathering, 310–312
  development of, 310
  diagnostic horizons, 319
  diagram of horizons, 5, 307, 315, 329,
    338, 339
  general explanation, 4–5, 310, 312
  horizon designations, 312–314
  inherited characteristics, 312
  of Alfisols, 329, 338

Soil profile [*cont.*]
  of catena, 339
  of Miami silt loam, 315
  of Mollisols, 307, 329, 338
  of peats, 360
  of prairie, 340
  of Spodosol, 329
  of Ultisols, 307
  subsoil vs. surface soil, 6
  theoretical profile, 313
  truncated, 314
Soil reaction (*see also* pH of soil)
  colloidal control of, 378
  correlations, 387–391
  major changes in, 379
  problems of in arid regions, 396
  relation to higher plants, 391
Soil separates (*see* Soil particles)
Soil series, 337
Soil solution (*see also* Capillary water)
  buffering of, 385
  concentration of, 33–34
  general nature of, 13, 33
  heterogeneity of, 381
  importance of buffering, 386–387
  ionic nature, 33–34
  major changes in pH, 379
  minor changes in pH, 38
  movement of, 179–180
  pH, controlled by colloids, 378
  pH, correlations of, 387–391
  pH range of, 34
  relation to capillary water, 178
  solutes of, 34
Soil structure
  defined, 55–56
  discussion of, 55–58
  genesis, 57
  granulation and its promotion, 58
  influence of adsorbed cations on, 59–60
  influence of cropping on, 55
  influence of tillage on, 60
  management of, 62
  of peat soils, 361
  puddled condition, 62, 397
  types of, diagram, 56
Soil subgroups, 336
Soil suborders, 335–336
Soil survey
  field map used, 343
  and food supply, 592
  maps and report, 344–346
  nature of, 343
  utilization, 345–347
Soil temperature, 266–275
  bare vs. cropped soils, 266–270
  control of, 274–275
  data, 272–274
  effect of color, 272
  effect of drainage, 232, 274–275
  effect of soil depth, 273
  factors governing equalization, 266
  influence on evaporation, 271
  influence on microorganisms, 127, 266
  influence on nitrification, 266
  influence on water vapor movement, 189

Soil temperature [*cont.*]
  influence on weathering, 281
  influenced by evaporation, 271
  influenced by water, 271, 275
  seasonal variations, 272–273
  surface and subsoil contrasts, 272, 273
Soil texture
  defined, 40
  influence on capillarity, 184
  influence on depth and interval of tile, 231, 232
  influence on erosion, 238
  influence on percolation, 62
  influence on soil nitrogen, 158
  light vs. heavy soil, 41–42
  soils named by, 45, 47
Soil type, 344
Soil water (*see also* specific names: Capillary water, Gravitational water, etc.)
  a dynamic solution, 13
  acquisition by plants, 197–198
  adhesion of, in soil, 166
  available, 195
  calculation of percentage, 175
  capillarity of, 178
  classification of, biological, 194
  classification of, physical, 192
  cohesion of in soil, 166
  consumptive use of, 195
  energy relations, 167
  erosion by, 234–245
  evaporation of, 186–189, 204, 271
  field capacity, 189
  forms of, 181–185
  function to plants, 18
  gravitational, 169, 192
  hydrogen bonding, 166
  hygroscopic, 191, 192
  hygroscopic coefficient, 191
  hysteresis, 175
  influence on aeration, 233, 254, 263
  influence on bacteria, 129
  influence on nitrification, 430
  influence on soil temperature, 232, 270, 271, 275
  influence on specific heat, 270
  interception of, by plants, 200, 201
  loss of, from soils, 200–252
  methods of measurement, 175–178
  moisture equivalent, 175
  movement, types of, 179, 181, 201
  of peat soils, 361
  optimum for plants, 194–195
  osmosis in, 168, 170–172
  percolation and leaching, 182, 201, 220–227
  polarity of, 165
  removal as drainage, 227
  retention by soil, 75, 189
  runoff of, 234
  saturated flow of, 182
  summary of regulation, 251–252
  surface tension, 166
  transpiration of, 204–207
  unavailable, 195

Soil water [*cont.*]
  unsaturated flow of, 184
  vaporization, 186–189, 204–207, 271
Soil water tension
  explanation of, 166, 172–173
  graphs for mineral soils, 174
  in terms of atmospheric pressure, 175
  in terms of bars, 175
  relation of, to relative humidity, 188
  table of equivalents, 173
  values of, 192
Solonchak soils, 323, 341
Solonetz soil, 323, 341
Solution (*see* Soil solution)
Specific gravity (*see* Particle density)
Specific heat of soil, 270
Spodosol soil order, 323, 327–330
Strip cropping and erosion control, 242–244
Strontium-90, 575
Structure of soil (*see* Soil structure)
Stubble mulch, 67, 213, 215, 216
Subgroups, soil, 321, 336
Suborders, soil, 333–334
Subsoil vs. topsoil
  bulk density, 50
  general comparison, 6
  organic matter, 155, 158
  pore space, 258
  structure, 55–57
Sulfur, 25, 146, 444–454
  added to soil in fertilizer, 445, 510, 515
  added to soil in rain, 446, 515
  adsorbed from air by soil, 515
  amount in mineral soils, 24
  amount in peat soils, 365, 366
  an essential element, 20, 21
  behavior in soils, 449
  cycle in soils, 449, 450
  fertility management, 453, 515
  forms in soil, 27, 29
  importance of, 444
  loss by erosion, 243
  loss in drainage, 226, 227
  organic, 449
  oxidation in soil, 146, 391, 451
  sources of, 446
  transfer to available forms, 33, 146, 449
  used as a fertilizer, 515
  used on alkali soil, 401
  used to increase soil acidity, 395–396, 401, 459
Summer fallow and moisture conservation, 216
Superphosphate
  ammoniated, 510, 511
  compared to rock phosphate, 512
  composition of, 510–511
  manufacture of, 510
Superphosphoric acid, 510–511
Surface tension, 166
Survey (*see* Soil survey)

Temperature (*see* Climate; Soil temperature)
Tension (*see* Soil water tension)
Terraces for erosion control, 245

Texture (*see* Soil texture)
Tile drains (*see* Drainage of soils)
Tillage
  contour, 243
  control of weeds by, 205, 214, 215
  influence on aeration, 68, 265–266
  influence on bulk density, 61
  influence on granulation, 60
  root pruning, 68, 214
  soil mulch and moisture conservation,
    210, 214
Tilth, 66–69
Trace elements
  amounts in soil, 23, 489
  as fertilizers, 501
  chelates of, 493
  deficiencies, 22, 376, 540
  effect of overliming on, 415, 484
  essential nutrients, 21–23, 484
  forms in soils, 489
  in fertilizer mixtures, 501, 516
  in peat soils, 370
  influence of pH on availability, 388, 389,
    490–493, 497
  ionic forms in soil, 38, 490, 496
  management of, 499
  other terms for, 21, 484
  oxidation and reduction of, 264, 491
  role of, 486
Transpiration by plants
  factors affecting, 204–207
  magnitude of, 207–208
Transpiration ratio
  defined, 208
  factors affecting, 209
  for different plants, 209
Trash mulch, 213, 215
Tropics
  potential for food production, 586, 589
  soils of, 431
Tundra soils, 326, 327

Ultisol soil order, 323, 331
Urea
  as a fertilizer, 505, 508
  in farm manure, 542
  loss of nitrogen from, by volatilization, 433

Vapor, water, movement of
  evaporation at soil surface, 187–189, 204,
    271
  from moist to dry soil zones, 188
  fundamental considerations, 188
Vapor pressure, 188
Vermiculite, a clay mineral, 80, 81, 85,
    87, 89, 90
Vertisol soil order, 323, 324–325

Volatilization
  of nitrogen, 145, 431–433, 542
  of water, 187–189, 204–208, 210–217, 271
Volume composition of soil, 9
Volume weight (*see* Bulk density)

Water (*see* Soil water)
Weathering
  a general case, 279
  acid clay, effect of, 283
  carbonation, 283
  chemical processes of, 282
  coloration due to, 283
  constructive, 4
  factors affecting, 285–286
  fresh rock to clay, 284
  hydration, 282
  hydrolysis, 282
  in action, 287
  mechanical forces of, 281
  oxidation, 283–284
  podzolization, 342
  processes classified, 280
  profile development by, 4, 310
  regolith developed by, 277, 310
  soil formation by, 310–312
  solution, 284
  synthesis of clay and humus by, 310
Weed control, 214, 215
Weight of soil (*see* Acre-furrow slice;
    Bulk density)
White alkali (*see* Alkaline and saline soils)
Wilting coefficient, 190–191
Wind
  influence on evaporation, 206
  influence on transpiration, 209
Wind erosion
  control of, 245–246
  extent of, 246
  in arid regions, 246–249
  in humid regions, 247
  loess deposits by, 246, 290
  on peat soils, 250, 369

Zinc
  amounts in soil, 23, 489
  an essential element, 20, 21, 484
  as pollutant, 563
  chelates of, 493
  deficiencies of, 488
  factors affecting availability, 490
  ionic form in soil, 35, 36, 495
  pH and availability, 388, 415, 491
  role of, 486, 487
  source of, 489
  used as a fertilizer, 489, 516
Zonal soil groups, 340–343